T0290928

Steel Connection Design by Inelastic Analysis

Steel Connection Design by Inelastic Analysis

Verification Examples per AISC Specification

Mark D. Denavit, Ali Nassiri, Mustafa Mahamid, Martin Vild, Halil Sezen and František Wald

Published by John Wiley & Sons, Inc., Hoboken, New Jersey.
Published simultaneously in Canada.

For general information on our other products and services or for technical support, please contact our Customer Care Department within the United States at (800) 762-2974, outside the United States at (317) 572-3993 or fax (317) 572-4002.

Wiley also publishes its books in a variety of electronic formats. Some content that appears in print may not be available in electronic formats. For more information about Wiley products, visit our web site at www.wiley.com.

Library of Congress Cataloging-in-Publication Data Applied for:

Paperback ISBN: 9781394222155

Cover Design: Wiley
Cover Images: © dvoevnore/Shutterstock, Courtesy of IDEA StatiCa

SKY10082566_082224

Contents

Introduction

While the use of finite element analysis in the design of a building structural system has been widely accepted in practice for decades, use of finite element analysis in the design of details and connections remains a novel approach. With Moore's law still in motion, the computational cost of such analyses is continually decreasing, but most engineers still calculate distinct failure modes using simple tools. Traditional hand-calculation-based approaches provide limited information about the behavior of the joint, and can be difficult to apply to complicated geometries or where interaction of limit states may occur. On the other hand, finite element analysis is not restricted by the number of connected members, geometry, or loading. Stresses and strains, as well as forces in bolts, are clearly visible. The deformed shape shows the real behavior of the joint as a whole under each investigated load case.

AISC 360-16 permits the use of computational methods based on nonlinear finite element analysis in Appendix 1, Design by advanced analysis. Structural engineers are responsible for producing safe designs, and must use only trusted tools to inform their judgment. To build trust, an approach to steel connection design by inelastic analysis using the component-based finite element method was subjected to a thorough validation and verification process by several expert teams. The results of that process are presented in this book.

Structure of the Book

Chapters 1–3, which provide general information on the design approach, were prepared by Martin Vild, a product owner at IDEA StatiCa, and Frantisek Wald, a Professor at the Czech Technical University in Prague, the Czech Republic. Chapters 4–11, which investigate several typical steel connection types that highlight various limit states, were prepared by Mark D. Denavit, an Associate Professor at the University of Tennessee, Knoxville, USA, and his student, Kayla Truman-Jarrell. Chapters 12–16, which investigate bracing connections, were prepared by Mustafa Mahamid, a Research Associate Professor at the University of Illinois Chicago, USA. Chapters 17–19, which investigate specific types of pinned, semi-rigid, and rigid connections, were prepared by a team from The Ohio State University, Columbus, Ohio, USA – Halil Sezen, Professor of Structural Engineering, Ali Nassiri, a Research Assistant Professor, and Baris Kasapoglu, a student of Halil Sezen. Martin Vild reviewed the whole text.

October 25, 2023

Martin Vild
Brno

1

Connection Design

1.1 Design Models

Structural steel connections are designed by analytical methods developed in past decades. Design models are the simplified version of reality and they are based on experimental campaigns. The small-scale tests on connections are typically not expensive, and a vast number of tests have been performed in the past to predict their behavior. Design tables for standardized connections were prepared by interpolation and even extrapolation based on tests. Databases of tests (Goverdhan, 1983; Abdalla and Chen, 1995; Mak and Elkady, 2021) have been collected and published. The collections are a valuable resource for learning about the behavior of typical joints, although some necessary data is missing. Curve-fitting models have been known since 1930. The mathematical formulas express the influence of geometrical and material parameters. They reproduce the behavior of connections that are similar in geometry, material, and loading well, but are restricted by their range of validity. Analytical models of connections need good engineering assumptions of internal forces and proper selection of components, which affect the resistance and stiffness.

Today, modeling is applied even for connections in seismic design. The analytical modeling of components of connections is well developed for connectors: bolts, welds, anchor bolts, etc. The complexity of finite element analyses has been studied extensively over the last twenty years. The procedures to achieve proper results in numerical simulations have been commonly accepted, together with the firm limits for applying numerical design calculations. The behavior of well-described and published components loaded by normal forces, shear forces, and their interaction was developed, based on numerical experiments, validated by physical experiments. The swift development of the computer-assisted design of steel structures, including complex structures, such as plated structures in bridges, excavators, and wind towers, clarified the design procedures of numerical models and their application in civil engineering.

1.2 Traditional Design Methods

The traditional design methods for steel connection design are described in AISC 360, *Specification for Structural Steel Buildings* (AISC, 2016), the *Steel Construction Manual* (AISC, 2017; 2022), and AISC Design Guides. There are two main approaches: Allowable Strength Design (ASD) and Load and Resistance Factor Design (LRFD). Both take advantage of inelastic behavior and are based on limit states design

Steel Connection Design by Inelastic Analysis: Verification Examples per AISC Specification, First Edition.
Mark D. Denavit, Ali Nassiri, Mustafa Mahamid, Martin Vild, Halil Sezen, and František Wald.

principles. Usually, several limit states apply in the determination of the nominal strength of a connection. The controlling limit state is the one that results in the least available strength. While the AISC *Specification* clearly defines the load-resistances of limit states, the actions on the individual components of the connection, i.e. load distribution in the connection, are left to the engineering judgment. Typical examples and procedures are provided in non-binding Design Guides and the *Steel Construction Manual.*

For most common connections, accurate equations have been developed. However, in many cases in engineering practice, design guidance is unavailable. Simplified models are used to estimate the behavior of connection elements. These models are typically based on equations developed for beams or columns, and the range of validity of such equations is disregarded; see Dowswell (2020).

1.3 Past and Present Numerical Design Calculations

The design models currently available for joint design are as follows (Wald et al., 2015a; Wald et al., 2021):

1. **Experimental models:** These are for contemporary design only. They provide value for experimentally reached resistance, which is never the expected design resistance. Design models are preferred as they reflect the need of designers for the safe prediction of joint behavior (Wald et al., 2015a). A plot of moment vs deformation capacity as obtained for an experimental model with respect to the experimental curve and design curve is shown in Figure 1.1.
2. **Curve-fitting model:** Curve-fitting procedures are based on experimental evidence and are still used to safely and economically design steel connections. A plot of moment vs. deformation capacity as obtained for a curve-fitting model with respect to the experimental curve and function curve is shown in Figure 1.2. Currently, this model is preferred only for hollow section joint design (Wald et al., 2015a).
3. **Analytical model:** Based on the resistance of connectors, such as welds, bolts, plates, and the estimated lever arm of internal forces, the resistance of the connection is predicted (Wald et al., 2018).

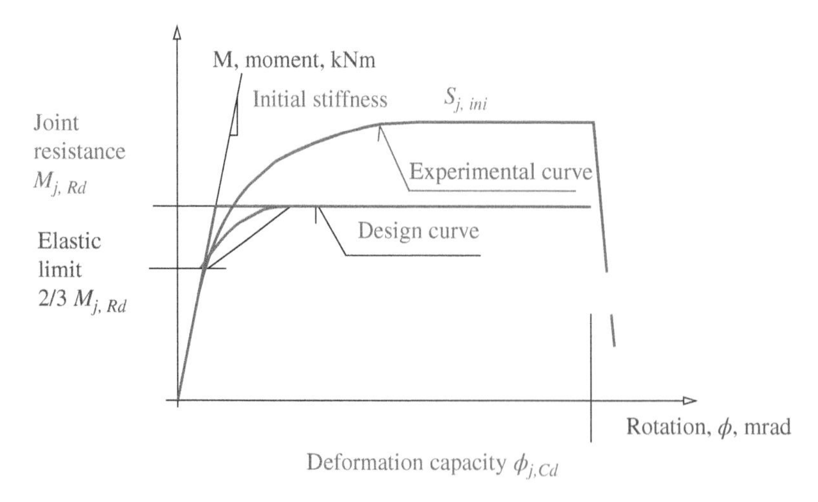

Figure 1.1 Plot of joint resistance vs deformation capacity of experimental model (Wald et al., 2015a).

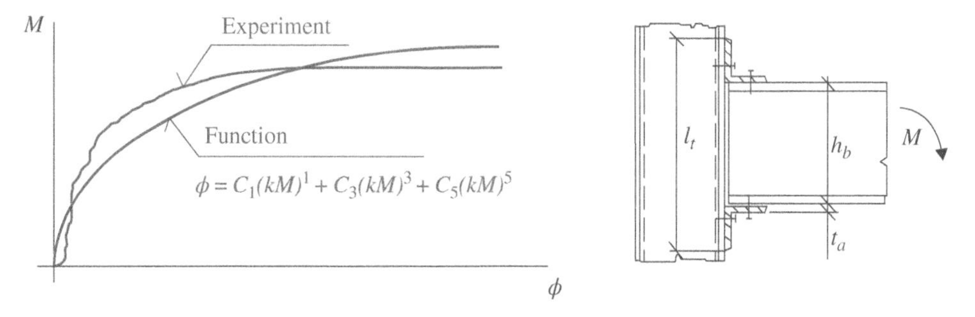

Figure 1.2 An example of curve-fitting models for steel connection (Wu and Chen, 1990).

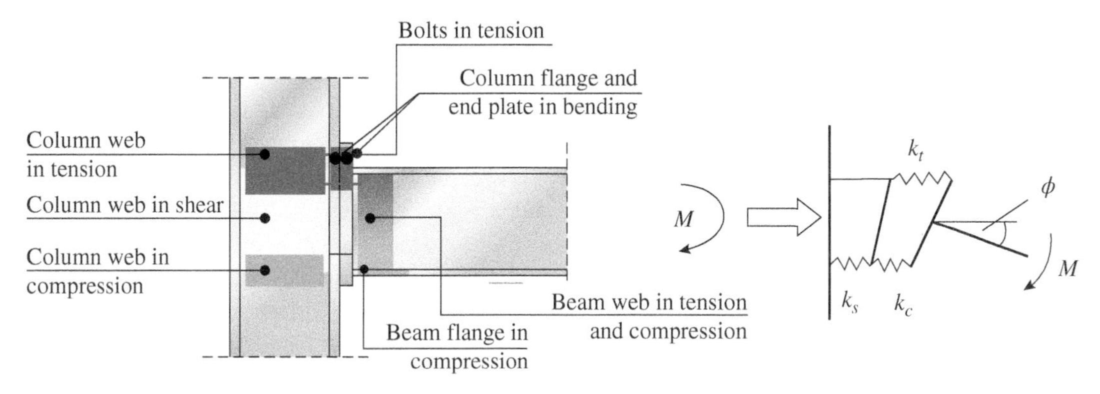

Figure 1.3 Components of a beam-to-column joint and a simple spring model (Block et al., 2012).

Also called the Component Method (CM), this method has been prepared for selected configurations, wherein not only the most essential design resistance for the connections of steel structures is predicted, but also the initial stiffness and the deformation capacity. This allows for the design of ductile structures. The procedure starts by decomposing a connection into components, followed by their description in terms of axial and shear force deformation behavior. After that, the components are grouped to examine connection moment–rotational behavior and classification/representation in a spring/shear model and application in global analyses (Wald et al., 2015a; Wald et al., 2019). The application of the CM to a beam-to-column joint and a simple spring model is shown in Figure 1.3.

The advantage of the CM is that it integrates the current experimental and analytical knowledge of the connection component's behavior: bolts, welds, end plates, flanges, anchor bolts, base plates, etc. This provides a very accurate prediction of the behavior in the elastic and ultimate loading levels. Verification of the model is possible using simplified calculations. The disadvantage of the component model is that the experimental evaluation of internal forces distribution is done only for a limited number of joint configurations. Also, in temporary scientific papers and background materials, typical components' descriptions are either absent or have low validity (Wald et al., 2015a; Wald et al., 2018).

4. **Numerical simulation:** Four decades ago, the Finite Element Analysis (FEA) of structural connections was treated by some researchers as a non-scientific matter. Two decades later, it was already a widely accepted addition or even necessarily an extension of experimental and theoretical work. Today,

computational analysis, particularly computational mechanics and fluid dynamics, is commonly used as an indispensable design tool and a catalyst for many relevant research fields (Godrich et al., 2015; Wald et al., 2021).

Finite element analyses (FEA) for connections have been used as research-oriented procedures since the 1970s. Their ability to express the real behavior of connections makes FEA a valid alternative to testing – a standard but expensive source of knowledge of connection behavior (Godrich et al., 2015; Wald et al., 2021).

The material model for numerical simulation uses a true stress-strain diagram, which is calculated from the experimental results of coupon tests considering the contraction of the sample during the inelastic stage of testing. Standard procedures with resistance factors for material/connections may be applied. A more advanced and accurate solution, which considers the accuracy of a model and material separately, gives a better and more economical solution to structural connections (Godrich et al., 2015; Wald et al., 2021). The stress-strain plots for several material models for steel used in numerical simulations and numerical design calculations are shown in Figure 1.4.

The complexity of FE modeling of structural steel connections has been studied extensively since the 1980s (see Krishnamurthy, 1978). Later, the procedures to reach proper results in scientifically-oriented FE models and the strict limits for the application of numerical design calculations were commonly accepted (see Bursi and Jaspart, 1997a; 1997b; Virdi, 1999).

Over the years, FEA has become a valuable tool for numerically modeling physical structures that are too complex for analytical solutions. Hence, the global analyses of steel structures are today carried out by FEA, and the traditional procedures are no longer widely used. Currently, rapid development in the software's ability to capture limit states in connection design by FEA has made thousands of experiments available for the validation process. In such a situation, the verification process performed

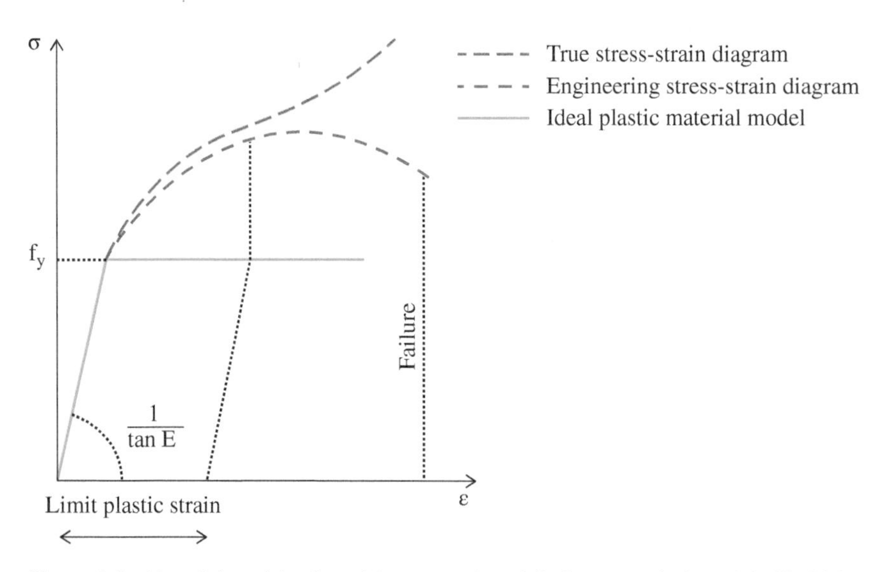

Figure 1.4 Material models of steel for research and design numerical models (Godrich et al., 2015; Kuříková et al., 2015).

through benchmark tests gains crucial importance. The source and the extent of such benchmark tests for the field of structural connections are yet to be established (Wald et al., 2018; Wald et al., 2019).

Since the 1970s, FEA for connections has been used for numerical simulations. Its ability to express the real behavior of connections makes numerical experiments a valid alternative to testing and a source of additional information about local stresses. The Validation and Verification (V&V) process of models is an integral part of the procedure (Bursi and Jaspart, 1997b), and the numerical simulations are based on the researcher's experiments (Wald et al., 2019).

A standard approach in numerical simulation is to perform an experiment and then create an advanced numerical model with fine meshing by using measured material properties and initial imperfections, often including residual stresses. The results of a numerical simulation must fit the experimental results closely. This process creates a validated numerical simulation, which may be used for further numerical experiments, in which the design material properties often are included. Creating a validated numerical simulation is very time-consuming and costly, yet it is still cheaper and more feasible than experiments (Šabatka et al., 2020).

Abaqus and ANSYS software packages are the preferred software tools for finite element analysis of steel connections to obtain numerical solutions, and the results can be used to validate the failure modes as observed from the experimental results. Figure 1.5 represents an Abaqus model of a steel connection. At the same time, Figure 1.6 ensures that the failure modes obtained from the numerical simulation are close to those obtained from the experimental setup results.

According to the von Mises yield criterion, a given material will only begin to fail if the maximum value of the distortion energy is at most the distortion energy required to yield the material in a tensile test. Von Mises stress is an equivalent stress value used to determine if a given material will begin to

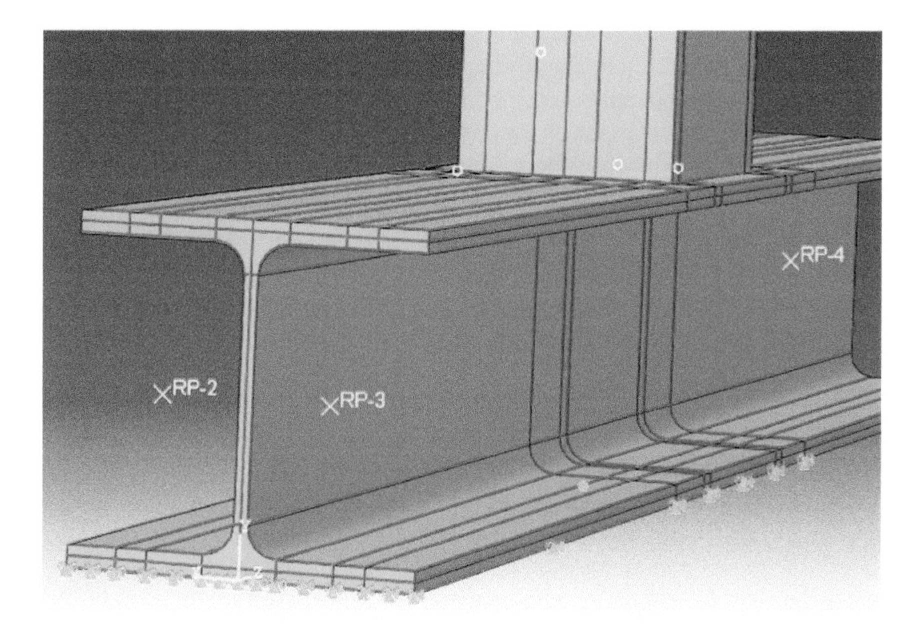

Figure 1.5 ABAQUS model of connection (Wald et al., 2015b).

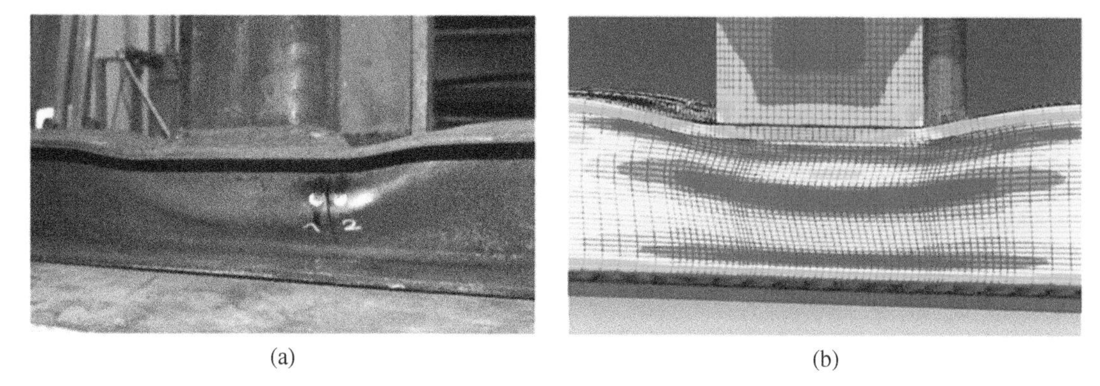

Figure 1.6 Well-predicted failure modes of chord web for experiment A1; (a) Failure in an experimental model; (b) von Mises stresses for failure in a numerical simulation (Wald et al., 2015b).

yield. In contrast, a given material will not yield if the maximum von Mises stress value does not exceed the yield strength of the material. Traditionally, von Mises stress has been used for ductile materials like metals (Thompson, 2023).

From a practical perspective, von Mises stress enables engineers to understand the performance of a complex loaded part with a simple uniaxial tensile test, from which we can extract performance parameters, such as yield strength, ultimate strength, and Young's modulus (Thompson, 2023).

5. **Numerical design calculation:** Numerical design calculations are used for direct load-resistance estimation of structures. The bilinear material curve with design strength is used and all resistance (LRFD) or safety (ASD) factors are applied. Numerical design calculations should reach the same or higher reliability as the AISC 360 *Specification*. The advantage is generally higher accuracy of models leading to smaller scatter of results, compared to experiments.

   Currently, several existing programs are based on the analytical spring model, CM, which is a perfectly fine solution. However, the major drawback here is that a new model must be put together from scratch for each topology. IDEA StatiCa came up with a new method called the Component-Based Finite Element Model (CBFEM), which is a synergy of the Component Method and the finite element method (Wald et al., 2015a; Wald et al., 2021). The numerical design calculations provided using the CBFEM have been extensively verified, and several studies have been published. It has been implemented in several commercial software packages such as IDEA StatiCa and Hilti PROFIS (Šabatka et al., 2020).

   Numerical design calculation for steel connections is carried out using IDEA StatiCa Connection. The software combines the finite element method with the component method, offering an alternative to conventional analytical models and the laborious CM. Contrary to numerical simulations, IDEA StatiCa software uses 2D shell elements for plates, whereas fasteners (welds, bolts, contacts, etc.) are represented by components with pre-defined properties based on experimental findings (Kuříková et al., 2021).

   CBFEM also provides code checks of failure modes that are very difficult to capture by finite element analysis alone, such as the crushing of concrete in compression or weld fracture. It removes the restrictions and most simplifications used in CM. The neutral axis and forces in components for any type of load combination are determined by the finite element method (Šabatka et al., 2020). The recommendation for design by FEA in structural steel is found in AISC 360-16, Appendix 1: Design by advanced analysis. However, this field is evolving fast, and the details are not strictly specified.

The modeling process in IDEAStatiCa is as follows:

1. Connections are divided into individual components, as shown in Figure 1.7 (IDEAStatiCa, n.d.b; Wald et al., 2021).
2. All steel plates are modeled as shell elements assuming ideal elastic-plastic material, as shown in Figure 1.8 (IDEA StatiCa, n.d.b).
3. Bolts are modeled as nonlinear springs, as shown in Figure 1.9 (IDEAStatiCa, n.d.b).

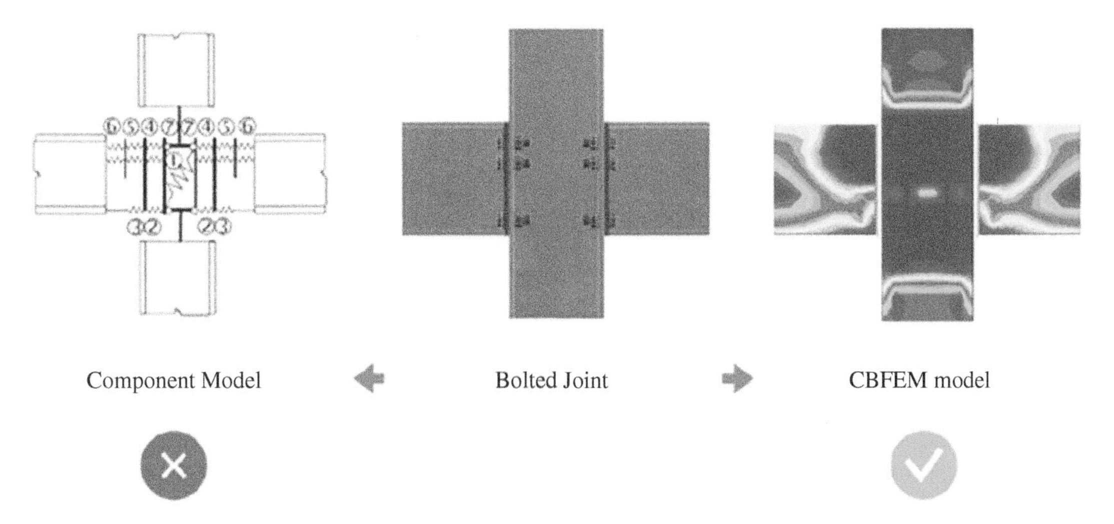

Figure 1.7 Components of a beam-to-column joint and a simple spring model (Block et al., 2012; IDEA StatiCa, n.d.b).

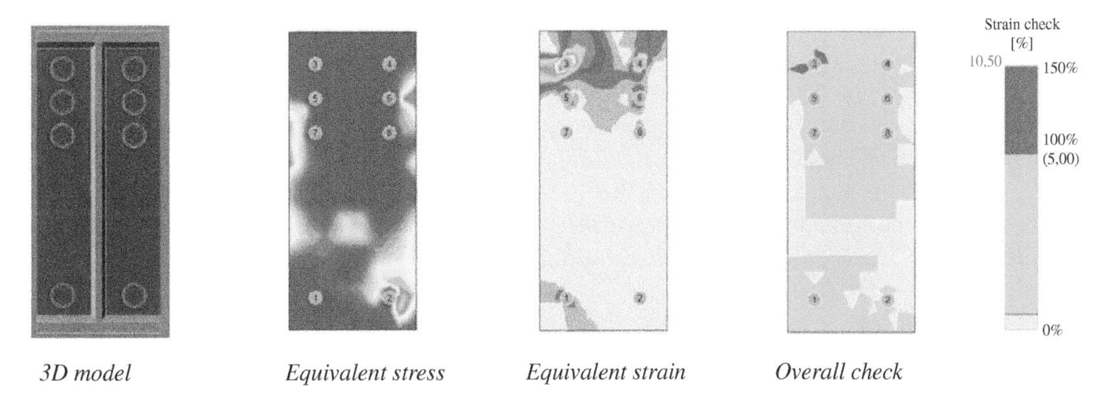

Figure 1.8 Steel plates as shell elements in CBFEM (IDEAStatiCa, n.d.b).

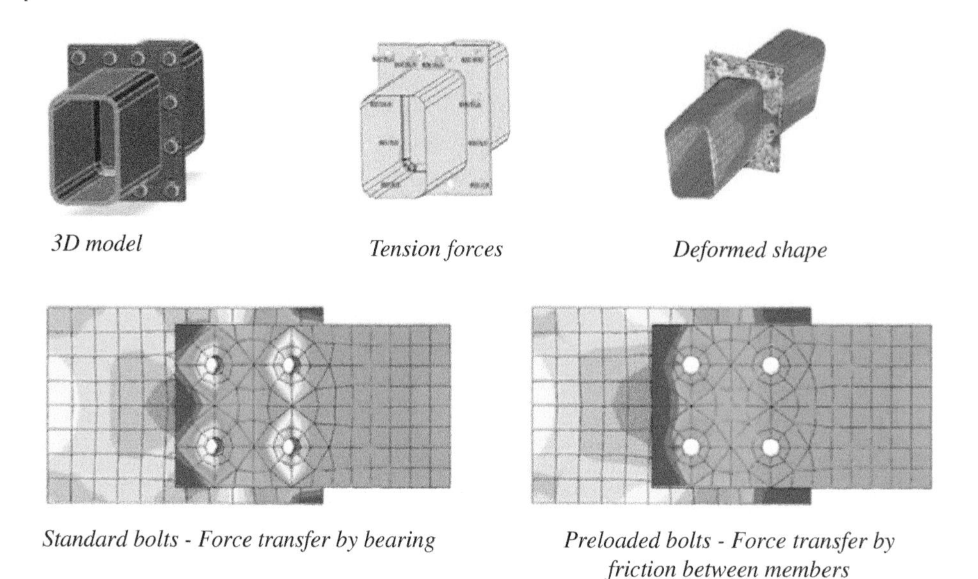

Figure 1.9 Standard bolts and preloaded bolts in CBFEM (IDEAStatiCa, n.d.b).

4. Welds are modeled as elastoplastic elements to allow for plastic stress redistribution, as shown in Figure 1.10 (IDEA StatiCa, n.d.b).

The finite element model is used to analyze internal forces in each of the components (IDEA StatiCa, n.d.b; Wald et al., 2021). Plates are checked for limit plastic strain – 5 % according to Cl. C.8(1) EN 1993-1-5 (CEN, 2006) and Cl. 8.1.5 in prEN 1993-1-14:2022 (CEN, 2022), which corresponds well also with recommendations in AISC 360-16 for rotation limit 0.02 rad for semi-rigid joints (IDEA StatiCa, n.d.b; Wald et al., 2021). Each component is checked according to specific formulas defined by the national code, similar to when using CM (IDEA StatiCa, n.d.b). Plates are decomposed into shell elements with 6 degrees of freedom at each node (3 rotations and 3 translations) and 5 layers. Materially nonlinear analysis is used to overcome challenges posed by the discontinuous regions in the models (IDEA StatiCa, n.d.b).

All plates of a beam cross-section have elements of a common size. The size of the generated finite elements is limited. The minimum element size is set by default to 8 mm and the maximum element size to 50 mm. Meshes on flanges and webs are independent of each other. The default number of finite elements is set to 12 elements per cross-section height, as shown in Figure 1.11 (Wald et al., 2021).

IDEAStatiCa Connection focuses on the actual design checking of steel connections, allowing the engineer to overcome the limitations of the current CM, on which all of the other existing connection software systems are based. It also provides a simple passed/failed status for all elements, which the general FEM tools simply do not have (IDEAStatiCa, n.d.a).

There is a big difference in user-friendliness and productivity compared to software such as Abaqus and ANSYS, where every single part of the procedure is very time-consuming. However, in terms of calculation results, they can be comparable. On the other hand, IDEA StatiCa Connection can only evaluate steel connections; the software is not as general as Abaqus or ANSYS (IDEA StatiCa, n.d.a).

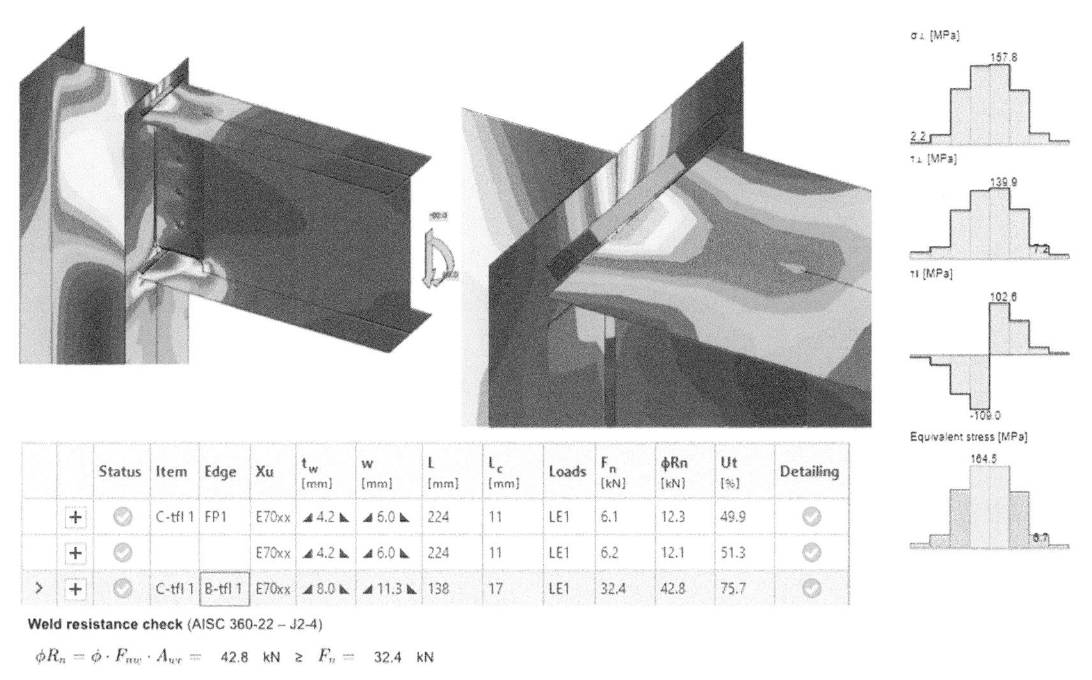

		Status	Item	Edge	Xu	t_w [mm]	w [mm]	L [mm]	L_c [mm]	Loads	F_n [kN]	ϕRn [kN]	Ut [%]	Detailing
	+	✓	C-tfl 1	FP1	E70xx	◢4.2◣	◢6.0◣	224	11	LE1	6.1	12.3	49.9	✓
	+	✓			E70xx	◢4.2◣	◢6.0◣	224	11	LE1	6.2	12.1	51.3	✓
>	+	✓	C-tfl 1	B-tfl 1	E70xx	◢8.0◣	◢11.3◣	138	17	LE1	32.4	42.8	75.7	✓

Weld resistance check (AISC 360-22 – J2-4)

$\phi R_n = \phi \cdot F_{nw} \cdot A_{we} =$ 42.8 kN ≥ $F_n =$ 32.4 kN

Where:

$F_{nw} =$ 413.1 MPa – nominal stress of weld material:

- $F_{nw} = 0.6 \cdot F_{EXX} \cdot (1 + 0.5 \cdot sin^{1.5}\theta)$, where:
  - $F_{EXX} =$ 482.6 MPa – electrode classification number, i.e. minimum specified tensile strength
  - $\theta =$ 64.1° – angle of loading measured from the weld longitudinal axis

$A_{we} =$ 138 mm^2 – effective area of weld critical element

$\phi =$ 0.75 – resistance factor for welded connections

Figure 1.10 Welds in CBFEM (IDEAStatiCa, n.d.b).

The component-based finite element method (CBFEM), which is further described in the chapters of this book, uses a bilinear material model for structural steel plates with negligible strain-hardening. The slope of the plastic branch is 1000× smaller than the slope of the elastic branch. Yield strength is multiplied by the resistance factor (LRFD) or divided by the safety factor (ASD). Bolts, welds, and anchors are modeled by a combination of nonlinear springs and special elements. Their capacity is determined by standard procedures with resistance (LRFD) or safety (ASD) factors.

1.4 Validation and Verification

Validation and Verification (V&V) studies are necessary to ensure the accuracy of the numerical models. They are used and created by both software developers and software users (see Kwasniewski, 2010; Kwasniewski and Bojanowski, 2015; Wald et al., 2021). prEN 1993-1-14:2022 (CEN, 2022) defines a formal procedure called System Response Quantity (SRQ), which differentiates between validation and

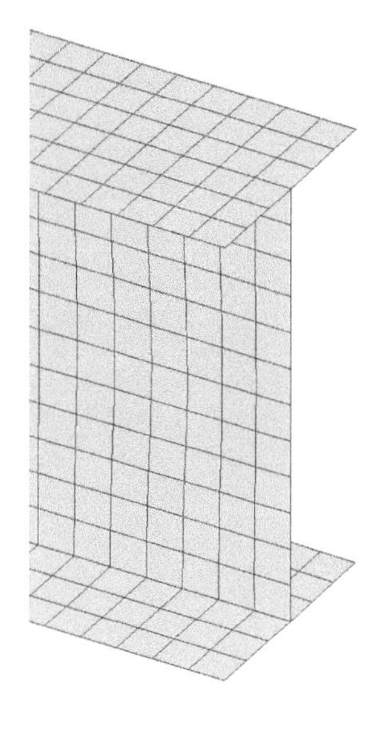

Figure 1.11 Mesh on beam with constraints between web and flange plate (Wald et al., 2021).

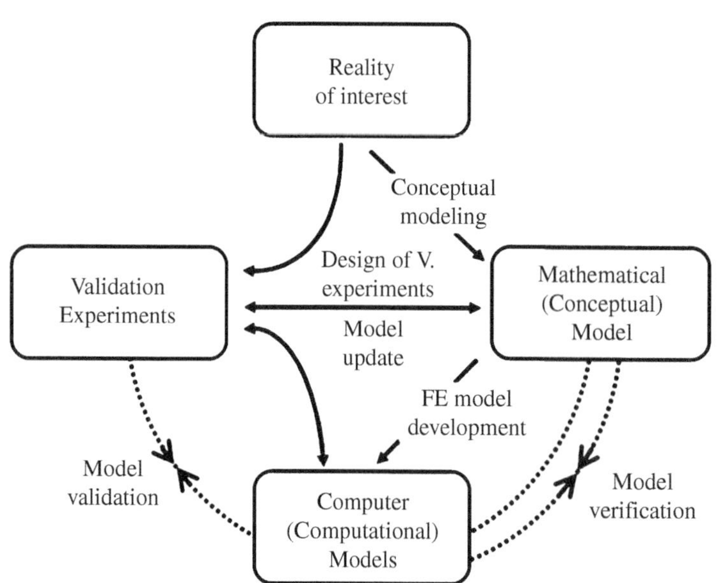

Figure 1.12 Relationship between modeling, verification, and validation (Kwasniewski and Bojanowski, 2015).

verification. Verification compares numerical solutions with highly accurate (analytical or numerical) benchmark solutions. Validation compares the numerical solution with the experimental data. The relationship between modeling, verification, and validation is shown in Figure 1.12.

The verification on the analyst's part is based on the test of agreement with the known correct results, if such are available. Most commercial codes, such as ANSYS, Abaqus (SIMULIA, 2011), and MIDAS, support lists of well-documented benchmark cases (tests). For example, Abaqus in three manuals provides a wide variety of benchmark tests (including 93 NAFEMS benchmarks), from simple one-element tests to complex engineering problems and experiments (validation benchmarks). These example problems, containing input files, are advantageous for a user not only as material for verification but also as a great help in individual modeling (Wald et al., 2014). Nevertheless, there is still a lack of benchmark studies for specific research areas, such as connection design.

The experimental data, which can be used for validation, should be treated separately and differently than benchmark solutions applied for verification. The reasons are unavoidable errors and uncertainties associated with the results of experimental measurements (Wald et al., 2015a; Wald et al., 2021). An error of measurement (calculation) can be defined as the result of measurement (calculation) minus the value of the measured (accurate solution) (ISO, 1993). As the accurate solution is usually unknown (eventually known for simplified cases), the user can only deal with estimates of errors. Uncertainty can be thought of as a parameter associated with the result of measurement (the solution) that characterizes the dispersion of the values that could be reasonably attributed to the measured.

The limitations of experimental validation increase the importance of verification, which is supposed to deliver evidence that mathematical models have been adequately implemented and that the numerical solution is correct with respect to the mathematical model.

The work of IDEA StatiCa and several cooperating universities on validation and verification of the CBFEM method and the results of IDEA StatiCa software, which is crucial for safety and reliability, can be found in (IDEA StatiCa, n.d.d). Validation and verification in CBFEM ensure that the finite element analysis of the model is approximately correct (IDEA StatiCa, n.d.d).

Verification compares a numerical model to an analytical method, most often incorporated in the building codes, for example, AISC 360-16 for steel structures in the United States. At the same time, validation compares results from a numerical model to the experimental results. Verifying CBFEM based on design examples from the building codes is relatively simple and preferable. The numerical model is often highly advanced, including material and geometrical nonlinearities. A benchmark case is prepared to help the users model and check the method's correctness and proper use (IDEA StatiCa, n.d.d).

Validation is a comparison of a precise numerical model to an experiment. The geometry and material properties are the same as those measured in the experiment. When the results – typically a load-displacement and stress-strain curves – of the numerical model are close to those of the experiment, the numerical model is validated.

Finally, the results of numerical simulations are compared to those of CBFEM. The results do not need to coincide perfectly; the CBFEM is typically much more straightforward, but the CBFEM results must be safe (Wald et al., 2015a; Wald et al., 2021). The material properties of the numerical model are then changed to nominal values, imperfections are increased according to manufacturing tolerances, and several sensitivity studies may be performed by changing the parameters, e.g. the thickness of plates and yield strength of the material. Analytical methods in codes are weighed down by simplifications, and the results between the *Specification* and the CBFEM for complicated connections may vary, especially at the boundaries of the range of validity. In that case, a comparison of CBFEM to an advanced model validated by experiments proves that CBFEM is safe, even though the resistances are higher than determined by the code (Wald et al., 2015a; Wald et al., 2021).

1.5 Benchmark Cases

A well-developed benchmark example should satisfy the following requirements. The problem considered should be relatively simple and easy to understand. For more complex problems, less reliable solutions can be provided. For example, with actual material properties of steel or concrete, only numerical solutions can be obtained.

Seeking simplicity, we should accept that a considered case may show little practical meaning. It is supposed to be used to verify computational models, not solve an engineering problem. Finding a good balance between simplicity and the practical meaning of the chosen benchmark case is difficult. A hierarchical approach is recommended where a set of problems is considered in benchmark studies, starting from simple cases with analytical solutions. Then, more complex problems, closer to practice, are investigated numerically. Such an approach gives more confidence in theobtained solutions.

The complete input data must be provided as part of the benchmark study. All assumptions must be clearly identified, such as material properties, boundary conditions, temperature distribution, loading conditions, and large/small deformations and displacements. All measurements and a detailed test procedure description should be provided for experimental examples. For numerical benchmark examples, a mesh density study should also be conducted. It should be shown that the provided results are within the range of asymptotic convergence. If possible, the recommended solution should be given as the estimate of the asymptotic solution, based on solutions for at least two succeeding mesh densities. For finite element calculations, complete procedures such as the Grid Convergence Index (GCI), based on the Richardson extrapolation, are recommended (Roache, 1998). During the development of benchmark studies, checking alternative numerical models, for example, using different codes or solid vs. shell finite elements (if possible), should also be considered. Such an approach increases the validity of the solution.

1.6 Numerical Experiments

A parametric study is a desired element of experimental work and an indispensable element of a numerical analysis. The cost of performing multiple experiments related to structural connections is usually small, but a probabilistic distribution of the system response is rarely available. However, in the case of simulated benchmark problems, the computational cost of running multiple instances of a simple numerical experiment with varying input parameters is competitive.

The variance of a system response depends on the variance in the input parameters but also on the range at which it is tested. Nonlinearity of the response must also be considered when designing the benchmark tests. The numerical experiments should be performed in the range where a reasonable variation in an input parameter causes a reasonable change in the system's response. Designing a benchmark test producing either a non-sensitive or overly sensitive response is undesirable. The sensitivity study for a system with multiple variable input parameters and multiple responses should be performed by regression analysis or variance-based methods.

Actually, the selection of the System Response Quantity (SRQ) (Kwasniewski, 2010) is important for both verification and validation. However, in both cases, it is subject to different limitations. In verification, SRQ means a quantity that describes the structure's response and is selected for comparison with the value obtained from the benchmark solution. A user is less limited here than in the case of validation, where the experimental data is always limited by the number of gauges and other instrumentation. The selection of the SRQ should reflect the main objective of the analysis. For mechanical structural response, we can choose between local and global (integral) quantities. Engineers are usually interested in stresses and internal forces, which are local quantities. They are subject to larger uncertainties, especially in the case of validation. More appropriate are global quantities such as deflection, which reflects the deformation of the whole, or a large part of the structure and its boundary condition.

1.7 Experimental Validation

As the experimental data is stochastic by nature and is always subject to some variation, it should be defined by a probability distribution. For complete comparison, the numerical results should also be presented in an analogous probabilistic manner, using a probability distribution, generated by repeated

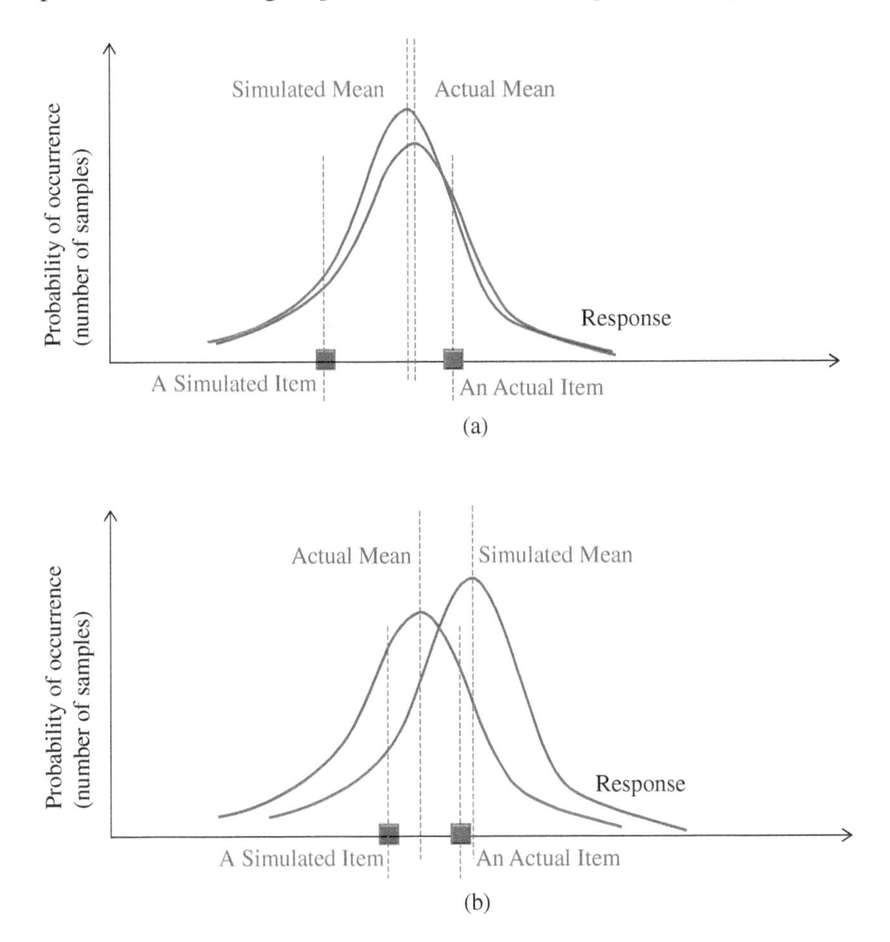

Figure 1.13 Example of calibration meaning unjustified shifting of the numerical results closer to the experimental data (Kwasniewski, 2010).

calculations with some selected input data varying, following prescribed distributions, so-called probability simulations. Such extensive calculations can be conducted automatically with the help of specialized optimization packages (e.g. LS-OPT®, HyperStudy®, or ModeFrontier®), which nowadays are often included in commercial computational systems.

For many authors working on the principles of validation and verification (Kwasniewski, 2010), the term calibration has a negative meaning and describes a practice that should be avoided in numerical modeling. Calibration here means unjustified modification of the input data applied to a numerical model to shift the numerical results closer to the experimental data. An example of erroneous calibration is shown in Figure 1.13, where at the beginning, it is assumed that the numerical model well reflects the experiment. However, due to some uncertainties associated with the experiment, the first numerical prediction differs from the first experimental result. Frequently, in such cases, the discrepancy between the experiment and the numerical simulation is attributable to some unidentified input parameter and not to a limitation of the software. Then, by hiding one error by introducing another, the calibration process itself is erroneous. Calibration, applied, for example, through variation of material input data, shifts the result closer to the experimental response but, at the same time, changes the whole numerical model whose probability is now moved away from the experimental one. Due to the calibration, the new numerical model may easily show poorer predictive capability. This fact is principally revealed for modified input data, for example, loading conditions. Figure 1.13(a) is not calibrated; Figure 1.13(b) is calibrated by introducing an error.

There is a situation when the calibration process actually makes sense. If a full stochastic description of the experimental data is known, a probabilistic analysis was performed for the simulation, and there is a difference between the means of the measured and simulated responses, then calibration of physical models may be necessary. Adjusting the model introduces a change in the response that brings the entire spectrum of results closer to the experimental data set. The calibration defined in that way is a much more complex process than just tweaking the models and must be confirmed on different simulated events.

References

Abdalla, K.M. and Chen, W.F. (1995). Expanded Database of Semi-Rigid Steel Connections. *Computer and Structures*, 56(4), 553–564.

AISC (2015). *Design Guide 29: Vertical Bracing Connections: Analysis and Design*. American Institute of Steel Construction, Chicago.

AISC 360 (2016). *Specification for Structural Steel Buildings*. American Institute of Steel Construction, Chicago.

AISC 360 (2017). *Steel Construction Manual*. American Institute of Steel Construction, Chicago.

AISC (2022). *Steel Construction Manual*, 16th Edition. American Institute of Steel Construction, Chicago. https://www.aisc.org/publications/steel-construction-manual-resources/16th-ed-steel-construction-manual/

Block, F., Burgess, I., Davison, B., and Plank, R. (2012). A Component Approach to Modelling Steelwork Connections in Fire: Behaviour of Column Webs in Compression. In *Structures 2004: Building on the Past, Securing the Future*, ASCE Structures Congress, 1–8. https://doi.org/10.1061/40700(2004)124

Bursi, O.S., and Jaspart, J.P. (1997a). Benchmarks for Finite Element Modelling of Bolted Steel Connections. *Journal of Constructional Steel Research*, 43(1), 17–42. https://doi.org/10.1016/S0143-974X(97)00031-X

Bursi, O.S., and Jaspart, J.P. (1997b). Calibration of a Finite Element Model for Isolated Bolted End-Plate Steel Connections. *Journal of Constructional Steel Research*, 44(3), 225–262. https://doi.org/10.1016/S0143-974X(97)00056-4

CEN (2005). *EN 1993-1-8:2005: Eurocode 3: Design of Steel Structures – Part 1-8: Design of Joints*. CEN, Brussels.

CEN (2006). *EN 1993-1-5: Eurocode 3: Design of Steel Structures – Part 1-5: Plated Structural Elements*. CEN, Brussels.

CEN (2022). *prEN 1993-1-14:2022: Eurocode 3: Design of Steel Structures – Part 1-14: Design Assisted by Finite Element Analysis*. CEN, Brussels.

Dowswell, B. (2020). Gusset Plates: The Evolution of Simplified Design Models. T.R. Higgins Presentation, American Institute of Steel Construction.

Dowswell, B. (2021). Analysis of the Shear Lag Factor for Slotted Rectangular HSS Members. *Engineering Journal*, 58(3), 77–89.

Godrich, L., Wald, F., Šabatka, L., Kurejkova, M., and Kabelac, J. (2015, September). Future Design Procedure for Structural Connections Is Component Based Finite Element Method. Paper presented at Nordic Steel Conference 2015, Tampere, Finland.

Goverdhan, A.V. (1983). *A Collection of Experimental Moment-rotation Curves and Evaluation of Prediction Equations for Semi-rigid Connections*. Vanderbilt University, Nashville, TN.

IDEA StatiCa (n.d.a). IDEA StatiCa Support Center: FAQ. https://www.ideastatica.com/support-center-faq

IDEA StatiCa (n.d.b). Structural Design of Steel Connection and Joints. Steel. https://www.ideastatica.com/steel

IDEA StatiCa (n.d.c). Theoretical Background of Steel Connections in IDEA StatiCa. https://www.ideastatica.com/support-center/general-theoretical-background

IDEA StatiCa (n.d.d). Verifications in IDEA StatiCa. https://www.ideastatica.com/support-center-verifications

ISO (1993). *Guide to the Expression of Uncertainty in Measurement*. ISO, Geneva.

Krishnamurthy, N. (1978). A Fresh Look at Bolted End-Plate Behaviour and Design. *AISC Engineering Journal*, 39–49.

Kuříková, M., Sekal, D., Wald, F., and Maier, N. (2021). Advanced Design of Block Shear Failure. *Metals*, 11, 1088. https://doi.org/10.3390/met11071088

Kuříková, M., Wald, F., Kabeláč, J., and Šabatka, L. (2015, November). Slender Compressed Plate in Component Based Finite Element Model. Paper presented at 2nd International Conference on Innovative Materials, Structures and Technologies. https://doi.org/10.1088/1757-899X/96/1/012050

Kwasniewski, L. (2010). Numerical Verification of Post-Critical Beck's Column Behavior. *International Journal of Non-Linear Mechanics*, 45(3), 242–255. https://doi.org/10.1016/j.ijnonlinmec.2009.11.007

Kwasniewski, L. and Bojanowski, C. (2015). Principles of Verification and Validation. *Journal of Structural Fire Engineering*, 6(1), 29–40. https://doi.org/10.1260/2040-2317.6.1.29

Mak, L. and Elkady, A. (2021). Experimental Database on Flush-End Plate Connections. *ASCE Journal of Structural Engineering*, 147(7). DOI: [10.1061/(ASCE)ST.1943-541X.0003064](https://ascelibrary.org/doi/abs/10.1061/%28ASCE%29ST.1943-541X.0003064).

Roache, P.J. (1998). Verification and Validation in Computational Science and Engineering. *Computing in Science Engineering*, Hermosa publishers, pp. 8–9.

Šabatka, L., Wald, F., Kolaja, D., and Pospíšil, M. (2020, August). Steel Connections. *STRUCTURE Magazine*.

SIMULIA. (2011). *Abaqus 6.11, Benchmarks Manual*. Dassault Systèmes.

Thompson, E. (2023, April 6). What Is Equivalent Stress? Blog. https://www.ansys.com/blog/what-is-equivalent-stress

Virdi, K.S. (1999). *Numerical Simulation of Semi-Rigid Connections by the Finite Element Method*. European Commission, Brussels.

Wald, F., Kabeláč, J., Kuříková, M., Perháč, O., Ryjáček, P., Minor, O., Pazmiño, M., Šabatka, L., and Kolaja, D. (2018). Advanced Procedures for Design of Bolted Connections. *IOP Conference Series: Materials Science and Engineering,* 419, 012044. https://doi.org/10.1088/1757-899X/419/1/012044

Wald, F., Burgess, I., Kwasniewski, L., Horová, K., and Caldová, E. (2014). *Benchmark Studies: Experimental Validation of Numerical Models in Fire Engineering*. CTU Publishing House, Prague.

Wald, F., Vild, M., Kuříková, M., Kožich, M., and Kabeláč, J. (2019). Component Based Finite Element Design of Seismically Qualified Joints. *Journal of Physics: Conference Series*, 1425(1), 012002. https://doi.org/10.1088/1742-6596/1425/1/012002

Wald, F. et al. (2015a). Connection Design by Component-Based Finite Element Method (Lecture 1 Beam-to-Column Moment Connection). http://steel.fsv.cvut.cz/merlion/

Wald, F. et al. (2015b). Connection Design by Component-Based Finite Element Method (Lecture 2 Joint of Hollow to Open Section). http://steel.fsv.cvut.cz/merlion/

Wald, F. et al. (2021). Component-Based Finite Element Design of Steel Connections. https://www.ideastatica.com/blog/new-cbfem-book-published-buy-it-online

Wu, F.-H., and Chen, W.-F. (1990). A Design Model for Semi-Rigid Connections. *Engineering Structures*, 12(2), 88–97. https://doi.org/10.1016/0141-0296(90)90013-I

2

The Component-Based Finite Element Method

2.1 Material Model

The most common material diagrams used in finite element modeling of structural steel are the ideal plastic, elastic model with strain hardening, and the true stress-strain diagram; see Figure 2.1.

The elastoplastic material with negligible strain hardening is used in CBFEM. The material behavior is based on the von Mises yield criterion. According to AISC 360-16, Appendix 1 (AISC, 2016), the yield strength is reduced by resistance factor ϕ (LRFD) or divided by the safety factor Ω (ASD). It is assumed to be elastic before reaching the yield strength, F_y. The slope of the plastic branch is $\tan(E/1000)$. That leads to negligible strain hardening and faster and reliable convergence. The ultimate limit state criterion for regions not susceptible to buckling is reaching a limiting value of the equivalent plastic strain. The value of 5% is recommended, according to Cl. C.8(1) EN 1993-1-5:2006 and Cl. 8.1.5 in prEN 1993-1-14 (CEN, 2022).

The limit value of the plastic strain is often discussed. In fact, ultimate load has low sensitivity to the limit value of the plastic strain when the elastic-plastic model is used. This is demonstrated in the following example of a beam-to-column joint. An open section beam S10 × 25.4 is connected to an open section column W8 × 40 and loaded by bending moment, as shown in Figure 2.2. The influence of the limit value of the plastic strain on the resistance of the beam is shown in Figure 2.3.

2.2 Plate Model and Mesh Convergence

2.2.1 Plate Model

Shell elements are recommended for modeling plates in numerical design calculations of structural connections. Four-node quadrangle shell elements with nodes at the corners are applied. Six degrees of freedom are considered in every node: 3 translations (u_x, u_y, u_z) and 3 rotations (ϕ_x, ϕ_y, ϕ_z). Deformations of the element are divided into membrane and flexural components.

The formulation of membrane behavior is based on the work of Ibrahimbegovic et al. (1990). Rotations perpendicular to the plane of the element are considered. Complete 3D formulation of the element is provided. The in-house stabilized variant of the Mindlin quad plate element with constant shear deformation along the edge is applied. The elements are inspired by MITC4; see Dvorkin and Bathe (1984).

Steel Connection Design by Inelastic Analysis: Verification Examples per AISC Specification, First Edition.
Mark D. Denavit, Ali Nassiri, Mustafa Mahamid, Martin Vild, Halil Sezen, and František Wald.

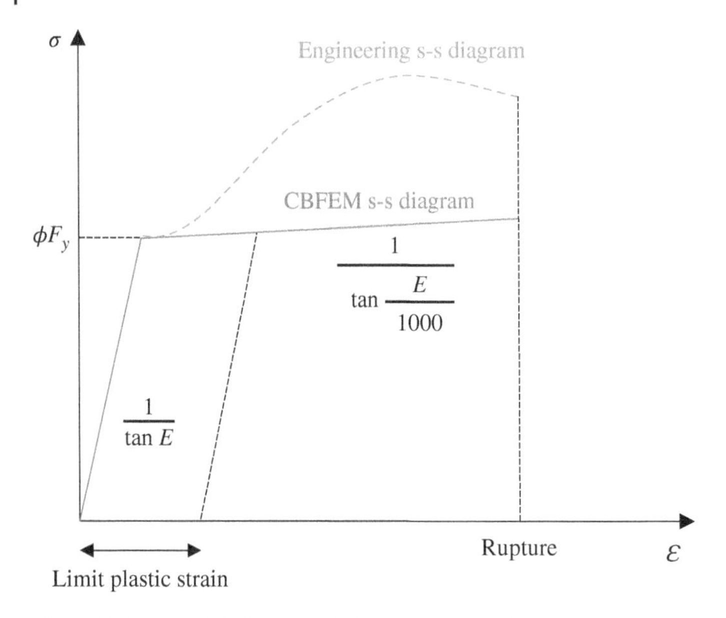

Figure 2.1 Material diagrams of steel in numerical models.

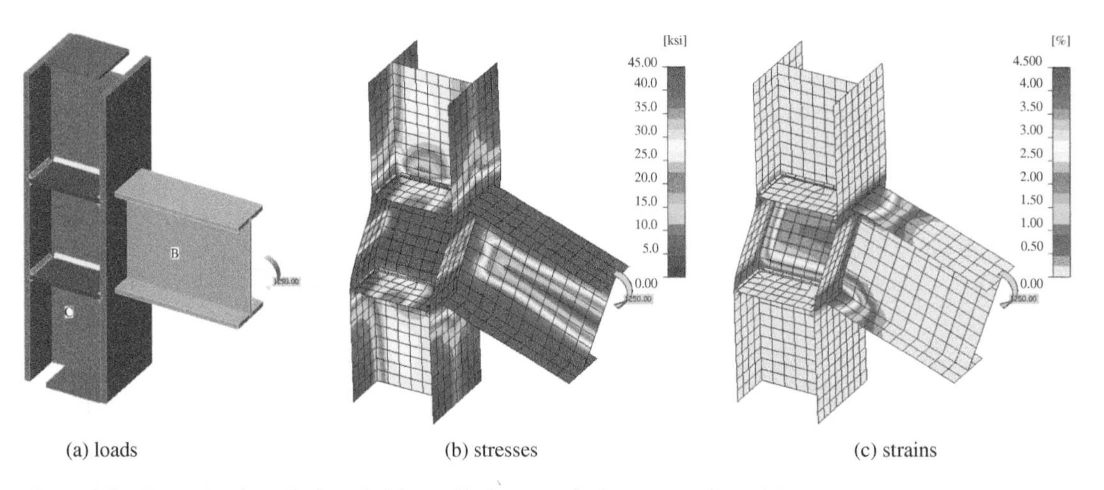

Figure 2.2 Example of prediction of ultimate limit state of a beam-to-column joint.

The shell is divided into five integration points along the thickness of the plate, and plastic behavior is analyzed at each point. This is called the Gaus–Lobatto integration. The nonlinear elastic-plastic stage of the material is analyzed in each layer, based on the known strains. The highest stresses and strains in a critical layer are used for plate failure evaluation.

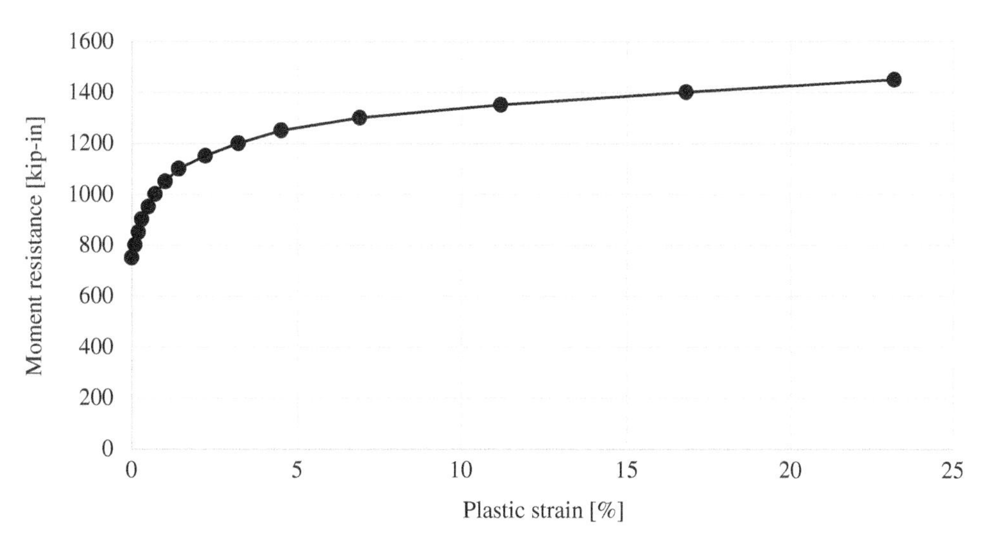

Figure 2.3 Influence of the limit value of plastic strain on the moment resistance.

2.2.2 Mesh Convergence

Ideally, the connection check should be independent of the element size, i.e. the difference in results for fine mesh and used mesh should be below 5%. Each plate is meshed separately, and multipoint constraints are used to connect the nodes of the plate meshes. This way, mesh generation on a separate plate is problem-free. Attention should be paid to complex geometries such as stiffened panels, T-stubs, and base plates. A sensitivity analysis considering mesh discretization should be performed for complicated geometries.

The size of the generated finite elements is limited. By default, the minimum element size is set to 0.315 in. (8 mm) and the maximum element size to 1.969 in. (50 mm). This setting is suitable for typical civil structural applications. Meshes on flanges and webs are independent of each other. The default number of finite elements is set to 12 elements per cross-section height, as shown in Figure 2.4.

The mesh of end plates is separate and independent of other connection parts. The default finite element size is set to 20 elements per cross-section height, as shown in Figure 2.5.

The following example of a beam-to-column joint shows the influence of mesh size on moment resistance. An open section beam W 10 × 26 is connected to an open section column W 8 × 40 and loaded by bending moment, as shown in Figure 2.6. The critical component is the column panel in shear. The number of finite elements along the cross-section height changes from 4 to 36, and the results are compared; see Figure 2.7. The dashed line represents a 5% difference. It is recommended to subdivide the cross-section height into 12 elements.

A mesh sensitivity study of a slender compressed end-plate stiffener is presented. The number of elements along the width of the stiffener is changed from 4 to 36. The first buckling mode and the influence of a number of elements on the buckling resistance and critical load are shown in Figure 2.8. The difference by 5% is displayed. It is recommended to use 12 elements along the stiffener width.

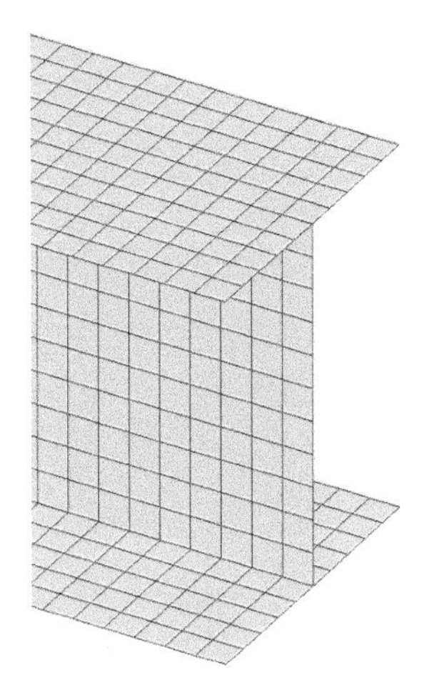

Figure 2.4 Mesh on beam web and flange plates.

Figure 2.5 Mesh on end plate, with eight elements along its width.

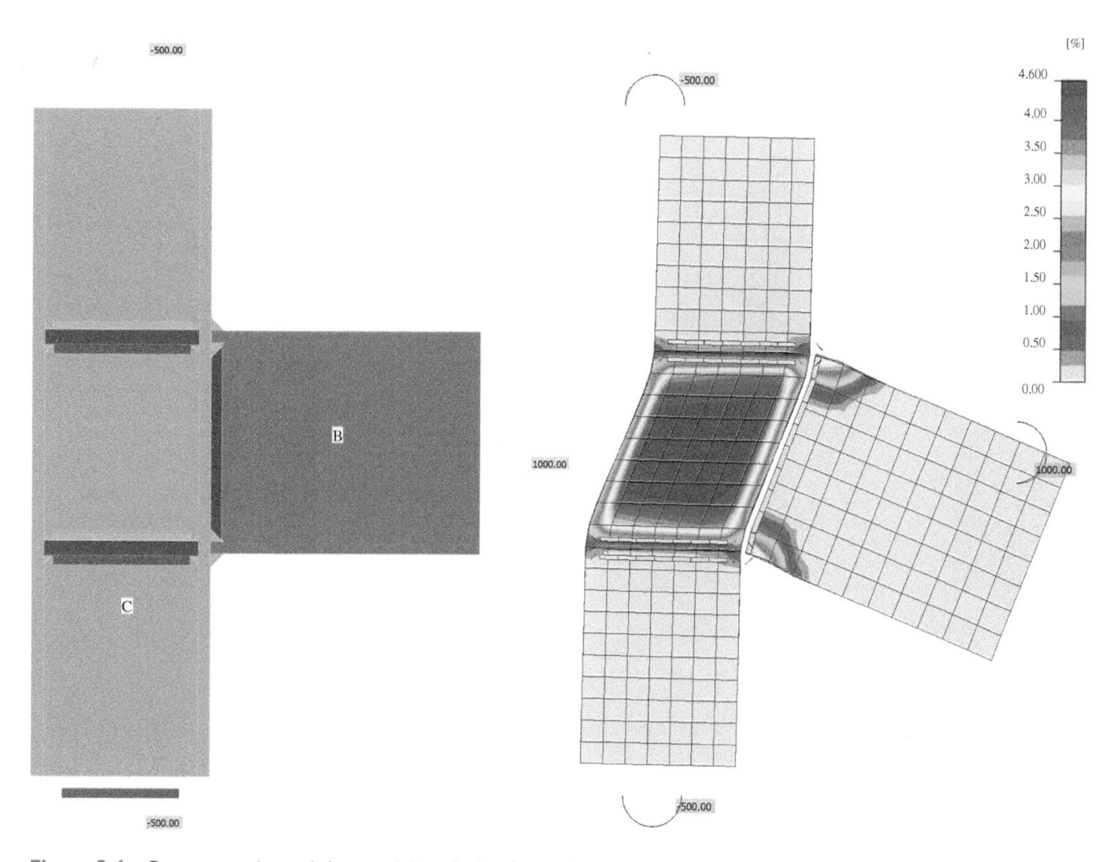

Figure 2.6 Beam-to-column joint model and plastic strains at ultimate limit state.

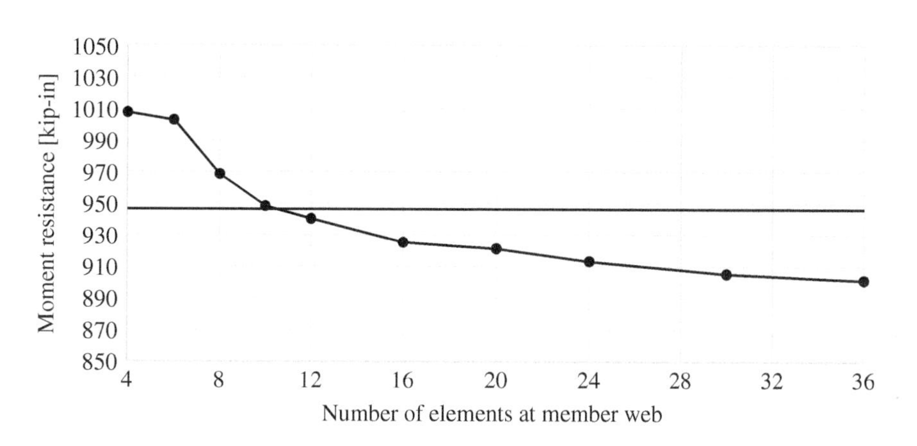

Figure 2.7 Influence of number of elements on the moment resistance.

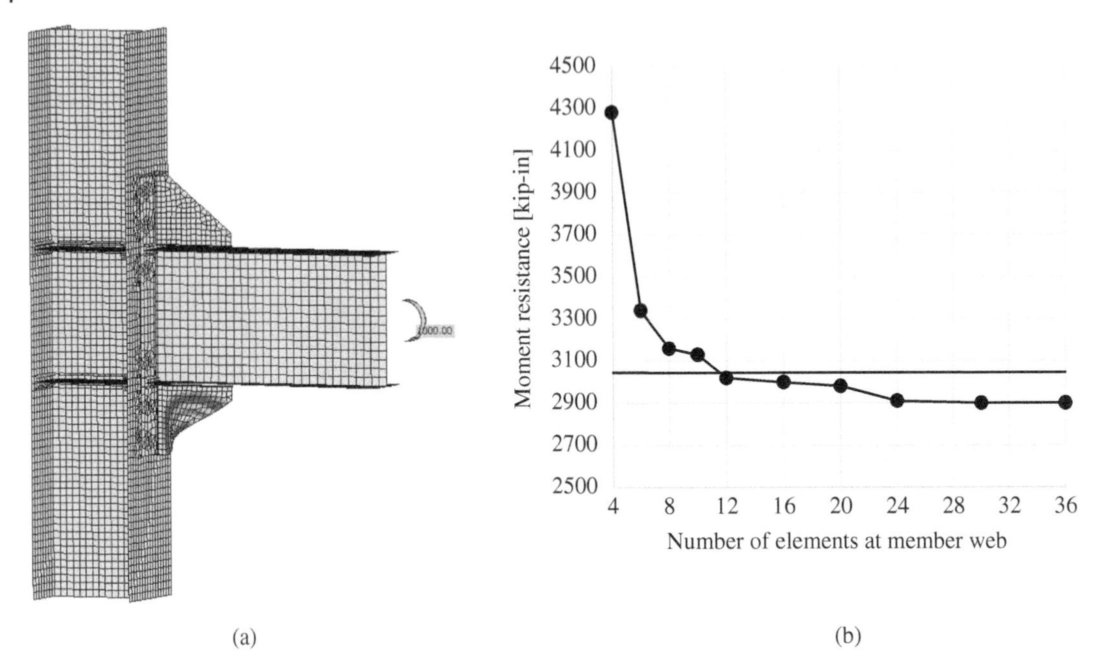

Figure 2.8 (a) First buckling mode; and (b) influence of number of elements along the stiffener on the moment resistance.

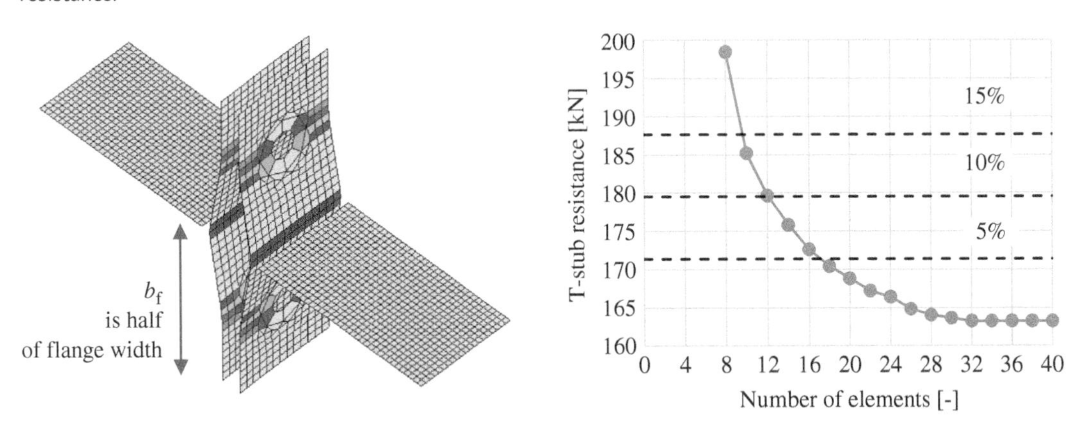

Figure 2.9 Influence of number of elements on the T-stub resistance.

A mesh sensitivity study of the T-stub in tension is presented. Half of the flange width is subdivided into 8–40 elements, and the minimum element size is set to 1 mm. The influence of the number of elements on the T-stub resistance is shown in Figure 2.9. The dashed lines represent the 5%, 10%, and 15% differences. It is recommended to use 16 elements on the half of the flange width.

A mesh sensitivity study of a uniplanar transverse plate circular hollow section T joint in compression is presented. The number of elements along the surface of the biggest circular hollow member is changed from 16 to 128. The influence of the number of elements on the joint resistance is shown in Figure 2.10. The dashed lines represent the 5% and 15% differences. It is recommended to use 64 elements along the surface of the circular hollow member.

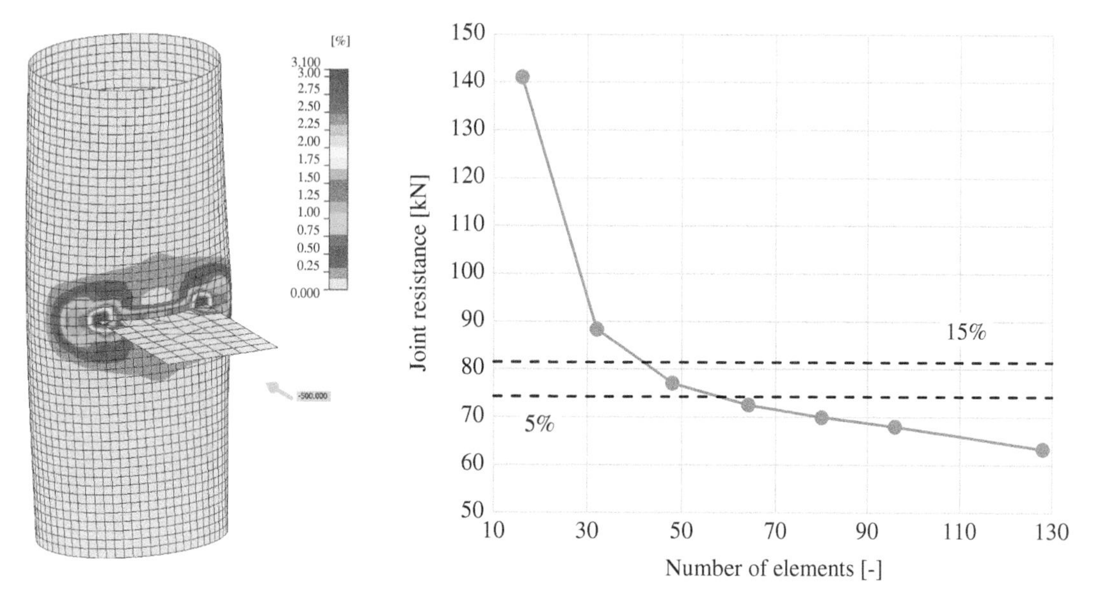

Figure 2.10 Influence of number of elements on the joint resistance of a CHS member.

A mesh sensitivity study of a uniplanar square hollow section X joint in compression is presented in Figure 2.11. The number of elements on the biggest web of a hollow member is changed from 4 to 24. The influence of the number of elements on the joint resistance is shown. The dashed lines represent the 5% and 15% differences. It is recommended to use 16 elements on the biggest web of a rectangular hollow member.

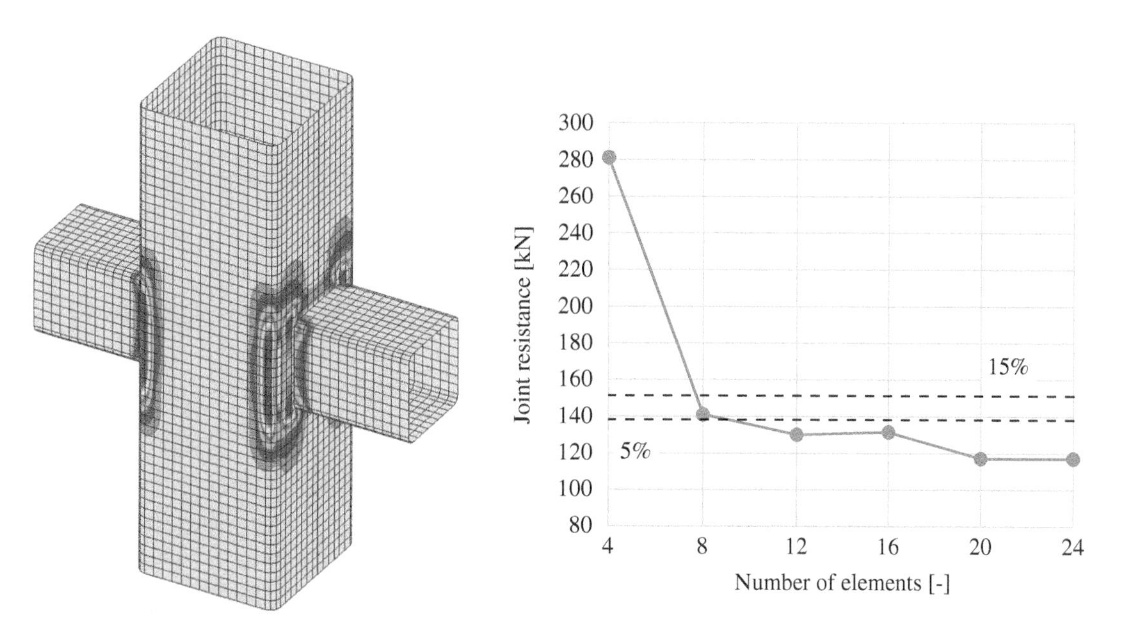

Figure 2.11 Influence of number of elements on the joint resistance of a rectangular hollow member.

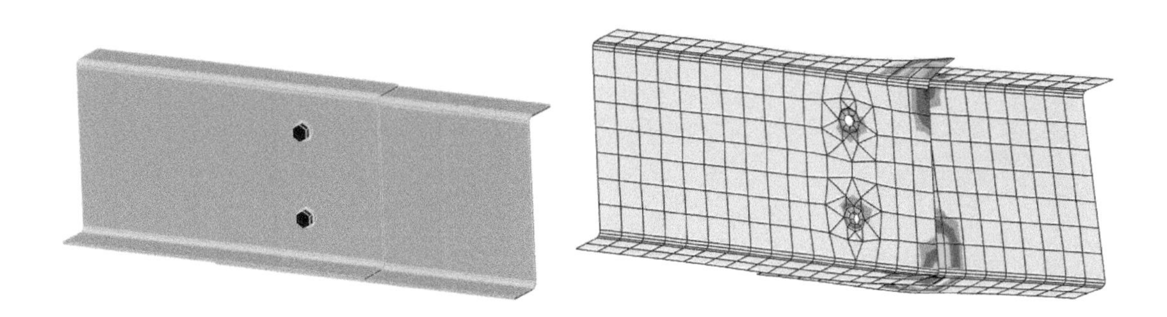

Figure 2.12 Example of contact stress of two overlapped Z sections for bolted connections of purlins.

2.3 Contacts

The standard penalty method is recommended for modeling a contact between plates. If penetration of a node into an opposite contact surface is detected, penalty stiffness is added between the node and the opposite plate. The penalty stiffness is controlled by a heuristic algorithm during nonlinear iteration to get better convergence. The solver automatically detects the point of penetration and solves the distribution of contact force between the penetrated node and the nodes on the opposite plate. It allows the contact between different meshes, see Figure 2.12. The advantage of the penalty method is the automatic assembly of the model. The contact between the plates significantly impacts the redistribution of forces in connection.

2.4 Welds

Several options for treating welds in numerical models exist; two approaches are described in this section.

2.4.1 Direct Connection of Plates

The first option for modeling the weld between plates is a direct merge of meshes by multipoint constraints, as shown in Figure 2.13. The load is transmitted to the opposite plate through force-deformation multipoint constraints (MPC) based on a Lagrangian formulation. The connection relates the finite element nodes of one plate edge to another plate. The nodes of the shell finite elements are not directly connected. The advantage of this approach is the ability to connect meshes with different densities.

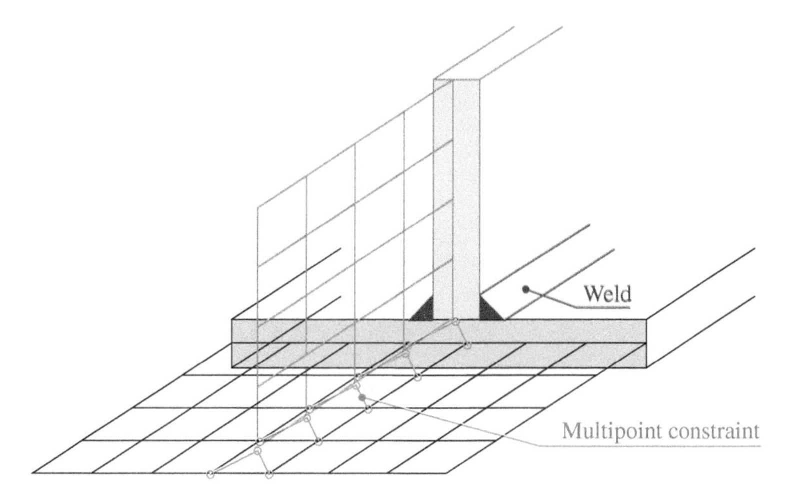

Figure 2.13 Constraint between mesh nodes.

The constraint allows modeling of the midline surface of the connected plates with the offset, which simulates the real plate thickness. This type of connection is used for full-penetration butt welds.

2.4.2 Weld with Plastic Redistribution of Stress

The load distribution in a weld is derived from the MPC, so the stresses are calculated in the throat section. This is important for stress distribution in a plate under the weld and for modeling the T-stubs. This model does not consider the stiffness of the weld, and the stress distribution is conservative. Stress peaks, which appear at the ends of plate edges, in corners, and rounding, govern the resistance along the whole weld length. To express the weld behavior, an improved weld model is applied. A special elasto-plastic element is inserted between the plates with the corresponding weld dimensions, as shown in Figure 2.14. The element considers the weld throat thickness, position, and orientation. Nonlinear material analysis is applied, and elasto-plastic behavior in the equivalent weld element is considered. The stress peaks are redistributed along the weld length.

The aim of the weld design models is not to capture reality perfectly. Residual stresses or weld shrinkage are ignored. Only three stress components appear on the weld element, $\sigma_\perp$, $\tau_\perp$, and $\tau_\parallel$; the normal stress along the weld length $\sigma_\parallel$ is ignored, as in design codes. The force and angle necessary for weld check, according to the AISC *Specification*, are derived from these stresses. The weld design models are verified for their resistance according to the relevant codes. An appropriate design weld model is selected for each design code. The resistances of regular welds, welds to unstiffened flange, long welds, and multi-oriented weld groups were investigated to select parameters of the design weld element. The plastic strain in steel plates is normalized to 5% to conform with the maximum plastic strain of plates.

2.4.3 Weld Deformation Capacity

The weld deformation capacity is determined to comply with the resistance of the welds to the unstiffened flange (Cl. 4.10 in EN 1993-1-8:2006) (CEN, 2006b), multi-oriented weld group

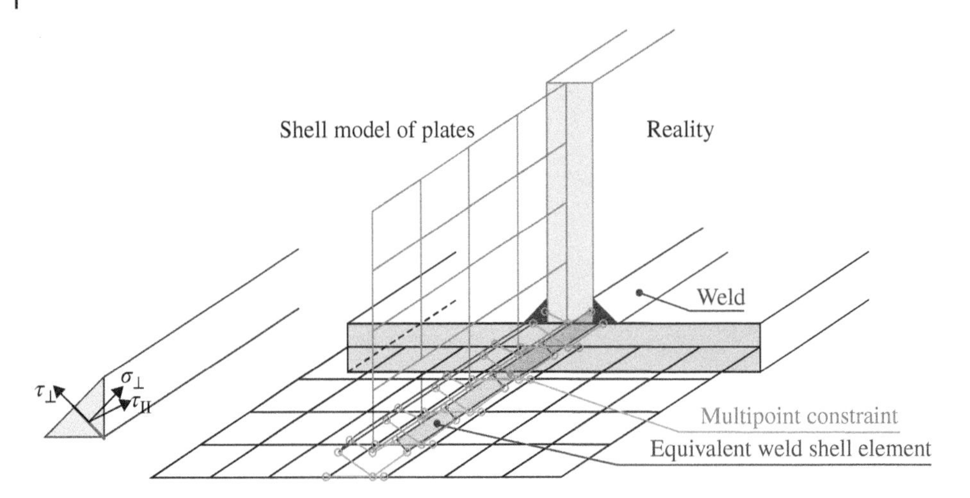

Figure 2.14 Constraint between weld element and mesh nodes.

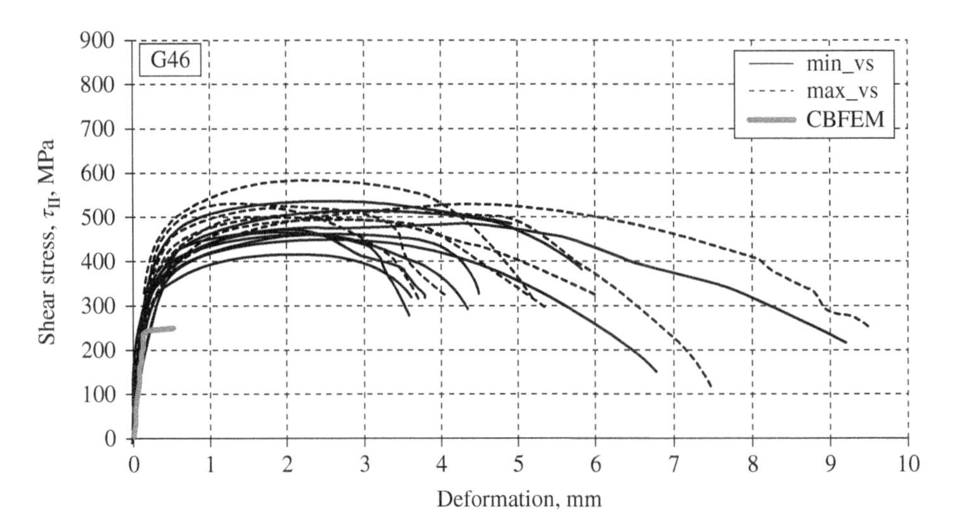

Figure 2.15 Longitudinal welds, stress-deformation diagrams of real tests from (Kleiner, 2018) compared to a model of a weld with plastic stress redistribution.

(AISC360-16, J2.4) (AISC, 2016), and long welds (AISC 360-16, J2.2b) (IDEA StatiCa, 2023a). This design deformation capacity was compared to sets of experiments from the literature for longitudinal welds (Kleiner, 2018) and transverse welds (Ng et al., 2002). The deformation capacity is, for simplicity, the same for longitudinal and transverse welds in CBFEM. This is not the case in reality. From Figures 2.15 and 2.16, it can be seen that the weld model with plastic redistribution of stress is very conservative for longitudinal welds and at the lower boundary for transverse welds. That confirms the model for strain compatibility of welds in AISC 360-10, Section J2-4 (AISC, 2010).

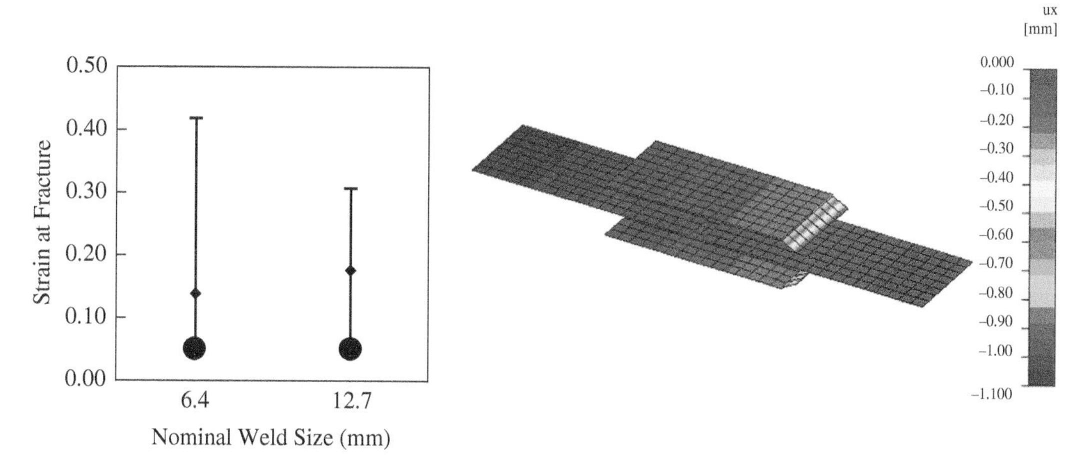

Figure 2.16 Transverse welds, strain at fracture, e.g. deformation divided by weld leg size, for tests of lapped splice joints in (Ng et al., 2002), compared to the model of a weld with plastic stress redistribution.

2.5 Bolts

In CBFEM, the bolt in tension and shear is modeled by a dependent nonlinear spring. The deformation stiffness of the shell element used for modeling the plates distributes the forces between the bolts and simulates the bearing on the plate adequately.

2.5.1 Tension

The spring of a bolt in tension is described by its initial deformation stiffness, design resistance, initialization of yielding, and deformation capacity. The initial stiffness is derived analytically according to Agerskov (1976) as:

$$D_{Lb} = \frac{L_s + 0.4 \cdot d_b}{E \cdot A_{ss}}$$

$$A_{PP} = \frac{0.75 \cdot D_H \cdot (L_w - D_H)}{D_{W1}^2 - D_{W2}^2}$$

$$A_{P1} = \frac{\pi}{4}\left(D_H^2 - D_{W1}^2\right)$$

$$A_{P2} = \frac{1}{2}\left(D_{W2}^2 - D_H^2\right)\tan^{-1} A_{PP}$$

$$A_P = A_{P1} + A_{P2}$$

$$D_{LW} = \frac{L_W}{EA_P}$$

$$k = \frac{1}{D_{LB} + D_{LW}}$$

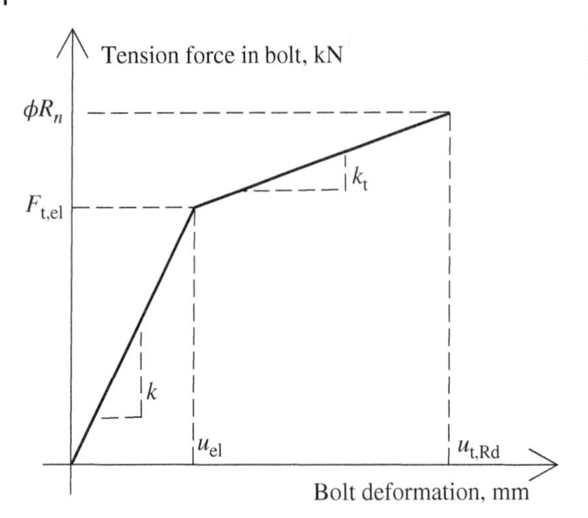

Figure 2.17 Load-deformation diagram of a bolt in tension.

where d_b is the bolt diameter, D_H is the bolt head diameter, D_{W1} and D_{W2} are the washer inner and outer diameter, respectively, L_W is the washer thickness, L_s is the bolt grip length, A_{ss} is the bolt gross area, and E is Young's modulus of elasticity.

The model corresponds well to experimental data (Gödrich et al., 2014). For the initialization of yielding and the deformation capacity, it is assumed that the plastic deformation occurs in the threaded part of the bolt shank only. The load-deformation diagram of the bolt is shown in Figure 2.17 and is derived as

$$F_{t,el} = \frac{\phi R_n}{c_1 \cdot c_2 - c_1 + 1}$$

$$k_t = c_1 \cdot k; c_1 = \frac{F_u - F_y}{\frac{1}{4} \cdot AE - F_y}$$

$$u_{el} = \frac{F_{t,el}}{k}$$

$$u_{t,Rd} = c_2 \cdot u_{el}; c_2 = \frac{A \cdot E}{4 \cdot F_y}$$

where k is the linear stiffness of the bolt, k_t the stiffness of bolt at the plastic branch, $F_{t,el}$ the limit force for linear behavior, A the percentage elongation after a fracture of a bolt, F_y the bolt yield strength, F_u the bolt tensile strength, ϕR_n the limit bolt resistance, and $u_{t,Rd}$ the limit deformation of a bolt.

2.5.2 Shear

The initial stiffness and the design resistance of a bolt in shear are defined by the following formulas:

$$k_{el} = \frac{1}{\frac{1}{k_{11}} + \frac{1}{k_{12}}}$$

$$k_{11} = \frac{8d_b^2 F_{ub}}{d_{M16}}$$

$$k_{12} = 15k_t d_b F_{up}$$

$$k_t = \min\left(2.5, \frac{1.5t_{min}}{d_{M16}}\right)$$

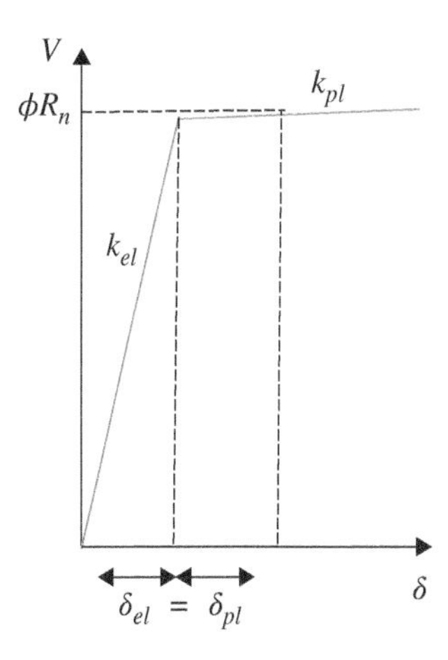

Figure 2.18 Load-deformation diagram of a bolt in shear.

$$k_{12} = 12 k_t d_b F_{up}$$

$$k_{pl} = \frac{k_{el}}{1000}$$

where d_b is the bolt diameter, F_{ub} is the bolt tensile strength, $d_{M16} = 16$ mm is the diameter of the reference bolt M16, F_{up} is the tensile strength of the connected plate, t_{min} is the minimum thickness of the connected plate.

The spring representing the bolt in shear has bi-linear force deformation behavior. Initialization of yielding is expected at $\phi R_{\mathrm{n,el}}$; see Figure 2.18.

$$\phi R_{\mathrm{n,el}} = 0.999\, \phi R_{\mathrm{n}}$$

The deformation at ultimate capacity is considered as:

$$\delta_{pl} = \delta_{el}$$

2.6 Interaction of Shear and Tension in a Bolt

A combination of shear and tension in a bolt is expressed in AISC 360-16, J3.7 (AISC, 2016) by a trilinear relation and checked as:

$$\phi R_n = \phi F'_{nt} A_b \quad \text{(LRFD)}$$

$$\frac{R_n}{\Omega} = \frac{F'_{nt} A_b}{\Omega} \quad \text{(ASD)}$$

$$1.3 F_{\mathrm{nt}} - \frac{F_{\mathrm{nt}}}{\phi F_{\mathrm{nv}}} f_{\mathrm{rv}} \leq F_{\mathrm{nt}} \quad \text{(LRFD)}$$

$$1.3 F_{\mathrm{nt}} - \frac{\Omega F_{\mathrm{nt}}}{F_{\mathrm{nv}}} f_{\mathrm{rv}} \leq F_{\mathrm{nt}} \quad \text{(ASD)}$$

where F_{nt} is the bolt nominal tensile strength, F_{nv} is the nominal bolt shear strength, f_{rv} is the required shear stress using LRFD or ASD combinations. A condition limiting the bolt resistance is shown in Figure 2.19.

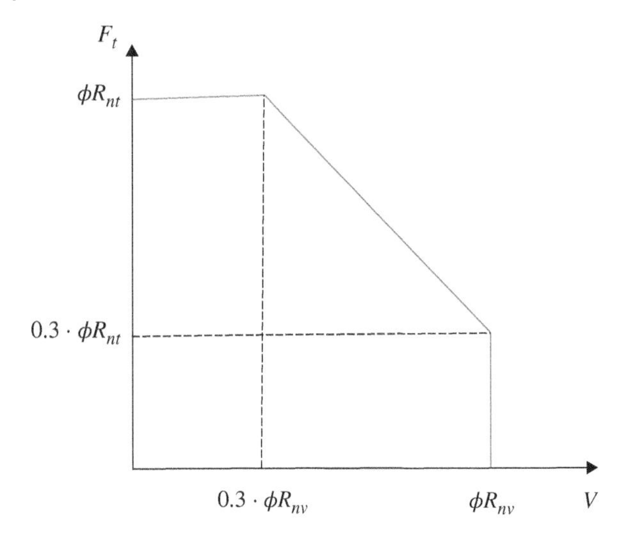

Figure 2.19 Limit condition of interaction of shear and tension in a bolt.

The tensile and shear behaviors of the bolt in a numerical model are represented by a bilinear spring model; see Section 2.5. The nonlinear spring shows special behavior in the interaction of the shear and tension. The shear and tension forces are presented as nonlinear functions of the shear and tension deformations composed of ruled surfaces; see Figure 2.20, where ULT is the bolt deformation at tensile resistance, ULS is the bolt deformation at shear resistance, R_{nt} is the bolt tensile resistance, R_{nv} is the

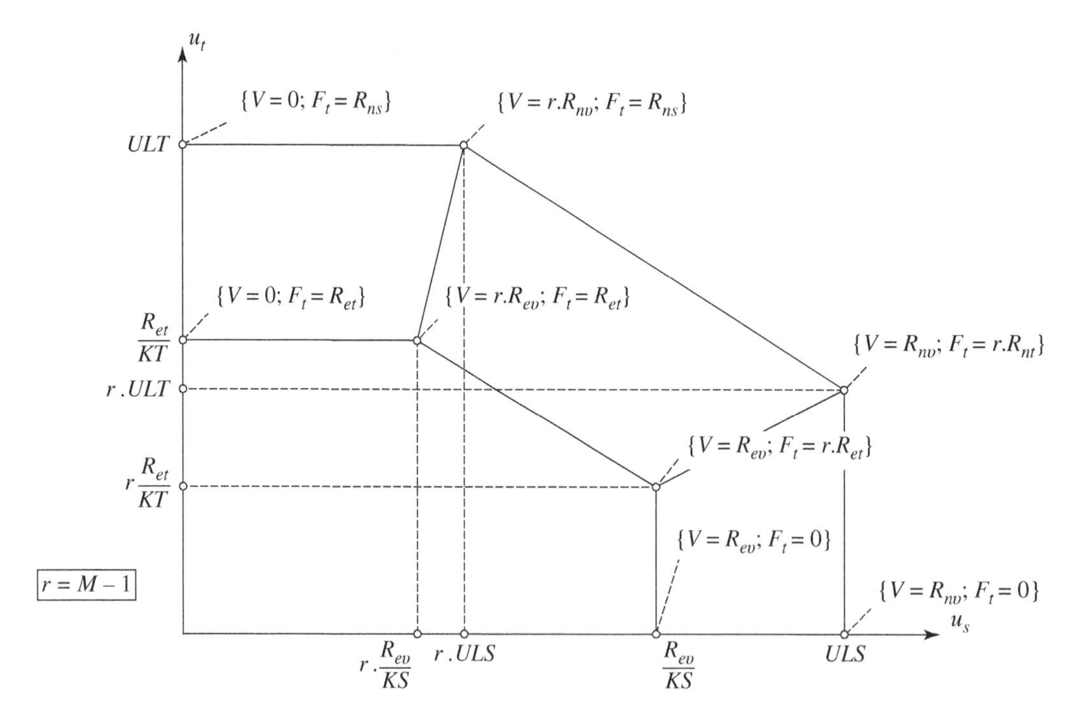

Figure 2.20 Bolt tension force as a function of deformation in shear and tension.

bolt shear resistance, $r = 0.3$. The functions consider the limit condition of interaction, shown above. It is clear that the bolt can be in four states: a linear behavior, a plastic state in tension, a plastic state in shear, and a plastic state in interaction of the tension and the shear.

2.7 High-Strength Bolts in Slip-Critical Connections

In connections with pretensioned bolts, the shear force is transferred by friction between the steel surfaces. Compared to snug-tight bolts, the friction is controlled by the preloading force. The final resistance is assured by bolt shearing and bolt bearing after the slippage of a bolt into a hole. In AISC 360-22, the pretension forces are given in Table J3.1 and the resistances are given in J3.9 and J3.10. The bolts are expected to be preloaded to at least 70% of their strength, F_u (Table J3.1 in AISC 360-22), and the bolt preloading force is

$$T_b = 0.7 F_u A_s$$

where F_u is the bolt tensile strength and A_s is the bolt area effective in tension. The bolt is expected to deform by 80% and the plate by 20%. If the external tensile force T_u is applied to the connection in the direction of a bolt, the slip resistance will be reduced by a factor

$$k_{sc} = 1 - \frac{T_u}{D_u T_b} \leq 1.0 \quad \text{(LRFD)}$$

$$k_{sc} = 1 - \frac{1.5\, T_u}{D_u T_b} \leq 1.0 \quad \text{(ASD)}$$

The bolt available slip resistance is determined by

$$\phi R_n = \phi \mu D_u h_f T_b \quad \text{(LRFD)}$$

$$\frac{R_n}{\Omega} = \frac{\mu D_u h_f T_b}{\Omega} \quad \text{(ASD)}$$

where μ is mean slip coefficient, $D_u = 1.13$ is a multiplier that reflects the ratio of the mean installed bolt pretension, h_f is factor for fillers. Each bolt and each shear plane is investigated separately, and only the most utilized is shown to the user.

In numerical design calculation, the component preloaded bolt is simulated either as a nonlinear spring using the preloading force and deformation or by restraints between surfaces, representing the friction, with a spring that includes a preloading force in a bolt. The CBFEM model is similar to the conventional, snug-tight bolt model for the bolt component represented by the preloading force and its slip deformation. The difference is in the transfer of shear forces to the plate. For a preloaded bolt, all the nodes of the bolt hole are loaded, while a snug-tight bolt contains gap elements, i.e. only the nodes in the direction of the shear are loaded. The shear characteristic of a preloaded bolt is shown in Figure 2.21. The initial linear shear stiffness is determined from the stiffness of the cylinder under the head of the bolt, and it is practically rigid. The shear force limit includes the external tensile load to the bolt per AISC 360-22, J3.10. The advantage of this simplified model is its computational stability and low demands of the FE model and the consistency of results with traditional calculation methods. The model does not consider the actual distribution of contact pressures between the plates and the history of preloading.

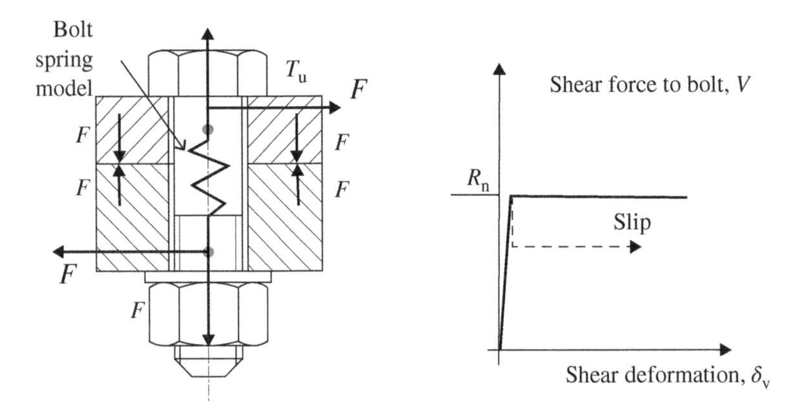

Figure 2.21 Shear characteristic of the preloaded bolt.

2.8 Anchor Bolts

2.8.1 Description

The anchor bolt is modeled with similar procedures as structural bolts. The bolt is fixed on one side to the concrete block. Its length L_b is taken according to EN 1993-1-8:2005 as a sum of the washer thickness t_w, the base plate thickness t_{bp}, the grout thickness t_g, and the free length embedded in concrete, which is expected as $8d$, where d is the bolt diameter. The stiffness in tension is calculated as $k = E\,A_s/L_b$. The load-deformation diagram of the anchor bolt is shown in Figure 2.22. Note that the failure modes related to concrete are brittle and may occur before any plastic deformation in anchor bolts.

The stiffness of the anchor bolt in shear is taken as the stiffness of the structural bolt in shear. The anchor bolt resistances are evaluated according to ACI 318-14 (IDEA StatiCa, 2023b). Both steel and

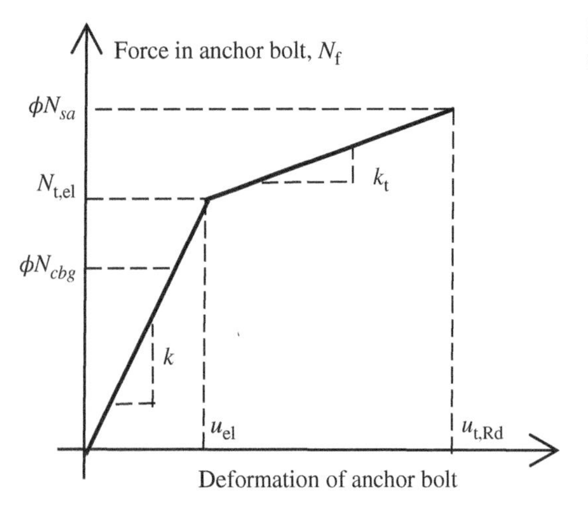

Figure 2.22 Load-deformation diagram of the anchor bolt.

concrete failure modes with general formulas are checked. However, some failure modes are determined by testing and only anchor manufacturers may provide the resistances. These are:

- Pull-out failure of post-installed mechanical fasteners
- Bond strength of adhesive anchors
- Concrete splitting failure during installation.

2.8.2 Anchor Bolts with Stand-Off

Anchors with stand-off can be checked as a construction stage before the column base is grouted or as a permanent state. The design guidelines for stand-off anchors are unclear. Therefore, an anchor with stand-off is designed as a bar element loaded by shear force, bending moment, and compressive or tensile force. Linear interaction is used for tension/compression and bending moment. The model contains two plastic hinges. The anchor is fixed on both sides; one side is $0.5 \times d$ below the concrete level, and the other side is in the middle of the thickness of the plate. The buckling length is conservatively assumed as twice the length of the bar element. Plastic section modulus is used. The forces in anchor with stand-off are determined using finite element analysis. Bending moment depends on the stiffness ratio of the anchors and base plate.

2.9 Concrete Block

2.9.1 Design Model

In the component-based finite element method (CBFEM), it is convenient to simplify the concrete block as 2D contact elements. The connection between the concrete and the base plate resists compression only. Compression is transferred via the Winkler-Pasternak subsoil model, which represents deformations of the concrete block. Tension force between the base plate and concrete block is carried by anchor bolts. Shear force is transferred by friction between a base plate and a concrete block, by shear lug, or by shearing and bending of anchor bolts. The resistance of bolts in shear is assessed analytically. Friction is modeled as a full single-point constraint in the plane of the base plate-concrete contact. In the shear lug model, the shear load is assumed to act in the mid-plane of the base plate, and it is resisted by concrete bearing stress at the whole area of the shear lug embedded in concrete. The bearing strength is based on a uniform bearing stress acting over the area of the shear lug. The area of the shear lug in the grout layer is assumed to be ineffective. There is a lever arm between the point of shear load application (base plate mid-plane) and the center of resistance (mid-depth of the shear lug embedded in concrete). This causes an additional bending moment that needs to be resisted by the bearing resistance of the concrete and the tension in the anchors.

2.9.2 Resistance

The resistance of concrete in 3D compression is determined based on AISC 360-16, J8, by calculating the design bearing strength of concrete $f_{p,max}$ under the compressed area A_1 of the base plate. The design bearing strength $\phi_c f_{p,max}$ and allowable bearing strength $\frac{f_{p,max}}{\Omega_c}$ are calculated as:

$$\phi_c f_{p,max} = \phi_c 0.85 f'_c \sqrt{\frac{A_2}{A_1}} \leq \phi_c 1.7 f'_c \quad \text{(LRFD)}$$

$$\frac{f_{p,max}}{\Omega_c} = \frac{0.85 f'_c \sqrt{\frac{A_2}{A_1}}}{\Omega_c} \leq \frac{1.7 f'_c}{\Omega_c} \quad \text{(ASD)}$$

where $\phi_c = 0.65$ is the resistance factor, $\Omega_c = 2.31$ is the safety factor, A_1 is the area of steel bearing on a concrete support, A_2 is the maximum area of the portion of the supporting surface that is geometrically similar to and concentric with the loaded area, and f_c' is the specified compressive strength of concrete.

The area where the concrete is in compression is taken from the results of FEA. A stress cut-off may be selected to ignore the compressed area with negligible compressive stress. This area in compression A_1 allows assessment of the resistance for a generally loaded column base of any column shape with any stiffeners. The average stress σ on the compressed area A_1 is determined as the compression force (including any compression due to bending and prying actions) divided by the compressed area. Check of the component is in stresses $\sigma \leq \phi_c f_{p,max}$ (LRFD) and $\sigma \leq \frac{f_{p,max}}{\Omega_c}$ (ASD).

This procedure of assessing the resistance of the concrete in compression is independent of the mesh of the base plate, as can be seen in Figure 2.23 and Figure 2.24. Two cases are investigated: loading by pure compression 600 kips, see Figure 2.23, and loading by a combination of compressive force 600 kips and bending moment 350 kip-ft, see Figure 2.24.

2.9.3 Concrete in Compression Stiffness

The stiffness of the concrete block may be predicted for the design of column bases as an elastic hemisphere. A Winkler-Pasternak subsoil model is commonly used for a simplified calculation of foundations. The stiffness of the subsoil is determined using the modulus of elasticity of the concrete and the effective height of the subsoil as

$$k = \frac{E_c}{(\alpha_1 + \upsilon) \cdot \sqrt{\frac{A_{eff}}{A_{ref}}}} \cdot \left(\frac{1}{\frac{h}{\alpha_2 \cdot d} + \alpha_3} + \alpha_4 \right)$$

where k is the stiffness in compression, E_c is the modulus of elasticity, v is the Poisson coefficient of the concrete foundation, A_{eff} is the effective area, A_{ref} is the reference area, d is the base plate width, h is the column base height, and α_i are coefficients. SI units must be used in the formula. The following values for coefficient were used, based on the results of research-oriented finite element models with concrete modeled by solid elements:

$$A_{ref} = 1\ \text{m}^2;\ \alpha_1 = 1.65;\ \alpha_2 = 0.5;\ \alpha_3 = 0,3;\ \alpha_4 = 1.0.$$

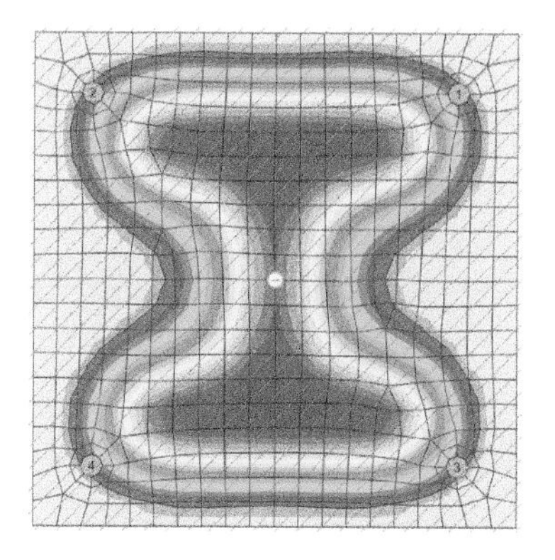

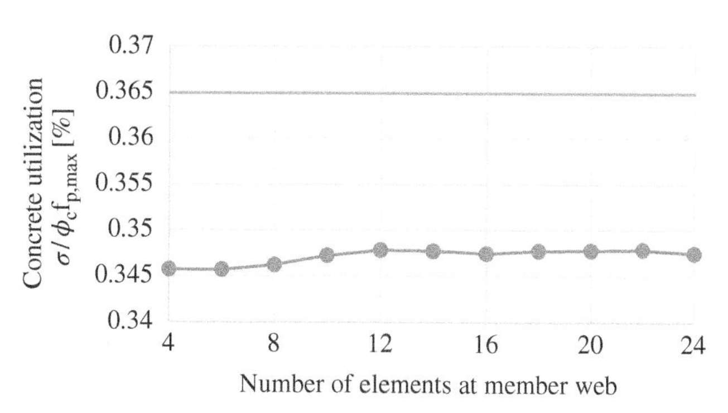

Number of elements	Concrete block utilization %
4	34.571
6	34.571
8	34.620
10	34.725
12	34.782
14	34.772
16	34.742
18	34.772
20	34.776
22	34.787
24	34.746

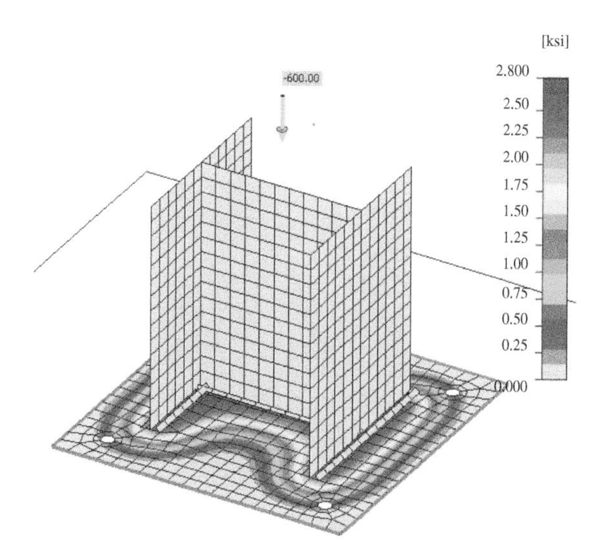

Figure 2.23 Influence of number of elements on prediction of resistance of concrete in compression in the case of pure compression.

2.10 Local Buckling of Compressed Internal Plates

In numerical simulations, the slender plates in compression considering plate geometrical imperfections, residual stresses, and large deformation during analyses may be designed according to EN 1993-1-5:2006 (CEN, 2006a) and prEN 1993-1-14. This should be précised according to the different plate/joint configurations. The FEA procedure naturally offers the prediction of the buckling load of the joint. The state-of-the-art design procedure for slender cross-sections according to the reduced stress method is described in Annex B of EN 1993-1-5:2006 (CEN, 2006a). It predicts the post-buckling resistance of joints. Critical buckling modes are determined using linear buckling analysis (LBA). In the first step, the minimum load amplifier for the design loads to reach the characteristic resistance value of the most critical

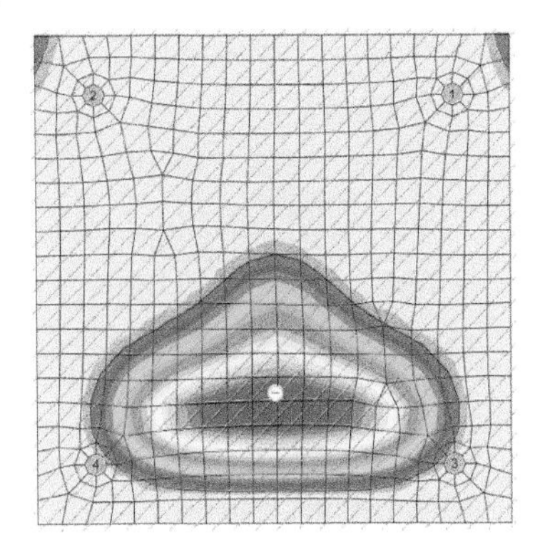

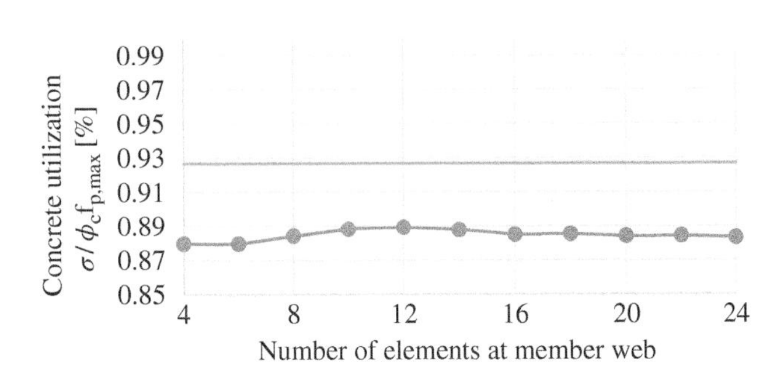

Number of elements	Concrete block utilization %
4	87.935
6	87.935
8	88.391
10	88.801
12	88.897
14	88.753
16	88.470
18	88.500
20	88.377
22	88.398
24	88.278

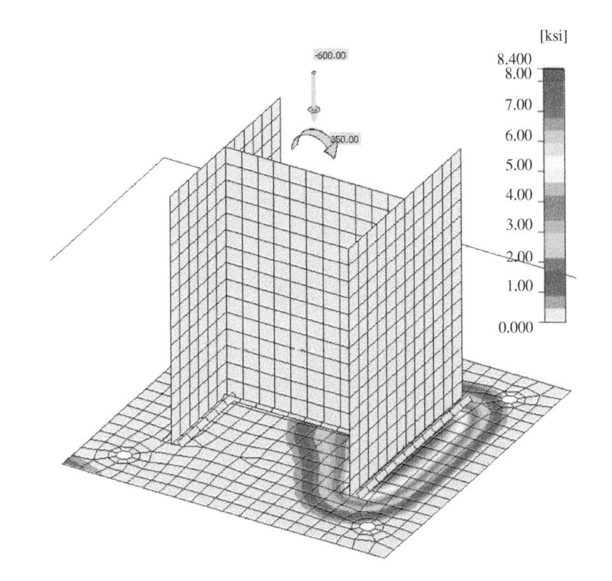

Figure 2.24 Influence of number of elements on prediction of resistance of concrete in compression in case of compression and bending.

point $\alpha_{ult,k}$ is obtained by MNA. The ultimate limit state is reached at 5% plastic strain. The critical buckling factor α_{cr} is determined by LBA and stands for the load amplifier to reach the elastic critical load under the complex stress field. Examples of critical buckling modes in steel joints are shown in Figure 2.25.

The load amplifiers are related to the non-dimensional plate slenderness, which is determined as follows

$$\bar{\lambda} = \sqrt{\frac{\alpha_{ult}}{\alpha_{cr}}}$$

The reduction buckling factor for internal plates restrained on all sides ρ is calculated according to EN 1993-1-5:2006, Annex B (CEN, 2006a). Conservatively, the lowest value from the longitudinal, transverse,

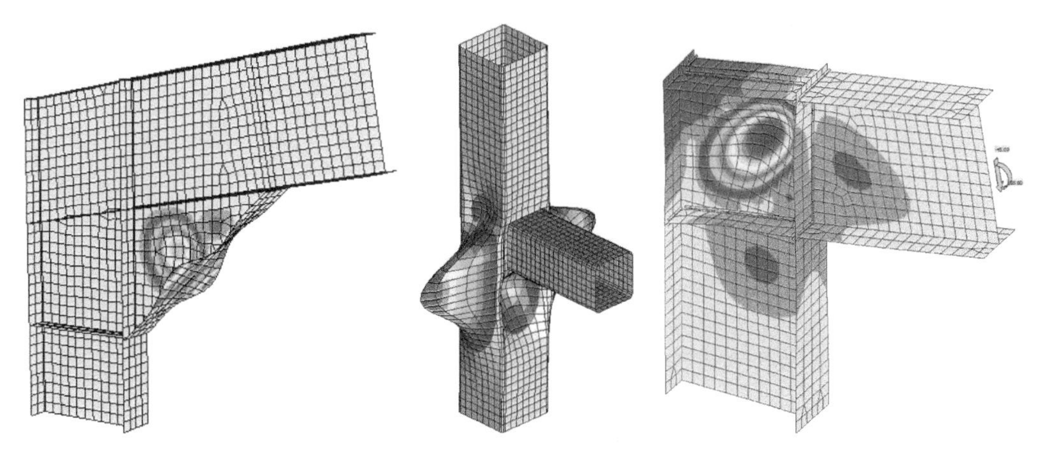

Figure 2.25 Examples of first buckling mode in CBFEM models.

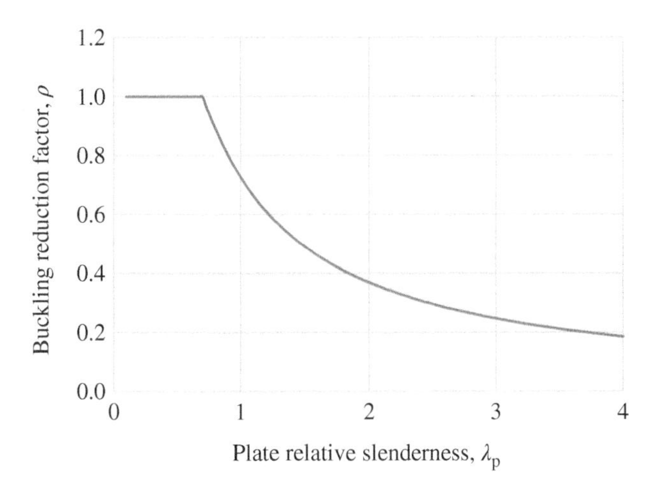

Figure 2.26 Buckling reduction factor ρ according to EN 1993-1-5:2006, Annex B (CEN, 2006a).

and shear stress is taken. Figure 2.26 shows the relation between the plate slenderness and the reduction buckling factor. In AISC 360-16 (AISC, 2016), the corresponding procedure is found in E7, Table E7.1.

Plate verification is based on the von Mises yield criterion and the reduced stress method. Buckling resistance is assessed as:

$$\phi \alpha_{ult} \rho \geq 1 \quad \text{(LRFD)}$$

$$\frac{\alpha_{ult} \rho}{\Omega} \geq 1 \quad \text{(ASD)}$$

where ϕ is the resistance factor and Ω is the safety factor. Based on the distribution of strains calculated in MNA and the critical buckling factor in LBA the buckling resistance is assessed without using second-order calculation and applying imperfections. It is recommended to check the buckling resistance

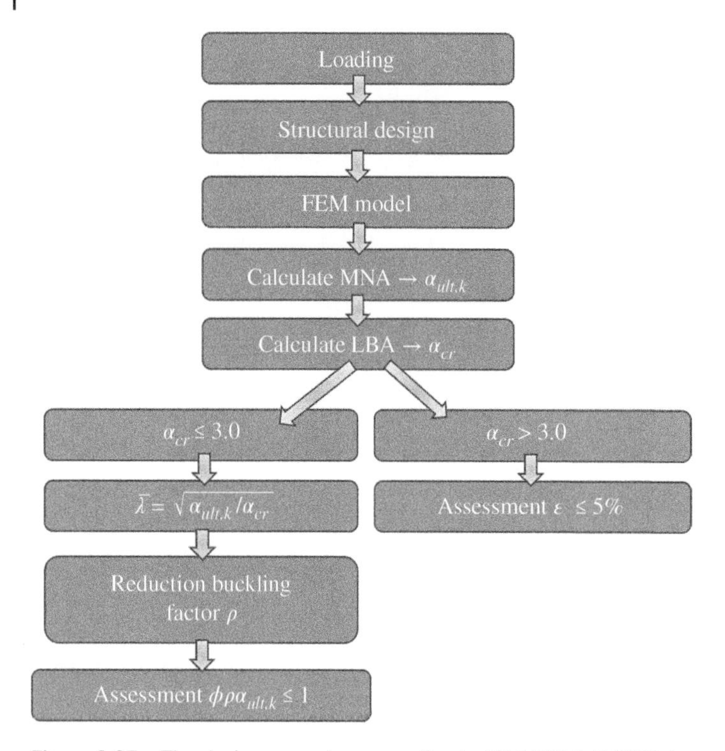

Figure 2.27 The design procedure according to EN 1993-1-5:2006 Annex B (CEN, 2006a).

for a critical buckling factor smaller than 3.0 for LRFD and smaller than 4.5 for ASD. Otherwise, the resistance is governed by reaching 5% plastic strain for plates with a smaller slenderness. The procedure is summarised in Figure 2.27.

2.11 Moment-Rotation Relation

The joint is generally three-dimensional. The moment–rotation curve for connection in a joint is evaluated for each member attached to the joint separately. One member is analyzed and all the others are supported. The procedure is divided into two steps. In the first step, the whole joint including all the fasteners is calculated and the rotation at the end of the analyzed member is noted. In the second step, all the members connected in node by rigid links are calculated, and the rotation at the end of the analyzed member is noted again. Then, the rotation of the second model is subtracted from the rotation of the first model. This way, the moment-rotation curve is independent of the length of the member stubs.

The modeling of the moment rotational curve may be documented on the behavior of a well-designed, portal frame eaves moment bolted connection, developed based on US best practice. The composition of the connection geometry of the bolted connection is demonstrated in Figure 2.28. The rafter of cross-section W14 × 38 is connected to column W12 × 65 by the full depth end plate of thickness ½ in. by 12 bolts 7/8 A490. The haunch is 27 in. long and 12 in. high with flange dimensions 3/8 × 4,725 in.

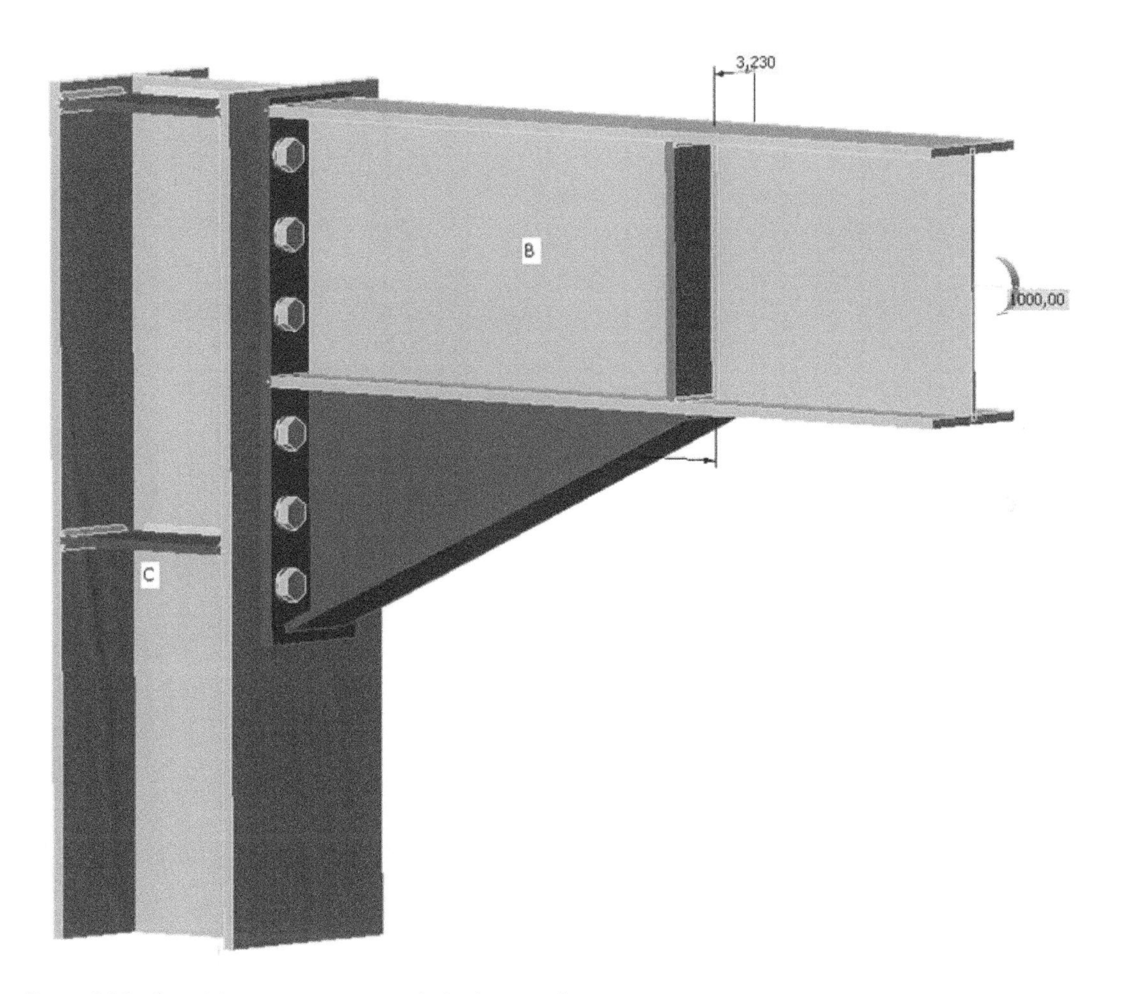

Figure 2.28 Portal frame eaves moment bolted connection.

The stiffeners are with the plate thickness of 9/16 in. The material of all plates is steel grade A992. The results of CBFEM analyses are shown in Figure 2.29. Plastic zones develop in connection, from first yielding under the bolt in tension through the development of full plasticity in the column web panel in shear, till reaching the 5% strain in the column web panel. After reaching this strain, the plastic zones propagate rapidly in the column web panel in shear, and for small steps in bending moment, the rotation of the joint rises significantly.

As is commonly known in well-designed connections, the plastification starts early; see Figure 2.29(a). The column web panel in shear brings in the deformation capacity of a connection and guides the nonlinear part of behavior; see Figure 2.29(e)–(f). Figure 2.30 demonstrates the fast development of yielding in the column web panel after reaching the 5% strain.

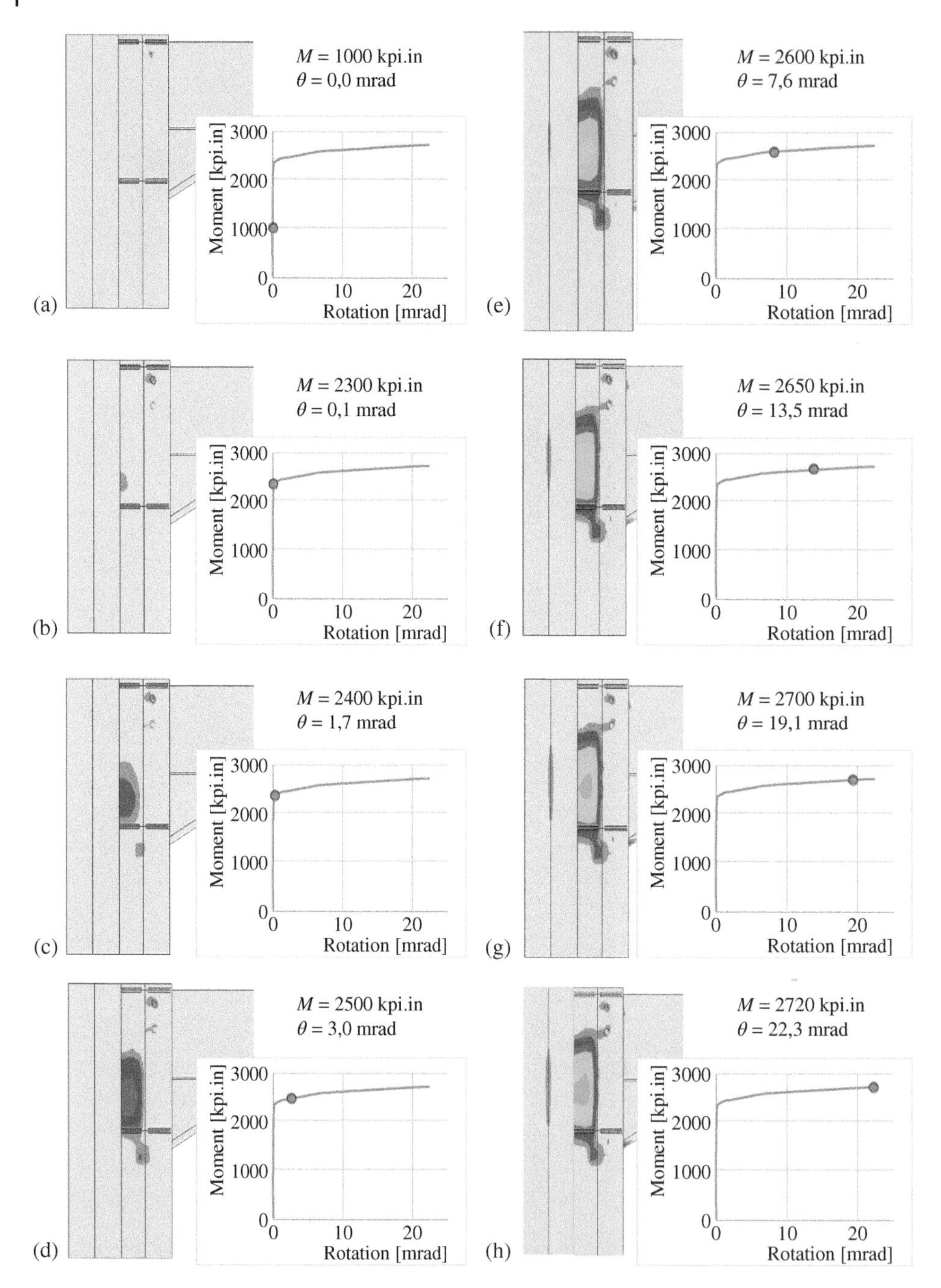

Figure 2.29 Development of plastic zones in connection by CBFEM analyses, from first yielding under the tensile bolt (a), through development of full plasticity in the column web panel loaded in shear (e)–(f), till reaching the 5% strain in panel (h).

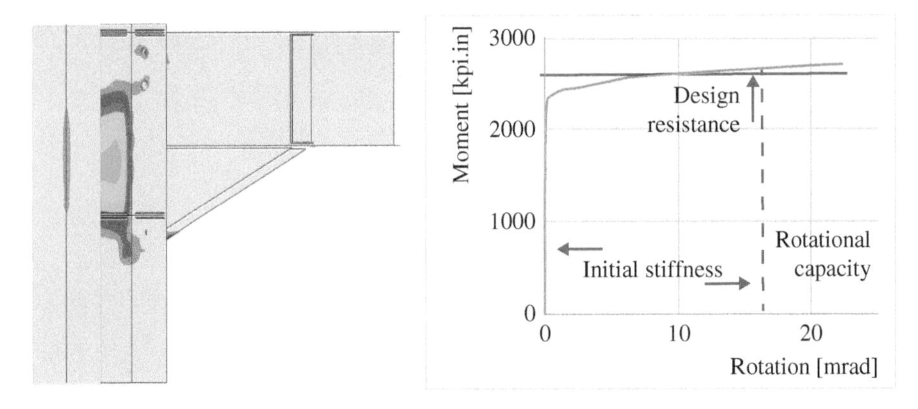

Figure 2.30 After reaching the 5% strain in the column web panel in shear, plastic zones propagate rapidly.

2.12 Bending Stiffness

The bending/deformation stiffness of structural joints has only an indirect influence on the resistance of the structure. The connection classification is included in AISC 360-16, Comm. B3.4 (AISC, 2016). Since it is a commentary, it is not binding and not described in depth. Analytically, the stiffness may be determined by the component method (CM) in EN 1993-1-8:2005. The prediction models using the CM show mostly higher values compared to experiments. The stiffness calculation by the CBFEM is different than the stress/strain analysis. The joint is usually three-dimensional. The bending/deformation stiffness for connection of a member is influenced by the shear deformation in the joint and the connection deformation of the particular member in the selected plane of the member. For joints connecting more members, the bending stiffnesses of the connections are analyzed separately in the planes, where they are loaded.

For the evaluation of the bending stiffness of a particular connection, it is assumed that members are supported at the ends, and only the analyzed member i has a free end; see Figure 2.31. The analyzed member i is loaded by a bending moment in plane yz coming from the global analyses. To learn the secant rotational stiffness $S_{\mathrm{j,i},yz}$ for rotation $\phi_{\mathrm{i},yz}$ the bending moment $M_{\mathrm{j,i},yz}$ is applied in a selected plane yz. The secant rotational stiffness $S_{\mathrm{j,s,i},yz}$ is derived from the formula:

$$S_{\mathrm{j,s,i},yz} = M_{\mathrm{j,i},yz}/\phi_{\mathrm{i},yz}$$

The total rotation $\phi_{\mathrm{j,i},yz}$ of the end section of analyzed member i in plane yz is derived on a model with one free/analyzed member. All the other members are fixed. The deformation of all members in the joint influences the calculated value. The effects of member deformation are removed by analyzing the substitute model of the joint, which is composed of members with appropriate cross-sections rigidly connected in node, as shown in Figure 2.32. By loading this substitute model by bending moment $M_{\mathrm{i},yz}$, the rotation $\phi_{\mathrm{j,ei,yz}}$ is reached. The rotation caused only by the construction of the connection is derived as:

$$\phi_{\mathrm{j,i,yz}} = \phi_{\mathrm{j,ti,yz}} - \phi_{\mathrm{j,ei,yz}}$$

Figure 2.31 FEA model for analysis of stiffness of a selected connected member.

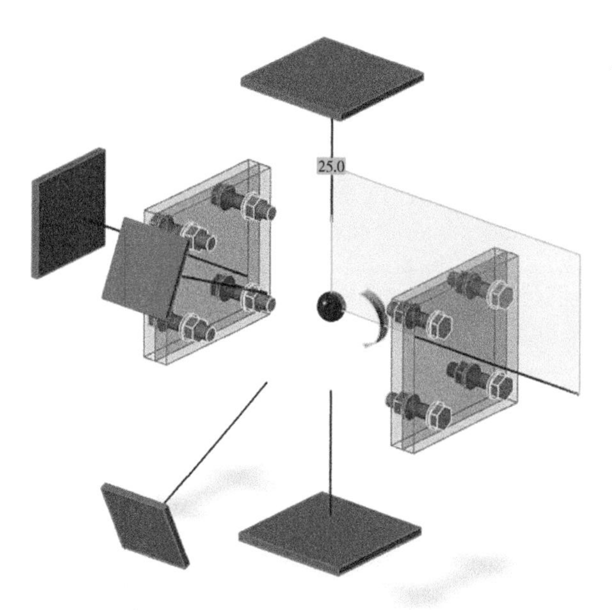

Figure 2.32 The substitute 1D model to eliminate the bending flexibility of the connected elements.

The secant stiffness of connection $\phi_{j,i,yz}$ is derived during its loading. The initial stiffness $\phi_{j,ini}$ is defined as elastic stiffness and is expected to be linear till 2/3 $M_{j,Rd}$; see Cl. 6.3.1 in EN 1993-1-8:2005. The calculation of the initial bending stiffness by the CBFEM is taken as a secant stiffness at 2/3 of the bending resistance ϕM_n. from the acting bending moment till 2/3 $M_{2/3Rd,i,yz}$ and corresponding rotation in connection $\phi_{2/3Rd,i,yz}$ as:

$$S_{j,ini,i,zy} \cong S_{j,s,2/3,i,zy} = \phi M_{2/3n,i,yz} / \phi_{2/3n,i,yz}$$

The values of the bending resistance $M_{j,Rd,i,yz}$ and initial stiffness $\phi_{j.ini,i,yz}$ are summarized in Figure 2.33.

The bending resistance $\phi M_{j,n,i,yz}$ of connection is in CBFEM evaluated by strain 5% in plates/sections or resistance of connectors, e.g. bolts, welds. Figure 2.34 shows the strain in the joint exposed to moments

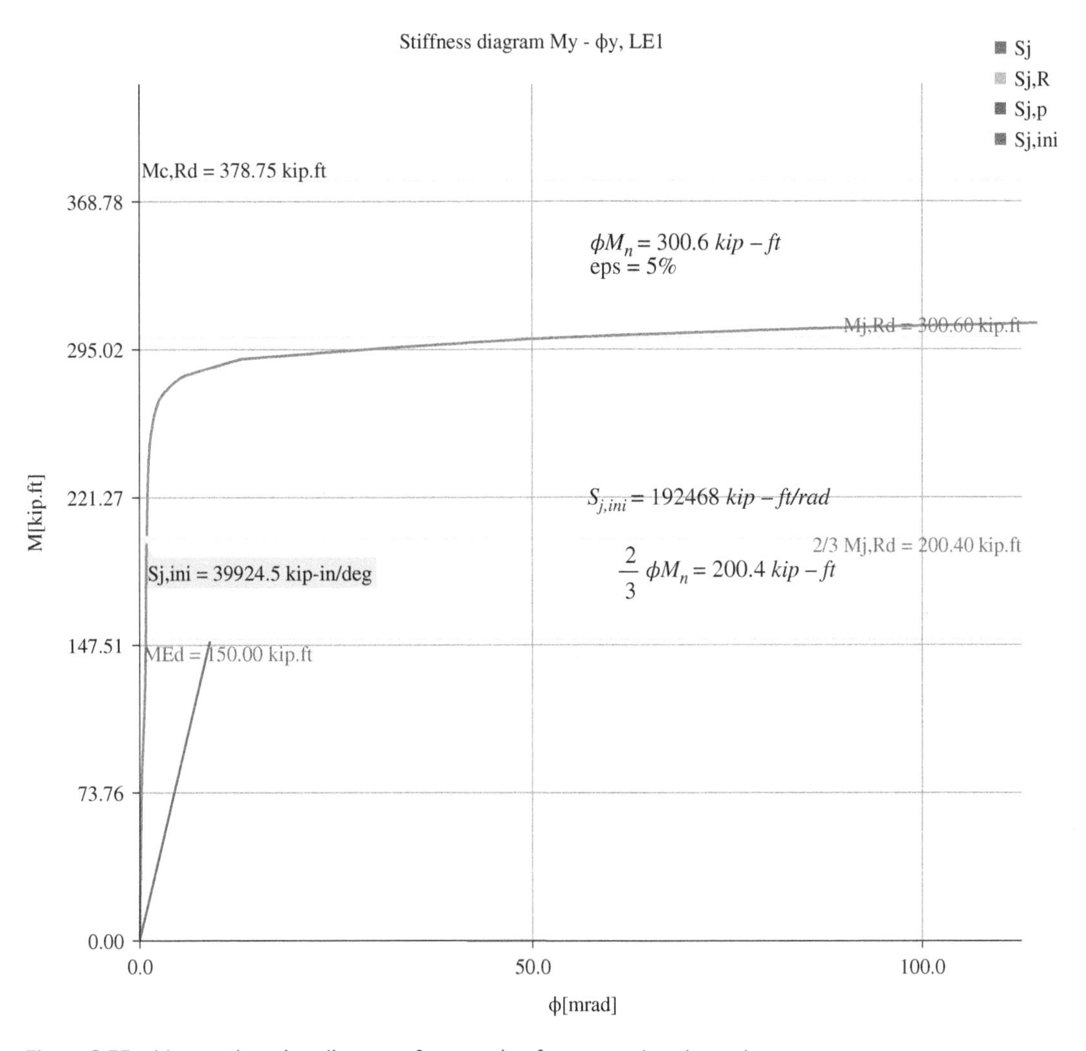

Figure 2.33 Moment/rotation diagram of connection for one analyzed member.

in the plane of the strong axis of the connected open I section beam from M_y = 25 kNm to 72 kNm. The maximum reached strain ε_{max} ranges from 0.2% to 22.7%.

2.13 Deformation Capacity

The deformation capacity/ductility δ_{Cd} belongs, together with resistance and stiffness, to the three basic parameters describing the behavior of connections. In moment-resistant connections, ductility is achieved by a sufficient rotation capacity ϕ_{Cd}. Sufficient ductility is essential especially in seismic regions but ductile connections are always a good design for avoiding brittle failures. The deformation/rotation

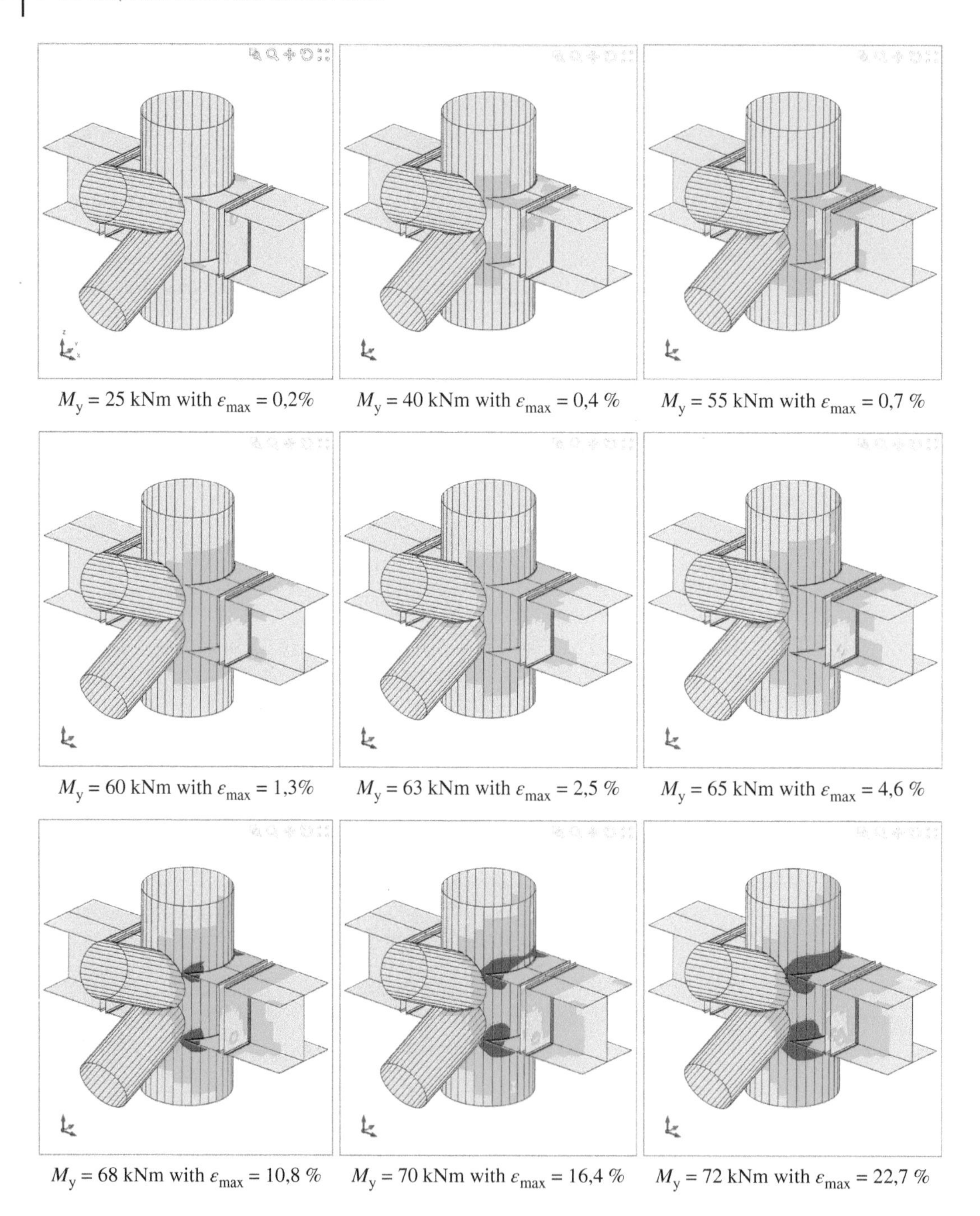

Figure 2.34 The maximum strains ε_{max} in joint during its loading by bending moment M_y in the plane of the strong axis of the connected open I section beam.

capacity is predicted for each connection in the joint separately. The prediction of deformation capacity δ_{Cd} of connections is currently not offered as a standardized procedure. Compared to the well-accepted methods for the determination of the resistance of many types of structural joints, there are no generally accepted standardized procedures for the determination of the rotation capacity.

The estimation of the rotation capacity is important for connections exposed to earthquakes; see (Gioncu and Mazzolani, 2002) and (Grecea et al., 2004), and extreme loading; see (Sherbourne and Bahaari, 1994; 1996a; 1996b). The deformation capacity of components has been studied for almost 30 years (Foley and Vinnakota, 1995). Faella et al. (2000) tested T-stubs and derived the analytical expressions for the deformation capacity. Kuhlmann and Kuhnemund (2000) performed tests on the column web subjected to transverse compression at different levels of axial compression force in the column. Da Silva et al. (2002) predicted deformation capacity at varying levels of axial force in the connected beam. Based on the test results combined with FE analysis, deformation capacities have been established for the basic components by analytical models by Beg et al. (2004). In this work, components are represented by nonlinear springs, and appropriately combined in order to determine the rotation capacity of the joint for the end plate connections, with an extended or flush end plate, and welded connections. For these connections, the most important components that may significantly contribute to the rotation capacity column were the web in compression, the column web in tension, the column web in shear, the column flange in bending, and the end plate in bending. Components related to the column web are relevant only when there are no stiffeners in the column that resist compression, tension or shear forces. The presence of a stiffener eliminates the corresponding component, and its contribution to the rotation capacity of the joint can be therefore ignored. End plates and column flanges are important only for end plate connections, where the components act as a T-stub, where also the deformation capacity of the bolts in tension is included. The questions and limits of the deformation capacity of connections of high-strength steel were studied by Girao et al. (2004).

For seismic regions, the current approach is not to trust any analytical and numerical models but to rely on experiments only. A set of prequalified connections with experimentally proven resistance and deformation capacity was developed and codified in AISC 358-22.

2.14 Connection Model in Global Analyses

A connection model for global analysis can be classified according to its bending stiffness as nominally pinned, rigid, or semi-rigid. Questions about the application of advanced models using FEA in global analyses are summarized in this chapter.

Connections of members modeled by 1D elements used for global analyses are modeled as massless points when analyzing a steel frame or a girder structure; see Figure 2.35. Equilibrium equations are assembled in joints, and after solving the whole structure, internal forces on the ends of beams are determined. In fact, the joint is loaded by those forces. The resultant of the forces from all members in the joint is zero – the whole joint is in equilibrium. The real shape of the joint is not known in the structural model. For better visualization of the CBFEM model, the end forces on 1D members are applied as loads on segment ends. Six components of the internal forces from the theoretical node are transferred to the outer end of segment – the values of forces are kept, but the moments are modified by the actions of the forces on the corresponding arms. Inner ends of the segments are not connected.

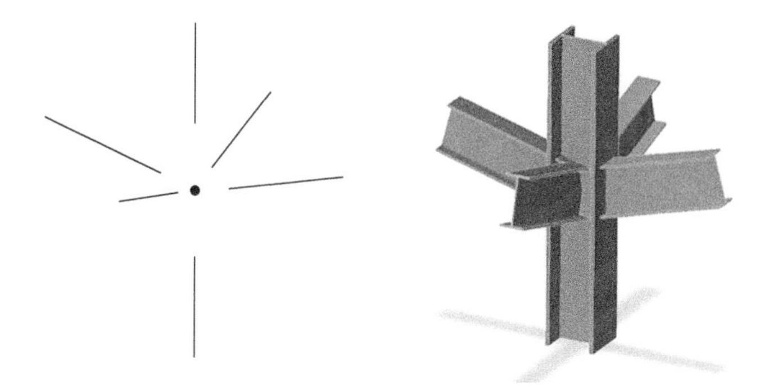

Figure 2.35 Theoretical (massless) joint and the real shape of the joint without modified member ends.

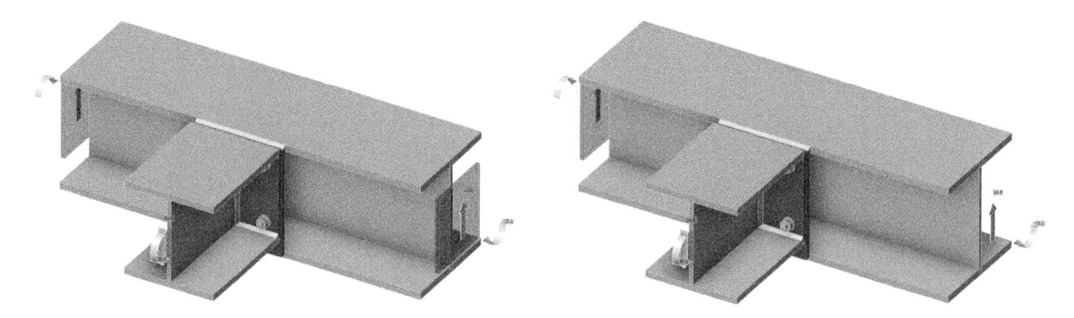

Figure 2.36 Simplified and advanced model of supports in CBFEM model of the joint.

Each node of a global FEM model is in equilibrium. The equilibrium requirement is correct, but the data input may be treated more sophisticatedly. One member in a connection is treated as bearing and the others as connected. If only a connection of the connected member is checked, keeping the equilibrium during the loading is unnecessary. Two modes of the load input are shown in Figure 2.36.

End forces of the frame analysis model are transferred to the outer ends of the member segments. Eccentricities of the members caused by the joint design are respected during the transfer. The analysis model created by the CBFEM corresponds to the real joint very precisely. On the other hand, the analysis of the internal forces is performed on the idealized global FEM model, where the individual beams are modeled using centerlines, and the joints are modeled using immaterial nodes, as shown in Figure 2.37.

The internal forces are analyzed using 1D members in global model. Figure 2.38 shows an example of the courses of the internal forces.

The effects caused by a member on a joint are important when designing the joint with its connections. The effects are illustrated in Figure 2.39.

Bending moment M and shear force V act in a theoretical joint. The point of the theoretical joint does not exist in the CBFEM model; thus, the load cannot be applied here. The model is loaded by actions M and V, which are transferred to the end of a segment in distance r

$$M_c = M - V \,.\, r$$

$$V_c = V$$

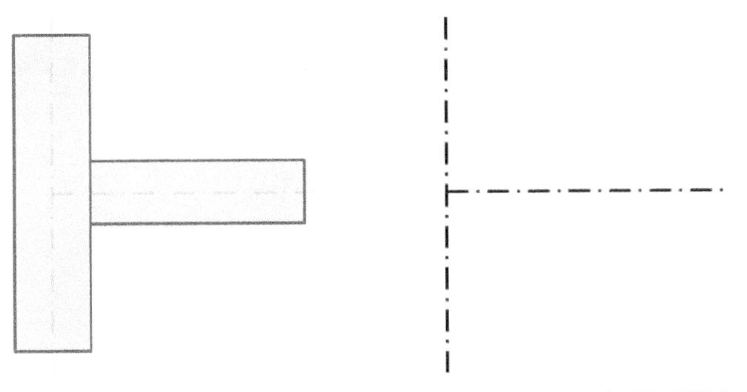

Figure 2.37 Joint of vertical column and horizontal beam.

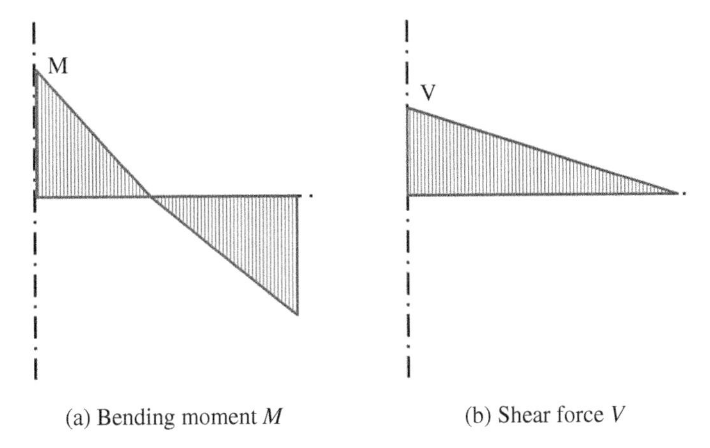

Figure 2.38 Internal forces on horizontal beam; M and V are the end forces of the joint.

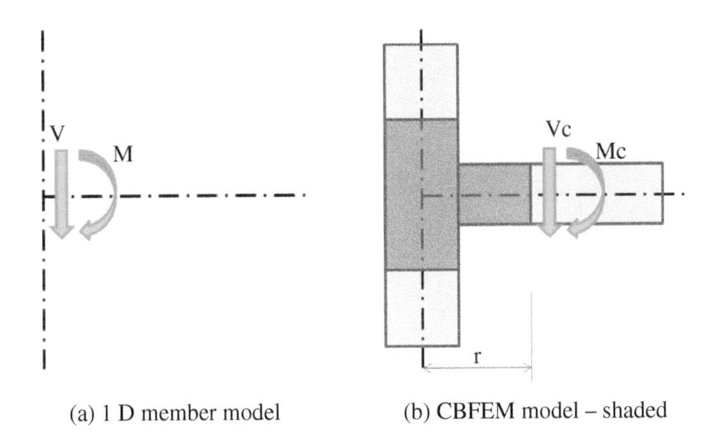

Figure 2.39 Load effects of a member on the connection.

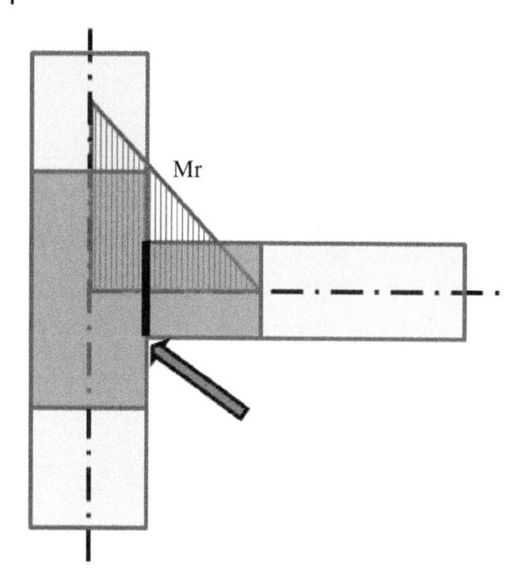

Figure 2.40 Course of bending moment in the CBFEM model; the arrow points to the real position of the connection.

In the CBFEM model, the end section of the segment is loaded by the moment M_c and the shear force V_c.

When designing the connection, its real position relative to the theoretical node should be determined and respected. The internal forces in the real connection's position often differ from the internal forces in the theoretical node. Thanks to the precise CBFEM model, the design is performed on reduced forces; see moment M_r in Figure 2.40.

Figure 2.41 illustrates that the position of a hinge in the theoretical 1D model differs from the real position in the structure. The global model does not correspond to reality. When applying the calculated internal forces, a significant bending moment is applied to the shifted joint, and the designed joint is

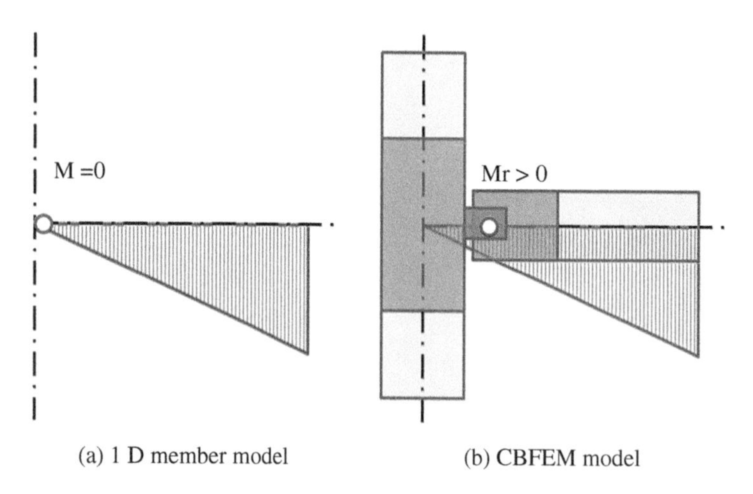

Figure 2.41 Position of the hinge in the global FEM model and in the real structure.

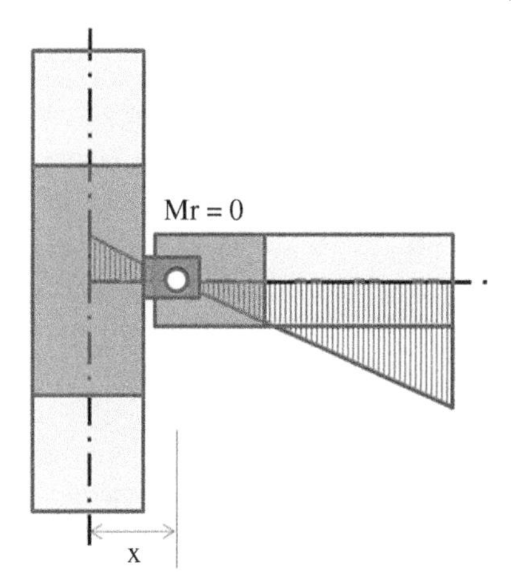

Figure 2.42 Position of the hinge in the real structure.

either overloaded or cannot be designed. The solution is simple – both models must correspond. Either the hinge in the 1D member model must be defined in the proper position, or the courses of the internal forces must be shifted to get the zero moment in the real position of the hinge, as shown in Figure 2.42.

References

Agerskov, H. (1976). High-Strength Bolted Connections Subject to Prying. *Journal of Structural Division*, 102(1), 161–175.

AISC 360-10 (2010). *Specification for Structural Steel Buildings*. AISC, Chicago.

AISC 360-16 (2016). *Specification for Structural Steel Buildings*. AISC, Chicago.

AISC 358-22 (2022). *Prequalified Connections for Special and Intermediate Steel Moment Frames for Seismic Applications*. AISC, Chicago.

American Concrete Institute (2014). *ACI 318-14: Building Code Requirements for Structural Concrete*. ACI, Farmington Hills, MI.

Beg, D., Zupančič, E., and Vayas, I. (2004). On the Rotation Capacity of Moment Connections. *Journal of Constructional Steel Research*, 60(3–5), 601–620.

CEN (2006a). *EN1993-1-5:2006, Eurocode 3: Design of Steel Structures – Part 1-5: Plated Structural Elements*. CEN, Brussels.

CEN (2006b). *EN 1993-1-8:2006, Eurocode 3: Design of Steel Structures – Part 1-8: Design of Joints*. CEN, Brussels.

CEN (2022). *prEN 1993-1-14:2022: Eurocode 3: Design of Steel Structures – Part 1-14: Design Assisted by Finite Element Analysis*. CEN, Brussels.

Da Silva, L., Lima, L., Vellasco, P., and Andrade, S. (2002). Experimental Behaviour of End-Plate Beam-to-Column Joints Under Bending and Axial Force, Database Reporting and Discussion of Results, Report on ECCS-TC10 Meeting in Ljubljana.

Dvorkin, E.N., and Bathe, K.J. (1984). A Continuum Mechanics Based Four Node Shell Element for General Nonlinear Analysis. *Engineering Computations*, 1(1), 77–88.

Faella, C., Piluso, V., and Rizzano, G. (2000). Plastic Deformation Capacity of Bolted T-Stubs: Theoretical Analysis and Testing. In F. Mazzolani (Ed.). *Moment Resistant Connections of Steel Frames in Seismic Areas: Design and Reliability*, E&FN Spon, London.

Foley, C.M. and Vinnakota, S. (1995). Toward Design Office Moment–Rotation Curves for End-Plate Beam-to-Column Connections. *Journal of Constructional Steel Research*, 35, 217–253.

Gioncu, V. and Mazzolani, F. (2002). *Ductility of Seismic Resistant Steel Structures*. Spon Press, London.

Girao, A.M.C., Bijlaard, F.S.K., and da Silva, L.S. (2004). Experimental Assessment of the Ductility of Extended End Plate Connections. *Journal of Engineering Structures,* 26, 1185–1206.

Gödrich, L., Wald, F., and Sokol, Z. (2014). Advanced modelling of end plate. In *Eurosteel 2014*, ECCS, Brussels, pp. 287–288.

Grecea, D., Stratan, A., Ciutina, A., and Dubina, D. (2004). Rotation Capacity of the Beam-to-Column Joints Under Cyclic Loading. *Connections in Steel Structures*. V.

Ibrahimbegovic, A., Taylor, R.L., and Wilson, E.L. (1990). A Robust Quadrilateral Membrane Element with Drilling Degrees of Freedom. *International Journal for Numerical Methods in Engineering*, 30(3), 445–457.

IDEA StatiCa (2023a). *CBFEM Weld Model: Validation and Verification*. https://www.ideastatica.com/support-center/cbfem-weld-model-validation-and-verification

IDEA StatiCa (2023b) *Check of Steel Connection Components (AISC)*. https://www.ideastatica.com/support-center/check-of-components-according-to-aisc

Kleiner, A. (2018). Beurteilung des Tragverhaltens von Flankenkehlnahtverbindungen aus normal- und höherfestem Baustahl unter Berücksichtigung statistischer Kriterien, PhD. thesis, Stuttgart University, p. 310. https://elib.uni-stuttgart.de/handle/11682/9861

Kuhlmann, U. and Kuhnemund, F. (2000). Rotation Capacity of Steel Joints: Verification Procedure and Component Tests. In C. C. Baniotopoulos and F. Wald (Eds.), *The Paramount Role of Joints into the Reliable Response of Structures*, Springer, Cham, pp. 363–372.

Ng, A.K.F., Driver, R.G., and Grondin, G.Y. (2002). *Behaviour of Transverse Fillet Welds*. Structural Engineering Report No. 245, University of Alberta, p. 317. https://era.library.ualberta.ca/items/08118758-edd0-4428-98cc-1760518c6999

Sherbourne, A.N. and Bahaari, M.R. (1994). 3D Simulation of End-Plate Bolted Connections. *Journal of Structural Engineering*, 120, 3122–3136.

Sherbourne, A.N. and Bahaari, M.R. (1996a). Simulation of Bolted Connections to Unstiffened Columns T-Stub Connections. *Journal of Constructional Steel Research*, 40, 169–87.

Sherbourne, A.N. and Bahaari, M.R. (1996b). 3D Simulation of Bolted Connections to Unstiffened Columns—II. Extended Endplate Connections. *Journal of Constructional Steel Research*, 40, 189–223.

3

Welded Connection

3.1 Fillet Weld in a Lap Joint

3.1.1 Description

The aim of this chapter is a verification of the component-based finite element method (CBFEM) of a fillet weld in a lap joint and the traditional method (TM) of AISC 360-16 (AISC, 2016). Two plates are connected in two configurations, namely with a transverse weld, and with a longitudinal weld. The length of the weld is the changing parameter in the study. The study covers long welds whose resistance is reduced due to stress concentration. The connection is loaded by a normal force.

3.1.2 Analytical Model

The fillet weld is the only component examined in the study. The welds are designed to be the weakest component in the connection. The weld is designed by load and resistance factor design (LRFD) according to AISC 360-16. The design strength of the fillet weld is determined by using Section J2.4 in AISC 360-16 (AISC, 2016). The available calculation methods for checking the strength of fillet welds are based upon the simplifying assumption that a fillet weld always fails in shear and the failure occurs along a plane through a throat section of a fillet weld, shown in Figure 3.1.

The design strength (ϕR_n) of welded joints shall be the lower values of the base material strength and the weld metal strength. The weld metal strength is determined according to the limit state of rupture, and the base material strength is determined according to the limit states of tensile rupture and shear rupture as follows:

For the base metal

$$R_n = F_{nBM} A_{BM}$$

For the filler metal

$$R_n = F_{nw} A_{we}$$

where A_{BM} is the cross-sectional area of the base metal, in.^{2} (mm^2), A_{we} is the effective area of the weld, in.^{2} (mm^2), F_{nBM} is the nominal stress of the base metal, ksi (MPa), F_{nw} is the nominal stress of the weld metal, ksi (MPa).

Steel Connection Design by Inelastic Analysis: Verification Examples per AISC Specification, First Edition.
Mark D. Denavit, Ali Nassiri, Mustafa Mahamid, Martin Vild, Halil Sezen, and František Wald.

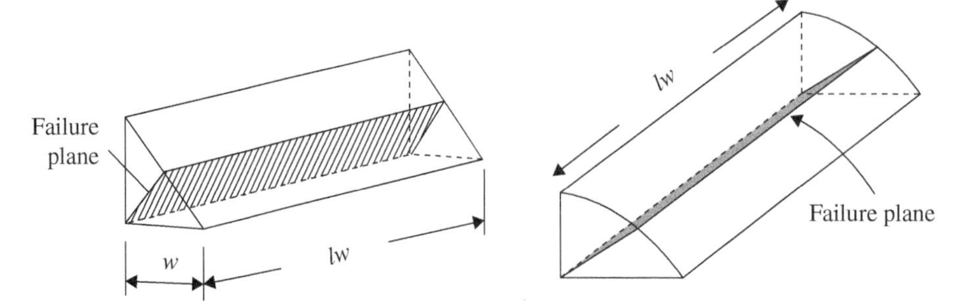

Figure 3.1 Failure plane through a throat section of a fillet weld.

The values of ϕ, Ω, and F_w and limitations thereon are given in Table J2.5 of AISC 360-16 (AISC, 2016).

For a linear weld group loaded through the center of gravity having a uniform leg size, the design strength is determined as follows:

$$\phi R_n = \phi 0.60 F_{EXX} \left(1.0 + 0.50 \sin^{1.5}(\theta)\right) A_{we}$$

where F_{EXX} is the filler metal classification strength, ksi (MPa), θ is the angle between the longitudinal direction of the force and the weld.

In the case of fillet weld groups concentrically loaded and consisting of elements with a uniform leg size that are oriented both longitudinally and transversely to the direction of applied load, the combined strength R_n, of the fillet weld group shall be determined as the greater of the following:

$$R_n = R_{nwl} + R_{nwt}$$

or

$$R_n = 0.85 R_{nwl} + 1.5 R_{nwt}$$

where R_{nwl} is the total nominal strength of the longitudinally loaded fillet welds, as determined in accordance with Table J2.5, kips (N), R_{nwt} is the total nominal strength of the transversely loaded fillet welds, as determined in accordance with Table J2.5 without the increase in Section J2.4 (b), kips (N).

If the length of a fillet weld is longer than 100 w, the reduction factor, β, is given by:

$$\beta = 1.2 - 0.002 \left(\frac{l_w}{w}\right) \leq 1{,}0$$

where l_w is the actual weld length, in. (mm), w is the size of weld leg, in. (mm). When the length of the weld exceeds 300 times the leg size, w, the effective length shall be taken as 180w.

An overview of the considered examples and the material properties is given in Table 3.1 and Table 3.2. The weld configurations T is for transverse, L for longitudinal weld; see the geometry in Figure 3.2. Steel grade was A36, and the filler metal was E60. The connection is symmetrical, and the plates are pulled out of the welded splice connection. The lengths of the transverse and longitudinal welds vary in this parametrical study.

Table 3.1 Overview of examples: Transverse welds.

Example	Material					Weld		Plate 1		Plate 2	
	Base metal (A36)		Filler metal (E60)								
	F_y	F_u	F_y	F_u	E	w	L	b_1	t_1	b_2	t_2
	[MPa]	[MPa]	[MPa]	[MPa]	[GPa]	[mm]	[mm]	[mm]	[mm]	[mm]	[mm]
T50	250	400	331	414	210	6	50	50	15	300	15
T60	250	400	331	414	210	6	60	60	15	300	15
T80	250	400	331	414	210	6	80	80	15	300	15
T100	250	400	331	414	210	6	100	100	15	300	15
T150	250	400	331	414	210	6	150	150	15	300	15
T200	250	400	331	414	210	6	200	200	15	300	15
T250	250	400	331	414	210	6	250	250	15	300	15
T300	250	400	331	414	210	6	300	300	15	300	15

Table 3.2 Overview of examples: Longitudinal welds.

Example	Material					Weld		Plate 1		Plate 2	
	Base metal (A36)		Filler metal (E60)								
	F_y	F_u	F_y	F_u	E	w	L	b_1	t_1	b_2	t_2
	[MPa]	[MPa]	[MPa]	[MPa]	[GPa]	[mm]	[mm]	[mm]	[mm]	[mm]	[mm]
L80	250	400	331	414	210	6	80	800	15	1000	15
L100	250	400	331	414	210	6	100	800	15	1000	15
L200	250	400	331	414	210	6	200	800	15	1000	15
L300	250	400	331	414	210	6	300	800	15	1000	15
L400	250	400	331	414	210	6	400	800	15	1000	15
L500	250	400	331	414	210	6	500	800	15	1000	15
L600	250	400	331	414	210	6	600	800	15	1000	15
L700	250	400	331	414	210	6	700	800	15	1000	15
L800	250	400	331	414	210	6	800	800	15	1000	15

3.1.3 Numerical Model

The weld component in CBFEM is described in Chapter 2, Section 2.4. Nonlinear elastic-plastic material is used for welds in this study. The stress peaks are redistributed along the longer part of the weld thanks to the plastic stress redistribution.

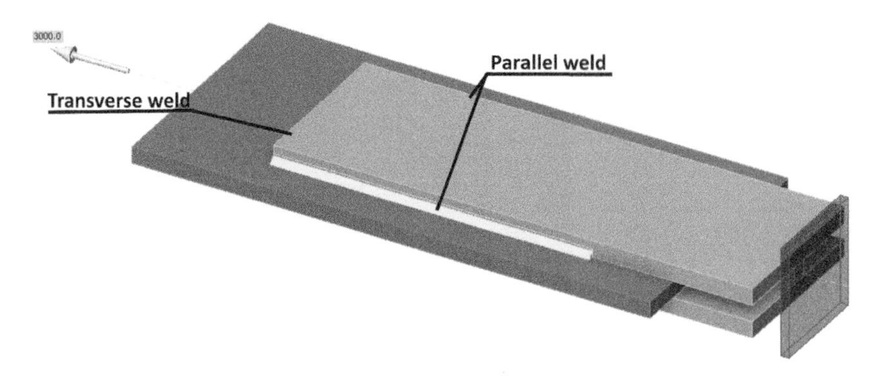

Figure 3.2 Specimen geometry.

Table 3.3 Comparison of CBFEM and TM: Transverse weld.

Example	Design strength		
	TM [kN]	CBFEM [kN]	Difference [CBFEM/ TM] [%]
T50	119	116	97
T60	142	140	99
T80	190	187	98
T100	237	234	99
T150	356	351	99
T200	474	469	99
T250	593	585	99
T300	711	700	98

3.1.4 Verification of Strength

The design weld strength calculated by CBFEM is compared with the results of TM. The results are presented in Table 3.3 and Table 3.4. The study is performed for one parameter, the length of the weld, in two weld configurations: transverse weld, and longitudinal weld. The sensitivity study of the weld length on the design strength is shown in Figure 3.3. The study shows good agreement for both weld configurations.

To illustrate the accuracy of the CBFEM model, the results of the study are summarized in Figure 3.4 comparing design resistance of CBFEM and TM. The results show that the difference between the two calculation methods in both weld configuration cases is less than 4%. For transverse welds, CBFEM provides very consistent results with 1–3% lesser strength.

Table 3.4 Comparison of CBFEM and TM: Longitudinal weld.

Example	Design strength		
	TM [kN]	CBFEM [kN]	Difference [CBFEM/TM] [%]
L80	253	256	101
L100	316	320	101
L200	474	476	100
L300	632	633	100
L400	790	788	100
L500	948	943	99
L600	1264	1269	100
L700	1581	1557	98
L800	1897	1828	96

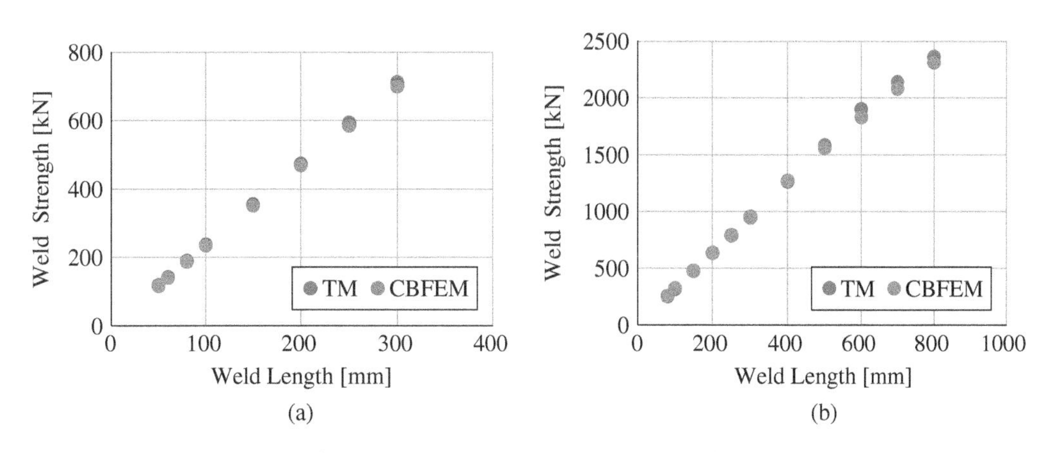

Figure 3.3 Sensitivity study of weld length (a) Transverse weld; (b) Longitudinal weld.

3.1.5 Benchmark Example

Inputs

Member 1 Iw45X300

- Welded from plates with thickness (t) = 15 mm
- Width (b) = 300 mm

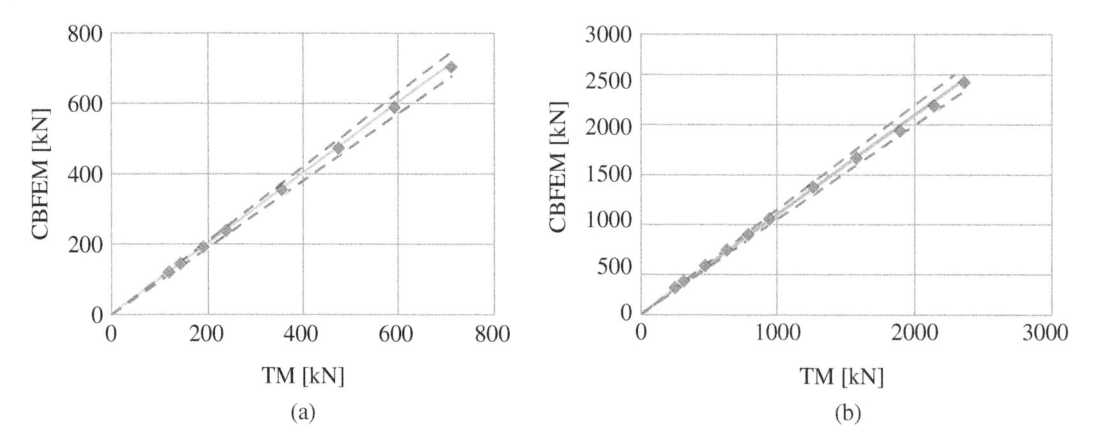

Figure 3.4 Comparison of CBFEM and TM (a) Transverse weld; (b) Longitudinal weld.

- Web is removed by opening manufacturing operation
- Steel A36

Member 2 Plate 15X300

- Thickness (t) =15 mm
- Width (b) = 300 mm
- Steel A36

Transverse fillet weld at both sides of Member 2; see Figure 3.5.

- Weld thickness (w) = 6 mm
- Filler metal = E60
- Weld length (L) = 300 mm

Output

- Design resistance in tension F_{Rd} = 700 kN

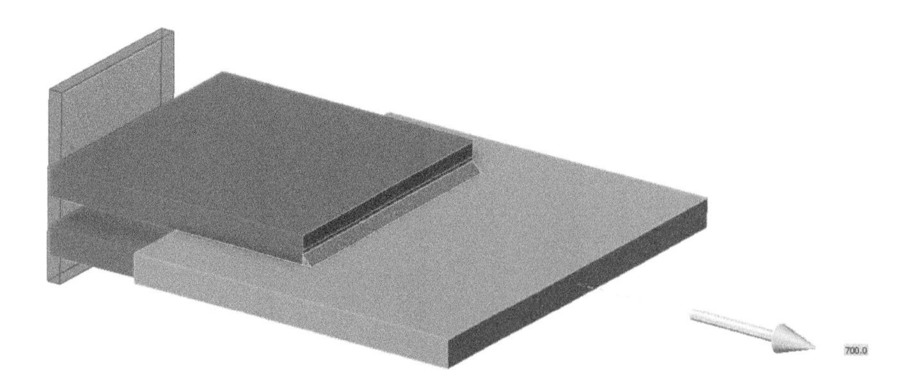

Figure 3.5 Benchmark example for the welded lap joint with transverse fillet welds.

3.2 Fillet Weld in a Cleat Connection

3.2.1 Description

In this section, the model of the fillet weld in a cleat connection calculated by the component-based finite element method (CBFEM) is compared with the traditional method (TM) of ANSI/AISC 360-16 (AISC, 2016). A cleat is welded to a plate and loaded by the normal force. The lengths of the welds are studied in a sensitivity study.

3.2.2 Investigated Cases

The fillet weld is the only component examined in the study. The welds are designed according to Chapter J.2 in AISC 360-16 (AISC, 2016) to be the weakest component in the joint. The design strength of the fillet weld is described in Section 3.1. An overview of the considered examples and the material is given in Table 3.5. The geometry of the joints with dimensions is shown in Figure 3.6.

3.2.3 Verification of Strength

The weld's strength calculated by CBFEM is compared with the results of TM; see Table 3.6. The study is performed for one parameter: length of the weld. The sensitivity study of weld length on design strength is shown in Figure 3.7(a).

To illustrate the accuracy of the CBFEM model, the results of the study are summarized in Figure 3.7(b) comparing design strength by CBFEM and TM. The results show that the numerical results are safer by 15–20%. This is caused by the eccentricity of the load imposed on the welds, which is ignored in the traditional method.

Table 3.5 Overview of examples.

Example	Material					Weld a		Weld b		Plate	
	Base metal (A36)		Filler metal (E60)								
	F_y [MPa]	F_u [MPa]	F_y [MPa]	F_u [MPa]	E [GPa]	w_a [mm]	L_a [mm]	w_b [mm]	L_b [mm]	b_p [mm]	t_p [mm]
L160X16	250	400	331	414	210	6	100	6	100	400	16
L160X16	250	400	331	414	210	6	110	6	110	400	16
L160X16	250	400	331	414	210	6	120	6	120	400	16
L160X16	250	400	331	414	210	6	130	6	130	400	16
L160X16	250	400	331	414	210	6	140	6	140	400	16
L160X16	250	400	331	414	210	6	150	6	150	400	16
L160X16	250	400	331	414	210	6	160	6	160	400	16

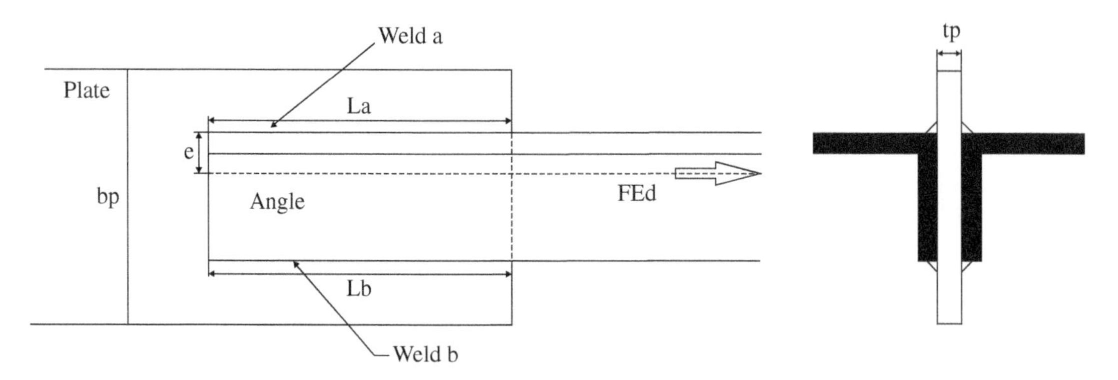

Figure 3.6 Geometry with dimensions.

Table 3.6 Comparison of CBFEM and TM.

Example	Design strength		
	TM [kN]	CBFEM [kN]	Difference [CBFEM/TM] [%]
L100X160X16	316	253	80
L110X160X16	348	285	81
L120X160X16	379	312	82
L130X160X16	411	342	83
L140X160X16	443	372	84
L150X160X16	474	401	85
L160X160X16	506	431	85

3.2.4 Benchmark Example

Inputs

Cleat

- Steel A36
- Cross-section = 2 Lt160/16
- Distance between cleats = 16 mm

Plate

- Thickness (t) = 16 mm
- Width (b) = 400 mm
- Steel A36

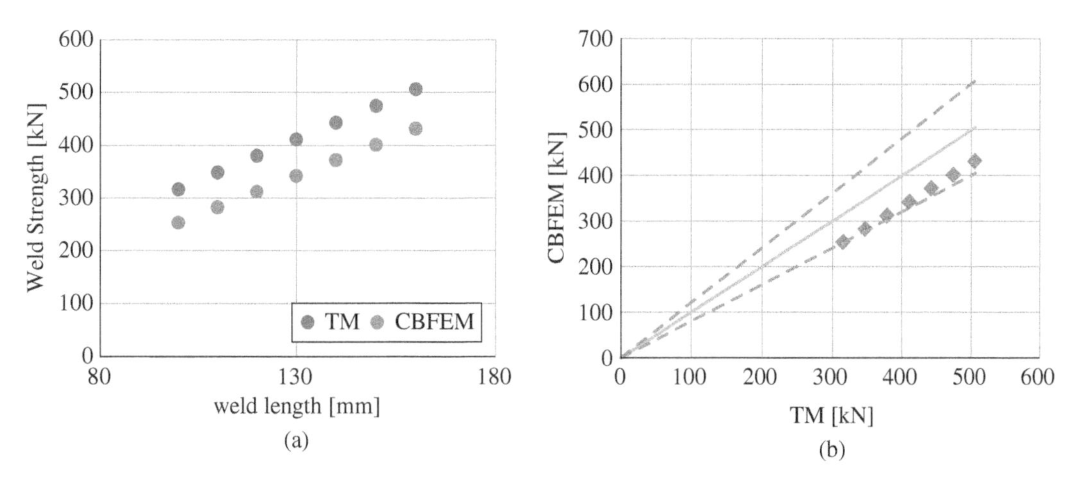

Figure 3.7 (a) Sensitivity study of weld length; (b) Comparison of CBFEM to TM.

Weld, parallel fillet welds, see Figure 3.8.

- Weld thickness (w) = 6 mm
- Filler metal = E60
- Weld length (L) = 160 mm

Outputs

- Design strength in tension F_{Rd} = 431 kN

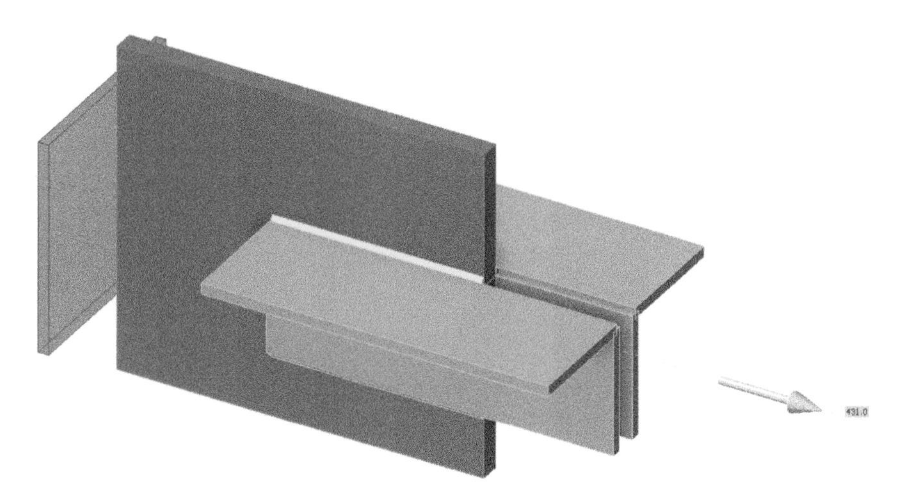

Figure 3.8 Benchmark example for the welded cleat connection with longitudinal fillet welds.

3.3 Fillet Weld of a Shear Tab

3.3.1 Description

In this section, the component-based finite element method (CBFEM) of a fillet weld in a shear tab connection is compared with the traditional method (TM). A shear tab is welded to an open section column HP 8X36. The height of the shear tab is changed from 50 to 200 mm. The plate/weld is loaded by shear force at 100 mm eccentricity from the weld.

3.3.2 Investigated Cases

The fillet weld is the only component examined in the study. The welds are designed according to Chapter J.2 in ANSI/AISC 360-16 (ASCI, 2016) to be the weakest component in the joint. The design strength of the fillet weld is described in Section 3.1. An overview of the considered examples and the material is given in Table 3.7. The geometry of the joints with dimensions is shown in Figure 3.9. Only the shear force is considered.

3.3.3 Comparison of Strength

The design strength calculated by CBFEM is compared with the results of the TM; see Table 3.8. The study is performed for one parameter: length of the weld, i.e. the height of the shear tab and one load case: shear force. The effect of additional bending moment due to the shear force is calculated by considering the elastic section modulus of the weld plane. The influence of the weld lengths on the design strength of the shear tab connection loaded by the shear force is shown in Figure 3.10(a).

To illustrate the accuracy of the CBFEM model, the results of the study are summarized in Figure 3.10(b) comparing the design resistance by the CBFEM and the TM. The results show that the difference between the two calculation methods is within 5% which is slightly more than the traditional

Table 3.7 Overview of examples.

Example	Material				Weld		Shear tab			Column
	Base metal (A36)		Filler metal (E60)							
	F_y	F_u	F_u	E	w	L	h_p	t_p	e	
	[MPa]	[MPa]	[MPa]	[GPa]	[mm]	[mm]	[mm]	[mm]	[mm]	Section
V50	250	400	414	210	6	50	50	20	100	HP8X36
V75	250	400	414	210	6	75	75	20	100	HP8X36
V100	250	400	414	210	6	100	100	20	100	HP8X36
V125	250	400	414	210	6	125	125	20	100	HP8X36
V150	250	400	414	210	6	150	150	20	100	HP8X36
V175	250	400	414	210	6	175	175	20	100	HP8X36
V200	250	400	414	210	6	200	200	20	100	HP8X36

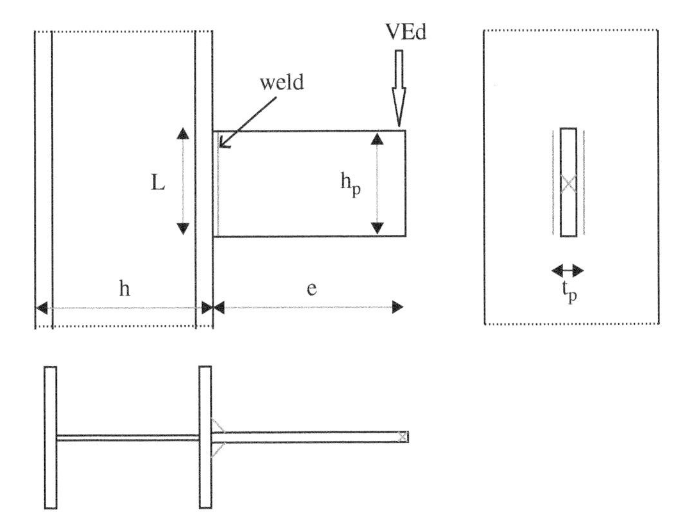

Figure 3.9 Geometry with dimensions.

Table 3.8 Comparison of CBFEM and TM.

Example	Design strength		
	TM [kN]	CBFEM [kN]	Difference [CBFEM/TM] [%]
V50	6.50	6.75	104
V75	14.50	15.00	103
V100	26.00	27.00	104
V125	40.25	41.75	104
V150	57.50	60.00	104
V175	77.50	80.75	104
V200	100.00	104.50	104

method. This is caused by the assumption of the elastic section modulus of the weld plane during the calculation of the bending moment resistance in the traditional method.

3.3.4 Benchmark Example

Inputs

Column

- Steel A36
- HP 8X36

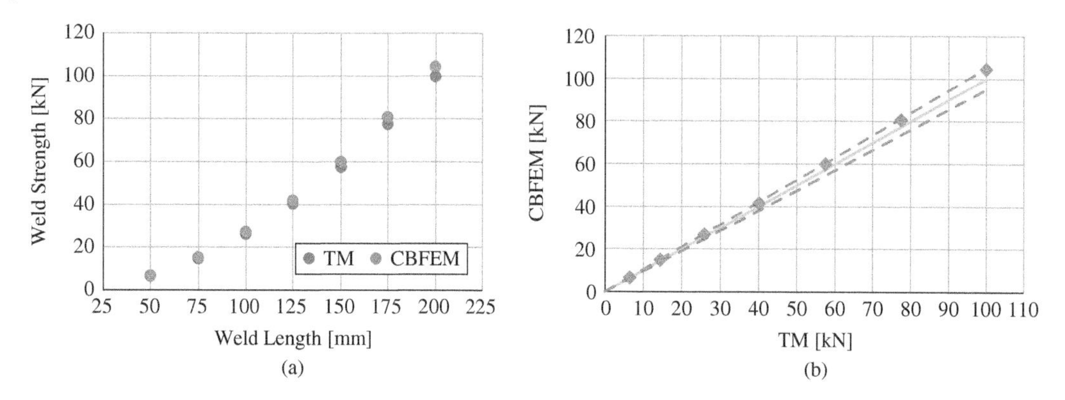

Figure 3.10 (a) Sensitivity study of weld length; (b) Comparison of CBFEM to TM.

Shear tab

- Thickness (t) = 20 mm
- Height (h) = 100 mm
- Load eccentricity to weld (e) = 100 mm

Weld, double fillet weld; see Figure 3.11.

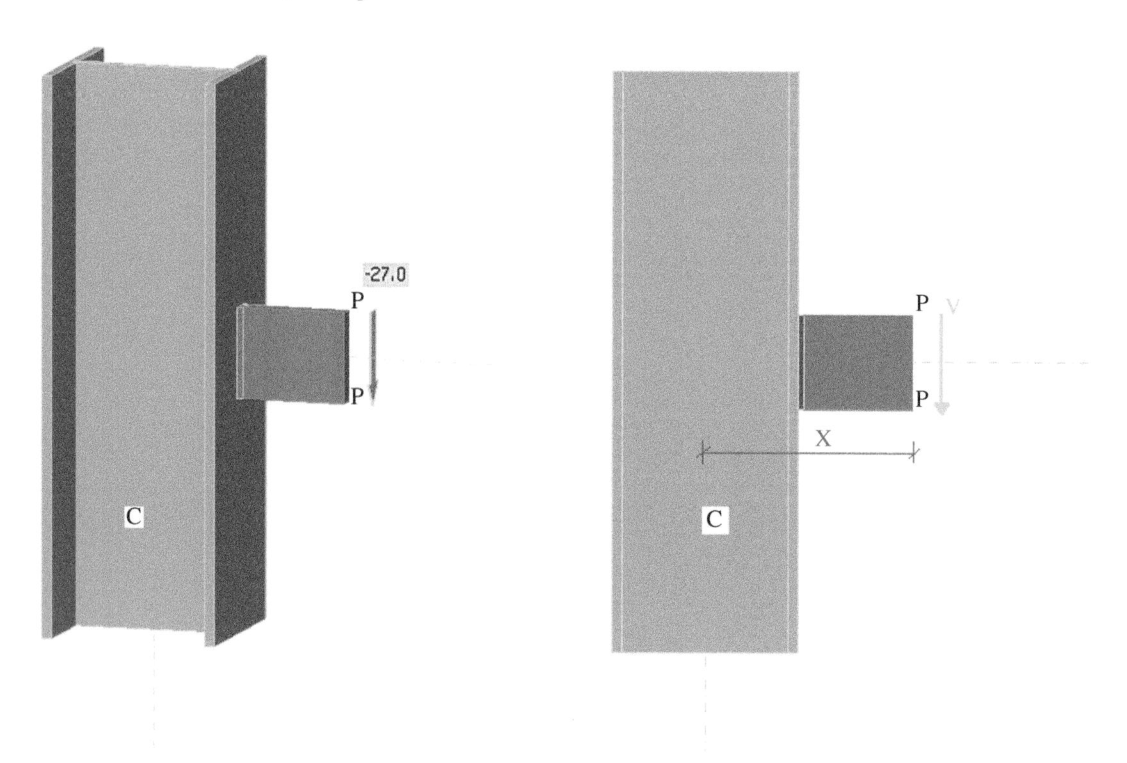

Figure 3.11 Benchmark example for the welded shear tab connection.

- Weld thickness, w = 6 mm
- Filler metal = E60

Loading

- Load magnitude (V) = 27 kN
- Load position (x) = 205 mm

Outputs

- Design resistance V_{Rd} = 27 kN

Reference

AISC 360-16 (2016). *Specification for Structural Steel Buildings*. AISC, Chicago.

4

T-Stub Connections

4.1 Description

A comparison between results from the component-based finite element method (CBFEM) and traditional calculation methods used in US practice for T-stub connections is presented in this chapter. A schematic of the connection investigated is presented in Figure 4.1. The limit states evaluated are slip, bolt tension and shear interaction strength, and flexural yielding of the flanges of the T-stub and the beam. The impact of prying action is considered.

For all connections investigated, the beam is a wide flange conforming to ASTM A992 (F_y = 50 ksi and F_u = 65 ksi) and the T-stub is built up from plates conforming to ASTM A572 Gr. 50 (F_y = 50 ksi and F_u = 65 ksi). Butt welds are used between the stem and the flange of the T-stub and between the tension member and the stem of the T-stub to simplify the evaluation. Each connection investigated had (8) 3/4 in. diameter bolts (i.e. 2 rows of 4 bolts) in standard holes with spacing, $s = 3$ in., edge distance $l_{eh} = 1.5$ in., and gage, $g = 5.5$ in.

The traditional calculations were performed in accordance with the provisions for load and resistance factor design (LRFD) in the AISC *Specification* (AISC, 2016) with prying action considered as described in Part 9 of the AISC *Manual* (AISC, 2017).

The CBFEM results were obtained from IDEA StatiCa Version 21.0. The maximum permitted loads were determined iteratively by adjusting the applied load input to a value that the program deems safe but if increased by a small amount (0.1 kip) the program would deem unsafe. DR type analyses can help identify the maximum permitted loads. However, some approximation is made in the evaluation of the joint design resistance, therefore, all results in this report are based on EPS type analyses.

4.2 Slip-Critical Connections

The first limit state investigated is slip. The configuration of this example matches that of Example J.5 of the AISC *Design Examples* v15.1 (AISC, 2019). Additional details of the connection include that the bolts are Group A (e.g. A325) with threads not excluded from the shear planes; the beam is a W18 × 175; the T-stub web thickness was $t_w = 0.75$ in.; the T-stub flange width was $b_f = 8.0$ in.; the T-stub flange thickness varied; and $\theta = 53.1°$. A three-dimensional view of one of the connections investigated is presented in Figure 4.2.

Steel Connection Design by Inelastic Analysis: Verification Examples per AISC Specification, First Edition.
Mark D. Denavit, Ali Nassiri, Mustafa Mahamid, Martin Vild, Halil Sezen, and František Wald.

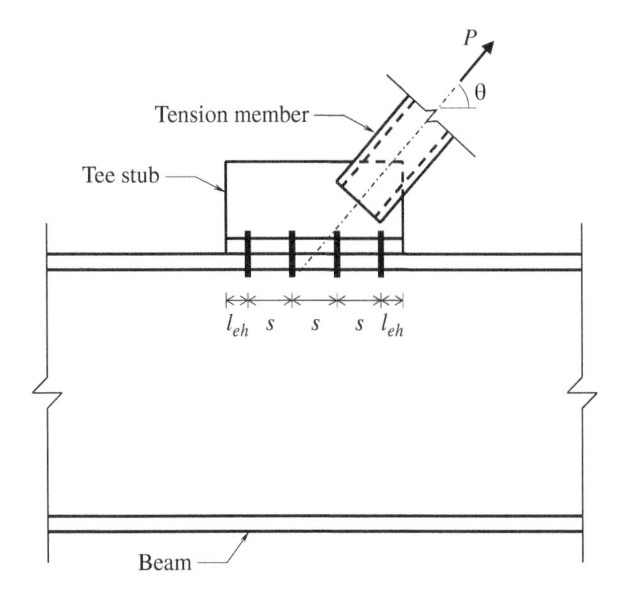

Figure 4.1 Schematic of the T-stub connection investigated in this chapter.

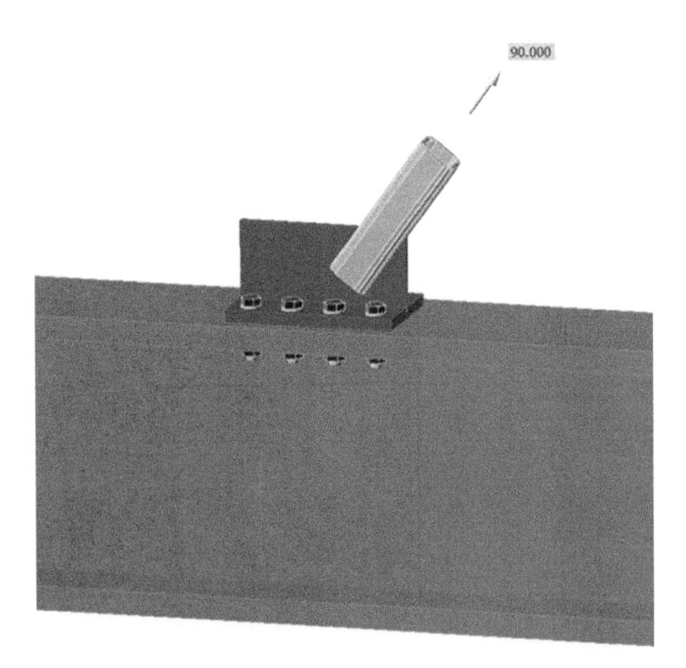

Figure 4.2 Three-dimensional view of the connection investigated.

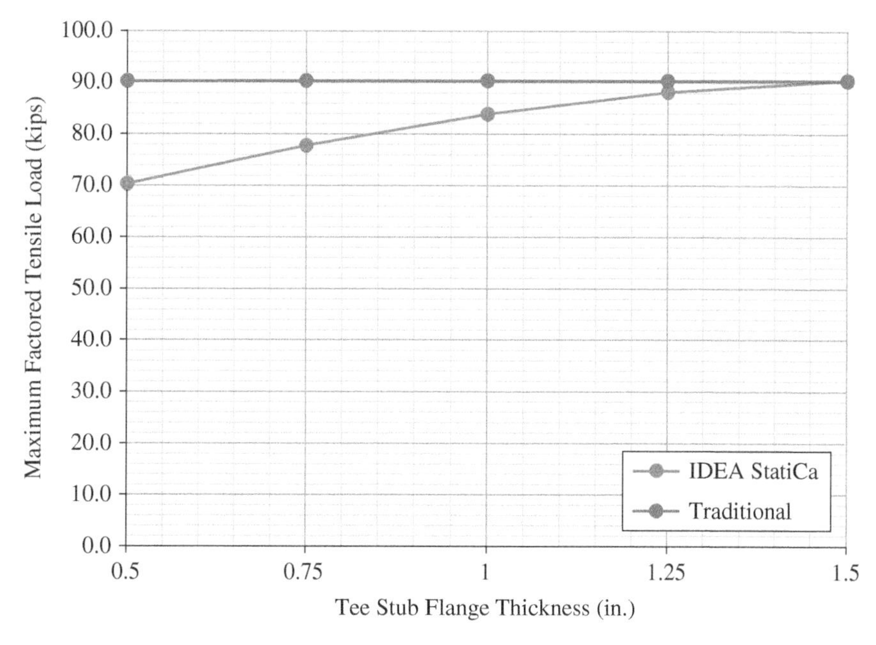

Figure 4.3 Design strength vs. T-stub flange thickness for slip-critical connections.

Calculations were performed for five T-stub flange thicknesses between 0.5 in. and 1.5 in. The maximum factored tensile load which could be applied to the connection is presented in Figure 4.3. For the traditional calculations, the maximum load does not vary with T-stub flange thickness except for the thinnest T-stub flange thickness where a slight reduction in maximum load is observed. Slip is the controlling limit state for all but the thinnest T-stub flange thickness which is controlled by the tensile strength of the bolts and the flexural yielding of the T-stub. For the CBFEM results, the maximum load varies continuously with the T-stub flange thickness.

The reason for the discrepancy can be identified through an examination of the detailed results provided by IDEA StatiCa. For this slip-critical connection, which is subjected to tension and shear, the provisions of Section J3.9 of the AISC *Specification* (AISC, 2016) apply. Specifically, a reduction factor, k_{sc}, which depends on the required tension force, is applied to the slip resistance. IDEA StatiCa includes the prying force in the required tension force used to compute k_{sc}. This is conservative as it is not required by the AISC *Specification* (AISC, 2016) and because prying forces do not reduce the clamping force which provides the slip resistance. For the thinnest flange investigated, IDEA StatiCa results in 23% less strength than the traditional calculations. For the thickest flange investigated, where prying action is prevented, IDEA StatiCa and the traditional calculations result in the same strength.

4.3 Prying Action

Prying action impacts the evaluation of plate bending strength and bolt strength. Part 9 of the AISC *Manual* (AISC, 2017) presents equations that account for prying action. The equations are presented in several forms for various design situations. In this work, plate bending is evaluated by comparing the thickness of the component (i.e. the T-stub flange or the beam flange) to t_{min} as given by Equation 9-19 of the AISC *Manual* and the bolt strength is evaluated by comparing the required bolt strength (i.e. $P\sin\theta$ divided by the number of bolts) to the available tensile strength, including the effects of prying action T_c as given by Equation 9-27 of the AISC *Manual*. Noting that LRFD was used for these analyses, t_{min} is calculated as:

$$t_{min} = \sqrt{\frac{4T_u b'}{\phi p F_u (1 + \delta\alpha')}}$$

$$b' = b - \frac{d_b}{2}$$

$$\delta = 1 - \frac{d'}{p}$$

$$\beta = \frac{1}{\rho}\left(\frac{B_c}{T_u} - 1\right)$$

$$\rho = \frac{b'}{a'}$$

$$a' = \left(a + \frac{d_b}{2}\right) \leq \left(1.25b + \frac{d_b}{2}\right)$$

If $\beta \geq 1$

$$\alpha' = 1$$

If $\beta < 1$

$$\alpha' = \min\left(1, \frac{1}{\delta}\frac{\beta}{1-\beta}\right)$$

where

B_c = available tension per bolt based on the limit state of tension or the combined limit states of tension and shear rupture = ϕr_n

F_u = specified minimum tensile strength of the connecting element

T_u = required tension force per bolt using LRFD load combinations = $P\sin\theta/n_b$

a = distance from bolt centerline to the edge of fitting

b = the distance from bolt centerline to the face of the stem of the T-stub

d_b = bolt diameter

d' = hole diameter

p = tributary length, based on yield line theory

ϕ = 0.9 (for plate bending)

Noting that LRFD was used for these analyses, T_c is calculated as:

$$T_c = B_c Q$$

If $\alpha' < 0$ (indicating the fitting has sufficient strength and stiffness).

$$Q = 1$$

If $0 \leq \alpha' \leq 1$ (indicating sufficient strength to develop full bolt available tension strength, but insufficient to prevent prying)

$$Q = \left(\frac{t}{t_c}\right)^2 (1 + \delta\alpha')$$

If $\alpha' > 1$ (indicating insufficient strength to develop full bolt tensile strength)

$$Q = \left(\frac{t}{t_c}\right)^2 (1 + \delta)$$

Note that the equation for α' for the determination of Q is different than the one used for the determination of t_{min}.

$$\alpha' = \frac{1}{\delta\,(1 + \rho)} \left[\left(\frac{t_C}{t}\right)^2 - 1\right]$$

$$t_c = \sqrt{\frac{4B_c b'}{\phi p F_u}}$$

where

t = thickness of the component

4.4 Prying of the T-Stub

The second investigation examines the strength of the T-stub and bolts. Same as the previous investigation, the bolts are Group A (e.g. A325) with threads not excluded from the shear planes; the beam is a W18 × 175; the T-stub web thickness was $t_w = 0.75$ in.; the T-stub flange width was $b_f = 8.0$ in.; the T-stub flange thickness varied; and $\theta = 53.1°$. Different from the previous investigation, the connections were not slip-critical.

Calculations were performed for eight T-stub flange thicknesses between 0.25 in. and 1.25 in. The maximum factored tensile load which could be applied to the connection is presented in Figure 4.4. As expected, for both the traditional calculation results and the IDEA StatiCa results, the maximum factored tensile load increases with the T-stub flange thickness until a plateau is reached where prying action is prevented. At the plateau, the strength of the connection is governed by the provisions of Section J3.7 of the AISC *Specification* (AISC, 2016) and results from the traditional calculations and IDEA StatiCa match. Where prying action impacts the strength of the connection, there are differences between the traditional calculations which follow the guidance of Part 9 of the AISC *Manual* (AISC, 2017) and IDEA StatiCa which explicitly models the connection using the CBFEM.

Typically, available flexural strength is computed based on the yield strength, F_y. The equations for prying action presented in Part 9 of the AISC *Manual* are based on the tensile strength, F_u, noting that using F_u in lieu of F_y provides better correlation with the available test data. Figure 4.5 presents the same data as Figure 4.4 but with the addition of traditional calculations using F_y in lieu of F_u. For T-stub flange thickness of 3/4 in. and 7/8 in., the use of F_y in the traditional calculation brings the strength closer to IDEA StatiCa (where strength is also based on F_y). For greater thicknesses, the bolt strength controls, so

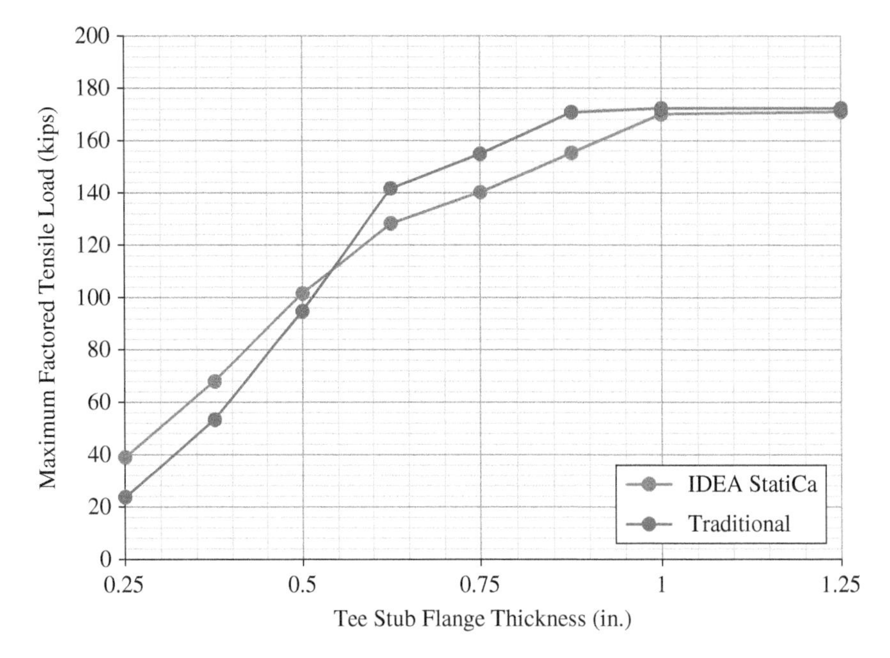

Figure 4.4 Design strength vs. T-stub flange thickness for bearing-type connections.

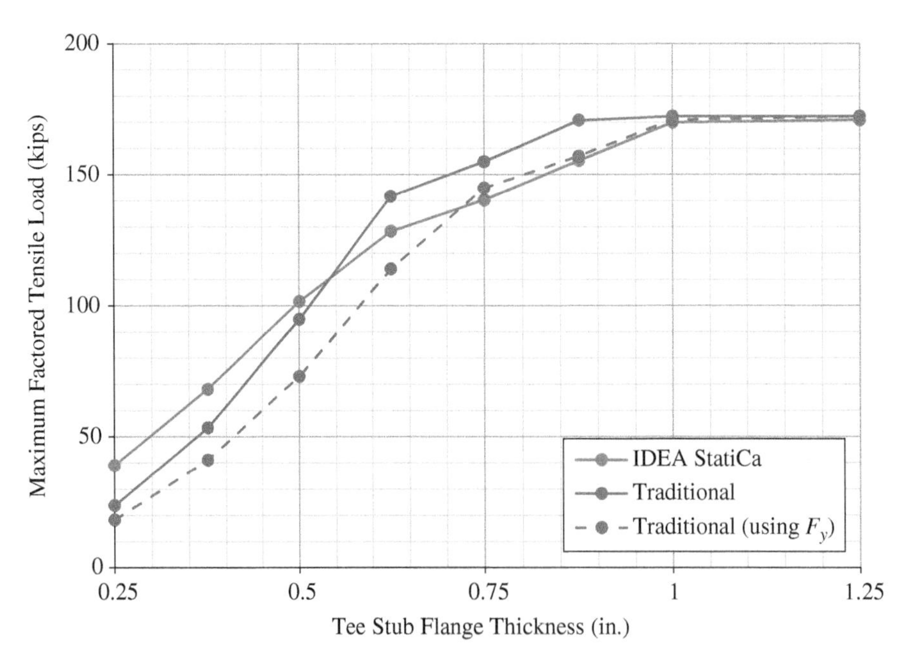

Figure 4.5 Design strength vs. T-stub flange thickness for bearing-type connections, including comparison to traditional calculations using F_y.

the choice of F_y or F_u does not impact the results. For smaller thicknesses, the use of F_y in the traditional calculations increases the discrepancy.

The discrepancy between results from the traditional calculations and IDEA StatiCa for prying action of thinner plates has been observed and investigated previously. Wald et al. (2020) compared traditional calculations to results of the component-based finite element method and to results from a research finite element model. The results showed that, while the component-based finite element method does result in greater strength than the traditional calculations for thinner plates, there remains a significant margin of safety when compared to the research model. The study by Wald et al. (2020) was expanded in this work by adding a comparison to the strength computed using the equations for prying action presented in Part 9 of the AISC *Manual*. Results, which are superimposed on existing results from Figure 5.1.5 from Wald et al. (2020), are presented in Figure 4.6. For thinner plates, the AISC results are close to that of the component method (CM).

The size of the finite elements used in IDEA StatiCa can impact the results. To investigate mesh sensitivity, analyses were repeated with four specific maximum element sizes: 2 in., 1 in., 0.5 in., 0.3 in. and compared against previous results using the "default" setting for maximum element size. The minimum element size was equal to 0.3 in. for all analyses except for those with a maximum element size of 0.3 in., in which case the minimum element size was set to 0.2 in. Results are plotted in Figure 4.7. Note that the results for maximum element sizes of 2 in. and 1 in. were the same as those for the default maximum element size and are excluded from the plot.

Smaller maximum element sizes decrease the maximum load which can be applied to the connection, according to IDEA StatiCa. The greatest differences are seen for the thinner plate. As a result, the IDEA

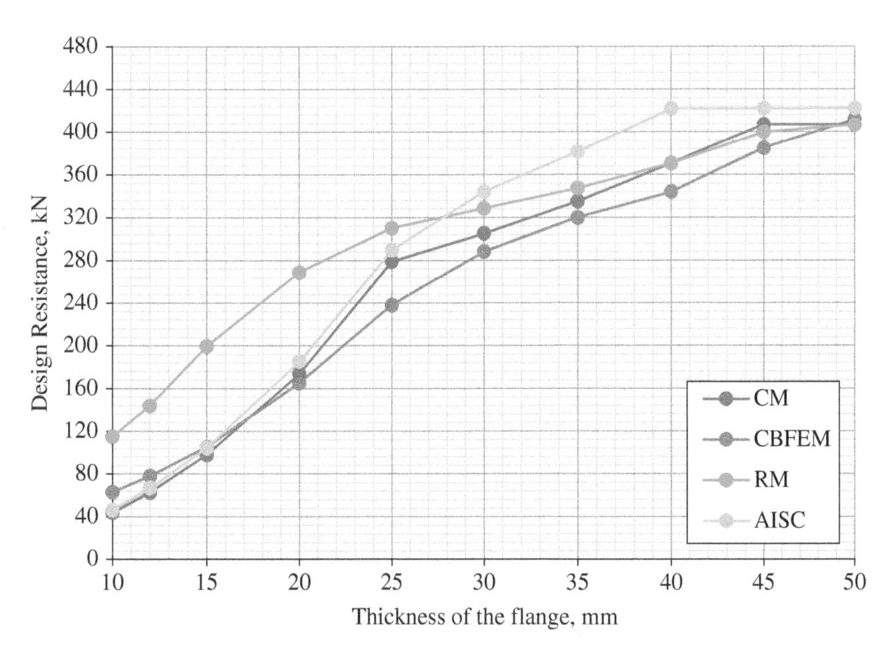

Figure 4.6 Sensitivity study of flange thickness; adapted from Figure 5.1.5 of Wald et al. (2020).

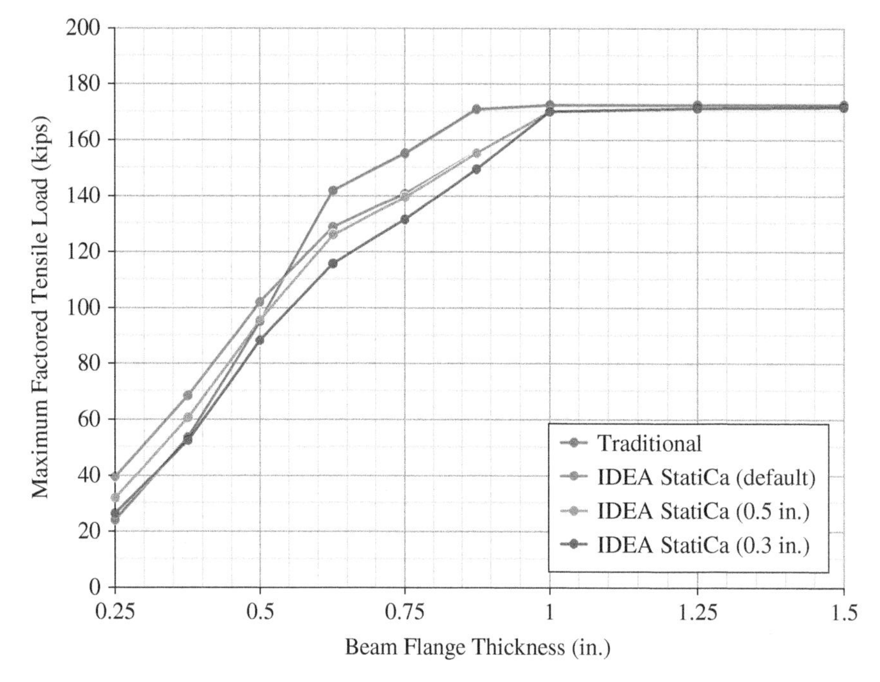

Figure 4.7 Design strength vs. T-stub flange thickness for bearing-type connections, including mesh sensitivity study.

StatiCa results with a maximum element size of 0.3 in. compares well to the results from the traditional calculations for the thinnest plates investigated.

4.5 Prying of the Beam Flange

The third investigation examines the strength of the beam flange and bolts. The beam flange was varied by selecting different beam sections. Six beam sections were selected for investigation as listed in Table 4.1. To accommodate the greater loads in this investigation, the bolts are Group B (e.g. A490) with threads not excluded from the shear planes; the T-stub flange width was $b_f = 8.0$ in.; the T-stub flange thickness was $t_f = 1.25$ in.; the T-stub web thickness was $t_w = 0.75$ in., and $\theta = 90°$. The connections were not slip-critical. The default mesh settings were used in IDEA StatiCa.

The maximum factored tensile load which could be applied to the connection is presented in Figure 4.8. As expected, for both the traditional calculation results and the IDEA StatiCa results, the maximum factored tensile load increases with the beam flange thickness until a plateau is reached where bending of the T-stub controls. Prying action impacts the strength of each of the connections in this investigation. For the traditional calculations, the guidance of Part 9 of the AISC *Manual* was adopted along with the

Table 4.1 Selected parameters.

Beam section	$t_{f,beam}$ (in.)	$b_{f,beam}$ (in.)
W18 × 175	1.59	11.4
W18 × 119	1.06	11.3
W18 × 97	0.870	11.1
W18 × 76	0.680	11.0
W12 × 40	0.515	8.01
W10 × 33	0.435	7.96

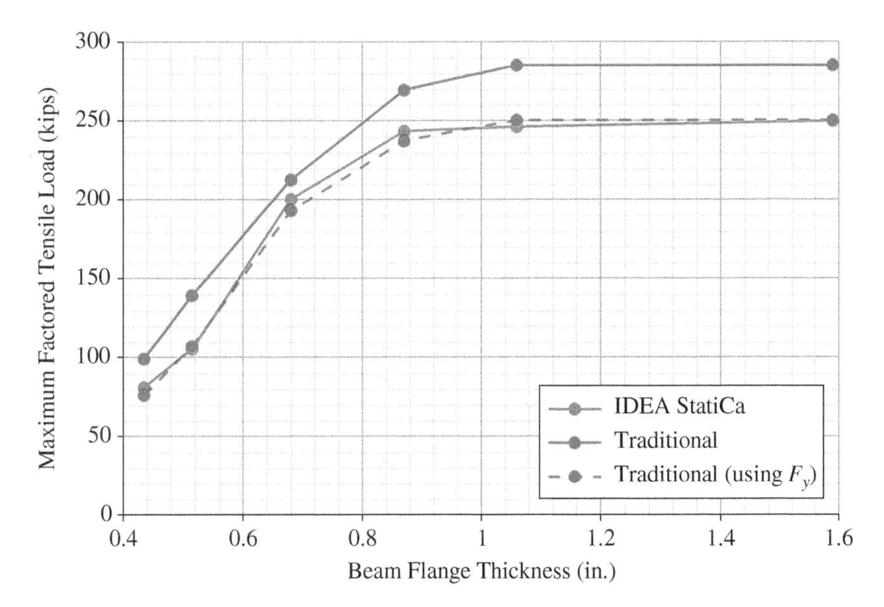

Figure 4.8 Design strength vs. beam flange thickness.

assumed yield-line pattern shown in Figure 4.9 (Dowswell, 2011). IDEA StatiCa explicitly models the connection using the CBFEM. The pattern of yielding observed from the CBFEM results (Figure 4.10) agreed with the assumed yield line used in the traditional calculations. IDEA StatiCa produced conservative results in comparison to the traditional calculations over the range investigated. As before, the IDEA StatiCa results were also compared to a variation of the traditional calculations where F_y was used in lieu of F_u. Use of F_y reduced the strength per the traditional calculations such that it matched closely with the IDEA StatiCa results.

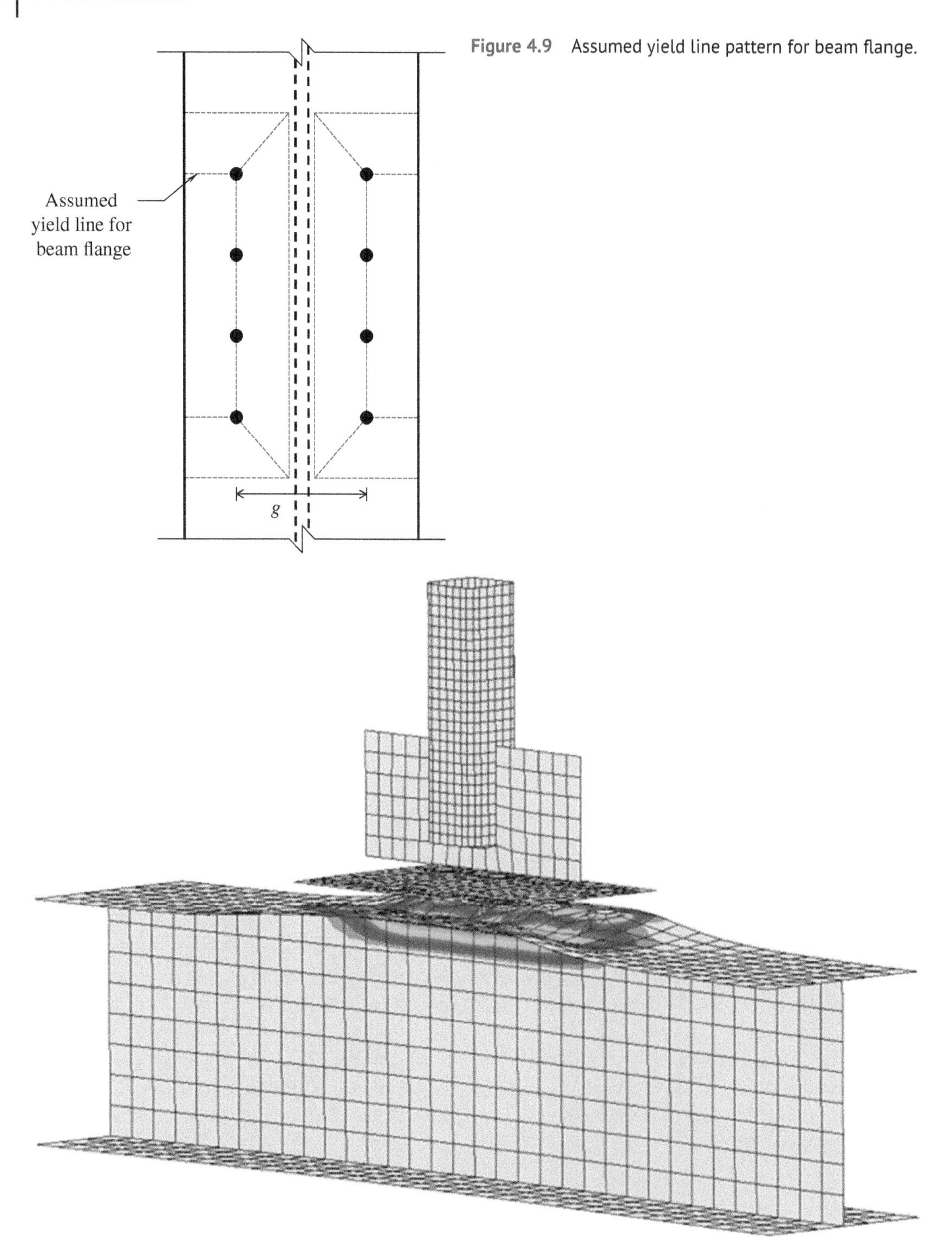

Figure 4.9 Assumed yield line pattern for beam flange.

Figure 4.10 Plastic strain for the connection with a W10 × 33 beam and t_f = 1.25 in. (deformation scale = 5).

4.6 Summary

This study compared the design of T-stub connections by traditional calculation methods used in US practice and IDEA StatiCa. Key observations from the study include:

- The available strength obtained from IDEA StatiCa agrees well with the traditional calculations with differences primarily on the conservative side.
- When evaluating slip-critical connections subjected to combined tension and shear, IDEA StatiCa conservatively considers only the tension in the bolts and not the contact pressure on the faying surfaces (i.e. the prying force) when determining the available strength.
- Some of the differences in connection strength are due to the equations for prying action presented in Part 9 of the AISC *Manual* being based on the tensile strength, F_u, while IDEA StatiCa limits the stress to the yield strength, F_y.
- IDEA StatiCa exhibited greater strength than the traditional calculations for the cases examined with thinner flanges. However, for these cases, there remains a significant margin of safety when compared to the results of a detailed finite element model.
- Some mesh dependence was observed. For cases where the plastic strain limit controlled, IDEA StatiCa exhibited reduced strengths when the mesh size was set smaller than the default.

References

AISC (2016). *Specification for Structural Steel Buildings*. American Institute of Steel Construction, Chicago.

AISC (2017). *Steel Construction Manual*, 15th Edition. American Institute of Steel Construction, Chicago.

AISC (2019). *Steel Construction Manual Design Examples*, v15.1. American Institute of Steel Construction, Chicago.

Dowswell, B. (2011). A Yield Line Component Method for Bolted Flange Connections. *Engineering Journal*, 48(2), 93–116.

Wald, F., Šabatka, L., Bajer, M., Jehlička, P., Kabeláč, J., Kožich, M., Kuříková, M., and Vild, M. (2020). *Component-Based Finite Element Design of Steel Connections*. Czech Technical University, Prague.

5

Beam-Over-Column Connections

5.1 Description

A comparison between results from the component-based finite element method (CBFEM) and traditional calculation methods used in US practice for beam-over-column connections is presented in this chapter. Connection limit states evaluated include beam web local yielding, beam web local crippling, HSS wall local yielding, HSS wall local crippling, cap plate bending, beam flange bending, and bolt tensile rupture. HSS member strength was also evaluated. A schematic of the beam-over-column connection investigated is presented in Figure 5.1.

The parameters of the connection change depending on the limit state being investigated. However, the typical connection has the following characteristics unless noted otherwise: (4) 3/4 in. diameter Group B (e.g. A490) bolts with spacing, $s = 11$ in. and gage, $g = 3.5$ in.; a W18 beam conforming to ASTM A992 ($F_y = 50$ ksi and $F_u = 65$ ksi); a 3/8 in. thick stiffener plate conforming to ASTM A36 ($F_y = 36$ ksi and $F_u = 58$ ksi); a 9 in. by 14 in. by 3/4 in. thick cap plate; and a HSS8x8 column conforming to ASTM A500 Gr. B ($F_y = 46$ ksi and $F_u = 58$ ksi).

The traditional calculations were performed in accordance with the provisions for load and resistance factor design (LRFD) in the AISC *Specification* (AISC, 2016) with prying action considered as described in Part 9 of the AISC *Manual* (AISC, 2017). The connections and method of evaluation were modeled after Example 4.1 of AISC Design Guide 24 (Packer et al., 2010). The axial load and moment are resolved to a force couple, the compression force is assumed to be centered at the face of the HSS, and the tensile force is assumed to be centered at the centerline of the bolts.

The CBFEM results were obtained from IDEA StatiCa Version 21.0. Loads were applied using the "Loads in Equilibrium" function to minimize the bending moment in the beam at the connection. For all analyses, the axial load was taken as constant, and the maximum permitted bending moment was determined iteratively by adjusting the applied load input to a value that met all limits; but if increased by a small amount (1 kip-in) would exceed the limits. Buckling analyses were performed and a limit of 3.00 on the buckling factor was enforced.

Steel Connection Design by Inelastic Analysis: Verification Examples per AISC Specification, First Edition.
Mark D. Denavit, Ali Nassiri, Mustafa Mahamid, Martin Vild, Halil Sezen, and František Wald.

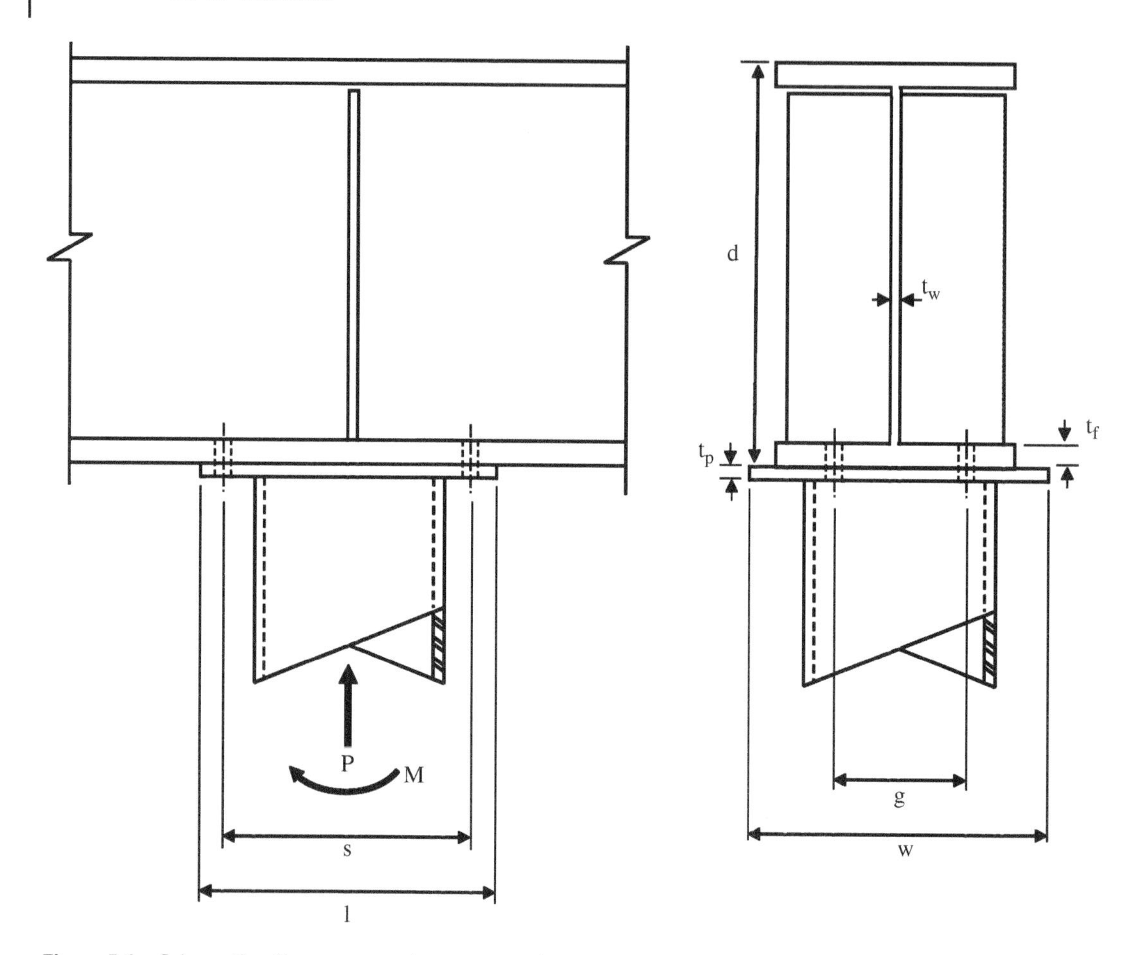

Figure 5.1 Schematic of beam-over-column connection.

5.2 HSS Column Local Yielding and Crippling

First, the limit states of local yielding and local crippling of the wall of the HSS column are investigated. Connections with five different beam sections (W18 × 35, W18 × 40, W18 × 46, W18 × 76, and W18 × 86) were analyzed. The beams have different flange thicknesses and thus distribute the load to the column differently. The cap plate conformed to ASTM A572 Gr. 50 (F_y = 50 ksi and F_u = 65 ksi). The column was a HSS8 × 8 × 3/16 which has a nominal moment strength of M_n = 580.5 kip-in and cross-sectional axial strength of P_n = 216.7 kips. The applied axial load was P_u = 45 kips for all analyses.

The maximum factored moment is presented in Figure 5.2. The limiting buckling factor of 3.00 controlled the strength of all connections in IDEA StatiCa. The strength increases slightly from 314 kip-in to 328 kip-in as the beam size increased and more evenly distributed the load to the wall of the HSS. An example of the buckling mode computed by IDEA StatiCa is presented in Figure 5.3.

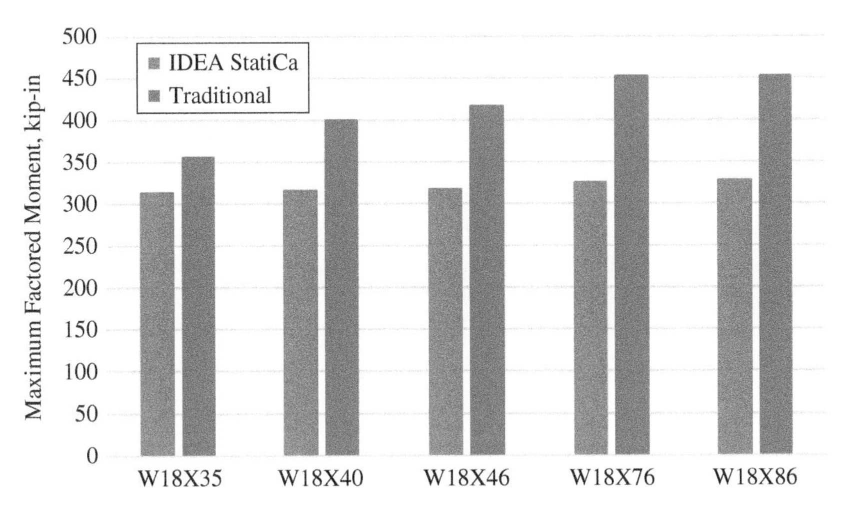

Figure 5.2 Comparison of results investigating HSS column local yielding and crippling.

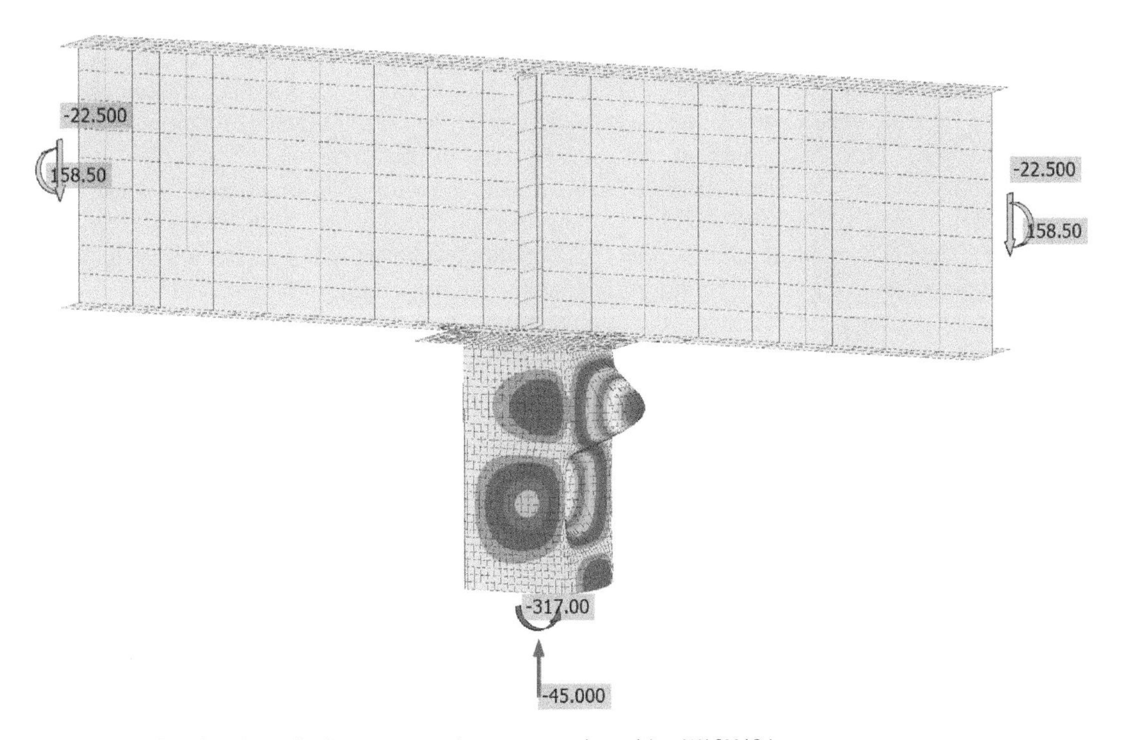

Figure 5.3 Buckled shape for beam-over-column connection with a W18X40 beam.

The strength according to the traditional calculations exhibited greater variation as the beam size increased, from 357 kip-in to 452 kip-in. HSS wall local yielding controlled for the connection with the W18 × 35 beam. HSS wall local crippling controlled for the connections with the W18 × 40 and W18 × 46 beams. HSS member strength controlled for the connections with the W18 × 76 and W18 × 86 beams.

These results indicate that limiting the buckling factor to 3.00 can be conservative. However, there was some indication that there was not significant reserve capacity beyond the buckling factor limit. Analyses in IDEA StatiCa were performed both with geometric nonlinearity on and with geometric nonlinearity off. Given that the boundary conditions were applied to the HSS member for this connection, geometric nonlinearity was on by default. Since the buckling factor limit controlled in all cases, there was no difference between the strength results between geometric nonlinearity on or off. However, for some cases and with geometric nonlinearity on, the strain increased rapidly with small increases in applied load shortly after the buckling limit was reached.

5.3 Beam Web Local Yielding and Crippling

Next, the limit states of local yielding and local crippling of the web of the wide flange beam are investigated. The beam for these analyses was a W18 × 40, but with the thickness of the web overridden to values of 0.30 in. 0.25 in. and 0.20 in. The connection was also analyzed with the beam's standard thickness of 0.315 in. Overriding the thickness allowed precise control of the thickness of the web in relation to other beam parameters. The cap plate conformed to ASTM A36 ($F_y = 36$ ksi and $F_u = 58$ ksi). The column was a HSS8 × 8 × 1/2 which has a nominal moment strength of $M_n = 1725$ kip-in and cross-sectional axial strength of $P_n = 621$ kips. The applied axial load was $P_u = 45$ kip for all analyses.

The maximum factored moment is presented in Figure 5.4. The controlling limit state for each analysis is presented in Table 5.1. The beam web local limit states controlled when the thickness was reduced significantly. The buckling mode computed by IDEA StatiCa for the analysis with beam web thickness of 0.20 in. is presented in Figure 5.5. For larger thicknesses, the tension side of the connection controlled with bending of the cap plate, bending of the beam flange, bolt tension, or a combination of these limit states controlling. Analyses were performed in IDEA StatiCa with geometric nonlinearity on and geometric nonlinearity off. Both sets of results are presented in Figure 5.4. There is only a small difference between the two.

When the beam web thickness is overridden to 0.20 in. or 0.25 in., local crippling of the beam web controls the strength per the traditional calculations. Buckling of the beam web controls the strength per IDEA StatiCa for the connection with beam web thickness of 0.20 in., but not for the connection with beam web thickness of 0.25 in. For both connections, IDEA StatiCa produces strengths greater than that from the traditional calculations. The discrepancy could be due to several factors. The traditional calculations do not account for the stiffener, which appears to influence the buckling mode (Figure 5.5). The finite element mesh in IDEA StatiCa may also be too coarse.

A mesh sensitivity study was performed to bring greater insight to the results. IDEA StatiCa analyses were repeated for each of the four connections presented in Figure 5.4 using different maximum element sizes. The analyses of this mesh refinement study were performed with geometric nonlinearity turned on. The results of the mesh refinement study are presented in Figure 5.6.

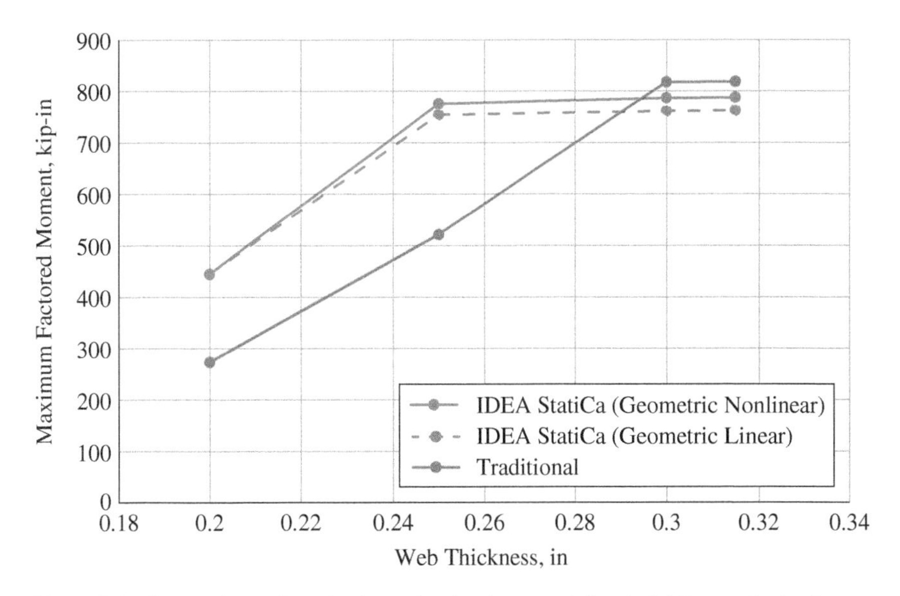

Figure 5.4 Comparison of results investigating beam web local yielding and crippling.

Table 5.1 Controlling limit state for results presented in Figure 5.4.

Web thickness (in.)	IDEA StatiCa	Traditional
0.200	Buckling (beam web)	Local crippling of the beam web
0.250	Plastic strain (cap plate)	Local crippling of the beam web
0.300	Plastic strain (cap plate)	Beam flange bending and bolt tension
0.315	Plastic strain (cap plate)	Beam flange bending and bolt tension

Overall, the results show a significant mesh dependence for this connection. The maximum factored moment capacity decreases as the size of the mesh decreases. Additionally, in some instances, the failure mode changes with refinement of the mesh. For the connections with web thicknesses of 0.25 in. and 0.30 in., the controlling limit state transitions from exceeding the strain limit in the cap plate at the default mesh size (1.969 in.) to exceeding the strain limit in the beam web for the decreased maximal element sizes. Note that cap plate bending was not expected to occur per the traditional calculations. The maximum element size also impacts the buckling results. For the connection with beam web thickness of 0.20 in., the buckling factor limit controls. The applied load at which the limit is reached decreases with mesh size and appears to converge at a maximum element size of 0.50 in.

Another potential reason for the discrepancy in results between the traditional calculations and IDEA StatiCa is the stiffener in the beam centered above the column. Since the stiffener is not located in line with the concentrated force (i.e. the wall of the column), it is not considered in the traditional calculations. The stiffener is included in the model and thus is considered by IDEA StatiCa.

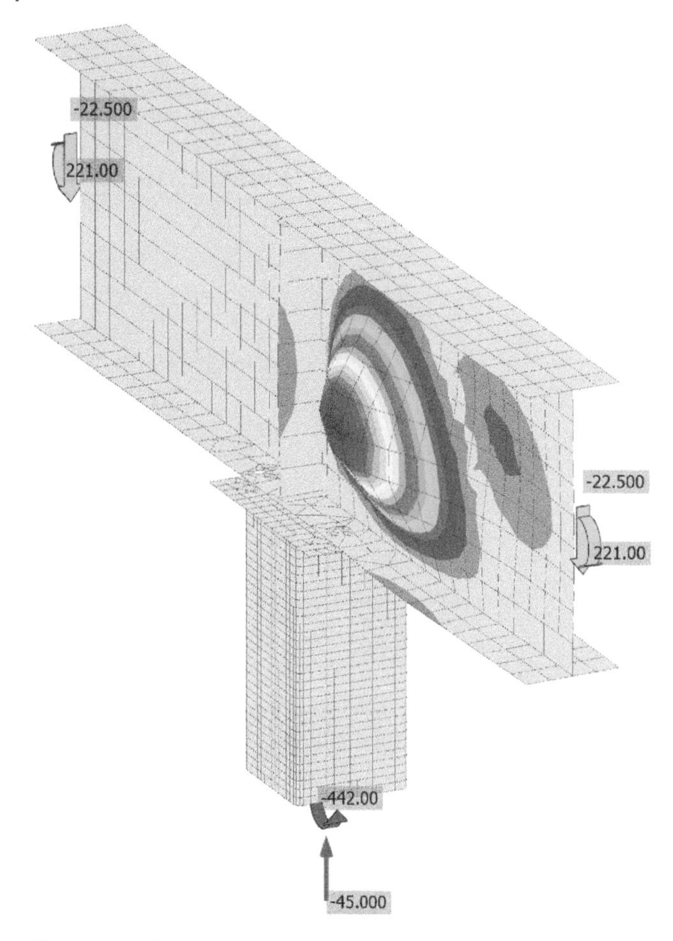

Figure 5.5 Buckled shape for beam-over-column connection with a W18X40 beam with web thickness overridden to 0.2 in.

Analysis of a simpler connection (Figure 5.7) was performed to evaluate the magnitude of the effect of a nearby stiffener. For this analysis, the beam was a W18x40 (A992) with web thickness overridden to $t_w = 0.25$ in. The beam was loaded by a 1 in. thick plate and 3/8 in. thick plate stiffeners were located 0.25 times the beam depth to 2 time the beam depth away from the centerline of the loading plate.

Analyses were performed to determine the maximum permitted applied load from IDEA StatiCa and Section J10 of the AISC *Specification* (AISC, 2016) for the limit states of web local yielding and web local crippling (Figure 5.8). The results of the traditional calculations do not consider the stiffener and do not vary with the position of the stiffener. Two results are shown for the traditional calculations. One where the k dimension (i.e. the distance from the outer face of the flange to the web toe of the fillet) was taken as the value of k listed in Part 1 of the AISC *Manual* (AISC, 2017) for the beam and one where the k dimension was taken as t_f, the thickness of the flange. IDEA StatiCa does not explicitly model the fillet

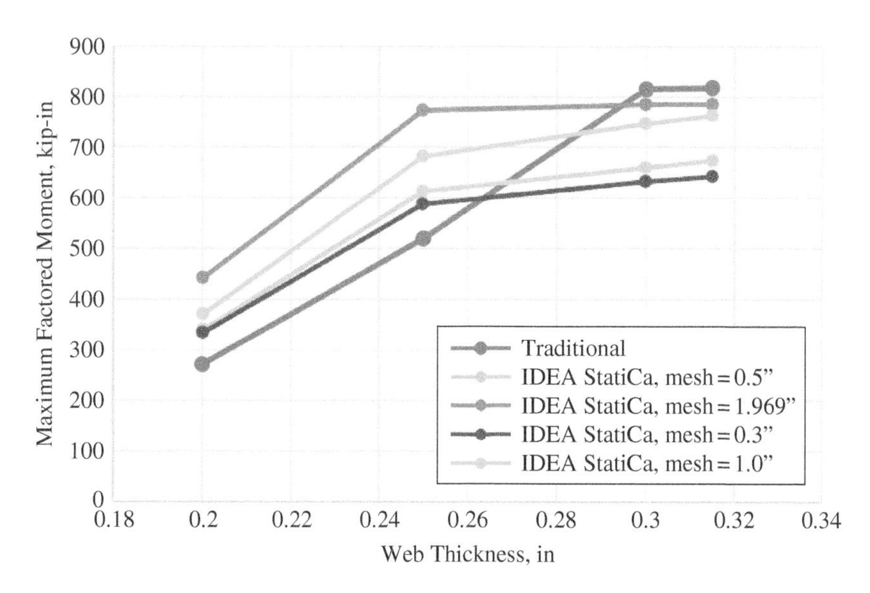

Figure 5.6 Comparison of results investigating beam web local yielding and crippling, showing mesh sensitivity study.

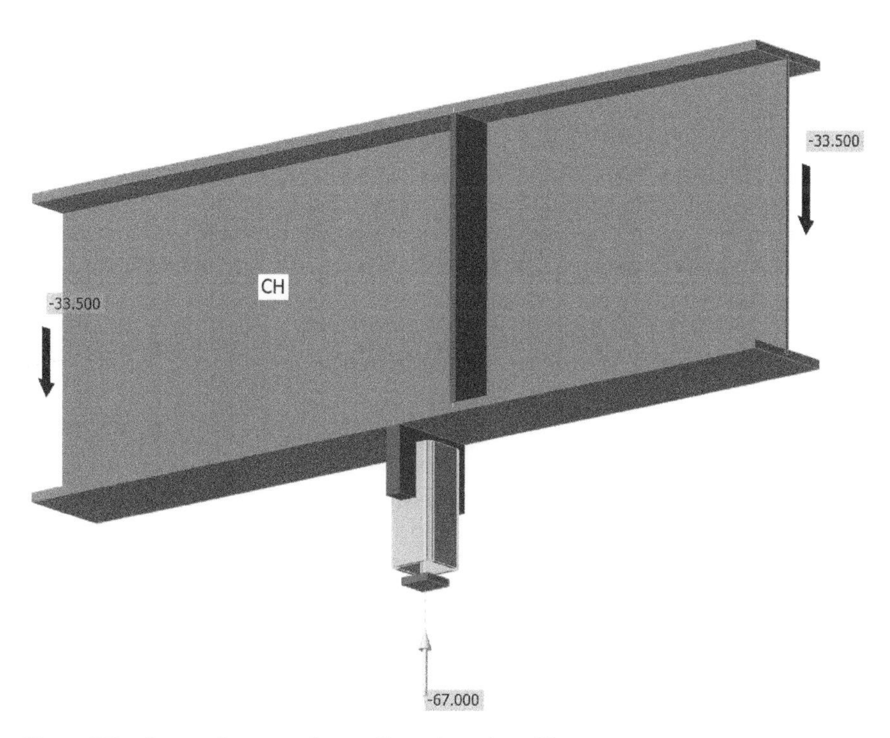

Figure 5.7 Connection to evaluate effect of nearby stiffener.

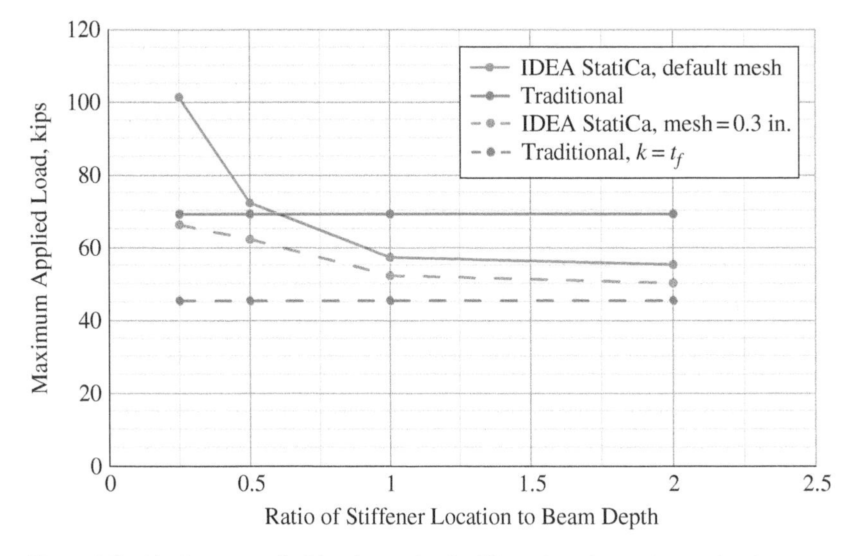

Figure 5.8 Maximum applied load vs ratio of stiffener location to beam depth.

of the wide flange shapes. Two results are also shown for IDEA StatiCa, one with the default mesh size and one with a mesh size of 0.3 in.

Local web yielding controls for the traditional calculations for all cases. The plastic strain limit controls for IDEA StatiCa when the stiffener is located one-quarter the beam depth from the applied load and the buckling limit controls otherwise. For the nearby stiffeners, IDEA StatiCa exhibits a greater strength than the traditional calculations. However, as the distance to the stiffener increases, the strength from IDEA StatiCa reduces, eventually settling below the strength from the traditional calculations. The strength from traditional calculations for $k = t_f$ is still lower, but this case is shown for informational purposes and not for direct comparison. Regardless, these results demonstrate that IDEA StatiCa captures the stiffening effect of nearby stiffeners and effect which contributed to the discrepancy in results shown in Figure 5.4.

5.4 Axial Compression/Bending Moment Interaction

Lastly, the variation of moment strength with the level of axial load is investigated. The traditional calculations utilize simple assumptions to convert the applied axial load and bending moment to a force couple. IDEA StatiCa computes the distribution of stresses explicitly. The beam for these analyses was a W18 × 35. The cap plate conformed to ASTM A572 Gr. 50 (F_y = 50 ksi and F_u = 65 ksi). The column was a HSS8 × 8 × 3/16 which has a nominal moment strength of M_n = 580.5 kip-in and cross-sectional axial strength of P_n = 216.7 kips.

An interaction diagram showing the maximum factored moment for each selected axial load is presented in Figure 5.9. The controlling limit state for each analysis is presented in Table 5.2. Analyses were performed in IDEA StatiCa with geometric nonlinearity on and geometric nonlinearity off. Both sets of results are presented in Figure 5.9. For most cases, where the buckling factor limit controlled, there is no difference between the two. Differences were noted for applied axial loads of 75 kips and 100 kips.

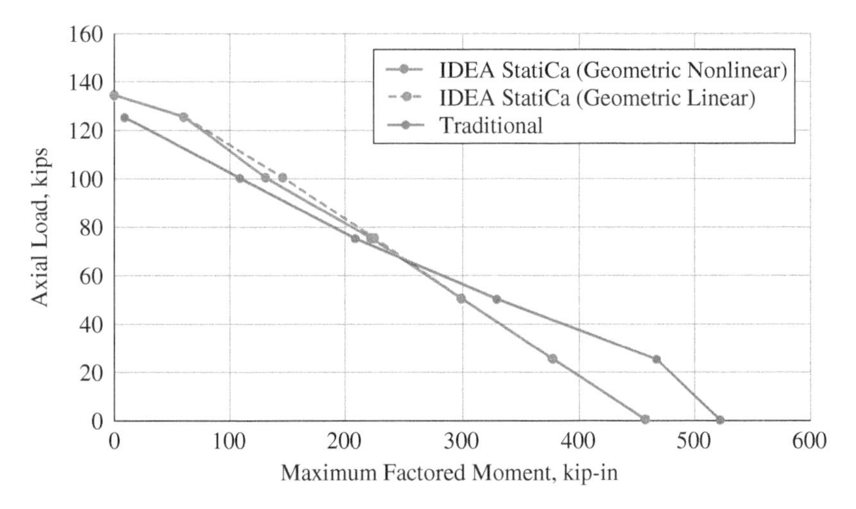

Figure 5.9 Comparison of results investigating axial compression / bending moment interaction.

Table 5.2 Controlling limit state for results presented in Figure 5.9.

Axial Load, (kips)	IDEA StatiCa (GMNA ON)	IDEA StatiCa (GMNA OFF)	Traditional
0	Buckling (HSS wall)	Buckling (HSS wall)	HSS member strength
25	Buckling (HSS wall)	Buckling (HSS wall)	Local yielding of HSS wall
50	Buckling (HSS wall)	Buckling (HSS wall)	Local yielding of HSS wall
75	Strain limit (HSS wall)	Buckling (HSS wall)	Local yielding of HSS wall
100	Limit point reached in analysis	Buckling (HSS wall)	Local yielding of HSS wall
125	Buckling (HSS wall)	Buckling (HSS wall)	Local yielding of HSS wall
134	Buckling (HSS wall)	Buckling (HSS wall)	n/a

For the connection with 75 kips applied axial load, when geometric nonlinearity was off, the buckling limit was reached at an applied moment of 225 kip-in. When geometric nonlinearity was on, the strain limit was reached at an applied moment of 222 kip-in. Important to note is that the strain limit was not reached gradually, rather, a large increase in strain (~3%) was noted for a small increase applied moment (1 kip-in) immediately prior to reaching the limit.

For the connection with 100 kips applied axial load, when geometric nonlinearity was off, the buckling limit was reached at an applied moment of 146 kip-in. When geometric nonlinearity was on, an applied load of 131 kip-in resulted in a buckling factor of 3.10 and maximum strain of 2.2%. For greater applied loads, the analysis was unable to complete, indicating that a limit point had been reached. The maximum factored moment was taken as the greatest applied moment for which the analysis completed 100%.

For both of these analyses, IDEA StatiCa provided greater strength than the traditional calculations. Further investigation is warranted to determine if an inelastic buckling analysis would be more appropriate or if other changes to the manner in which this connection is evaluated are necessary.

5.5 Summary

This study compared the design of beam-over-column connections by traditional calculation methods used in US practice and IDEA StatiCa. Key observations from the study include:

- The available strength obtained from IDEA StatiCa agrees well with the traditional calculations with differences primarily on the conservative side.
- For the cases examined, limiting the buckling factor to 3.00 was found to be an effective and conservative means of limiting the effects of geometric nonlinearity and considering elastic stability limit states.
- IDEA StatiCa considers the effect of nearby stiffeners which impacts the strength web local limit states.
- Some mesh dependence was observed. IDEA StatiCa exhibited reduced strengths when the mesh size was set smaller than the default.

References

AISC (2016). *Specification for Structural Steel Buildings*. American Institute of Steel Construction, Chicago.

AISC (2017). *Steel Construction Manual*, 15th Edition. American Institute of Steel Construction, Chicago.

Packer, J., Sherman, D., and Lecce, M. (2010). *Hollow Structural Section Connections. Design Guide 24*. American Institute of Steel Construction, Chicago.

6

Base Plate Connections

6.1 Description

A comparison between results from the component-based finite element method (CBFEM) and traditional calculation methods used in US practice for base plate connections is presented in this chapter. Three loading conditions are evaluated: concentric axial compressive load, shear load, and combined axial compressive load and moment. A schematic of the column to base plate connection investigated is shown in Figure 6.1.

The traditional calculation methods are based upon the recommendations presented in AISC Design Guide 1 (Fisher and Kloiber, 2006). The recommendations presented in this guide are based on simplifying assumptions of the behavior of the base plate that can lead to highly conservative results if redistribution of bearing stress is possible after base plate yielding or unconservative results if the tensile forces in the anchor rods are underestimated. In particular, the assumption of uniformly distributed bearing stress (i.e. rigid base plate) is often inaccurate since the flexibility of the base plate results in a non-uniform stress distribution (Fitz et al., 2018). Accordingly, results from traditional calculations based on alternative assumptions that are less conservative will also be presented. In both cases, the calculations were performed in accordance with the provisions for load and resistance factor design (LRFD) in the AISC *Specification* (AISC, 2016). The ACI *Code* (ACI, 2019) also includes provisions relevant to the strength of base plate connections. However, concrete limit states other than concrete bearing strength were avoided in this study and the provisions for concrete bearing strength in the ACI *Code* are identical to those in the AISC *Specification*.

The CBFEM results were obtained from IDEA StatiCa Version 22.1. The maximum permitted loads were determined iteratively by adjusting the applied load input to a value that the program deems safe but if increased by a small amount (e.g. 1 kip) the program would deem unsafe. DR type analyses can help identify maximum permitted loads. However, some approximation is made in the evaluation of the joint design resistance, therefore, all results in this report are based on EPS type analyses.

Steel Connection Design by Inelastic Analysis: Verification Examples per AISC Specification, First Edition.
Mark D. Denavit, Ali Nassiri, Mustafa Mahamid, Martin Vild, Halil Sezen, and František Wald.

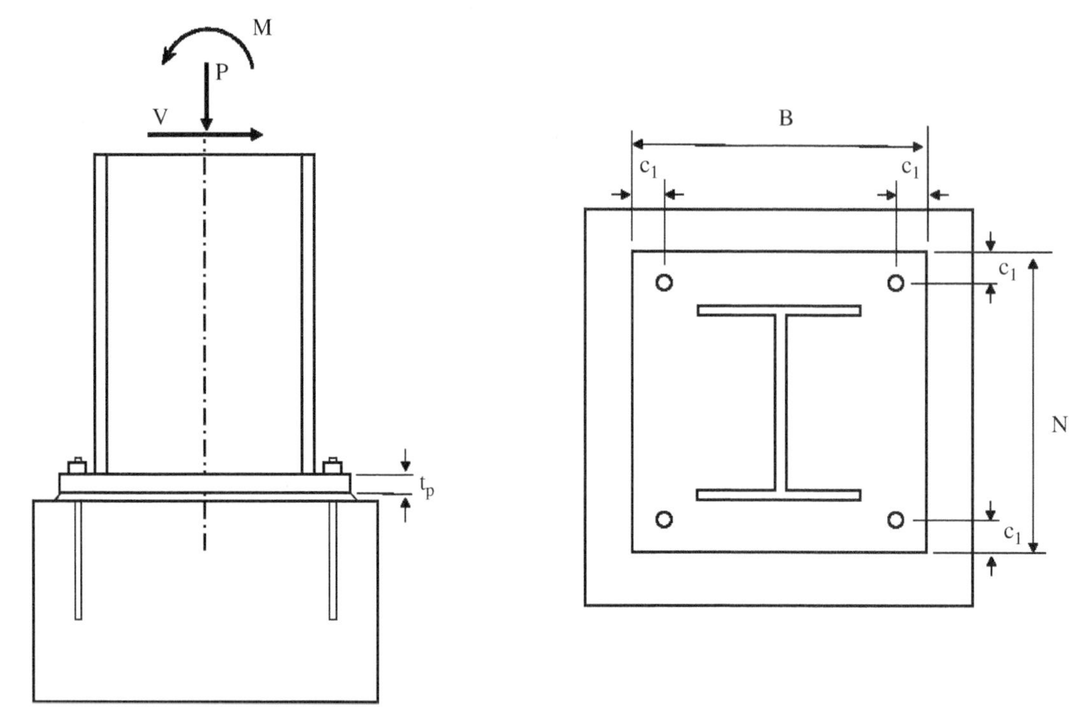

Figure 6.1 Schematic of base plate connection showing the wide flange column. Base plate for the HSS column is similar.

6.2 Concentric Axial Compressive Load

First, base plates subjected to concentric axial compressive load are investigated. Limit states evaluated for this loading condition are concrete crushing and flexural yielding of the base plate. Two cases are examined, one with a rectangular HSS column and one with a wide flange column.

For the case with the rectangular HSS column, the column section was a HSS10x4x5/8 (ASTM A500 Gr. C, $F_y = 50$ ksi) and the plate was square with plan dimensions of 12 in. by 12 in., thickness varying from 0.25 in. to 2.50 in., and steel conforming to ASTM A36 ($F_y = 36$ ksi). Anchor rods were 3/4 in. diameter (ASTM F1554 Gr. 36, $F_y = 36$ ksi) and had an edge distance of $c_1 = 1$ in. The holes for the anchor rods were 1-5/16 in. in diameter in accordance with the recommendations of Table 14-2 of the AISC *Manual* (AISC, 2017). The base plate was assumed to bear directly on the concrete ($f'_c = 4$ ksi). The plan area of the concrete was large, such that the maximum permitted bearing strength would apply (i.e. $\sqrt{A_2/A_1} \geq 2$). A three-dimensional view of the base plate connection is shown in Figure 6.2.

The maximum factored axial compressive loads that can safely be applied to the base plate connection as determined from IDEA StatiCa and traditional calculations are presented in Figure 6.3. For thick base plates, i.e. $t_p \geq 2.25$ in., the traditional results and those from IDEA StatiCa are nearly identical. For these cases, bearing controls the strength and the whole area of the base plate is in contact with the concrete. The small difference in strength between the results of the traditional method and IDEA StatiCa

Figure 6.2 Three-dimensional view of the base plate with HSS column.

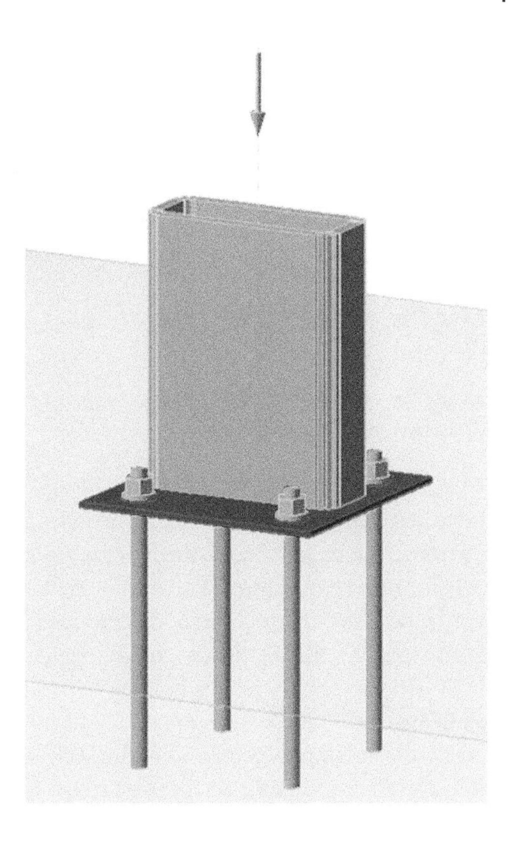

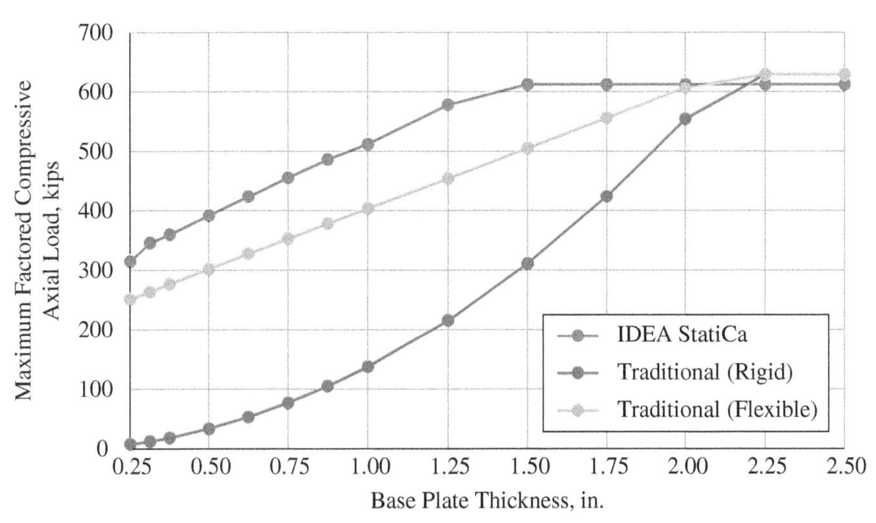

Figure 6.3 Maximum factored compressive axial load vs. plate thickness for base plate with HSS column.

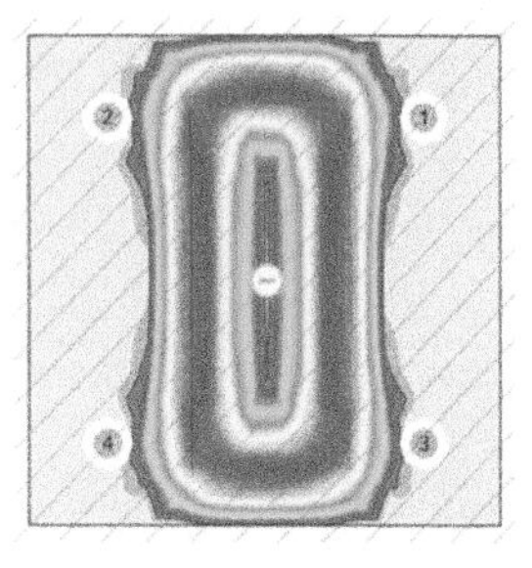

(a) $t_p = 0.25$ in. and $P = 317$ kips

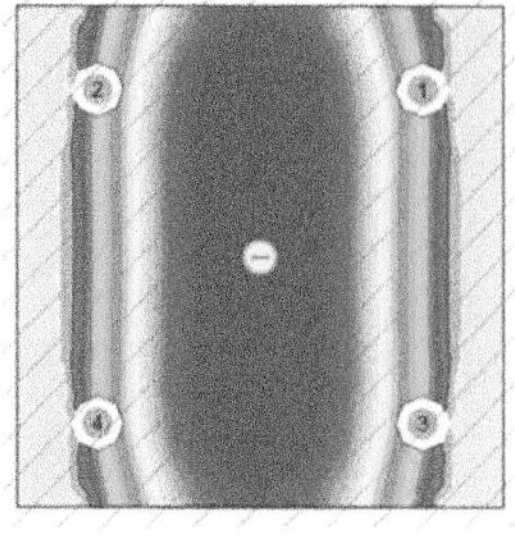

(b) $t_p = 0.75$ in. and $P = 458$ kips

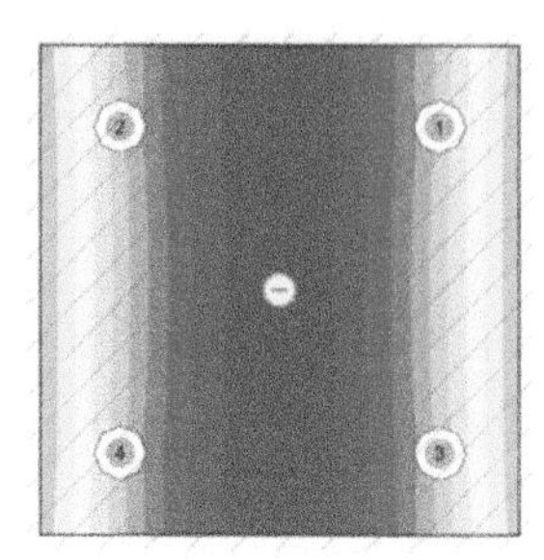

(c) $t_p = 2.00$ in. and $P = 614$ kips

Figure 6.4 Bearing stress distribution from IDEA StatiCa for base plate with HSS column. Hatching indicates area A_2 and extends beyond the view.

is because IDEA StatiCa considers the holes for the anchor rods when computing the bearing area while the reduction in area due to the holes is typically ignored in the traditional method.

For thinner base plates, results from the traditional calculations and IDEA StatiCa differ significantly. For these cases, the traditional calculations are controlled by base plate bending while the controlling limit state in IDEA StatiCa is concrete crushing. The uniformly distributed bearing stress assumed in AISC Design Guide 1 results in large flexural demands in the base plate. However, the base plate, especially when thin, is flexible and will deform, resulting in a distribution of bearing stresses concentrated below the column as shown in Figure 6.4. Yielding of the base plate further increases the flexibility of the base plate and limits the bearing stress at the extremities of the base plate. This behavior is modeled explicitly in IDEA StatiCa. Thus, while yielding of the base plate occurs, the plastic strain in the base plate never reaches the 5% limit and concrete strength controls.

To explore the differences further, the traditional calculations were repeated with assumptions more consistent with a flexible base plate. The assumed stress distribution for these alternative traditional calculations is shown in Figure 6.5. The bearing stress is uniform, but only over a portion of the base plate. The magnitude of the bearing stress is equal to the maximum permitted by the AISC *Specification* (AISC, 2016) (i.e. $\phi 1.7 f_c'$ noting that the plan area of the concrete is large). The width of the bearing area is dependent on the applied load and bearing stress. For these calculations, the location of the yield lines was the same as recommended in AISC Design Guide 1. While this alternative assumption regarding the distribution of bearing stress is different than that presented in the guide, it still complies with the AISC *Specification*. Another way of interpreting the alternative bearing stress assumption is that portions of the base plate that are in excess of what is needed for concrete bearing are ignored.

The maximum factored axial compressive loads computed using the alternative traditional calculations are presented in Figure 6.3. Use of the alternative bearing stress assumption provides strengths that are much higher than those using the assumptions of AISC Design Guide 1. Given that both sets of assumptions are valid, this indicates that assuming uniform bearing stress over the entire base plate is conservative for base plates that are oversized for bearing. The strengths from IDEA StatiCa are still greater than the strengths from the traditional calculations using the alternative assumption. The reason for this is that the bearing stress distribution in IDEA StatiCa is not uniform (see Figure 6.4). The stresses

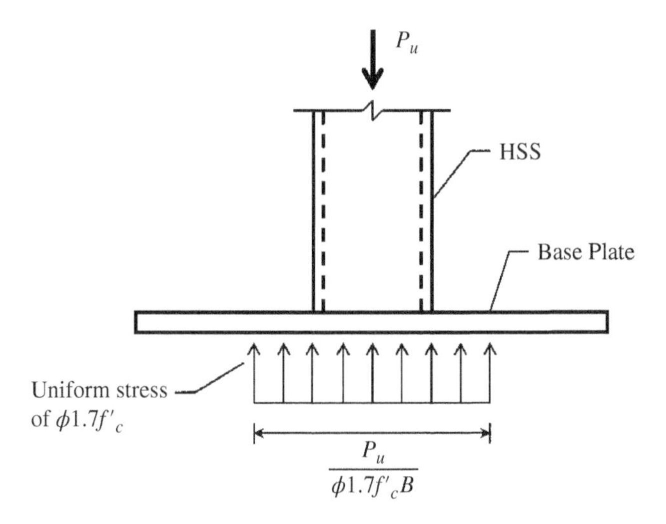

Figure 6.5 Assumed bearing stress distribution for the traditional (flexible) calculations for base plate with HSS column.

are concentrated near the column, thus placing less flexural demand on the plate. While this behavior is physically realistic, it is difficult to capture with hand calculations.

The geometry of the HSS base plate connection makes calculation of flexural demands in the base plate with more realistic assumptions of the bearing stress distribution simple. Such calculations are more difficult with wide flange columns, but the uniform bearing stress distribution assumption is similarly conservative. To explore this, additional analyses were performed with a W12 × 120 (ASTM A992, F_y = 50 ksi) column on a square base plate with plan dimensions of 18 in. by 18 in., thickness varying from 0.25 in. to 3.00 in., and steel conforming to ASTM A36 (F_y = 36 ksi). Anchor rods were 3/4 in. diameter (ASTM F1554 Gr. 36, F_y = 36 ksi) and had an edge distance of c_1 = 1.5 in. The holes for the anchor rods were 1-5/16 in. in diameter in accordance with the recommendations of Table 14-2 of the AISC *Manual* (AISC, 2017). The base plate was assumed to bear directly on the concrete (f'_c= 4 ksi). The plan area of the concrete was large, such that the maximum permitted bearing strength would apply (i.e. $\sqrt{A_2/A_1} \geq 2$).

The maximum factored axial compressive loads that can safely be applied to the base plate connection, as determined from IDEA StatiCa and traditional calculations are presented in Figure 6.6. For thick base plates, i.e. $t_p \geq 2.25$ in., the traditional results and those from IDEA StatiCa are nearly identical. Just as for the HSS column base plate, the difference is due to different handling of the holes for the anchor rods in the computation of the bearing area.

Also, as for the HSS column base plate, a significant difference in strength is noted for the thinner base plates. A major source of the difference is the uniform bearing stress over the entire base plate assumed in the traditional calculations. An alternative approach to the traditional calculations, based on European practice, is to assume a uniform bearing stress over only a portion of the base plate. The portion of the base plate subject to bearing stress is the column cross-section extended out by dimension c, as shown in Figure 6.7.

In European practice, the dimension c is based on a cantilever beam analogy as the maximum uniformly loaded length that can support the bearing stress without yield. A value for the dimension c can be

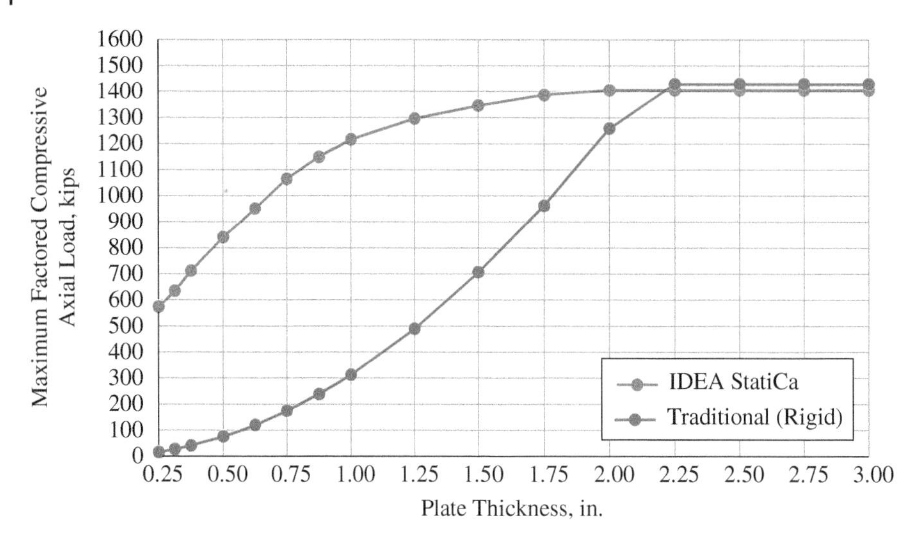

Figure 6.6 Maximum factored compressive axial load vs. plate thickness for base plate with wide flange column.

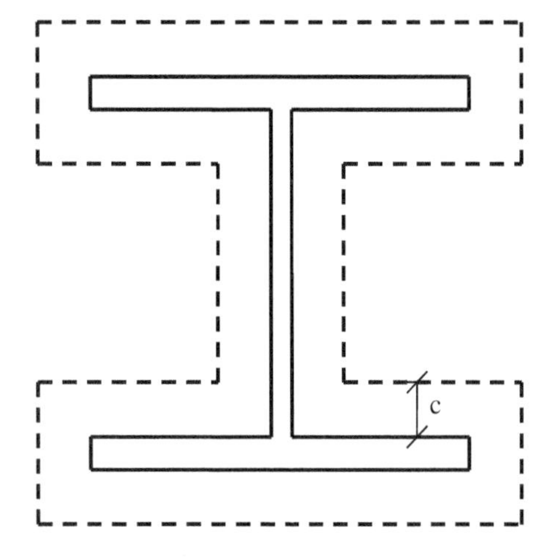

Figure 6.7 Assumed bearing area for the traditional (flexible) calculations for base plate with wide flange column.

determined by applying this concept to this example and calculations used in US practice. The cantilever beam analogy is shown in Figure 6.8. The uniform bearing stress F_p is equal to 1.7 times the concrete compressive strength, given that the plan area of the concrete is large in this example (i.e. $\sqrt{A_2/A_1} \geq 2$). The design bearing stress is $\phi F_p = 1.105f'_c$ after applying the resistance factor for concrete crushing of 0.65. The resulting required moment strength at the support for a unit width of the cantilever is

$$M_u = 1.105f'_c\frac{c^2}{2}$$

Figure 6.8 Cantilever beam analogy for determination of dimension c.

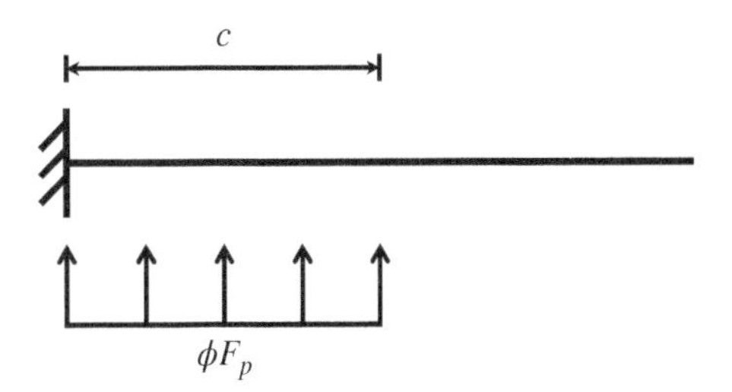

The available moment strength for the limit state of flexural yielding for a unit width of the cantilever is

$$\phi M_n = 0.9F_y \frac{t_p^2}{4}$$

Equating the required and available moment strengths (i.e. $M_u = \phi M_n$) results in an equation for c as a function of plate thickness.

$$c = 0.638t_p\sqrt{\frac{F_y}{f_c'}}$$

For the material strengths used in this example, $F_y = 36$ ksi and $f'_c = 4$ ksi, the value of c is $1.91t_p$ for a ratio of $c/t_p = 1.91$.

Steenhuis et al. (2008) evaluated the relative stiffness of the base plate and concrete foundation and recommended a ratio of $c/t_p = 1.5$. Another potential value for the ratio is $c/t_p = 2.5$, based on the 2.5:1 slope for the spread of load that is assumed in other aspects of steel design, e.g. the web local yielding provisions of Section J10.2 of the AISC *Specification* (AISC, 2016).

The strengths of the base plate using the three different c/t_p ratios are shown with the IDEA StatiCa results and the traditional calculation results using a rigid base plate assumption in Figure 6.9. For the thinner base plates, the alternative bearing stress distribution permits maximum factored loads greater than when using the assumptions of AISC Design Guide 1. The strengths are closer to the strengths from IDEA StatiCa, but IDEA StatiCa still shows greater strength. There are two main reasons for this. First, the base plate does not behave as a cantilever between the flanges of the column. Using a bearing stress distribution based on the cantilever beam analogy in this region between the flanges is conservative. Second, IDEA StatiCa does not use a uniform bearing stress, even within the bearing area.

The distribution of bearing stress in IDEA StatiCa results from the relative stiffness of the base plate and the concrete foundation. The bearing stress is greatest directly below the column web and flanges and decreases away from these elements as shown in Figure 6.10. Thus, the distribution of bearing stress is not uniform as assumed in the cantilever beam analogy. Also, the peak bearing stress can exceed the uniform bearing stress used in design since IDEA StatiCa evaluates the utilization ratio based on the average bearing stress in the bearing area. The bearing area is defined in IDEA StatiCa as the area with a bearing stress greater than a fraction of the maximum bearing stress. The fraction, called the stress cut-off ratio, is taken as 0.1 by default, but can be set by the user in the code setup menu. Using a different stress cut-off ratio yields different results. The maximum factored compressive axial load per IDEA StatiCa using a stress cut-off ratio of 0.4 is shown in Figure 6.9.

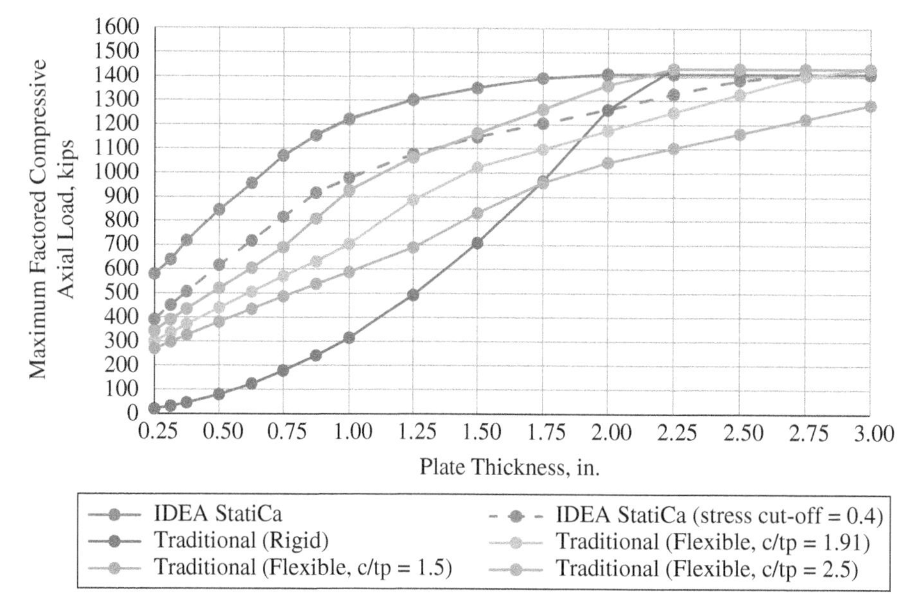

Figure 6.9 Maximum factored compressive axial load vs. plate thickness for base plate with wide flange column including traditional calculations with flexible base plate.

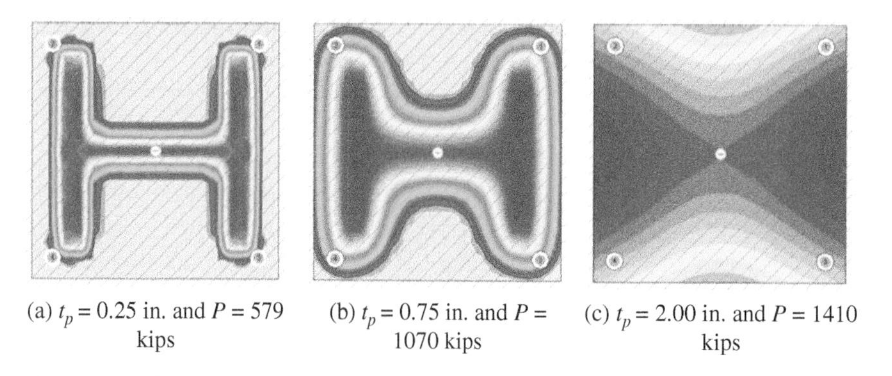

Figure 6.10 Bearing stress distribution from IDEA StatiCa for base plate with WF column. Hatching indicates area A_2 and extends beyond the view.

To further investigate the different approaches, results from traditional calculations and IDEA StatiCa are compared to experimental results. Shelson (1957), Hawkins (1968), and DeWolf and Sarisley (1980) each performed physical experiments subjecting base plate connections to axial compression load. The specimens tested by Shelson (1957) had square base plates with side dimensions ranging from 1.41 in. to 3 in. and a solid square "column" with a side dimension of 1 in. The yield strength of the plate was not reported and was assumed to be 40 ksi. The specimens tested by Hawkins (1968) had square base plates with side dimensions ranging from 3.9 in. to 4 in. and a solid square "column" with side dimensions ranging from 2 in. to 2.28 in. The specimens tested by DeWolf and Sarisley (1980) had rectangular base plates with dimensions 7 in. by 9 in. and a 4 in. by 4 in. by 0.5 in. thick steel tube column. None of the specimens had anchors.

The experimentally reported strength is compared to the strength per the traditional calculations assuming a rigid base plate, traditional calculations assuming a flexible base plate with $c/t_p = 1.5$, IDEA StatiCa results using a stress cut-off ratio of 0.1, and IDEA StatiCa results using a stress cut-off ratio of 0.4. In IDEA StatiCa, the solid square columns were modeled with square tubes with thickness of approximately 3/8 times the side dimension. For this comparison to the experimental results, strengths were computed without resistance factors. The resistance factor for concrete crushing cannot be adjusted in IDEA StatiCa's code setup menu, so the analyses were performed to a utilization ratio of $1/\phi = 1/0.65 = 153.8\%$. Additionally, due to limitations on the input, the concrete blocks in the specimens by DeWolf and Sarisley (1980) were modeled in IDEA StatiCa as 12 in. by 10 in. instead of 12 in. by 10.5 in. as they were physically.

A total of 43 specimens were compared as detailed in Table 6.1. Calculated strength results are listed in Table 6.2. For each calculation method, the strength is listed in units of kips as well as normalized by the experimentally reported strength. The normalized strength results are also presented graphically in Figure 6.11.

The results from the comparison to experimental results show many of the same trends as identified previously. The use of a uniform bearing stress over the whole base plate for base plates that are oversized for bearing is clearly conservative. The average normalized strength for the traditional calculations with a rigid plate assumption is 0.278. The use of a flexible plate assumption within traditional calculations improves the accuracy of the calculation while still being conservative. The average normalized strength

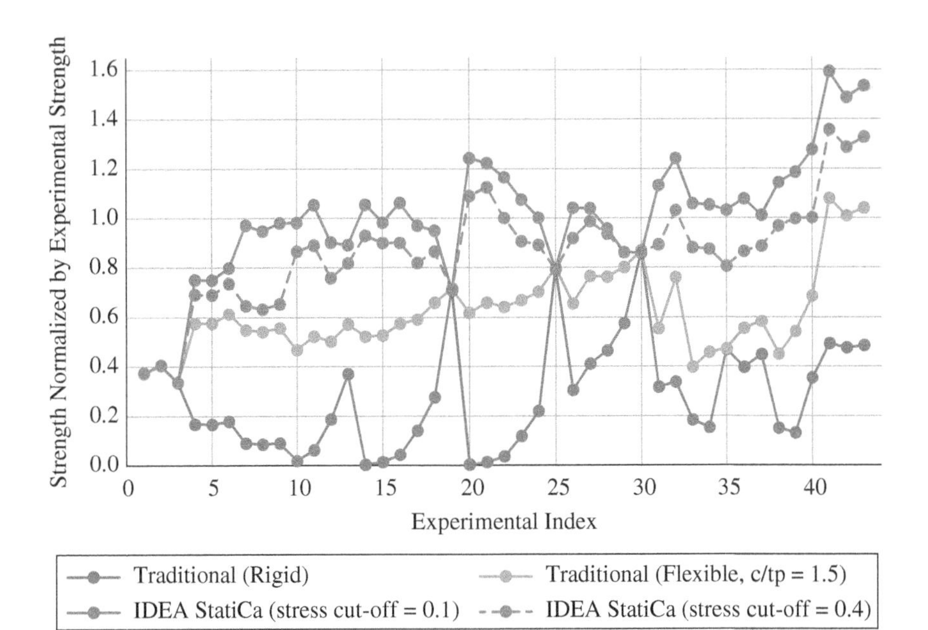

Figure 6.11 Experimental tests on concentric axial compressive loaded base plate connections: normalized strength results.

Table 6.1 Experimental tests on concentric axial compressive loaded base plate connections: specimen details.

Specimen	Column	Plate	t_p (in.)	$F_{y,plate}$ (ksi)	Concrete block	f'_c (ksi)	P_{exp} (kips)
Shelson (1957)							
4	1 in. square	1.41 in. square	0.25	40	8 in. square	6.53	59
5	1 in. square	1.41 in. square	0.25	40	8 in. square	6.17	51.4
6	1 in. square	1.41 in. square	0.25	40	8 in. square	6.24	62.7
10	1 in. square	2 in. square	0.25	40	8 in. square	6.54	59.2
11	1 in. square	2 in. square	0.25	40	8 in. square	6.58	59.6
12	1 in. square	2 in. square	0.25	40	8 in. square	6.54	55.6
19	1 in. square	3 in. square	0.25	40	8 in. square	6.48	61.6
20	1 in. square	3 in. square	0.25	40	8 in. square	6.67	64.2
21	1 in. square	3 in. square	0.25	40	8 in. square	6.55	61.4
Hawkins (1968)							
A1	2.13 in. square	4 in. square	0.060	49	6 in. square	2.75	40
A2	2.13 in. square	4 in. square	0.120	40	6 in. square	2.75	41.6
A3	2.13 in. square	4 in. square	0.250	38	6 in. square	2.75	57.9
A4	2.13 in. square	4 in. square	0.350	41	6 in. square	2.75	62.1
B1	2.25 in. square	4 in. square	0.030	50	6 in. square	4.49	60.1
B2	2.25 in. square	4 in. square	0.060	49	6 in. square	4.49	64.2
B3	2.25 in. square	4 in. square	0.120	40	6 in. square	4.49	68.1
B4	2.25 in. square	4 in. square	0.250	38	6 in. square	4.49	87.3
B5	2.25 in. square	4 in. square	0.350	41	6 in. square	4.49	94.7
B6	2.25 in. square	4 in. square	1.000	40	6 in. square	4.49	127.8
C1	2.25 in. square	4 in. square	0.030	50	6 in. square	6.02	68
C2	2.25 in. square	4 in. square	0.060	49	6 in. square	6.02	68.8
C3	2.25 in. square	4 in. square	0.120	40	6 in. square	6.02	81.5
C4	2.25 in. square	4 in. square	0.250	38	6 in. square	6.02	103.3
C5	2.25 in. square	4 in. square	0.350	41	6 in. square	6.02	119
C6	2.25 in. square	4 in. square	1.000	40	6 in. square	6.02	153.5
D1	2 in. square	3.9 in. square	0.350	46	6 in. square	4.19	77.7
D2	2.28 in. square	3.9 in. square	0.350	46	6 in. square	4.19	79.4
D3	2 in. square	3.9 in. square	0.487	40	6 in. square	4.19	86.1
D4	2.28 in. square	3.9 in. square	0.487	40	6 in. square	4.19	95.7
D5	2.28 in. square	3.9 in. square	0.618	38	6 in. square	4.19	96
E1	2.28 in. square	5.9 in. square	0.498	75	10 in. square	4.19	155.6

Table 6.1 (Continued)

Specimen	Column	Plate	t_p (in.)	$F_{y,plate}$ (ksi)	Concrete block	f'_c (ksi)	P_{exp} (kips)
E2	2.28 in. square	5.9 in. square	0.727	38	10 in. square	4.19	158
F1	2.28 in. square	6 in. square	0.250	90	8 in. square	3.02	79
F2	2.28 in. square	5.9 in. square	0.350	40	8 in. square	3.02	84.4
F3	2.28 in. square	5.9 in. square	0.498	75	8 in. square	3.02	105.2
F4	2.28 in. square	5.9 in. square	0.618	42	8 in. square	3.02	107.4
F5	2.28 in. square	5.9 in. square	0.727	38	8 in. square	3.02	119.1
G1	2.28 in. square	6 in. square	0.250	90	8 in. square	4.19	97
G2	2.28 in. square	5.9 in. square	0.350	40	8 in. square	4.19	99
G3	2.28 in. square	5.9 in. square	0.618	42	8 in. square	4.19	120.4
DeWolf and Sarisley (1980)							
1	4 in. by 4 in. tube	9 in. by 7 in.	0.884	38.8	12 in. by 10.5 in.	3.36	155
5	4 in. by 4 in. tube	9 in. by 7 in.	0.884	36.2	12 in. by 10.5 in.	3.04	150
9	4 in. by 4 in. tube	9 in. by 7 in.	0.884	35.4	12 in. by 10.5 in.	3.01	144

for the traditional calculations with a flexible plate assumption ($c/t_p = 1.5$) is 0.621 with only the three specimens tested by DeWolf and Sarisley (1980) exhibiting a normalized strength greater than 1.0.

IDEA StatiCa provides greater strengths than either of these methods, however, it is based on realistic assumptions of behavior and bearing strength checks are performed in accordance with the AISC *Specification*. With a stress cut-off ratio of 0.1 (the default value), the average normalized strength is 0.999. While IDEA StatiCa computes strengths greater than the experimental results for several specimens, most notably those tested by DeWolf and Sarisley (1980), the IDEA StatiCa results are computed without resistance factors, which, when applied, supply a margin of safety. The use of a stress cut-off ratio of 0.4 was shown to better match hand calculations (e.g. Figure 6.9) and is more conservative than use of the default stress cut-off ratio. The average normalized strength from IDEA StatiCa using a stress cut-off ratio of 0.4 is 0.865. Engineers who desire additional conservatism may adjust this factor in the code setup menu.

6.3 Shear Load

Base plates subjected to shear load are investigated in this section. The transfer of shear from a base plate to the concrete can occur through several mechanisms, including friction, bearing of the base plate or a shear lug against the concrete, and shear in the anchor rods. This study investigates only the mechanism of shear in the anchor rods.

As noted in AISC Design Guide 1, the design of anchor rods for shear depends on the connection details and corresponding load path. Holes in base plates for anchor rods typically have a larger tolerance than bolt holes to allow for misalignment of the rods during setting. Recommended sizes for anchor rod holes

Table 6.2 Experimental tests on concentric axial compressive loaded base plate connections: results.

Specimen	Traditional (rigid)		Traditional (flexible, $c/t_p = 1.5$)		IDEA StatiCa (stress cut-off = 0.1)		IDEA StatiCa (stress cut-off = 0.4)	
Shelson (1957)								
4	22.1	0.374	22.1	0.374	22.1	0.374	22.1	0.374
5	20.9	0.406	20.9	0.406	20.9	0.406	20.9	0.406
6	21.1	0.336	21.1	0.336	21.1	0.336	21.1	0.336
10	10.0	0.169	34.0	0.575	44.5	0.751	41.0	0.693
11	10.0	0.168	34.3	0.575	44.7	0.751	41.2	0.691
12	10.0	0.180	34.0	0.612	44.5	0.800	41.0	0.738
19	5.6	0.091	33.7	0.548	59.8	0.971	39.9	0.648
20	5.6	0.088	34.7	0.541	60.9	0.948	40.7	0.633
21	5.6	0.092	34.1	0.555	60.2	0.980	40.2	0.654
Hawkins (1968)								
A1	0.8	0.020	18.7	0.468	39.3	0.983	34.7	0.867
A2	2.6	0.063	21.7	0.523	43.9	1.054	37.1	0.891
A3	10.9	0.188	29.1	0.502	52.3	0.903	44.0	0.760
A4	23.0	0.370	35.5	0.571	55.4	0.892	50.8	0.819
B1	0.2	0.004	31.3	0.522	63.4	1.055	55.9	0.930
B2	0.9	0.014	33.8	0.527	63.1	0.983	57.8	0.901
B3	3.0	0.044	39.0	0.573	72.3	1.062	61.4	0.901
B4	12.4	0.142	51.5	0.590	84.8	0.971	71.6	0.820
B5	26.2	0.277	62.3	0.658	89.9	0.950	81.9	0.865
B6	91.6	0.717	91.6	0.717	90.7	0.710	90.7	0.710
C1	0.2	0.003	42.0	0.618	84.6	1.244	74.1	1.090
C2	0.9	0.013	45.3	0.659	84.1	1.223	77.3	1.124
C3	3.0	0.037	52.3	0.642	94.9	1.165	81.5	1.000
C4	12.4	0.120	69.1	0.669	111.0	1.075	93.6	0.906
C5	26.2	0.221	83.6	0.702	119.1	1.001	106.1	0.891
C6	122.8	0.800	122.8	0.800	121.6	0.792	121.6	0.792
D1	23.7	0.306	51.0	0.656	80.9	1.041	71.4	0.918
D2	32.7	0.411	60.8	0.765	82.4	1.038	78.3	0.986
D3	40.0	0.464	65.6	0.762	82.4	0.958	80.7	0.937
D4	55.0	0.575	76.7	0.801	82.4	0.861	82.4	0.861
D5	83.3	0.868	83.3	0.868	82.4	0.859	82.4	0.859
E1	49.4	0.318	86.0	0.553	176.1	1.132	139.0	0.893

Table 6.2 (Continued)

Specimen	Traditional (rigid)		Traditional (flexible, $c/t_p = 1.5$)		IDEA StatiCa (stress cut-off = 0.1)		IDEA StatiCa (stress cut-off = 0.4)	
E2	53.4	0.338	120.1	0.760	196.3	1.242	163.0	1.031
F1	14.6	0.185	31.4	0.398	83.6	1.058	69.6	0.881
F2	13.0	0.154	38.6	0.457	88.9	1.053	73.8	0.875
F3	49.4	0.470	49.6	0.471	108.5	1.031	84.9	0.807
F4	42.6	0.397	59.5	0.554	115.7	1.077	92.9	0.865
F5	53.4	0.448	69.3	0.582	120.4	1.011	105.6	0.887
G1	14.6	0.151	43.6	0.449	110.9	1.144	94.0	0.969
G2	13.0	0.131	53.5	0.541	117.5	1.186	98.8	0.998
G3	42.6	0.354	82.5	0.685	153.8	1.278	120.7	1.002
DeWolf and Sarisley (1980)								
1	76.4	0.493	167.2	1.079	246.7	1.591	210.5	1.358
5	71.3	0.475	151.3	1.008	223.2	1.488	193.1	1.287
9	69.7	0.484	149.8	1.040	221.0	1.535	191.2	1.328
Average		0.278		0.621		0.999		0.865
St. Dev.		0.217		0.167		0.258		0.212

For each method, strength is listed in units of kips and as normalized by the experimentally reported strength.

in base plates are presented in Table 14-2 of the AISC *Manual* (AISC, 2017). To avoid slip and transfer the shear to all the anchor rods equally, a setting plate can be installed below the base plate or plate washers can be installed above the base plate (and below the anchor rod nuts). Once the setting plate or plate washers are welded to the base plate; the shear will be transferred uniformly to each of the anchor rods. However, if plate washers are used, bending of the anchor rod within the base plate should be considered in design.

IDEA StatiCa does not consider bending of the anchor rod within the base plate. A series of analyses were performed to demonstrate the effect of this bending. The analyses were performed with a W12 × 120 (ASTM A992, $F_y = 50$ ksi) column on a square base plate with plan dimensions of 18 in. by 18 in., thickness varying from 0.25 in. to 2.50 in., and steel conforming to ASTM A36 ($F_y = 36$ ksi). Anchor rods were 3/4 in. diameter (ASTM F1554 Gr. 36, $F_y = 36$ ksi) with threads not excluded from the shear plane and had an edge distance of $c_1 = 1.5$ in. The holes for the anchor rods were 1-5/16 in. in diameter in accordance with the recommendations of Table 14-2 of the AISC *Manual* (AISC, 2017). The base plate was assumed to bear on a 2 in. thick grout pad (mortar joint) above the concrete ($f'_c = 4$ ksi). The plan area of the concrete was large, such that edge effects did not need to be considered. The shear was applied with the point of zero moment at the top of the base plate.

Maximum factored shear loads from IDEA StatiCa and traditional calculations are presented in Figure 6.12. The IDEA StatiCa results are nearly constant with a maximum factored shear load of

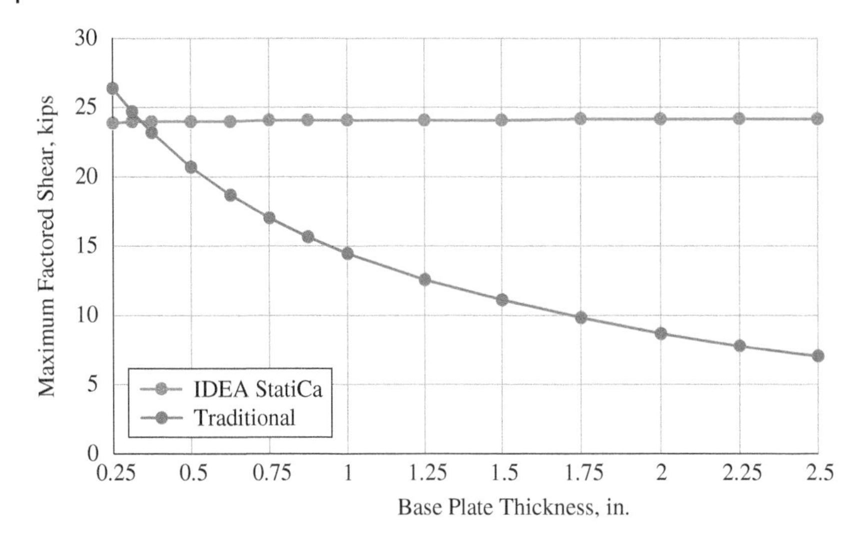

Figure 6.12 Maximum factored shear load vs. plate thickness.

24 kips. This value is the available shear strength of the four anchor rods with a 0.8 reduction factor applied as required by the ACI *Code* (ACI, 2019) for base plates with grout pads. This strength is appropriate when a setting plate is used, or the anchor rod holes do not have a large tolerance. However, if plate washers are used, the strength reduces with increasing base plate thickness. The traditional calculations were performed following the procedure laid out in Example 4.11 of AISC Design Guide 1, including a lever arm for bending of half the distance from the center of the plate washer to the top of the grout. As recommended in AISC Design Guide 1, the 0.8 reduction factor for grouted base plates defined in the ACI *Code* (ACI, 2019) was not applied. For this case, the traditional approach per AISC Design Guide 1 results in lower maximum factored shear than IDEA StatiCa for base plates 3/8 in. and thicker. If using base plates with welded plate washers or other details that enable significant bending of the anchor rods within the base plate, it is recommended that checks be performed outside of IDEA StatiCa.

6.4 Combined Axial Compressive Load and Moment

Base plates subjected to combined axial compressive load and moment are investigated in this section. The limit states evaluated for this loading condition are concrete bearing, flexural yielding of the base plate, tensile yielding of the anchor rod, and member strength.

The analyses were performed with a W12 × 120 (ASTM A992, $F_y = 50$ ksi) column on a square base plate with plan dimensions of 20 in. by 20 in., thickness varying from 0.5 in. to 2.50 in., and steel conforming to ASTM A36 ($F_y = 36$ ksi). Anchor rods were 1 in. diameter (ASTM F1554 Gr. 55, $F_y = 55$ ksi) embedded a sufficient depth in the concrete such that tensile strength of the anchor rod controlled over all concrete tensile failure modes. The anchor rods had an edge distance of $c_1 = 2$ in. The holes for the anchor rods were 1-7/8 in. in diameter in accordance with the recommendations of Table 14-2 of the AISC *Manual* (AISC, 2017). The base plate was assumed to bear on a 2 in. thick grout pad (mortar joint) above the

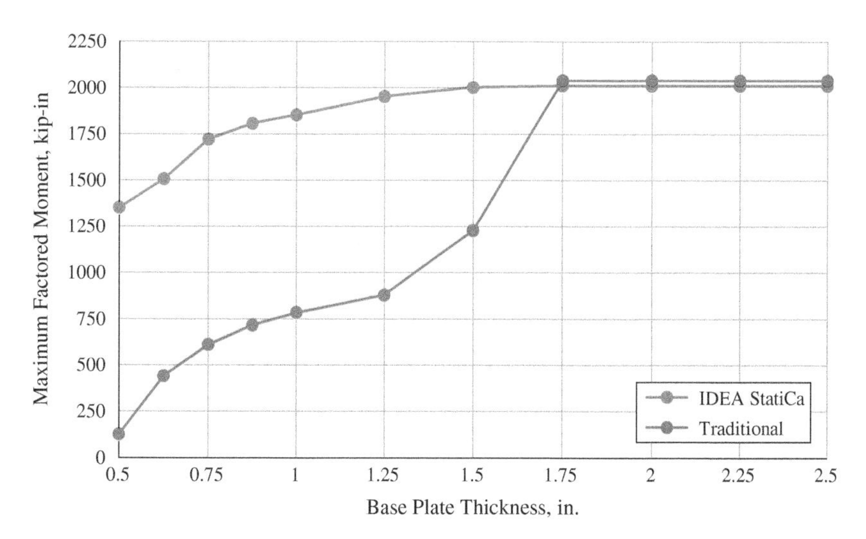

Figure 6.13 Maximum factored moment vs. plate thickness for base plate with 100 kips axial compressive load.

concrete (f'_c= 4 ksi). The plan area of the concrete was large, such that edge effects did not need to be considered and the maximum permitted bearing strength would apply (i.e. $\sqrt{A_2/A_1} \geq 2$).

The applied axial compressive load was held constant at 100 kips and the maximum bending moment that can be applied concurrently was determined. The maximum factored bending moment is presented in Figure 6.13. For IDEA StatiCa, the plastic strain limit on the tension side of the base plate controlled the strength of the connection with a 0.5 in. thick base plate. For the connection with a 0.625 in. thick base plate, an interesting limit state of concrete crushing controlled as the corners of the base plate on the tension side were bent down into the concrete by the anchors, as shown in Figure 6.14. The tensile strength of the anchors was reached at approximately 5% greater applied moment. The tensile strength of the anchors controlled for all other connections (i.e. $t_p \geq 0.75$ in.). With the traditional calculations, flexural yielding of the base plate on the compression side controlled the strength of the connections with plate thickness of 1.5 in. and less and tensile strength of the anchor rod controlled otherwise.

Where base plate bending controlled the traditional calculations, the maximum permitted factored moments were less for the traditional method than for IDEA StatiCa. The reason for this result is similar to that for the base plates subjected to concentric axial load, specifically, that the assumed distribution of bearing stress is conservative and does not account for increased flexibility of the base plate upon yield. Traditional calculation methods have been developed for evaluating flexible base plates subject to axial compression and bending and have been compared to IDEA StatiCa in other studies (https://www.ideastatica.com/support-center/column-base-open-section-column-in-bending-to-strong-axis).

Conversely, where the tensile resistance of the anchor rod controlled the traditional calculations, the maximum permitted factored loads were slightly greater for the traditional method than for IDEA StatiCa. The available tensile strength of the anchor rods is slightly greater for the traditional calculations since it is based on recommendations from AISC Design Guide 1 while IDEA StatiCa is based on provisions from the ACI Code. The two approaches also vary in the assumed bearing stress distribution resulting in

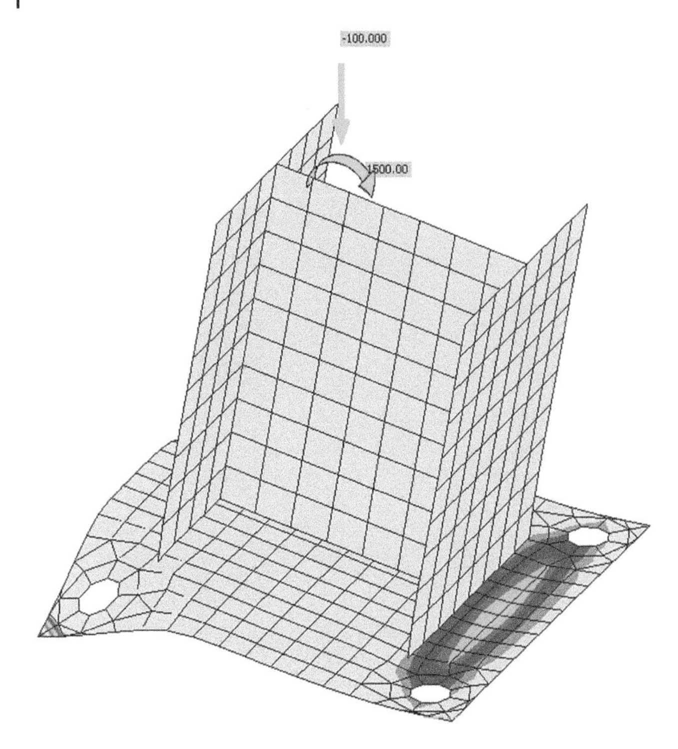

Figure 6.14 Deformed shape (scale factor = 5) and concrete bearing stress for base plate connection with 0.625 in. thick base plate. Note bearing stresses at corners of the tension side of the base plate.

a slightly different lever arm for the force couple formed between the anchor rod and the centroid of the bearing force.

6.5 Summary

This study compared the design of base plate connections by traditional calculation methods used in US practice and IDEA StatiCa as well as experimental results. Key observations from the study include:

- For thick base plates that better conform to the rigid base plate assumption, IDEA StatiCa provides strengths that are comparable to the traditional calculations presented in AISC Design Guide 1.
- For thinner base plates, where flexural yielding of the base plate due to bearing stresses controls, IDEA StatiCa can provide significantly greater strengths than the traditional calculations since the distribution of bearing stresses is calculated explicitly and redistributes upon the initiation of yielding of the base plate.
- IDEA StatiCa correctly calculates the shear strength of anchor rods but neglects the potential reductions in shear strength due to bending of the anchor rod within the base plate that can occur in certain base plate configurations (e.g. base plates with welded plate washers).

References

ACI (2019). *Building Code Requirements for Structural Concrete and Commentary*. American Concrete Institute, Farmington Hills, MI.

AISC (2016). *Specification for Structural Steel Buildings*. American Institute of Steel Construction, Chicago.

AISC (2017). *Steel Construction Manual*, 15th Edition. American Institute of Steel Construction, Chicago.

DeWolf, J.T. and Sarisley, E.F. (1980). Column Base Plates with Axial Loads and Moments. *Journal of the Structural Division*, 106(11), 2167–2184.

Fisher, J. and Kloiber, L. (2006). *Base Plate and Anchor Rode Design*, 2nd Edition. Design Guide 1, American Institute of Steel Construction, Chicago.

Fitz, M., Appl, J., and Geibig, O. (2018). Comprehensive Base Plate and Anchor Design Based on Realistic Behavior – New Design Software Based on Realistic Assumptions. *Stahlbau*, 87(12), 1179–1186. [In German] https://doi.org/10.1002/stab.201800036

Hawkins, N.M. (1968). The Bearing Strength of Concrete Loaded Through Flexible Plates. *Magazine of Concrete Research*, 20(63), 95–102.

Shelson, W. (1957). Bearing Capacity of Concrete. *Journal Proceedings*, 54(11), 405–414.

Steenhuis, M., Wald, F., Sokol, Z., and Stark, J. (2008). Concrete in Compression and Base Plate in Bending. *Heron*, 53(1/2), 51–68.

7

Bracket Plate Connections

7.1 Description

A comparison between results from the component-based finite element method (CBFEM) and traditional calculation methods used in US practice for bracket plate connections are presented in this chapter. Both bolted and welded bracket plates are considered. The focus of this investigation is the strength of the eccentrically loaded bolt and weld groups that connect the bracket plates to the column flanges.

The instantaneous center of rotation method is the primary method described in the AISC *Manual* (AISC, 2017a) for computing the strength of eccentrically loaded bolt and weld groups. Details of the method differ between bolt and weld groups; however, the general approach is the same. The force in each individual bolt or segment of weld is assumed to act perpendicular to a line that passes through the individual component and the common center of rotation. The magnitude of the force in each component is based on equations representing the load deformation relationship. For welds, the load deformation relationship considers the direction of force with respect to the longitudinal axis of the weld. The center of rotation is typically found by using an iterative process and is known to be valid when static equilibrium is achieved (i.e. the sum of the forces and moments equals zero). In practice, calculations using the instantaneous center of rotation method are completed using tabulated solutions for common bolt and weld groups provided in Parts 7 and 8 of the AISC *Manual*.

7.2 Bolted Bracket Plate Connections

A schematic of the bolted bracket plate connection investigated is presented in Figure 7.1 and Figure 7.2. The parameters change depending on the limit state being investigated. However, the typical connection has the following characteristics unless noted otherwise: bracket plate thickness of 5/8 in., ASTM A572 Grade 50 conforming steel for the plates (F_y = 50 ksi and F_u = 65 ksi), horizontal and vertical edge distances of $l_{eh} = l_{ev}$ = 2.25 in., gage of g = 5.5 in., and 6 bolts in each vertical row with a spacing of s = 3 in. The bolts are 7/8 in. diameter A325 with threads not excluded from the shear plane and in standard holes. The column is a W12×106 conforming to ASTM A992 steel (F_y = 50 ksi and F_u = 65 ksi). The properties of the bolt group match those of Example II.A-24 of the AISC Design Examples (AISC, 2019).

Steel Connection Design by Inelastic Analysis: Verification Examples per AISC Specification, First Edition.
Mark D. Denavit, Ali Nassiri, Mustafa Mahamid, Martin Vild, Halil Sezen, and František Wald.

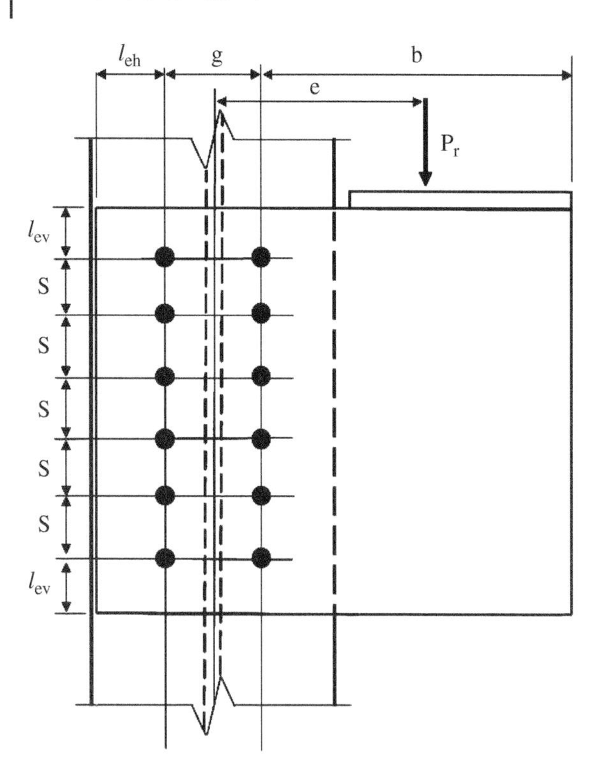

Figure 7.1 Schematic of bolted bracket plate connection.

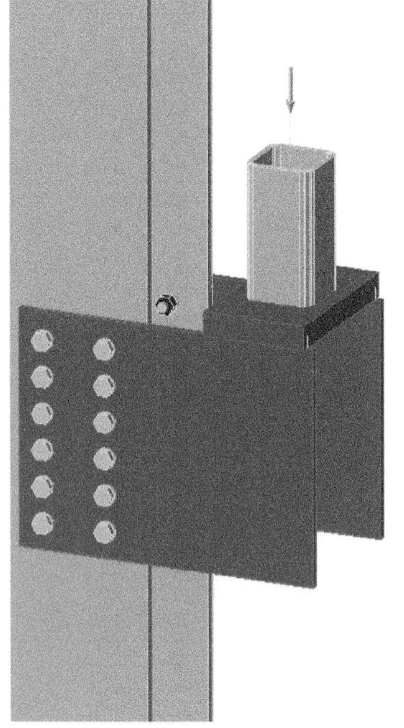

Figure 7.2 IDEA StatiCa model of bolted bracket plate connection.

The traditional calculations are performed following the provisions for load and resistance factor design (LRFD) in the AISC *Specification* (AISC, 2016). The limit states evaluated are shear rupture of the bolt, bearing, tearout, and slip.

7.3 Bolt Shear Rupture

The first investigation explores how the bolt utilization percentage varies with applied load. For one value of eccentricity, $e = 16$ in., the applied load was varied from 0 to 200 kips and the bolt utilization percentage as reported by IDEA StatiCa was recorded. The results are presented in Figure 7.3. The relationship

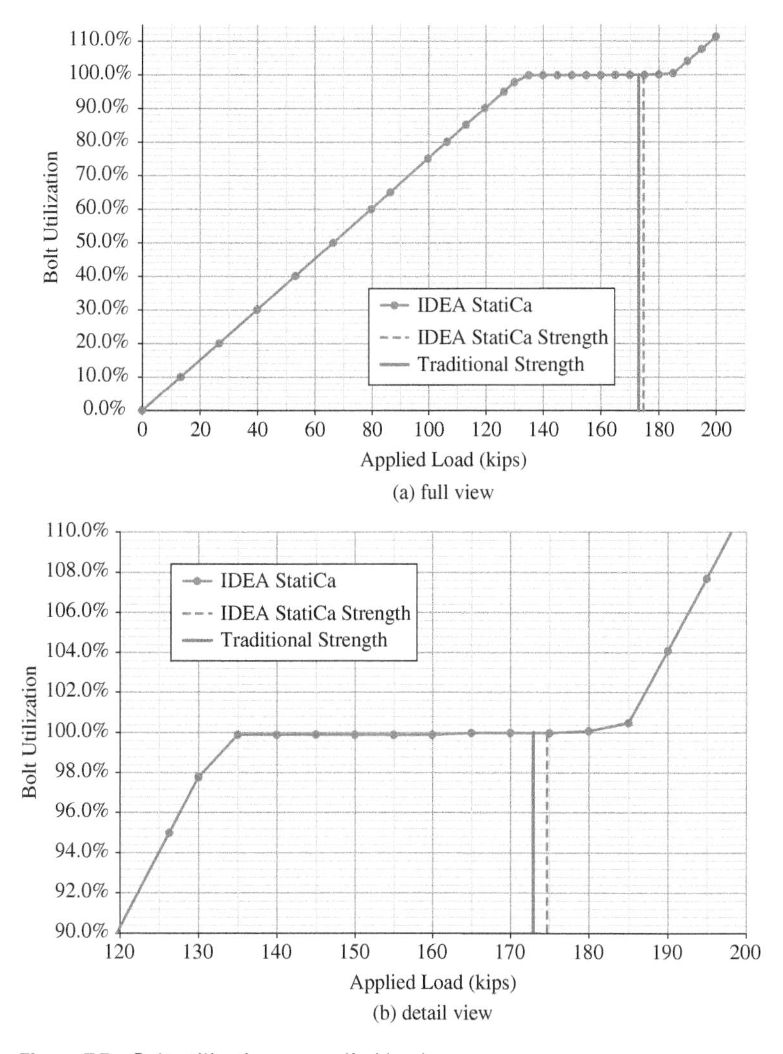

Figure 7.3 Bolt utilization vs. applied load.

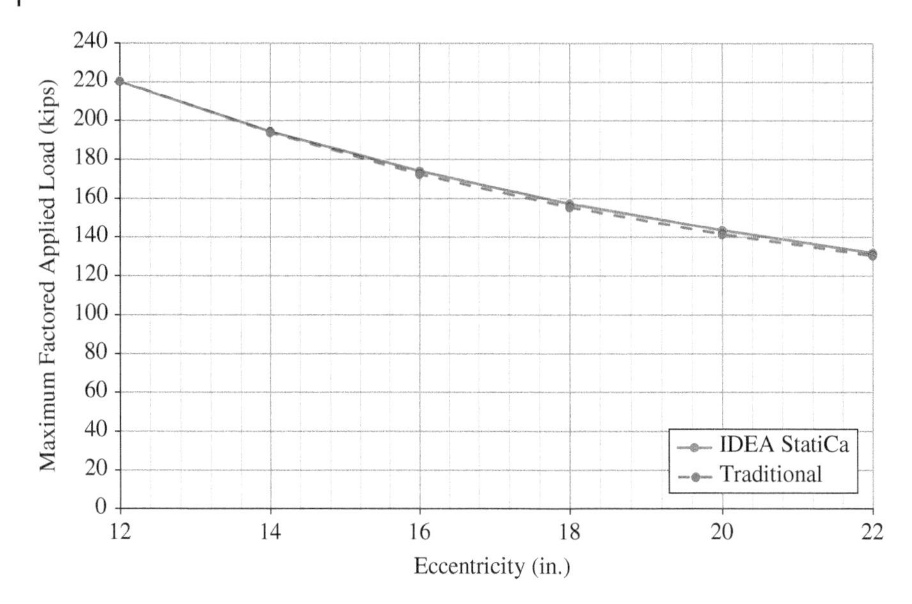

Figure 7.4 Maximum factored applied load vs. eccentricity.

between applied load and bolt utilization percentage is essentially linear until an applied load of roughly 135 kips, at which point the bolt utilization percentage plateaus at near 100% until an applied load of roughly 185 kips, at which point the bolt utilization percentage again increases linearly. The load at which failure of the bolts indicated by IDEA StatiCa (i.e. with a red "x" in the results display) occurs later in the plateau, at an applied load of 174.7 kips. The strength of this connection per the traditional calculations is 172.6 kips.

These strength results for the same connection and a range of values of eccentricity are presented in Figure 7.4. As expected, the maximum permitted applied load decreases with eccentricity. The results from IDEA StatiCa are in close agreement with the traditional calculations.

7.4 Additional Bolt Groups

Additional bolt groups are investigated in this section. The connections investigated are like those investigated in the previous section but the first has a larger gage ($g = 8$ in.) and the second has only two bolts in each vertical row ($g = 5.5$ in., $s = 6$ in.). A larger column size (W14×132) was used with the connection with the larger gage to ensure minimum edge distance requirements were satisfied. The results for the larger gage are presented in Figure 7.5 and the results for the connection with two bolts in each vertical row are presented in Figure 7.6. As before, the IDEA StatiCa results are in close agreement with the traditional calculations.

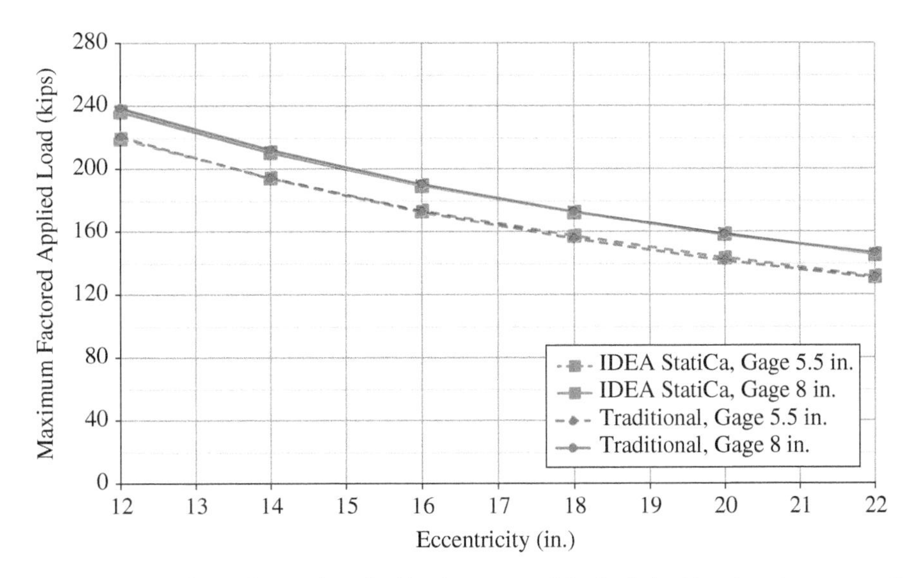

Figure 7.5 Maximum factored applied load vs. eccentricity for bolted bracket plate connections with two different values of bolt gage.

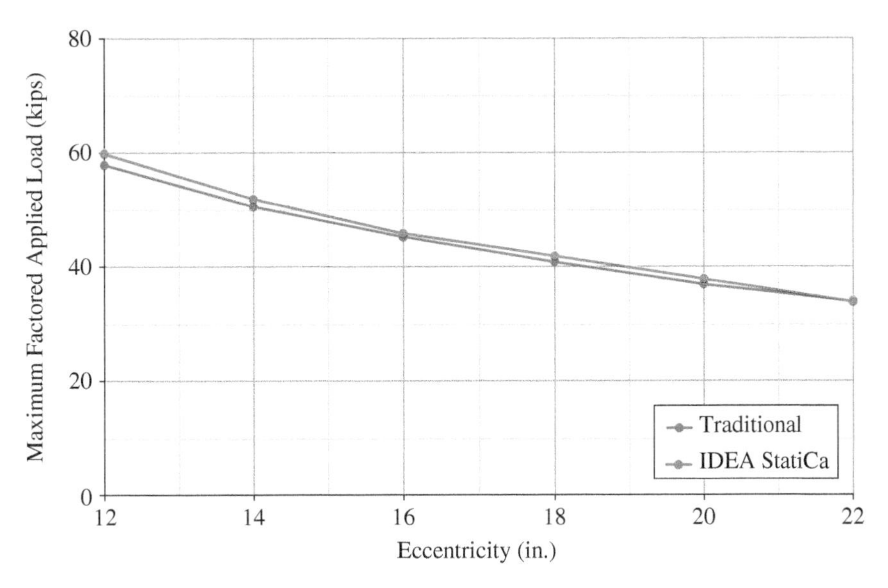

Figure 7.6 Maximum factored applied load vs. eccentricity for bolted bracket plate connection with two bolts in each vertical row.

7.5 Tearout

A disadvantage of the instantaneous center of rotation method is that the tabulated solutions assume that the bolts all have the same strength. The bolts in an eccentrically loaded bolt group may not all have the same strength if the edge distances are small and tearout controls over bearing or bolt shear rupture. This is additionally challenging for the traditional calculations since, when using the tabulated solutions, the direction of force for each bolt is not known and thus the clear distance, a key factor in tearout strength, cannot be accurately determined. When evaluating eccentrically loaded bolt groups with small edge distances, engineers often employ the "poison bolt method" whereby the strength of all the bolts is set equal to the lowest possible strength (i.e. that computed from the lowest possible clear distance). In IDEA StatiCa, tearout strength is computed individually for each bolt based on the computed direction of force.

A comparison between IDEA StatiCa results and results from traditional calculations using the poison bolt method are shown in Figure 7.7. The connection for this comparison is like that described in Section 7.2 but with a bracket plate thickness of 3/8 in. and varying horizontal edge distance, l_{eh}. The edge distance varies between 1.125 in., the minimum edge distance per Table J3.4 of the AISC *Specification* (AISC, 2016), and 2.25 in., a value at which bolt shear rupture will control over tearout. The results show close agreement, indicating that IDEA StatiCa is appropriately considering the effects of tearout in eccentrically loaded bolt groups.

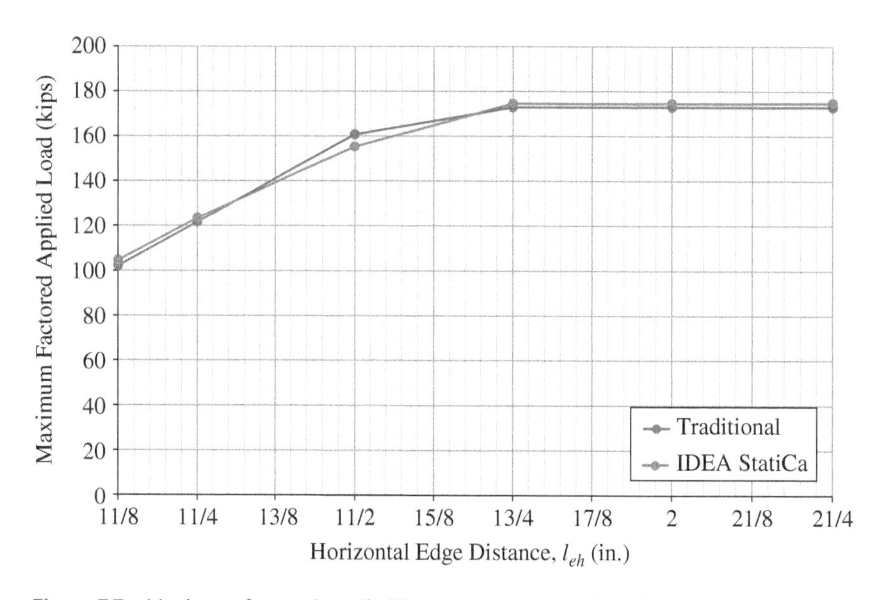

Figure 7.7 Maximum factored applied load vs. horizontal edge distance.

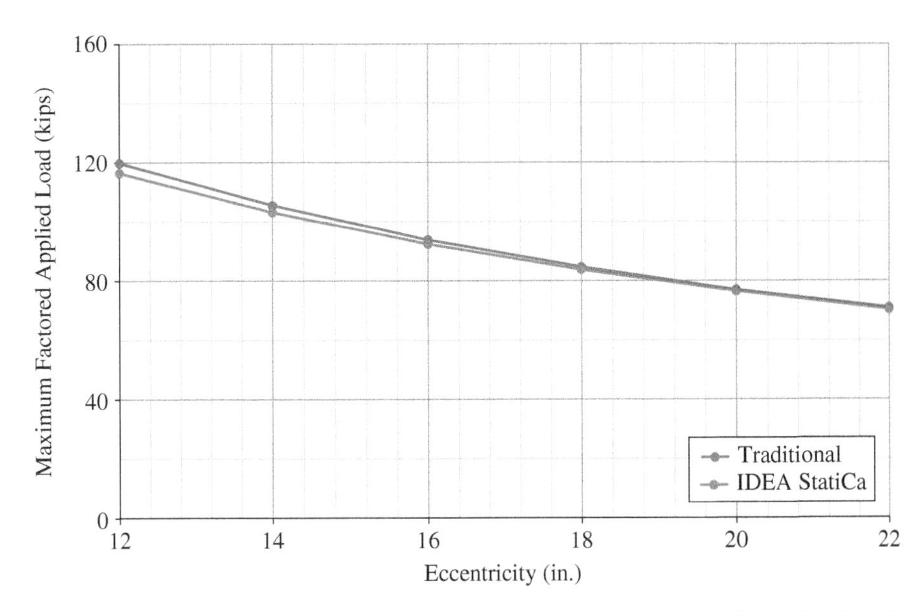

Figure 7.8 Maximum factored applied load vs. eccentricity for slip-critical bolted bracket plate connection.

7.6 Slip Critical

The instantaneous center of rotation method is also applicable to slip-critical connections even though the mechanics of force transfer are different than those assumed in the method. The results of a comparison using the same connection parameters as for the connection explored in Section 7.3 but for a slip-critical connection are presented in Figure 7.8. The average difference between the IDEA StatiCa results and traditional US methods is about 1.5%.

7.7 Welded Bracket Plate Connections

A schematic of the welded bracket plate connection investigated is presented in Figure 7.9 and an image of the IDEA StatiCa model is presented in Figure 7.10. The parameters of the connections investigated are as follows: plate thickness of 9/16 in., ASTM A572 conforming steel for the plates ($F_y = 50$ ksi and $F_u = 65$ ksi), 3/8 in. fillet welds with E70XX weld metal, weld length, $l = 10$ in., and an aspect ratio of either $k = 0.5$ or $k = 0.3$. The column is a W8×40 conforming to ASTM A992 steel ($F_y = 50$ ksi and $F_u = 65$ ksi). The properties of the weld group match those of Example II.A-26 of the AISC *Design Examples* (AISC, 2019). The traditional calculations are performed following the provisions for load and resistance factor design (LRFD) in the AISC *Specification* (AISC, 2016). Only the limit state of weld rupture is evaluated.

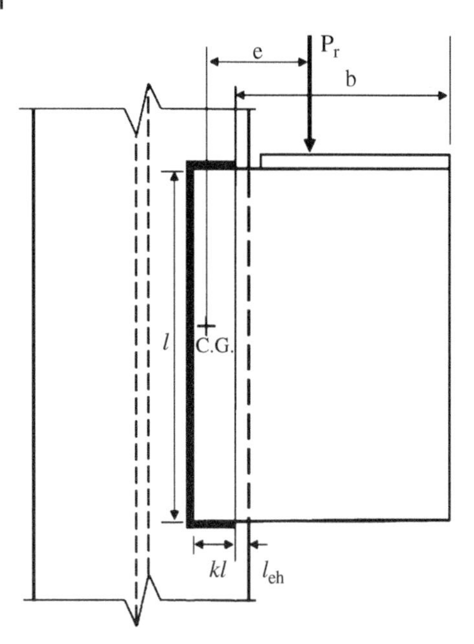

Figure 7.9 Schematic of welded bracket plate connection.

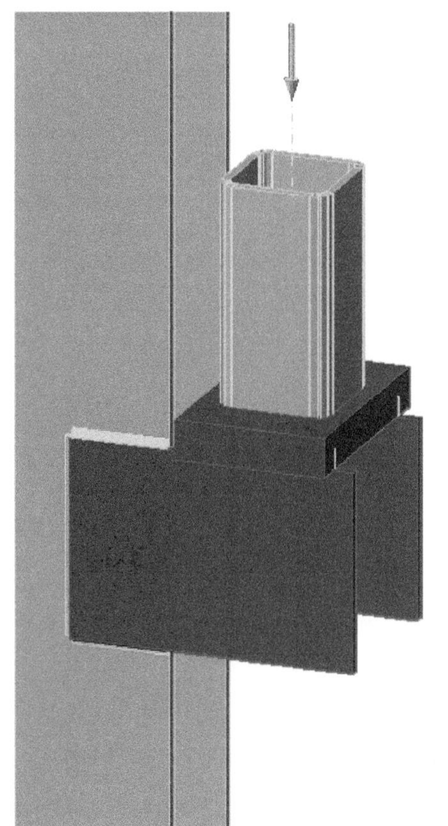

Figure 7.10 IDEA StatiCa model of welded bracket plate connection.

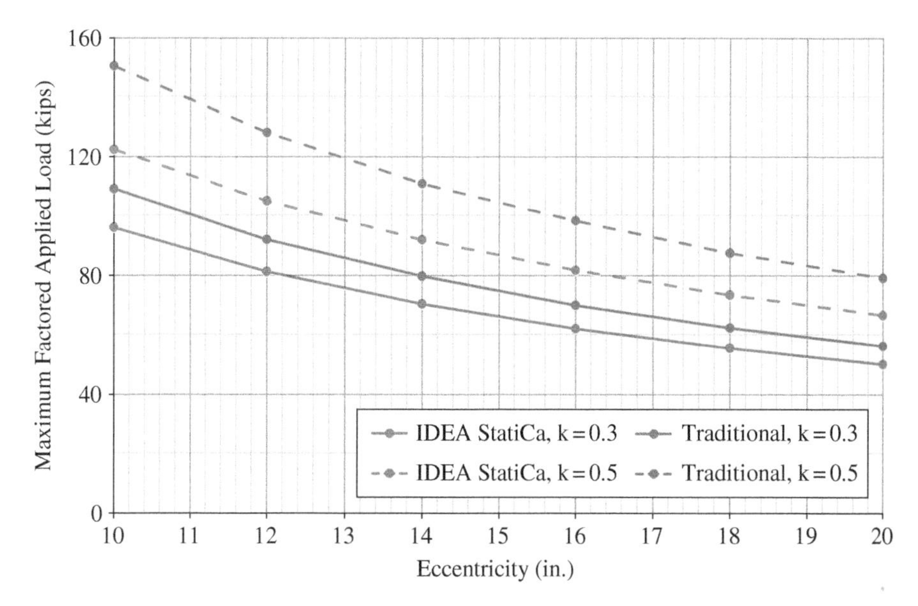

Figure 7.11 Maximum factored applied load vs. eccentricity for welded bracket plate connections.

The strength of the connections per IDEA StatiCa and the traditional calculations for a range of eccentricities are presented in Figure 7.11. As expected, and like the bolted connections, the maximum permitted applied load decreases with eccentricity. The results show a relatively uniform level of conservatism for IDEA StatiCa as compared to traditional US practice. The case with $k = 0.5$ exhibits an average difference of approximately 17%, whereas the case with $k = 0.3$ exhibits an average difference of approximately 12%.

7.8 Summary

This study compared the design of bracket plate connections by traditional calculation methods used in US practice and IDEA StatiCa. Key observations from the study include:

- The available strength of bolted bracket connections per IDEA StatiCa agrees very well with traditional calculations per the instantaneous center of rotation method.
- Eccentrically loaded bolt groups may exhibit a plateau during which IDEA StatiCa shows a bolt utilization of near 100% for a range of applied loads. The applied load at which IDEA StatiCa indicates failure was taken as the limit in this study and compares well to the traditional calculations.
- IDEA StatiCa detects the clear distance for each bolt individually for consideration of tearout, resulting in appropriate reductions in strength when edge distances are small.

- The available strength of welded bracket connections per IDEA StatiCa was found to be conservative in comparison to the traditional calculations using the instantaneous center of rotation method for the cases examined.

References

AISC (2016). *Specification for Structural Steel Buildings*. American Institute of Steel Construction, Chicago.

AISC (2017). *Steel Construction Manual*, 15th Edition. American Institute of Steel Construction, Chicago.

AISC (2019). *Companion to the AISC Steel Construction Manual*, Vol. 1: *Design Examples* Version 15.1. American Institute of Steel Construction, Chicago.

8

Single Plate Shear Connections

8.1 Description

A comparison between results from the component-based finite element method (CBFEM) and traditional calculation methods used in US practice for single plate shear connections is presented in this chapter. A schematic of the connection investigated is presented in Figure 8.1.

The traditional calculation methods used in this work are based upon the recommendations presented in Part 10 of the AISC *Manual* (AISC, 2017). Two approaches for the design of single plate shear connections are presented in Part 10 of the AISC *Manual*. The first, for "conventional" configurations, offers some simplifications if certain dimensional limitations are met. The second, for "extended" configurations, is more broadly applicable but without the simplifications permitted for design of conventional configurations. Specifically, conventional configurations must have a single vertical row of between 2 and 12 bolts, the distance between the bolt line and the weld line, a, must be equal to or less than 3.5 in., bolts must be in standard holes or short-slotted holes transverse to the member reaction, the vertical edge distance, l_{ev}, must satisfy the minimum edge distance requirements of Table J3.4 of the AISC *Specification* (AISC, 2016), the horizontal edge distance, l_{eh}, must be greater than or equal to $2d$, where d is the bolt diameter, and either the thickness of the plate, t_p, or the thickness of the beam web, t_w, must satisfy maximum thickness requirements.

The primary simplification to design for connections that meet these requirements is that the bolt group strength may be evaluated as follows: bolt shear strength checked using the eccentricity listed in Table 10-9 of the AISC *Manual* (AISC, 2017) and bearing and tearout checked, assuming the reaction is applied concentrically. This simplification avoids the need to consider tearout in an eccentrically loaded bolt group. For extended configuration calculations, where tearout is considered when determining the strength of the eccentrically loaded bolt group, two different methods are employed. The first method is a commonly used conservative approximation known as the "poison bolt" method. In this method the strength of the eccentrically loaded bolt group is obtained by identifying the smallest possible strength for any of the bolts for any direction of force, then utilizing that value of strength in conjunction with a value of C from the tables in Part 7 of the AISC *Manual*. The values of C listed in the tables are computed from the instantaneous center of rotation (IC) method. The second method is to use the modified instantaneous center of rotation method developed by Denavit et al. (2021), in which tearout is considered explicitly within the iterative procedure for determining the strength of the bolt group.

Steel Connection Design by Inelastic Analysis: Verification Examples per AISC Specification, First Edition.
Mark D. Denavit, Ali Nassiri, Mustafa Mahamid, Martin Vild, Halil Sezen, and František Wald.

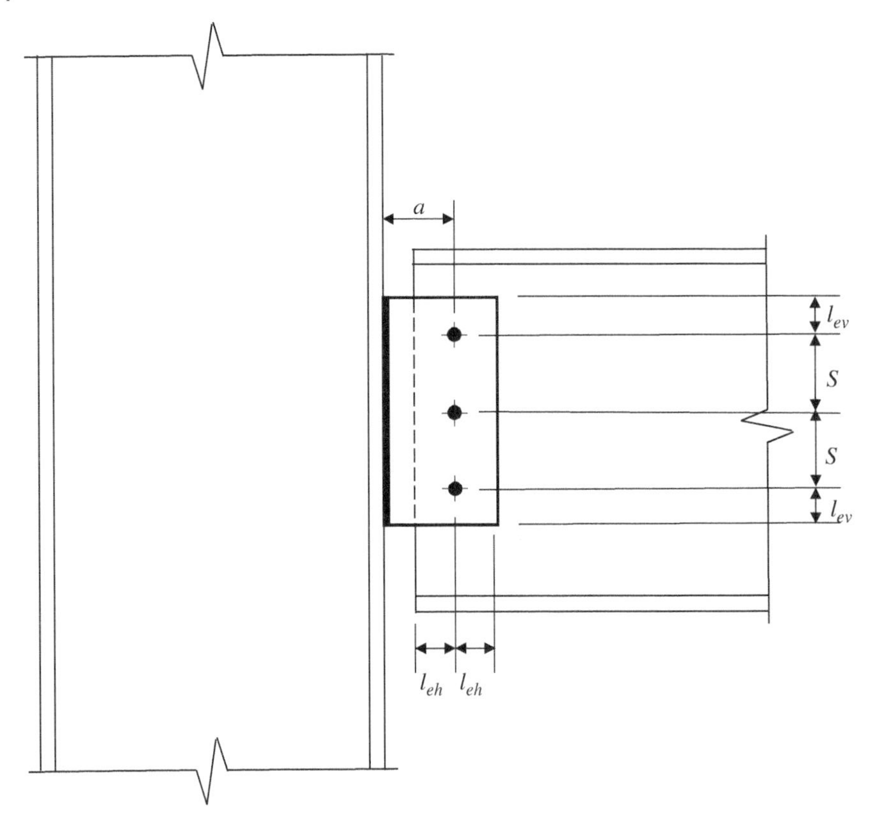

Figure 8.1 Schematic of single plate shear connection.

Beyond bolt group strength, shear yielding of the plate, shear rupture of the plate, block shear rupture of the plate, and weld shear are also checked for conventional configurations. Additional checks for extended configurations include those for flexural rupture, plate interaction strength, and plate buckling.

All traditional calculations were performed in accordance with the provisions for load and resistance factor design (LRFD) in the AISC *Specification* (AISC, 2016).

The CBFEM results were obtained from IDEA StatiCa Version 21.0. An example model is shown in Figure 8.2. The maximum permitted loads were determined iteratively by adjusting the applied load input to a value that the program deems safe but if increased by a small amount (e.g. 0.1 kip) the program would deem unsafe. In all models, the supported beam was assigned a "N-Vz-My" model type to ensure in-plane behavior. Unless noted otherwise, forces were defined such that the point of zero moment was located at the weld line, matching the assumption of the design methods presented in Part 10 of the AISC *Manual* (AISC, 2017).

8.2 Bolt Group Strength

First, connections where the strength of the bolt group controls the strength of the connection are investigated. For these comparisons the column is a W14x90, and the supported beam, which frames into the

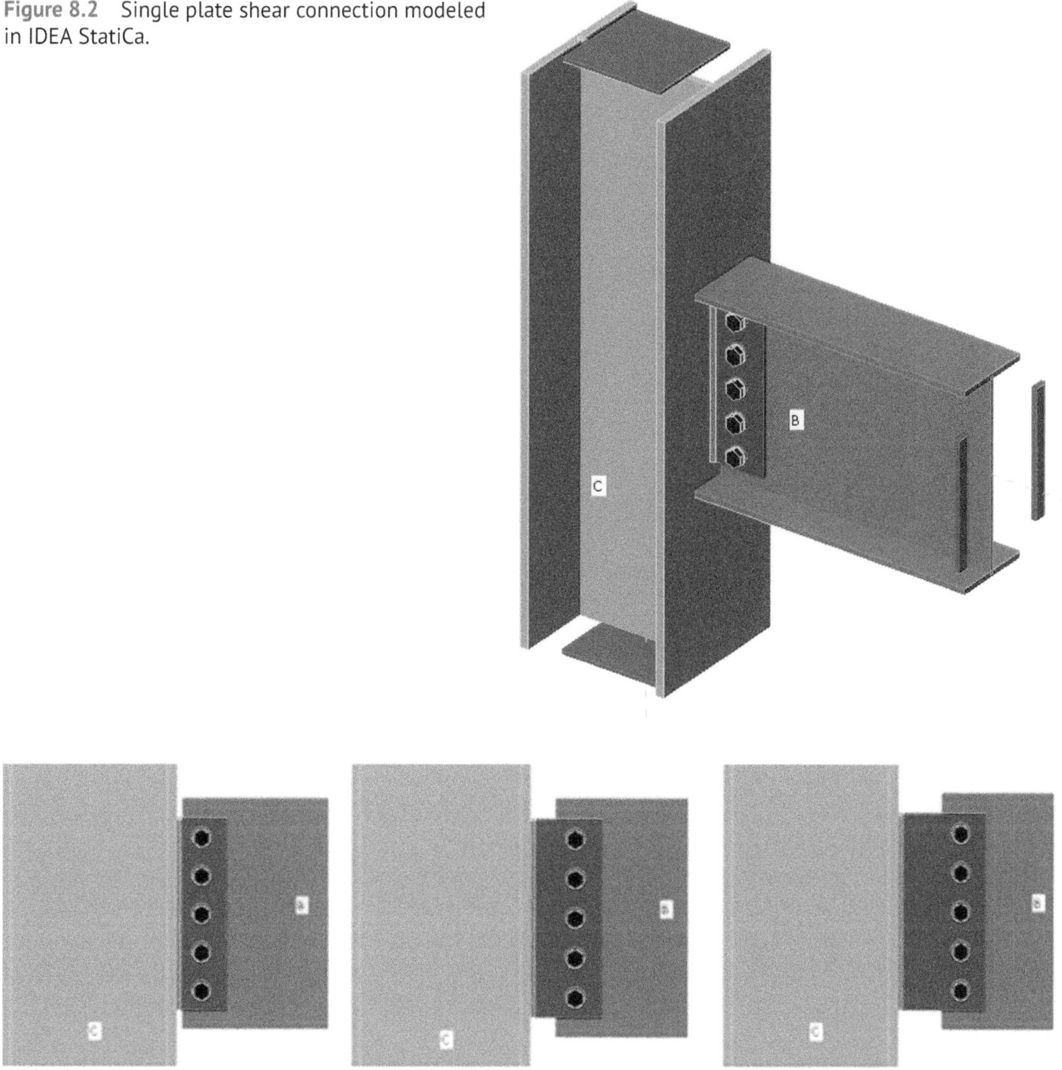

Figure 8.2 Single plate shear connection modeled in IDEA StatiCa.

Figure 8.3 Variation of a in the IDEA StatiCa model.

flange of the column, is a W18x50. Both conform to ASTM A992 (F_y = 50 ksi, F_u = 65 ksi). The plate is 15 in. tall (s = 3 in., l_{ev} = 1.5 in.), ½ in. thick, and conforms to ASTM A36 (F_y = 36 ksi, F_u = 58 ksi). Each vertical row of bolts has (5) ¾in. diameter A325 bolts with threads not excluded from the shear plane and horizontal edge distance, l_{eh} = 2.0 in. The weld was a 5/16 in. fillet weld on both sides in accordance with the $(^5/_8)t_p$ rule noted in Part 10 of the AISC *Manual* (AISC, 2017). The distance from the weld line to the bolt line, a, was varied from 2 in. to 5 in. (Figure 8.3). Note that this connection satisfies the requirements for the conventional configuration when $a \leq 3.5$ in.

Variation of the shear capacity of the connections with the distance a is presented in Figure 8.4. Bolt shear rupture was the controlling limit state for all values of a and all methods of calculation. The IDEA

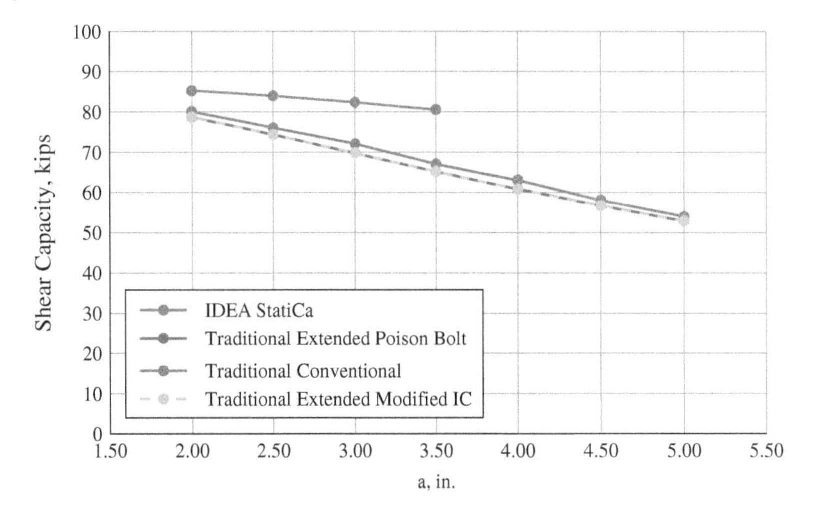

Figure 8.4 Shear capacity of single plate shear connection with respect to *a*.

StatiCa results match well with the traditional calculations for the extended configuration. Where applicable, the traditional calculations for the conventional configuration give somewhat greater shear capacity. The reason for this is that a reduced eccentricity of $a/2$ is permitted to be assumed for conventional configurations per Table 10-9 of the AISC *Manual* (AISC, 2017). The eccentricity of the bolt group is taken as a for the extended configuration calculations. The eccentricity of the bolt group is also equal to a for IDEA StatiCa because the point of zero moment was defined to be at the weld line. The poison bolt method and the modified IC method provide the same results indicating that tearout did not control for any bolt (i.e. the plate and beam web were sufficiently thick and the bolt spacing and edge distances were sufficiently large).

Variation of the shear capacity with the distance a is presented in Figure 8.5 for connections with the same properties as previously described but with two vertical rows of bolts (Figure 8.6) and l_{eh} = 1.5 in. The horizontal spacing between vertical rows of bolts was 3 in. These connections are extended configuration, regardless of the value of a, given that they have more than one vertical row of bolts. Again, bolt shear rupture was the controlling limit state for all values of a and all methods and the IDEA StatiCa results match well with the traditional calculations.

8.3 Plate Thickness

Varying the plate thickness allows for a wider range of limit states to control, including bearing and tearout at the bolt holes and shear yielding and rupture of the plate. For these comparisons the column is a W14x90 and the supported beam, which frames into the flange of the column, is a W18x130. Both conform to ASTM A992 (F_y = 50 ksi, F_u = 65 ksi). The plate is 14 in. tall (s = 3 in., l_{ev} = 1 in.) and conforms to ASTM A572 Gr. 50 (F_y = 50 ksi, F_u = 65 ksi). The thickness of the plate varies from 3/16 in. to 3/4 in. in these analyses. There is one vertical row of (5)3/4 in. diameter A490 bolts with threads not excluded from the shear plane and horizontal edge distance, l_{eh} = 1.5 in. Fillet welds were provided on both sides

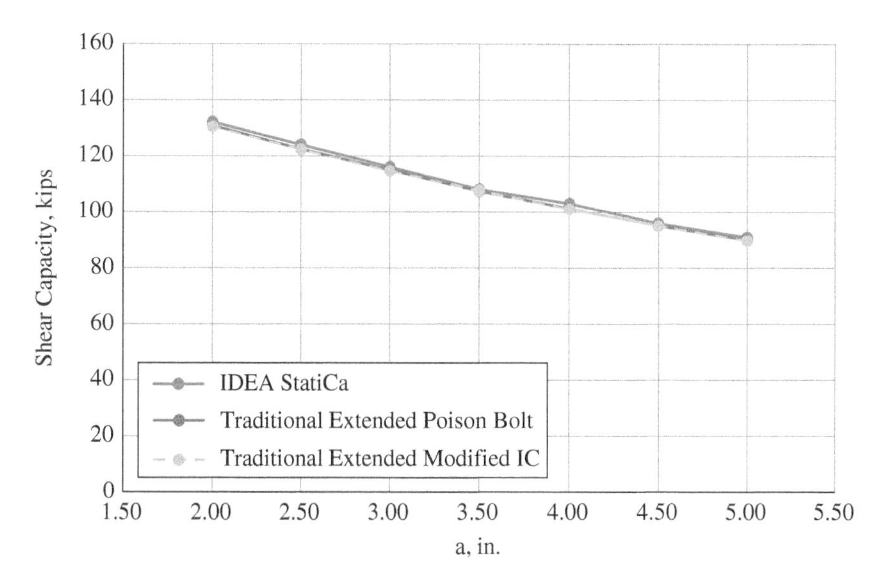

Figure 8.5 Shear capacity of extended configuration with two rows of bolts with respect to *a*.

Figure 8.6 Extended configuration with two rows of bolts modeled in IDEA StatiCa.

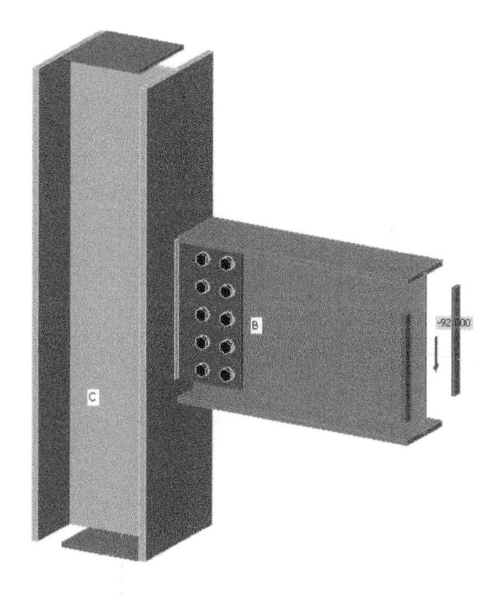

of the plate with size varying with plate thickness in accordance with the $(^5/_8)t_p$ rule noted in Part 10 of the AISC *Manual* (AISC, 2017). The distance from the weld line to the bolt line, a, was 3.0 in. These connections satisfy the requirements for the conventional configuration for plate thicknesses less than or equal to 7/16 in.

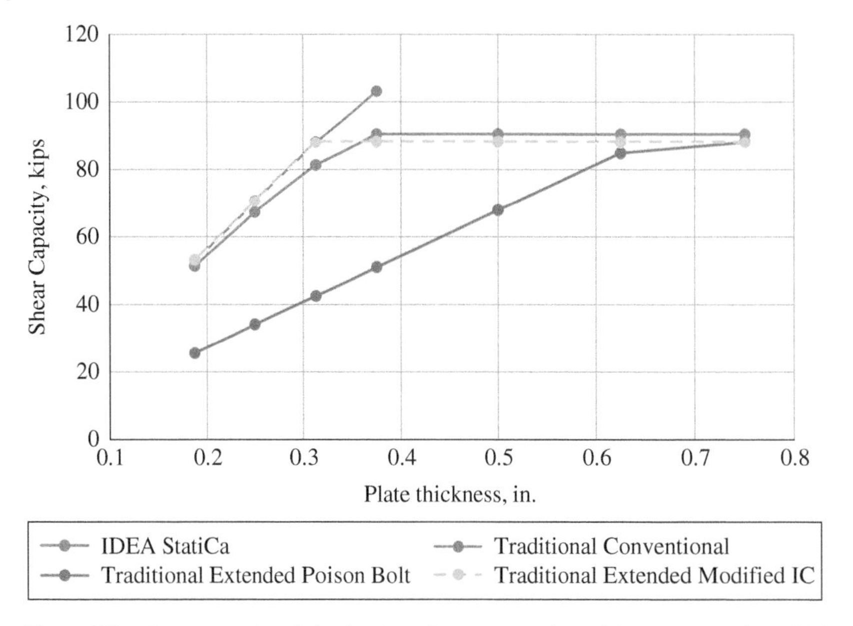

Figure 8.7 Shear capacity of single plate shear connection with respect to plate thickness.

Variation of the shear capacity of the connections with plate thickness is presented in Figure 8.7 with the controlling limit states presented in Table 8.1. The most notable result is that the traditional calculations for the extended configuration using the poison bolt method show far lower strengths than the other methods. The poison bolt method, in which the lowest possible strength for any bolt is taken as the strength of every bolt, can be highly conservative. However, it is used in practice for the evaluation of eccentrically loaded bolt groups where tearout may control. For this connection, the strength of all the bolts is based on the tearout strength of the bottom bolt using an edge distance of $l_{ev} = 1$ in. resulting in a clear distance $l_c = 0.594$ in. In IDEA StatiCa and the modified IC method, the strength of each individual bolt is based on the clear distance in the direction of force for that individual bolt. For example, at the

Table 8.1 Controlling limit state for results presented in Figure 8.7

Plate thickness (in.)	IDEA StatiCa	Traditional conventional	Traditional extended (poison bolt)	Traditional extended (modified IC)
3/16	Plate strain	Plate shear rupture	Bolt group	Plate shear rupture
1/4	Plate strain	Plate shear rupture	Bolt group	Plate shear rupture
5/16	Plate strain	Plate shear rupture	Bolt group	Bolt group
3/8	Bolt shear rupture	Bolt shear rupture	Bolt group	Bolt group
1/2	Bolt shear rupture	n/a	Bolt group	Bolt group
5/8	Bolt shear rupture	n/a	Bolt group	Bolt group
3/4	Bolt shear rupture	n/a	Bolt group	Bolt group

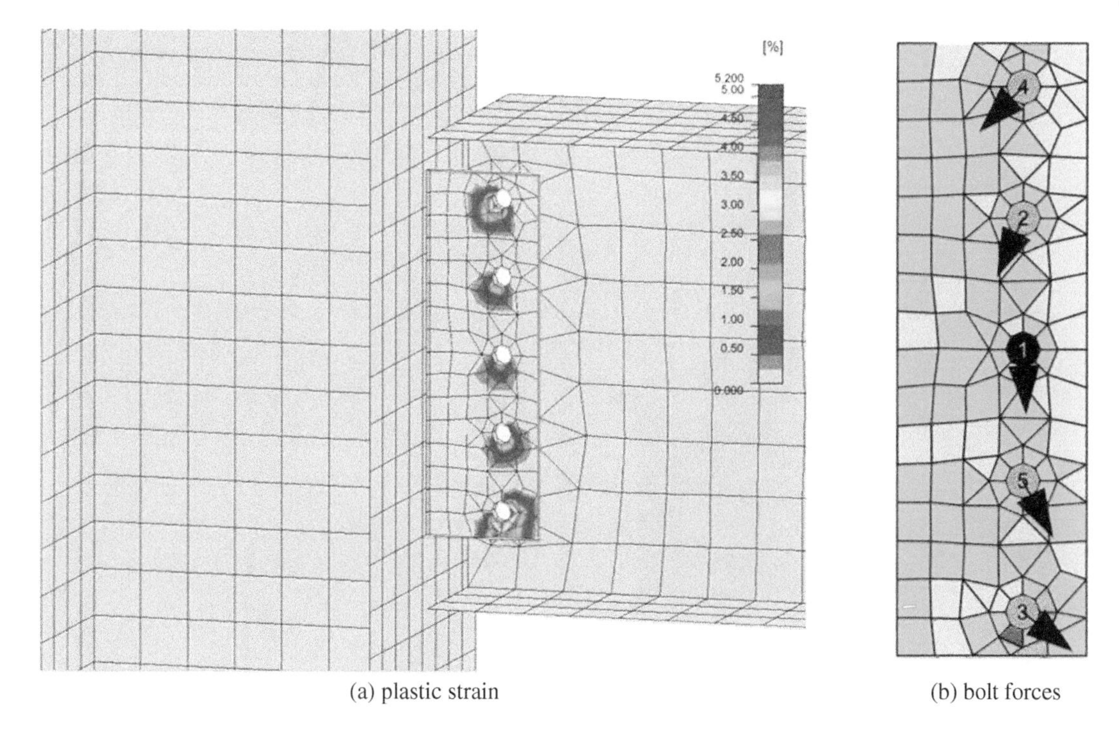

(a) plastic strain (b) bolt forces

Figure 8.8 Detailed results for connection with ¼ in. plate thickness.

limiting shear capacity of the connection with ¼ in. thick plate, the clear distance for the bottom bolt computed by IDEA StatiCa is $l_c = 1.240$ in. based on the angle of load in the bolt (Figure 8.8(b)). Tearout strength is proportional to clear distance, so the strength of the bolts per IDEA StatiCa is significantly greater than assumed in the poison bolt method.

For the connections with the thinner plates, the plate controlled in both IDEA StatiCa and the traditional calculations (other than those using the poison bolt method). However, in IDEA StatiCa, plastic strains were concentrated at the holes of the top and especially the bottom bolts (Figure 8.8). This contrasts with the assumed shear rupture failure plane used in the traditional calculations (i.e. a vertical line through the center of the bolts). Despite the differences in behavior, the resulting shear strength was close with IDEA StatiCa providing slightly lower shear capacities for the connections with the thinner plates.

8.4 Other Framing Configurations

Single plate shear connections are used for a variety of framing configurations. This section investigates two additional configurations, one where the supported beam frames into the web of a column and another where the supported beam frames into the web of a girder.

For the case of the supported beam framing into the web of a column (Figure 8.9), the column is a W27x114, and the supported beam is a W18x50. For the case of the supported beam framing into the web of a girder (see Figure 8.11), the girder is a W21x55, and the supported beam is a W18x46. All wide flange

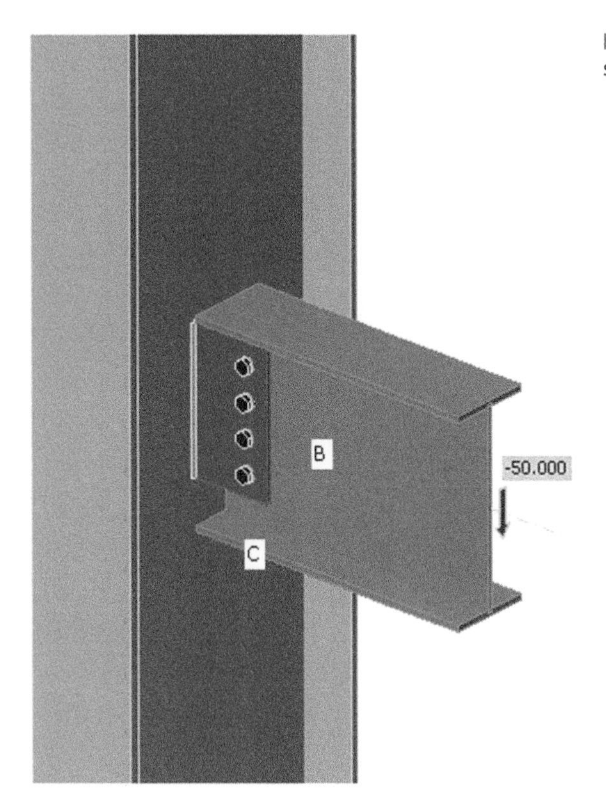

Figure 8.9 IDEA StatiCa model of single plate shear connection welded to weak axis of column.

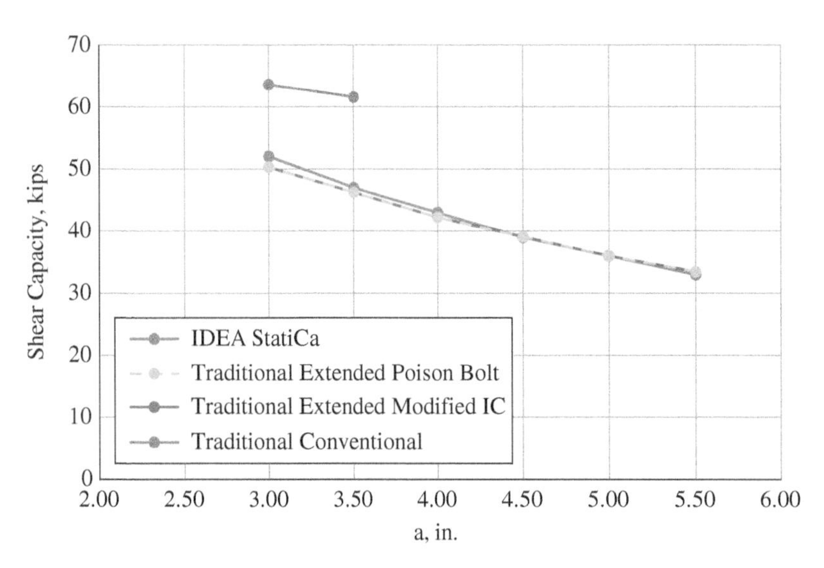

Figure 8.10 Shear capacity of single plate shear connection welded to weak axis of column with respect to *a*.

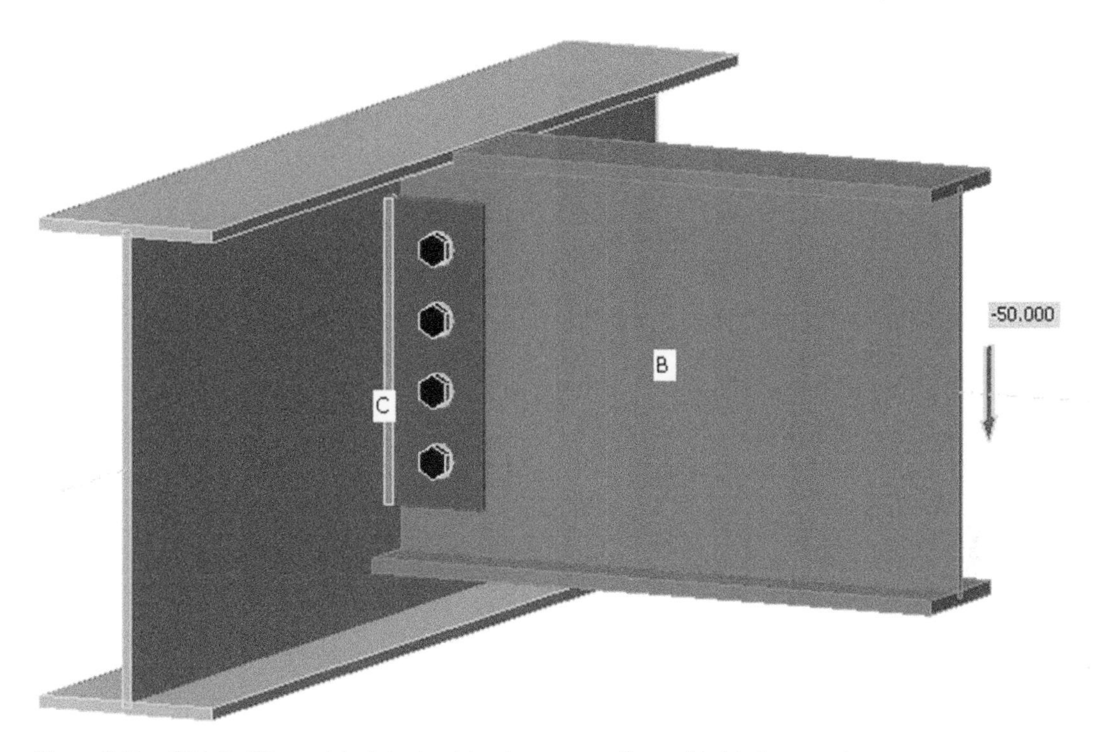

Figure 8.11 IDEA StatiCa model of single plate shear connection welded to beam web.

shapes conform to ASTM A992 (F_y = 50 ksi, F_u = 65 ksi). For both cases, the plate is 13 in. tall (s = 3 in., l_{ev} = 2 in.), 3/8 in. thick, and conforms to ASTM A36 (F_y = 36 ksi, F_u = 58 ksi). The connections have a single vertical row of (4) ¾ in. diameter A325 bolts with threads not excluded from the shear plane and horizontal edge distance, l_{eh} = 2 in. The weld was a 5/16 in. fillet weld on both sides of the plate. The distance from the weld line to the bolt line, a, was varied from 3 in. to 5.5 in.

Variation of the shear capacity of the connections with the distance a is presented in Figure 8.10 for the case of the supported beam framing into the web of a column and Figure 8.12 for the case of the supported beam framing into the web of a girder. Bolt shear rupture was the controlling limit state for all values of a and all methods in both framing configurations. The capacity determined from IDEA StatiCa agrees with that from the traditional calculations.

8.5 Location of the Point of Zero Moment

The design methodology for single plate shear connections in Part 10 of the AISC *Manual* (AISC, 2017) presumes that the location of the point of zero moment is at the weld line. Accordingly, all the IDEA StatiCa analyses thus far in this document have utilized an equivalent assumption for the position on the member from the node where the load is applied, X. However, other choices of the location of the point

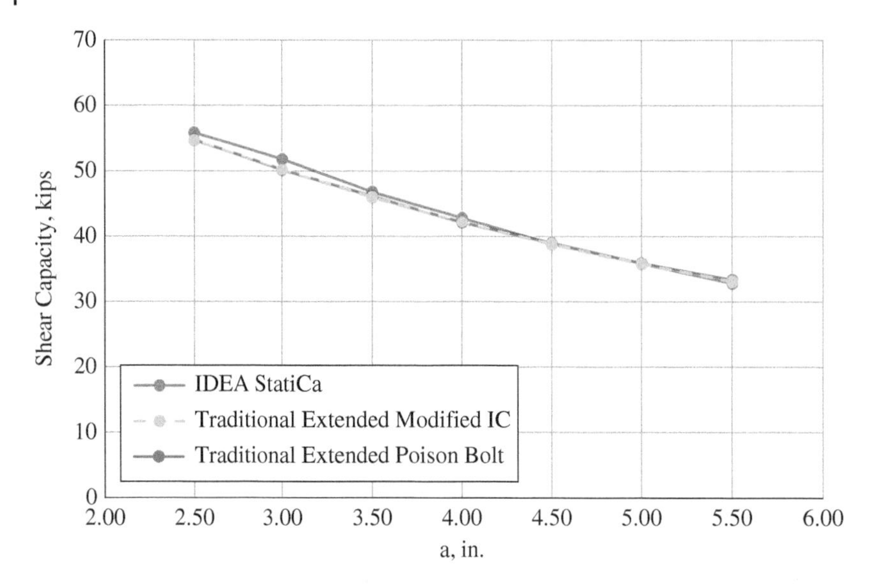

Figure 8.12 Shear capacity of single plate shear connection welded to beam web with respect to *a*.

of zero moment could be made, especially if the choice is made consistently with the location of the pin in the structural analysis model of the frame.

Analyses were performed to investigate the impact of the location of the point of zero moment. For these analyses, the column is a W14x90, and the supported beam, which frames into the flange of the column, is a W18x143. Both conform to ASTM A992 (F_y = 50 ksi, F_u = 65 ksi). The plate is 14 in. tall (s = 3 in., l_{ev} = 1 in.), 3/8 in. thick, and conforms to ASTM A572 Gr. 50 (F_y = 50 ksi, F_u = 65 ksi). There is one vertical row of (5) ¾ in. diameter A490 bolts with threads excluded from the shear plane and horizontal edge distance, l_{eh} = 1 in. Fillet welds were provided on both sides of the plate with size varying with plate thickness in accordance with the $(^5/_8)t_p$ rule noted in Part 10 of the AISC *Manual* (AISC, 2017). The distance from the weld line to the bolt line, a, was 9 in.

Variation of the shear capacity with the distance X (measured from the centerline of the column to the location of the point of zero moment) is presented in Figure 8.13. The controlling limit state per IDEA StatiCa was bolt tearout for $x \leq 16$ in. and weld resistance for greater values of X. The controlling limit states for the traditional calculations using the modified IC method were bolt group strength for $x <$ 17 in. and shear rupture of the plate for greater values of X. The controlling limit state for the traditional calculations using the poison bolt method was bolt group strength for all values of X. It is interesting to note that the IDEA StatiCa results were near those from the poison bolt method for this comparison. For these cases, the direction of the force in the controlling bolt is near that of the worst-case condition used in the poison bolt method (Figure 8.14).

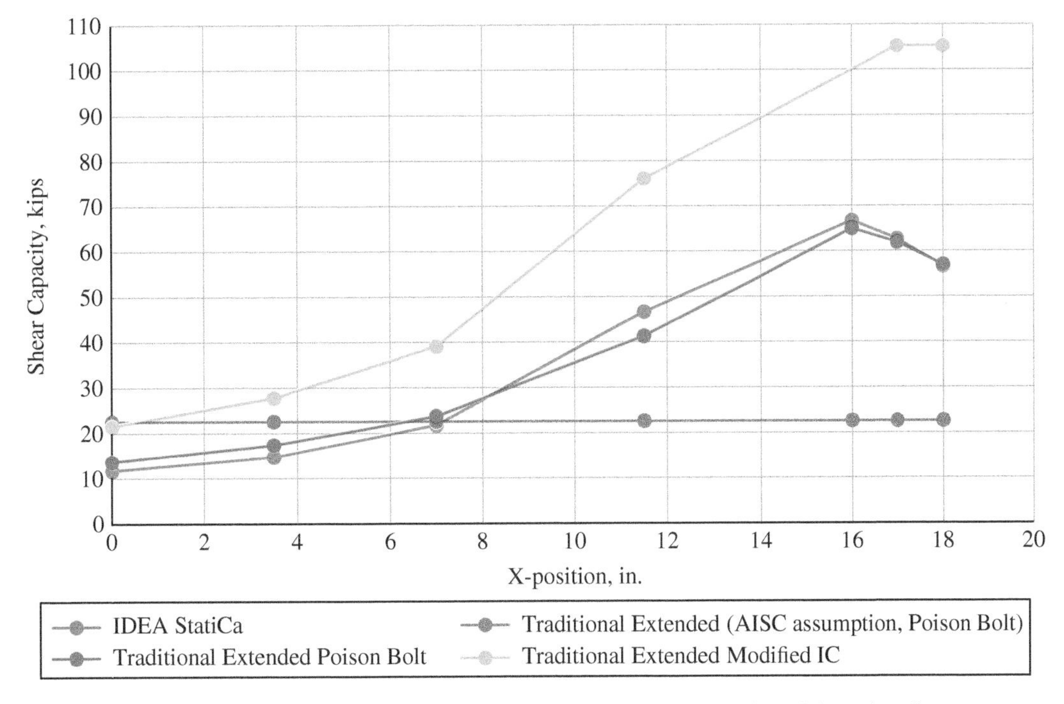

Figure 8.13 Shear capacity of single plate shear connection versus the location of the point of zero moment.

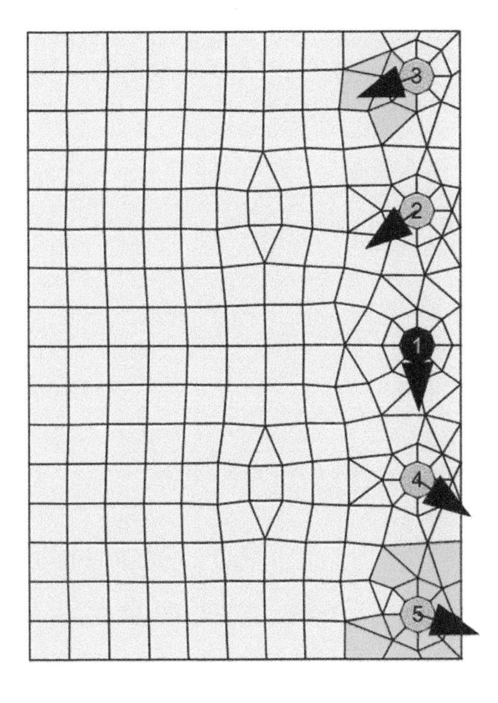

Figure 8.14 Detailed results for connection of the position of the point of zero moment located at the weld line.

8.6 Stiffness Analysis

In addition to strength requirements, single plate shear connections must also satisfy rotation capacity requirements. Section B3.4a of the AISC *Specification* (AISC, 2016) states that "a simple connection shall have sufficient rotation capacity to accommodate the required rotation determined by the analysis of the structure." For the traditional calculations, this requirement is satisfied by maximum plate and beam web thickness limitations described in Part 10 of the AISC *Manual* (AISC, 2017). With IDEA StatiCa, this requirement can be satisfied by performing a stiffness analysis.

The rotation capacities from a series of analyses on connections with varying plate thickness are presented in Figure 8.15. For these analyses, the column is a W14x90, and the supported beam, which frames into the flange of the column, is a W18x130. Both conform to ASTM A992 (F_y = 50 ksi, F_u = 65 ksi). The plate is 15 in. tall (s = 3 in., l_{ev} = 1.5 in.) and conforms to ASTM A572 Gr. 50 (F_y = 50 ksi, F_u = 65 ksi). There is one vertical row of (5) 7/8 in. diameter A325 bolts with threads not excluded from the shear plane and horizontal edge distance, l_{eh} = 1.5 in. Fillet welds were provided on both sides of the plate with size varying with plate thickness in accordance with the $(^5/_8)t_p$ rule noted in Part 10 of the AISC *Manual* (AISC, 2017). The distance from the weld line to the bolt line, a, was 3 in. These connections satisfy the requirements for the conventional configuration and rotation capacity since all plate thicknesses are less than or equal to 1/2 in. (AISC *Manual*, Table 10-9).

The analyses were performed using the "ST" (stiffness) analysis type. Unlike the previous analyses, these models were loaded with bending moments about the major axis of the beam. The rotational capacity was independent of the magnitude of applied load.

Per Section B3.4a of the AISC *Specification* (AISC, 2016), the required rotation capacity is determined from the structural analysis and depends on the framing and loads. A value of 0.03 rad or 30 mrad is commonly accepted as a reasonable upper bound for beam end rotation and the plate thickness limitations of Part 10 of the AISC *Manual* (AISC, 2017) were calibrated to meet this upper bound (Muir and

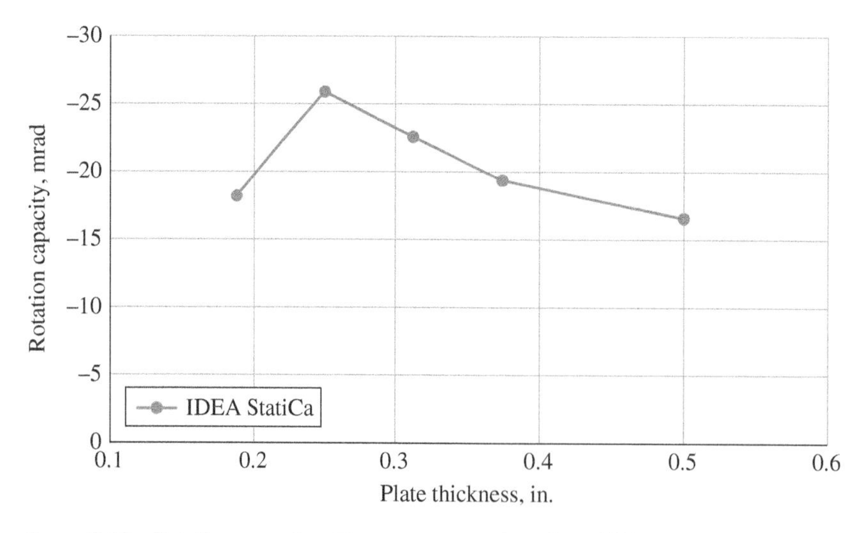

Figure 8.15 Rotation capacity with respect to varying plate thickness.

Thornton, 2011). The rotation capacities shown in Figure 8.15 are less than 30 mrad, despite meeting the plate thickness requirements. The values may still be acceptable to a wide range of cases which have less beam end rotation than the upper bound, however, it is also possible that the stiffness analysis in IDEA StatiCa is not fully capturing the ductility of the connections.

8.7 Summary

This study compared the design of single plate shear connections by traditional calculation methods used in US practice and IDEA StatiCa. Key observations from the study include:

- The available strength of single plate shear connections per IDEA StatiCa agrees well with traditional calculations using the method for extended configurations.
- The available strength per IDEA StatiCa was found to be conservative in comparison to the traditional calculations using the method for conventional configurations, which assumes a reduced eccentricity in some cases.
- IDEA StatiCa detects the clear distance for each bolt individually for consideration of tearout, resulting in appropriate reductions in strength when edge distances are small.
- IDEA StatiCa allows the investigation of different assumed locations of the point of zero moment.
- Stiffness analysis in IDEA StatiCa can be used to evaluate the rotation capacity requirements of AISC *Specification* Section B3.4a. However, the results were found to be conservative in comparison to the design rules presented in the AISC *Manual* for the cases examined.

References

AISC (2016). *Specification for Structural Steel Buildings.* American Institute of Steel Construction, Chicago.

AISC (2017). *Steel Construction Manual*, 15th Edition. American Institute of Steel Construction, Chicago.

Denavit, M.D., Franceschetti, N., and Shahan, A. (2021). *Investigation of Bearing and Tearout of Steel Bolted Connections.* Final Research Report to the American Institute of Steel Construction, Chicago.

Muir, L.S. and Thornton, W.A. (2011). The Development of a New Design Procedure for Conventional Single-Plate Shear Connections. *AISC Engineering Journal*, 48(2), 141–152.

9

Extended End-Plate Moment Connections

9.1 Description

A comparison between results from the component-based finite element method (CBFEM) and traditional calculation methods used in US practice for extended end-plate moment connections (Figure 9.1) is presented in this chapter.

The traditional calculation methods used in this work for non-seismic connections are based upon the recommendations presented in AISC Design Guide 4 (Murray and Sumner, 2003), in addition to the requirements for load and resistance factor design (LRFD) in the AISC *Specification* (AISC, 2016a). The traditional calculation methods used in this work for seismic (i.e. capacity designed) connections are based upon AISC *Prequalified Connections for Special and Intermediate Steel Moment Frames for Seismic Applications* (AISC, 2016b), hereafter referred to as AISC 358. For both seismic and non-seismic connections, these references include minimum end-plate thickness and column flange thickness limitations that are not directly based on the applied loads. These limits are intended to avoid prying action and ensure that the connection is fully restrained. For non-seismic connections, it is permitted to use thinner plates and column flanges if prying action is considered, for example, using the recommendations from Dowswell (2011). However, the minimum thickness limits were enforced for all traditional calculations in this study.

The limit states evaluated in the traditional calculations include tensile rupture of the bolts, flexural yielding of the end plate and column flange (via thickness limitations), shear yielding and rupture of the end plate, local column limit states (i.e. web local yielding, web local crippling, and web compression buckling), column web panel-zone yielding, and bolt shear limit states (i.e. bolt shear rupture, bearing, tearout – note, only the shear strength of the compression bolts was considered). For simplicity, all welds were modeled as butt welds and their strength was not evaluated in the traditional calculations.

The CBFEM results were obtained from IDEA StatiCa Version 21.0. Example models are shown in Figure 9.2. The maximum permitted loads were determined iteratively by adjusting the applied load input to a value that the program deems safe but if increased by a small amount (e.g. 1 kip-in.) the program would deem unsafe. In contrast to the traditional calculations, the influence of prying was evaluated in IDEA StatiCa and results shown include cases with prying action. The rigidity of the connection was evaluated using stiffness analyses (i.e. "ST" analysis type).

Steel Connection Design by Inelastic Analysis: Verification Examples per AISC Specification, First Edition.
Mark D. Denavit, Ali Nassiri, Mustafa Mahamid, Martin Vild, Halil Sezen, and František Wald.

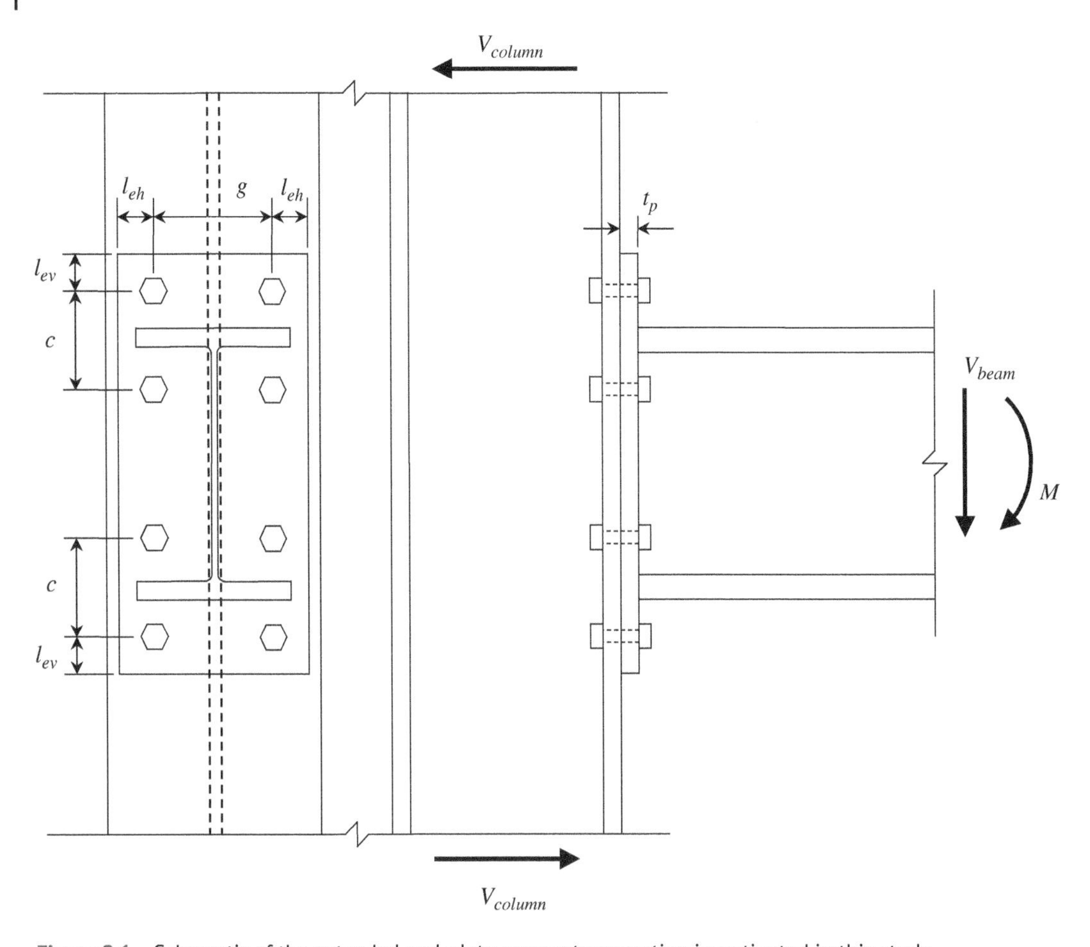

Figure 9.1 Schematic of the extended end-plate moment connection investigated in this study.

9.2 End-Plate Thickness

First, the impact of the end-plate thickness on the behavior and strength of the connection is investigated. For these comparisons, the beam is a W21 × 68, and the column is a W14 × 193. Both conform to ASTM A992 (F_y = 50 ksi, F_u = 65 ksi). The column was selected to be large (t_f = 1.44 in.) and provided with 5/8 in. thick stiffeners (i.e. continuity plates) to ensure that the controlling limit state was not in the column. The end plate has a depth of 29 in., a width of 9.5 in., and thickness varies from 3/8 in. to 2.5 in. All plate material (i.e. end plate and stiffeners) conforms to ASTM A572 Gr. 50 (F_y = 50 ksi, F_u = 65 ksi). The connection has four bolts near each beam flange (8 total bolts) and the end plate is not stiffened. This configuration is commonly referred to as a four-bolt unstiffened, 4E, configuration. The bolts are 1 1/8 in. diameter A325 with a horizontal gage of g = 5.5 in. and vertical spacing of c = 4.5 in. The vertical distance from the centerline of the bolts to the edge of the end plate is l_{ev} = 2 in.

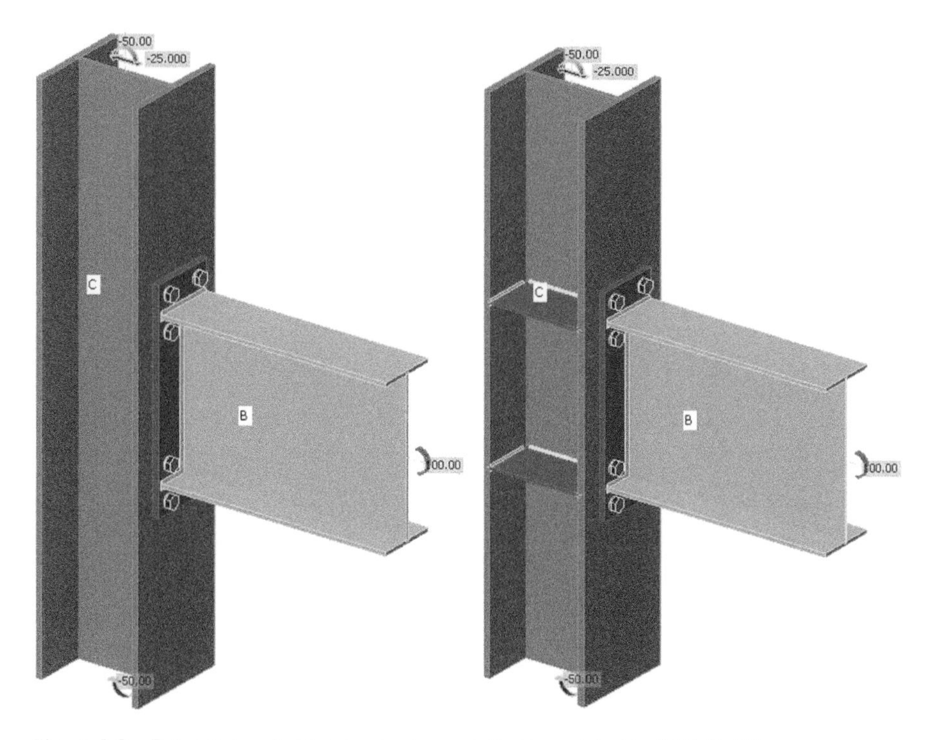

Figure 9.2 Extended end-plate moment connections modeled in IDEA StatiCa.

In IDEA StatiCa, loads were applied using the "loads in equilibrium" option. The moments applied at the top and the bottom of the column were each equal to half the moment applied to the beam. A shear load of 25 kips was also applied to the column (V_{column} = 25 kips, see Figure 9.1). For simplicity, no shear was applied to the beam (V_{beam} = 0 kips, see Figure 9.1).

Variation of the maximum applied moment with end-plate thickness is presented in Figure 9.3. The controlling limit state for each thickness is presented in Table 9.1. Results are not shown for the traditional calculations for end-plate thicknesses less than 1 in. since thinner plates did not meet the minimum thickness requirements to avoid prying action. The controlling limit state from the traditional calculations for the connections that met the end-plate thickness requirement was tensile rupture of the bolts. As a result, the maximum applied moment does not vary with end-plate thickness.

Variation of maximum applied moment with end-plate thickness is seen in the IDEA StatiCa results. For very thin plates ($t \leq 0.5$ in.), the plastic strain in the end plate controls the design. Otherwise, the bolt tension controls the design. The maximum applied moment increases with increasing end-plate thickness over the entire range investigated. The increase in maximum applied moment is rapid for thin plates as increasing thickness directly increases the flexural yielding strength of the end plate. The increase in maximum applied moment is more gradual when the bolt tension controls. For end-plate thicknesses 1.25 in. and greater, the maximum applied moment for IDEA StatiCa exceeds that of the traditional calculations. The reason for this is that the traditional calculations assume that the contact force at the interface between the column flange and the end plate is centered about the beam flange, whereas IDEA

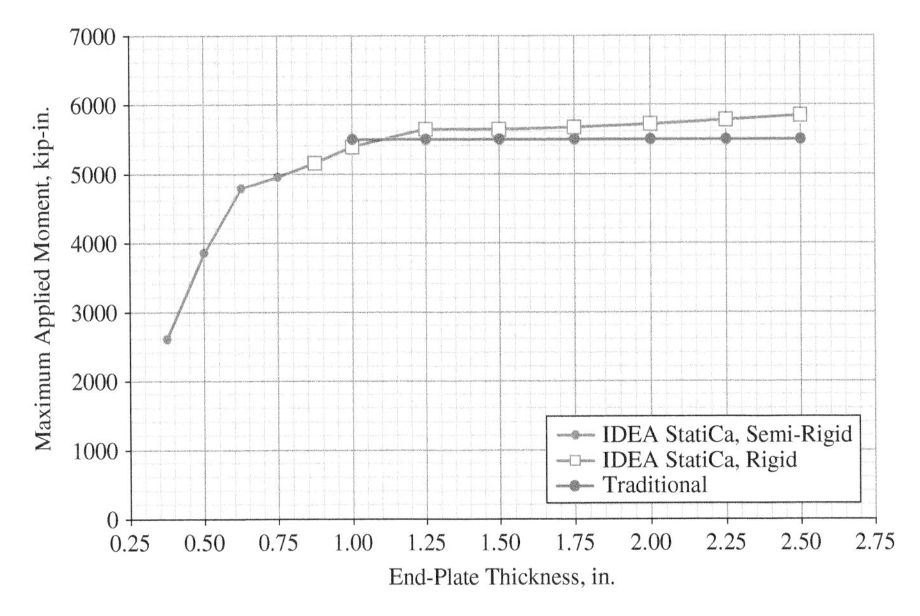

Figure 9.3 Maximum applied moment vs. end-plate thickness.

Table 9.1 Controlling limit state for results presented in Figure 9.3.

End-plate thickness (in.)	IDEA StatiCa	Traditional
0.375	Plastic strain (end plate)	N/A
0.500	Plastic strain (end plate)	N/A
0.625	Bolt tension	N/A
0.750	Bolt tension	N/A
0.875	Bolt tension	N/A
1.000	Bolt tension	Bolt tension
1.250	Bolt tension	Bolt tension
1.500	Bolt tension	Bolt tension
1.750	Bolt tension	Bolt tension
2.000	Bolt tension	Bolt tension
2.250	Bolt tension	Bolt tension
2.500	Bolt tension	Bolt tension

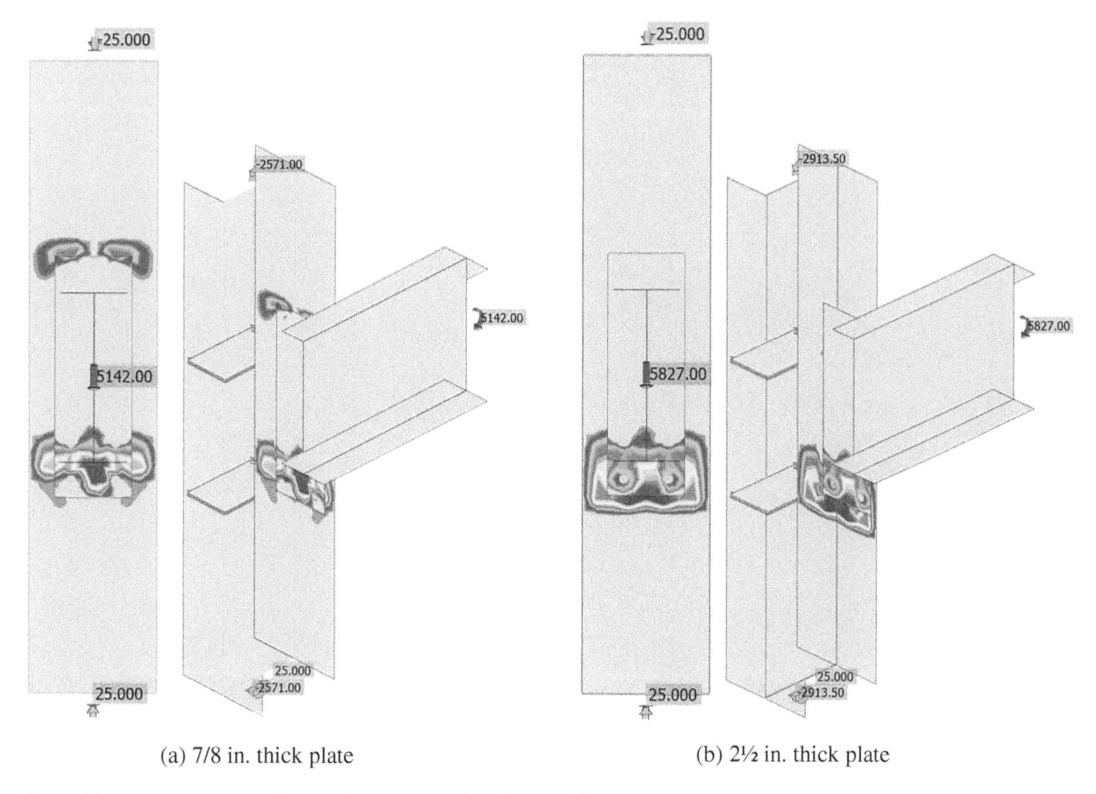

(a) 7/8 in. thick plate (b) 2½ in. thick plate

Figure 9.4 Contact stress for results presented in Figure 9.3.

StatiCa models the contact pressure explicitly. As the end-plate thickness increases, the portion of the end plate that extends past the beam flange is stiffer and more capable of resisting contact pressure, shifting the compressive force below the beam bottom flange (Figure 9.4). Therefore, while the tensile capacity of the bolts is no different between IDEA StatiCa and the traditional calculations, the lever arm of the couple is greater for IDEA StatiCa resulting in greater moment capacity.

For each end-plate thickness, the presence of prying action and the rigidity of the connection were determined by IDEA StatiCa. A connection was assumed to have prying action if there was contact stress on the tension side of the connection. For example, as shown in Figure 9.4, prying action was observed for the connection with a 7/8 in. thick plate, but not for the connection with the 2½ in. thick plate. There was no prying action for end-plate thicknesses of 1 in. and greater. This agrees with the corresponding minimum thickness limitation of the traditional calculations. Connections with end-plate thicknesses of 7/8 in. and greater were determined to be fully restrained (i.e. rigid) by a stiffness analysis in IDEA StatiCa, indicating that the minimum thickness limitation of the traditional calculations also provides a good indirect check of the rigidity of the connection for this case.

Adding stiffeners to the end plate changes the behavior of the connection. Contact stresses are presented in Figure 9.5. Variation of the maximum applied moment with end-plate thickness is presented in

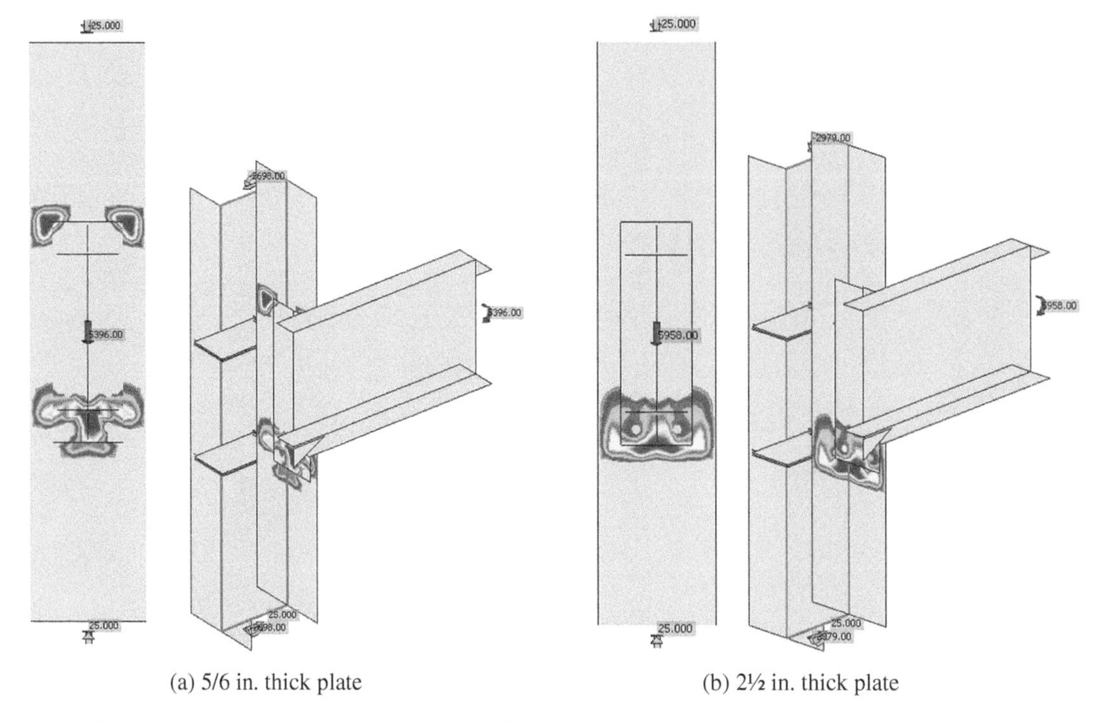

(a) 5/6 in. thick plate (b) 2½ in. thick plate

Figure 9.5 Contact stress for results presented in Figure 9.6 (with end-plate stiffener).

Figure 9.6 for the same connections investigated previously but with end-plate stiffeners added. The IDEA StatiCa results presented in Figure 9.3 for connections without stiffeners are included in Figure 9.6 for reference. The stiffeners were ½ in. thick, 3.5 in. wide, 6.5 in. long and placed on both flanges of the beam. Plate material for the stiffeners conformed to ASTM A572 Gr. 50 (F_y = 50 ksi, F_u = 65 ksi).

For the traditional calculations, adding stiffeners changes the yield line pattern for the flexural strength of the end plate, decreasing the minimum thickness. However, the addition of stiffeners did not change the strength of the connection, which remained controlled by the tensile rupture of the bolts, since the compression force is assumed to be centered with the flange regardless of the end plate stiffness. A review of recent research has confirmed that the compression force does shift below the bottom flange with the addition of end plate stiffeners and a method of accounting for the shift in design has been proposed (Landolfo et al., 2018).

For IDEA StatiCa, adding stiffeners increased the maximum applied load. The controlling limits were the same as those presented in Table 9.1. The increase in maximum applied loads was greatest for end-plate thicknesses between 5/8 in. and 1 in. where the bolt tension controlled, and the stiffeners helped decrease the prying action and increase the lever arm of the force couple.

The preceding analyses all used a relatively large column to ensure column limit states did not control. The column for the following analyses was smaller, a W14 × 109. Other aspects of the connections, including column stiffener thickness, beam, end plate, and bolts remained the same. The end plate for these analyses was unstiffened.

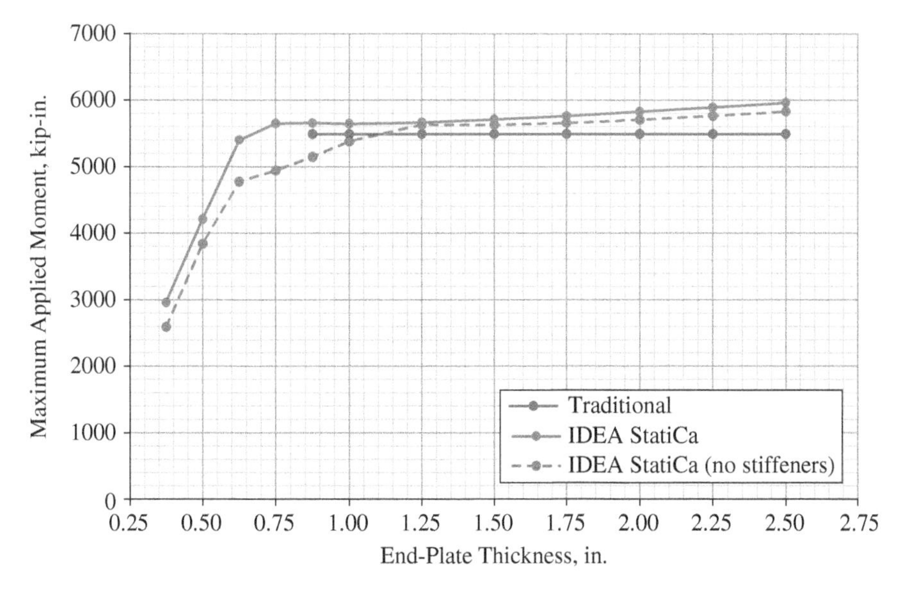

Figure 9.6 Maximum applied moment vs end-plate thickness.

Variation of the maximum applied moment with end-plate thickness is presented in Figure 9.7. The controlling limit state for each thickness is presented in Table 9.2. Multiple lines are plotted in Figure 9.7 for both IDEA StatiCa and the traditional calculations.

For the traditional calculations, results are plotted for when the effect of inelastic panel-zone deformations on the frame stability is not accounted for in the frame analysis and for when the effect is accounted

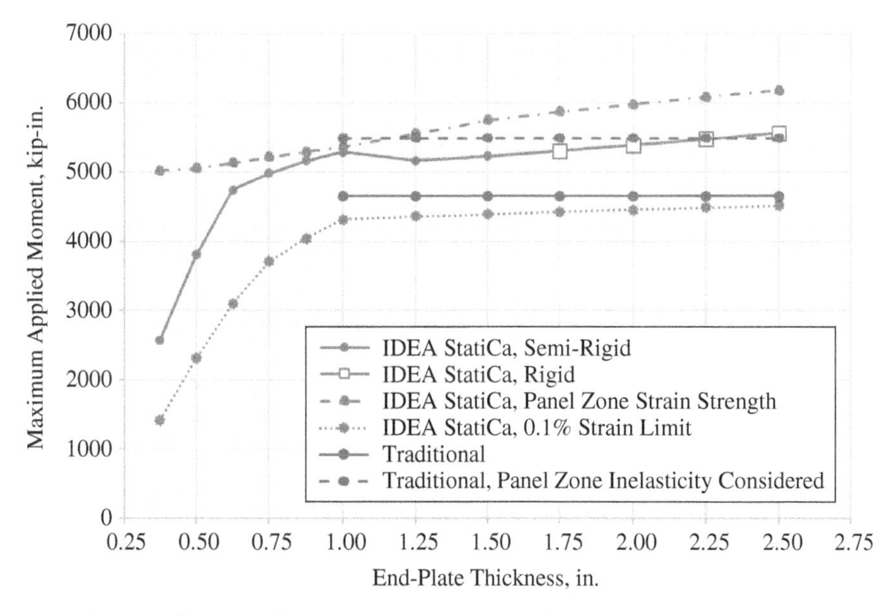

Figure 9.7 Maximum applied moment vs. end-plate thickness.

Table 9.2 Controlling limit state for results presented in Figure 9.7.

End-plate thickness (in.)	IDEA StatiCa	Traditional[1]	Traditional[2]
0.375	Plastic strain (end plate)	N/A	N/A
0.500	Plastic strain (end plate)	N/A	N/A
0.625	Bolt tension	N/A	N/A
0.750	Bolt tension	N/A	N/A
0.875	Bolt tension	N/A	N/A
1.000	Bolt tension	Panel zone shear	Bolt tension
1.250	Bolt tension	Panel zone shear	Bolt tension
1.500	Bolt tension	Panel zone shear	Bolt tension
1.750	Bolt tension	Panel zone shear	Bolt tension
2.000	Bolt tension	Panel zone shear	Bolt tension
2.250	Bolt tension	Panel zone shear	Bolt tension
2.500	Bolt tension	Panel zone shear	Bolt tension

[1]Effect of inelastic panel-zone deformations on frame stability not accounted for in the analysis.
[2]Effect of inelastic panel-zone deformations on frame stability accounted for in the analysis.

for in the frame analysis. Panel-zone yielding affects overall the frame stiffness and can significantly increase the second-order effects. If inelasticity of the panel-zone is not accounted for in the analysis to determine the required strengths of the frame, the AISC *Specification* (AISC, 2016a) limits the panel-zone behavior to the elastic range. If inelasticity of the panel-zone is accounted for when determining the required strengths of the frame, additional inelastic shear strength of the panel zone is recognized.

In the case where the panel-zone inelasticity is not accounted for in the analysis, the panel-zone shear strength controls the strength of the connection, with a maximum applied moment of 4,649 kip-in. In the case where the panel-zone inelasticity is accounted for in the analysis, the tensile strength of the bolts controls the strength of the connection with a maximum applied moment of 5,490 kip-in. (note the maximum applied moment for panel-zone yielding is only slightly higher at 5,495 kip-in.).

The controlling limit for IDEA StatiCa is the plastic strain limit in the end plate for very thin end plates ($t \leq 0.5$ in.) and bolt tension otherwise. The maximum applied moment is greater for IDEA StatiCa than it is for the traditional calculations. The controlling limit states also differ, so additional analyses were performed to quantify the applied moment at which the plastic strain limit is reached for the column web panel zone, shown in Figure 9.8 for a plate thickness of 1.25 in. These values are plotted as a dashed line in Figure 9.7 (note the bolt strength limits were exceeded for these analyses).

IDEA StatiCa captures the panel-zone yielding limit state although with a greater strength than permitted by the AISC *Specification* (AISC, 2016a) when the effect of inelastic panel-zone deformations on frame stability is accounted for in the analysis. Connections can be designed in IDEA StatiCa to limit yielding of the panel-zone simply by enforcing a plastic strain limit less than 5%. For example, the maximum applied load for the connection with a 1.75 in. thick end plate to have nearly elastic behavior (i.e. the limit of 0.1%

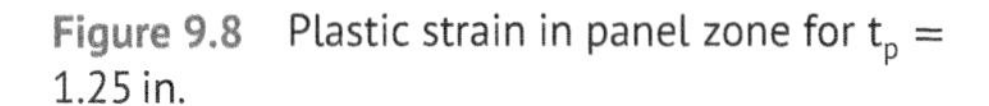

Figure 9.8 Plastic strain in panel zone for t_p = 1.25 in.

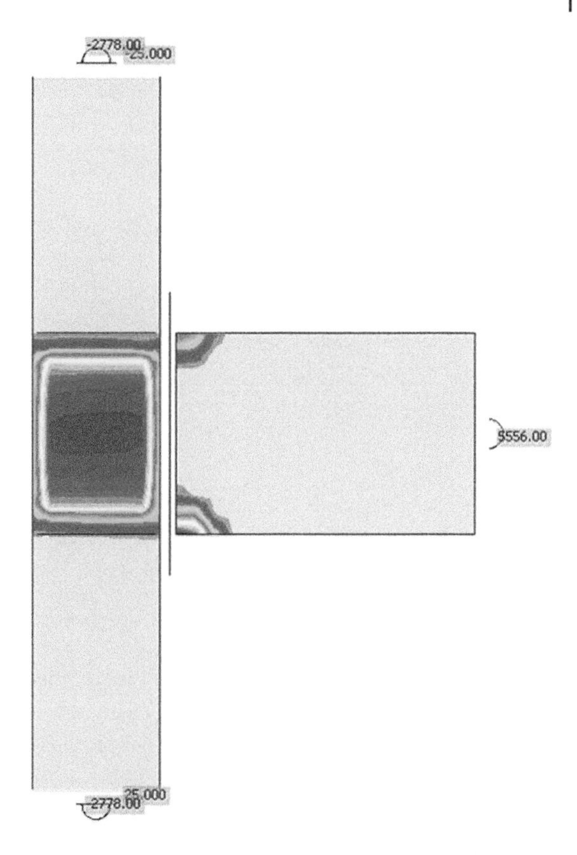

plastic strain) of the column web is 4,418 kip-in. which compares well to the maximum applied moment of 4,649 kip-in from the traditional calculations when the effect of the inelastic panel-zone deformations on frame stability is not accounted for in the analysis.

Interestingly, prying action is identified and the connection is classified as partially restrained (semi-rigid) in IDEA StatiCa for end-plate thicknesses up to 1.5 in. The traditional calculations permit end-plate thickness as low as 1 in. with the assumption of no prying.

9.3 Vertical Bolt Spacing

Thickness is not the only parameter that impacts the behavior of the end plate. As the vertical distance between the centerlines of the bolts increases, so does the pitch (distance from face of beam flange to centerline of nearer bolt). Generally, the smallest possible bolt pitch is most economical (Murray and Sumner, 2003), however, larger values may be necessary for constructability or other reasons.

A series of analyses is performed with varying vertical bolt spacing. For these comparisons, the beam is a W21 × 55, and the column is a W14 × 109. Both conform to ASTM A992 (F_y = 50 ksi, F_u = 65 ksi). The end plate has a depth of 28.5 in., a width of 9.0 in., a thickness of 1 in., and conforms to ASTM A572 Gr. 50 (F_y = 50 ksi, F_u = 65 ksi). The connection has four bolts near each beam flange (8 total bolts) and the

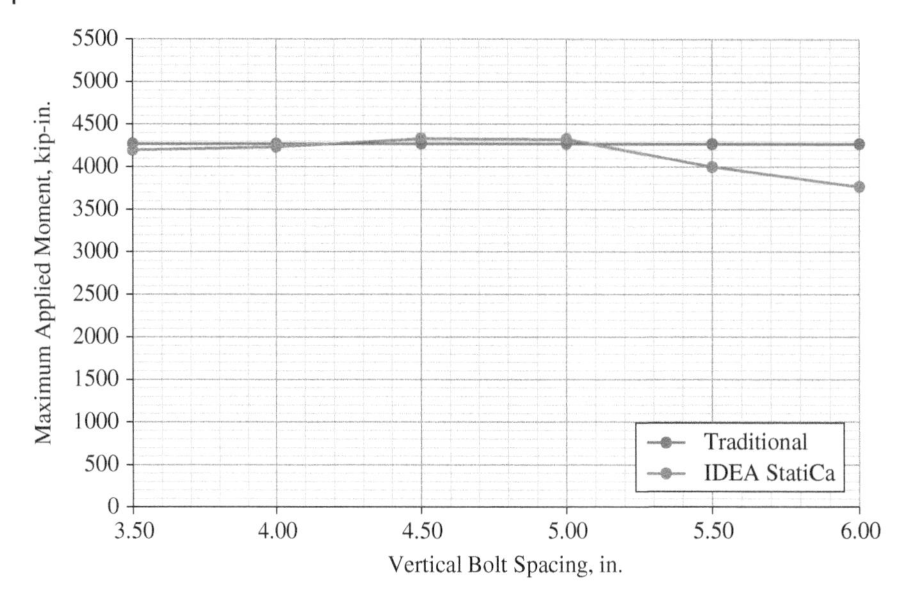

Figure 9.9 Maximum applied moment vs. vertical bolt spacing.

end plate is not stiffened. The bolts are 1 in. diameter A325 with a horizontal gage of 5.5 in. The vertical spacing between the bolts varies from 3.5 in. to 6 in. and the distance from the centerline of the bolts to the edge of the end plate varies from 2.5 in. to 1.25 in. The centroid of the bolt group was held constant. Loads were applied as described in the previous section, including 25 kips of shear in the column.

Variation of the maximum applied moment with vertical bolt spacing is presented in Figure 9.9. The controlling limit state for both the traditional calculations and IDEA StatiCa was tensile rupture of the bolt in all cases. For vertical bolt spacing less than or equal to 5 in., there is close agreement between the traditional calculations and IDEA StatiCa. For larger vertical bolt spacing, the maximum applied load from IDEA StatiCa decreases. The maximum applied load from the traditional calculations is constant over the entire range. The reason for the discrepancy is prying action. The plate thickness meets the minimum thickness requirement of the traditional calculations to assume no prying action. However, prying action is observed in the IDEA StatiCa results for vertical bolt spacing of 5.5 in. and 6 in., decreasing the maximum applied moment.

9.4 Capacity Design

Extended end-plate moment connections are one of the connection types that are prequalified for use in special and intermediate steel moment frames (AISC, 2016b). However, they are only prequalified if they meet the limitations and were designed according to the highly prescriptive procedure of AISC 358. The design criteria of AISC 358 are intended to ensure that inelastic deformation of the connection is achieved by beam yielding.

Use of IDEA StatiCa in lieu of the design procedure specified in AISC 358 is not permitted for demonstrating conformance with the requirements for beam-to-column connections for special and intermediate steel moment frames. However, IDEA StatiCa does have the capability to perform capacity design and produce comparable results.

For capacity design in IDEA StatiCa, specific elements are designated as dissipative components. The stress strain response of these components is overridden to be based on that of the expected material strengths and to include strain hardening. Then, loads are applied corresponding to the maximum probable load effects. For the extended end-plate moment connection, the beam is the dissipative component and the maximum probable load effects are computed per AISC 358.

In this investigation, a series of connections are capacity designed according to the AISC 358 procedure and IDEA StatiCa to compare results. Note that the default resistance factors were overridden in IDEA StatiCa to match those specified in AISC 358. The beam is varied from a W18 × 35 to a W18 × 60, the column is a W14 × 211. All wide flange shapes conform to ASTM A992 (F_y = 50 ksi, R_y = 1.1, F_u = 65 ksi). The end plate conforms to ASTM A572 Gr. 50 (F_y = 50 ksi, F_u = 65 ksi) and has a depth of 28 in. The width of the plate was 7 in. for the W18 × 35, W18 × 40, and W18 × 46 beams and 8.5 in. for the W18 × 50, W18 × 55, and W18 × 60 beams. The thickness of the end plate was selected during the design process. A four-bolt unstiffened, 4E, configuration was used with A490 bolts. The diameter of the bolts was selected during the design process. The horizontal gage was 5.5 in., the vertical bolt spacing was 5.5 in, and the vertical distance from the centerline of the bolts to the edge of the end plate was l_{ev} = 2 in.

The applied moment and applied beam shear for each beam section are listed in Table 9.3. The applied beam shear was based on an assumed beam shear force resulting from gravity loads of 30 kips and beam length (between column centerlines) of 30 ft. The loads were applied at the "X-position" (i.e. the distance from the centerline of the column to assumed plastic hinge location). A shear load of 30 kips was also applied to the column. Interestingly, the plastic strains in the beam reached a maximum of approximately 30% in these analyses. However, this high level of plastic strain does not breach any limits since the beam is classified as the dissipative component.

The designed end-plate thickness and bolt diameter are shown as a function of the beam weight in Figure 9.10 and Figure 9.11, respectively. One design is shown for each beam size for the traditional calculations since the AISC 358 procedure inhibits prying action and results in a unique efficient design. Two designs are shown for each beam size for IDEA StatiCa. With the capability of explicitly considering prying action in IDEA StatiCa, a range of efficient designs are possible, depending on the relative priority

Table 9.3 Applied loads for capacity design example.

Beam section	Applied moment, kip-in	Applied beam shear, kip	X-position (in.)
W18 × 35	4,206	55.8	16.70
W18 × 40	4,959	60.4	16.80
W18 × 46	5,737	65.2	16.90
W18 × 50	6,388	69.2	16.85
W18 × 55	7,084	73.4	16.90
W18 × 60	7,780	77.7	16.95

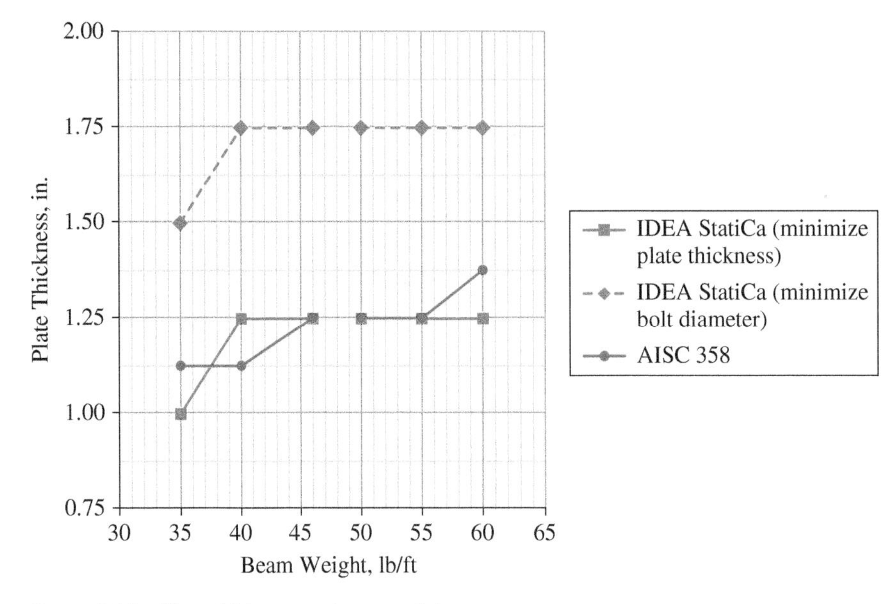

Figure 9.10 Plate thickness vs. beam weight.

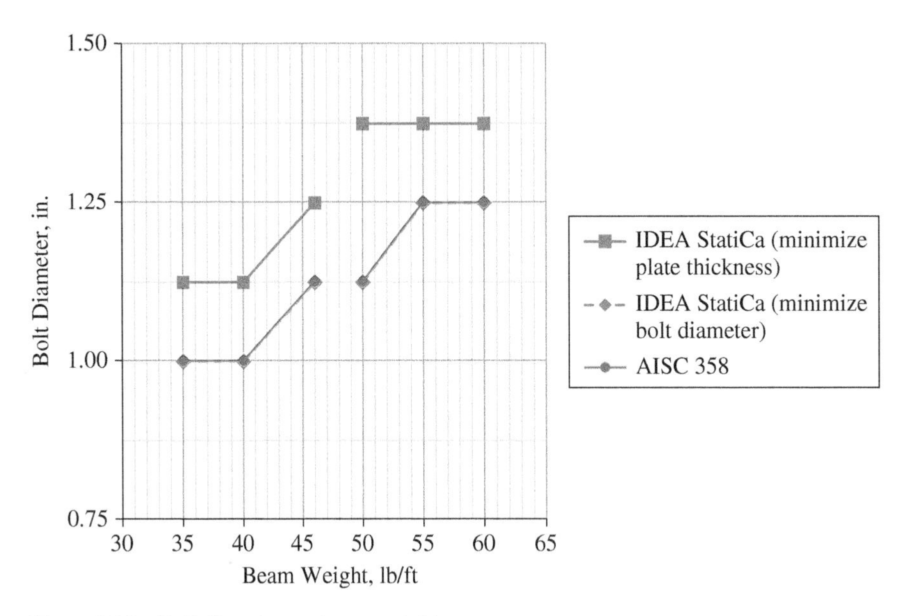

Figure 9.11 Bolt diameter vs. beam weight.

of bolt diameter and plate thickness. An informal optimization was performed to determine one design where the plate thickness was minimized and another where the bolt diameter was minimized.

When the bolt diameter is minimized, the resulting bolt diameter is the same between the traditional calculations and IDEA StatiCa, but the plate thickness is greater for the IDEA StatiCa design. The thicker plates are needed in IDEA StatiCa to eliminate the effect of prying action and minimize the demand on the bolts.

When the plate thickness is minimized, the resulting plate thickness for the IDEA StatiCa design is approximately the same as for the traditional calculations with some designs the same, some with one size thicker plate, and some with one size thinner plate. The bolts for the IDEA StatiCa design for these cases are larger than required per the traditional calculations due to the increased demands from prying action.

These results indicate that the modeling assumptions built into IDEA StatiCa result in a more conservative evaluation of prying action than the traditional calculations and, accordingly, IDEA StatiCa provides a conservative design of these two components of the extended end-plate moment connection.

9.5 Summary

This study compared the design of extended end plate moment connections using traditional calculation methods used in US practice and IDEA StatiCa. Key observations from the study include:

- IDEA StatiCa results in available strengths for extended end plate moment connections that are similar to the traditional calculations.
- Differences in strength are mainly due to prying action and the distribution of bearing stress, both of which are addressed with simplified assumptions in the traditional calculations, but explicitly modeled in IDEA StatiCa.
- Using default parameters, web panel-zone strength from IDEA StatiCa is similar to the strength from the AISC *Specification* when the effect of inelastic panel-zone deformations on frame stability is accounted for in the analysis to determine the required strengths. The lower strength given in the AISC *Specification* for when the effect of inelastic panel-zone deformations on frame stability is not accounted for in the analysis to determine the required strengths can be achieved by adjusting the plastic strain limit in IDEA StatiCa.
- The capacity design capabilities of IDEA StatiCa allow for the selection of the bolt diameter and the plate thickness that are conservative with respect to the procedure defined in AISC 358.

References

AISC (2016a). *Specification for Structural Steel Buildings*. American Institute of Steel Construction, Chicago.

AISC (2016b). *Prequalified Connections for Special and Intermediate Steel Moment Frames for Seismic Applications*. American Institute of Steel Construction, Chicago.

Dowswell, B. (2011). A Yield Line Component Method for Bolted Flange Connections. *AISC Engineering Journal*, 48(2), 93–116.

Landolfo, R., D'Aniello, M., Costanzo, S., Tartaglia, R., Demonceau, J., Jaspart, J., Stratan, A., Jakab, D., Dubina, D., Elghazouli, A., and Bompa, D. (2018). Equaljoints PLUS – Volume with Information Brochures for 4 Seismically Qualified Joints. Paper presented at European Convention for Constructional Steelwork (ECCS), Brussels, Belgium.

Murray, T.M. and Sumner, E.A. (2003). *Extended End-Plate Moment Connections: Seismic and Wind Applications*, 2nd Edition. Design Guide 4. American Institute of Steel Construction, Chicago.

10

Bolted Wide Flange Splice Connections

10.1 Description

A comparison between results from the component-based finite element method (CBFEM) and traditional calculation methods used in US practice for bolted wide flange splice connections (Figure 10.1) is presented in this chapter.

The traditional calculation methods used in this work are based upon the requirements for load and resistance factor design (LRFD) in the AISC *Specification* (AISC, 2016). The limit states evaluated in the traditional calculations include shear rupture, bearing, and tearout for the strength of the bolts; tensile yielding, tensile rupture, block shear rupture, and compression yielding for the strength of the splice plates; and tensile yielding, tensile rupture, compression yielding, and flexural yielding for the wide flange members. Deformation at the bolt hole at service load was assumed to be a design consideration. Slip was evaluated for some connections.

The CBFEM results were obtained from IDEA StatiCa Version 22.1. Example models are shown in Figure 10.2. The maximum permitted loads were determined iteratively by adjusting the applied load input to a value that the program deems safe but if increased by a small amount (e.g., 1 kip) the program would deem unsafe.

For the comparisons presented in this study, the upper column was always a W14 × 159 and the lower column was either a W14 × 159 or W14 × 370. All wide flange shapes were assumed to conform to ASTM A992 (F_y = 50 ksi, F_u = 65 ksi). The splice connection was based on Table 14-3 of the AISC *Manual* (AISC, 2017). A total of (24) 7/8 in. diameter A490 bolts (threads not excluded from shear planes) were used in the connection (6 for each of the splice plate to column flange connections). The connection was not slip-critical, unless noted otherwise. There was no gap between the columns. The connections were evaluated both with consideration of contact bearing (Muir, 2015) and ignoring the contact bearing. The bolt spacing was s = 3 in. and the vertical edge distances were l_{ev1} = 1.5 in. and l_{ev2} = 1.75 in., for a total length of 18.5 in. for the splice plate. The splice plate was 14 in. wide. The bolt gage was g = 11.5 in., and the horizontal edge distance was l_{eh} = 1.25 in. for some cases as recommended by Table 14-3. In other cases, the bolt gage was g = 8 in., and the horizontal edge distance was l_{eh} = 3 in. to avoid block shear rupture. The thickness of the splice plate varied in the analysis. Table 14-3 recommends a plate thickness of 0.5 in. for W14 × 159 columns. The splice plates were assumed to conform to ASTM A36 (F_y = 36 ksi, F_u = 58 ksi).

Steel Connection Design by Inelastic Analysis: Verification Examples per AISC Specification, First Edition.
Mark D. Denavit, Ali Nassiri, Mustafa Mahamid, Martin Vild, Halil Sezen, and František Wald.

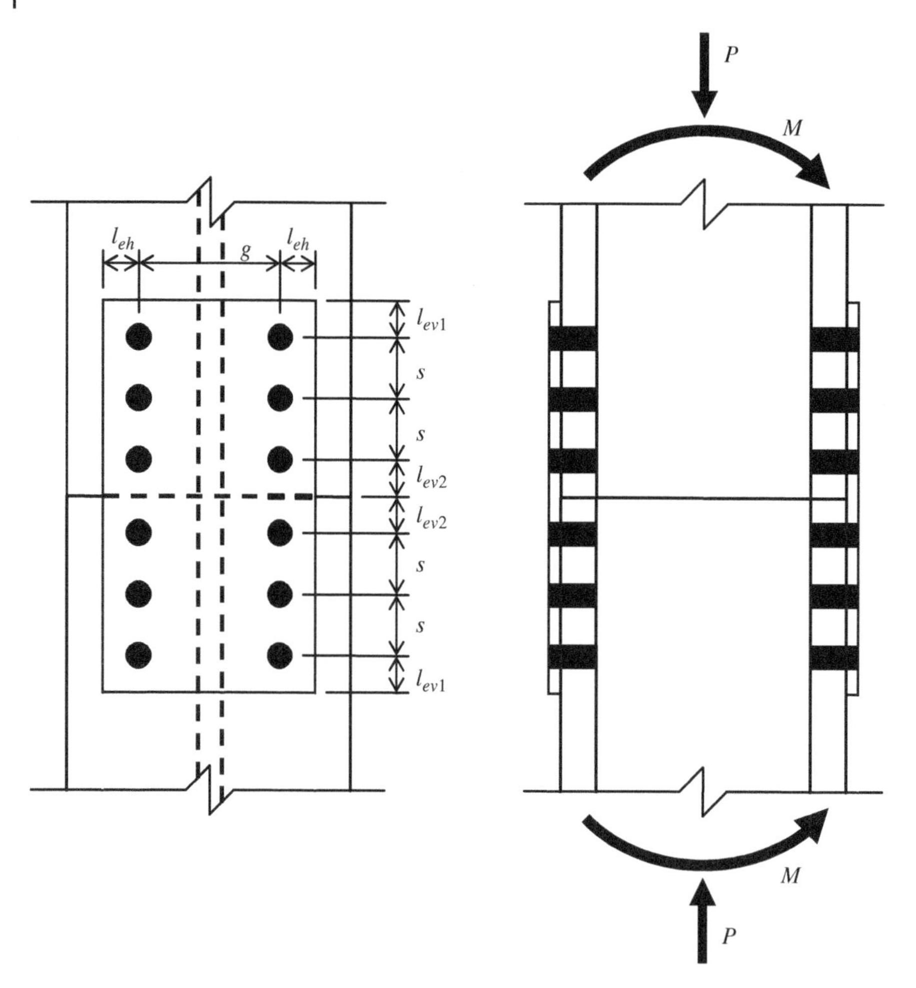

Figure 10.1 Schematic of bolted wide flange splice connection.

This column splice design is most appropriate for connecting two axially loaded columns through contact bearing. This study investigates cases where contact bearing is ignored, as well as cases where the columns are loaded in tension or with combined major-axis flexure. The same splice connection was used throughout the study for uniformity and ease of comparison; however, different connections would likely be more efficient for cases with significant tension or combined flexure.

10.2 Axial Loading

First, the strength of the connection under axial loading is investigated for the case of equal depth columns and bolt gage, $g = 11.5$ in. Variation of the maximum applied axial compression load with splice plate thickness is presented in Figure 10.3. The strength of the connection is much greater with contact

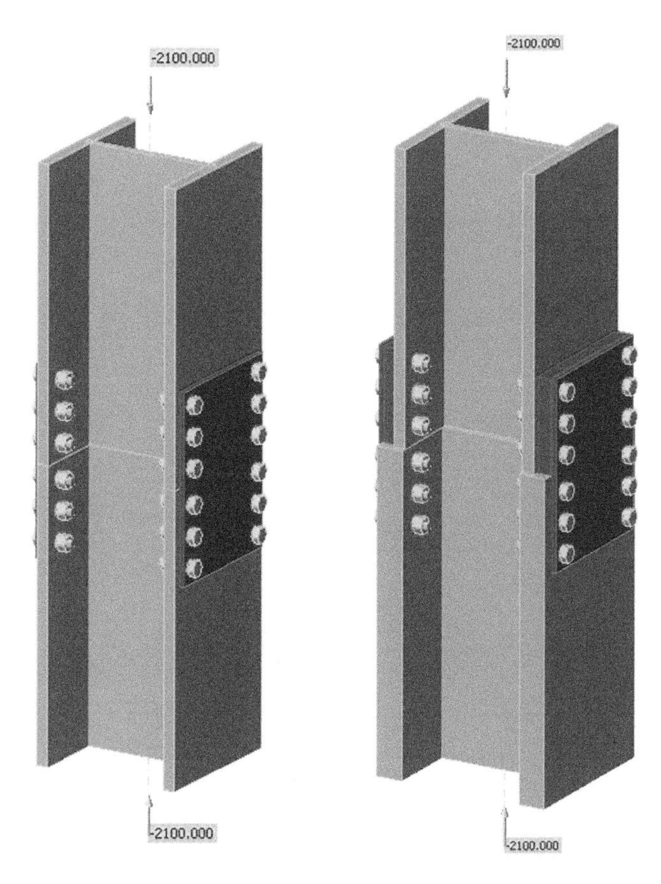

Figure 10.2 Bolted wide flange splice connections modeled in IDEA StatiCa.

bearing than without. With contact bearing, the plastic strain limit in the column web controlled for IDEA StatiCa and compression yielding of the column controlled for the traditional calculations. The bolts and splice plates are essentially unstressed in these analyses; hence strength did not vary with splice plate thickness. IDEA StatiCa gives a maximum permitted load that is approximately 4% greater than the traditional calculations, primarily due to the small amount of strain hardening assumed in the model and small differences in the cross-sectional area of the wide flange (i.e. IDEA StatiCa does not model the fillets and a portion of the area at each junction of the web and flange is double counted).

Without contact bearing the load is transferred from one wide flange to the other through the bolts and splice plates. For the thinnest splice plate (i.e. 3/8 in.), the plastic strain limit in the splice plate controlled for IDEA StatiCa and compression yielding of the splice plate controlled for the traditional calculations. Note that $L_c/r \leq 25$ for the splice plate when the effective length factor is taken as 0.65 for a fixed-fixed condition, so no stability reduction was applied. The maximum applied load was 309 kips for IDEA StatiCa and 340 kips for the traditional calculations. IDEA StatiCa gives a smaller maximum applied load because the stress in the splice plate is concentrated near the bolt holes. For all other splice

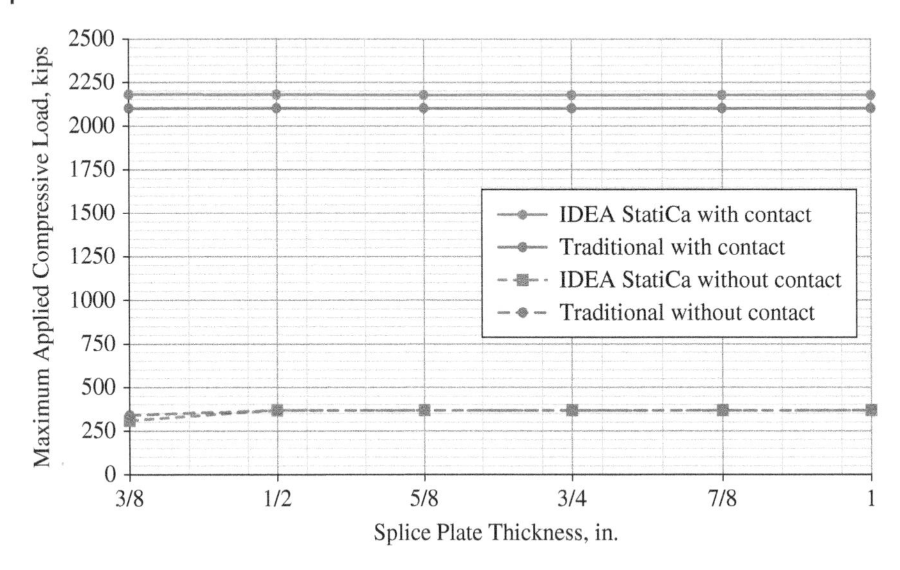

Figure 10.3 Maximum applied compressive load vs. splice plate thickness.

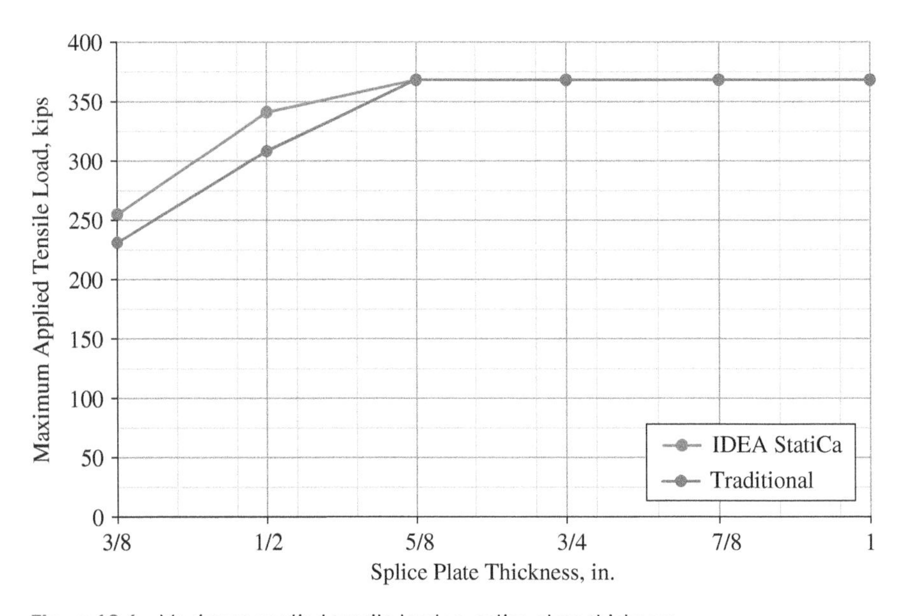

Figure 10.4 Maximum applied tensile load vs. splice plate thickness.

plate thicknesses, bolt shear rupture controlled for both IDEA StatiCa and the traditional calculations, and the maximum applied load was identical.

Variation of the maximum applied axial tensile load with splice plate thickness is presented in Figure 10.4. The IDEA StatiCa analyses were performed with and without bearing contact operations, however, the results from the two cases were identical.

For the connections with thicker splice plates (5/8 in. thick or more), bolt shear rupture controlled for both IDEA StatiCa and the traditional calculations. The maximum applied load was the same for the two methods. For the connections with thinner splice plates, tearout controlled the strength per IDEA StatiCa and block shear rupture of the splice plate controlled the strength per the traditional calculations. The mismatch in controlling limit states is resolved by refining the mesh in IDEA StatiCa. For the connection with 1/2 in. thick splice plates and the default mesh (maximum mesh size of 1.969 in.), tearout controls with a maximum applied tensile load of 341 kips. For a maximum mesh size of 1 in., the plastic strain limit is reached in the splice plates at an applied load of 338 kips. Further refinement to a maximum mesh size of 0.25 in., yields a maximum applied load of 328 kips with plastic strain of the splice plates controlling. The pattern of plastic strain for this connection is consistent with a block shear rupture failure (Figure 10.5). Even with the refined mesh, IDEA StatiCa gives a greater maximum applied load than the traditional calculations. For the connection with 1/2 in. thick splice plates, the maximum applied load per the traditional calculations is 308 kips.

Researchers have noted that the block shear rupture provisions in the AISC *Specification* (AISC, 2016) can be conservative in comparison to physical test data and have proposed alternative equations to better predict block shear rupture strength (Teh and Deierlein, 2017). Their proposed equation for the nominal strength for block shear rupture, $R_n = F_u A_{nt} + 0.6F_u A_{ev}$, utilizes an effective shear area, A_{ev}, equal to the average of the gross and net shear areas currently used in the AISC *Specification* (i.e. $A_{ev} = (A_{gv} + A_{nv})/2$). The available strength for block shear rupture for the connection with 1/2 in. thick splice plates using this equation is 391 kips, thus other limit states would control. If the equation proposed by Teh and Deierlein (2017) is accurate, then the IDEA StatiCa results would be conservative.

To further explore the behavior of this connection under tensile load, the analysis was re-run using a bolt gage of g = 8 in. Block shear rupture does not control the tensile strength of the splice plate with

Figure 10.5 Plastic strain in splice plate at 328 kips applied load for connection with 1/2 in. thick splice plate and maximum mesh size of 0.25 in.

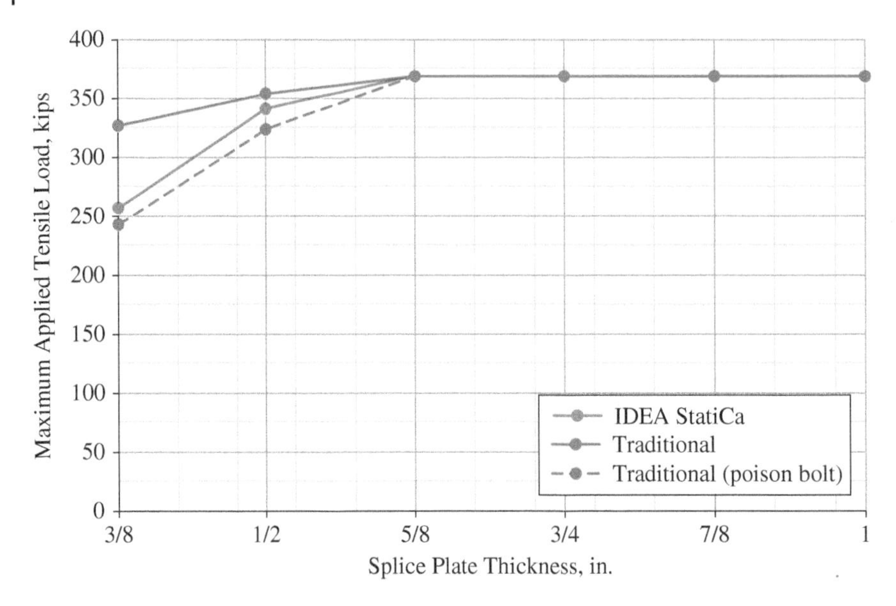

Figure 10.6 Maximum applied tensile load vs. splice plate thickness (bolt gage, $g = 8$ in.).

this value of g. Variation of the maximum applied axial tensile load with splice plate thickness for this case is presented in Figure 10.6. The IDEA StatiCa results are essentially the same as for the case with the larger bolt gage. For IDEA StatiCa and the connections with the two thinnest splice plates (i.e. 3/8 in. and 1/2 in.), the controlling limit state was tearout with only the bolts at the extreme ends of the splice reaching 100% utilization (Figure 10.7). Bolt shear rupture controlled for the other connections in IDEA StatiCa with all bolts reaching 100% utilization. For the traditional calculations, the strength of the bolt group controlled for all cases. However, the maximum applied load for the traditional calculations was

Figure 10.7 Display indicating bolt utilization at 256 kips applied load for connection with 3/8 in. thick splice plate.

greater than IDEA StatiCa for the connections with the two thinnest splice plates. For the traditional calculations, the effective strength of each bolt in the group is evaluated and summed to obtain the strength of the bolt group. Therefore, some of the bolts are controlled by tearout while others are controlled by shear rupture, but they all contribute their peak strength to the bolt group. In IDEA StatiCa, the bolts are all modeled with the same stiffness, so they all experience roughly the same load in this connection. For the thinner plates, tearout controls the strength of the extreme bolts and they achieve their strength first before the remaining bolts can achieve their strength. This is akin to the poison bolt method, which is more commonly used for eccentrically loaded bolt groups in traditional calculations. Using the poison bolt method in this case yields strength results more like those from IDEA StatiCa.

10.3 Axial Loading with Unequal Column Depths

When the columns to be connected have different depths, filler plates are used to pack out the depth of the smaller column and create a flush surface for the splice plates. Filler plates can be developed or undeveloped. Developed filler plates have an additional attachment to the column beyond the splice plates. Undeveloped filler plates do not have an additional attachment. The AISC *Specification* (AISC, 2016) requires reductions in shear and slip strength for bolted connections with undeveloped fillers.

The results presented in this section are for a splice connection with a W14 × 159 upper column and W14 × 370 lower column. The difference in depth between these two shapes is 2.90 in., therefore it was assumed that the total thickness of filler plates was 1.45 in., which was achieved with two plies, one 1-1/4 in. thick and another 3/16 in. thick.

Variation of the maximum applied axial compression load with splice plate thickness is presented in Figure 10.8. The bolt gage was taken as $g = 11.5$ in. for this case as would be typical for a column splice. With contact bearing, the results are essentially the same as for the case with equal depth columns and

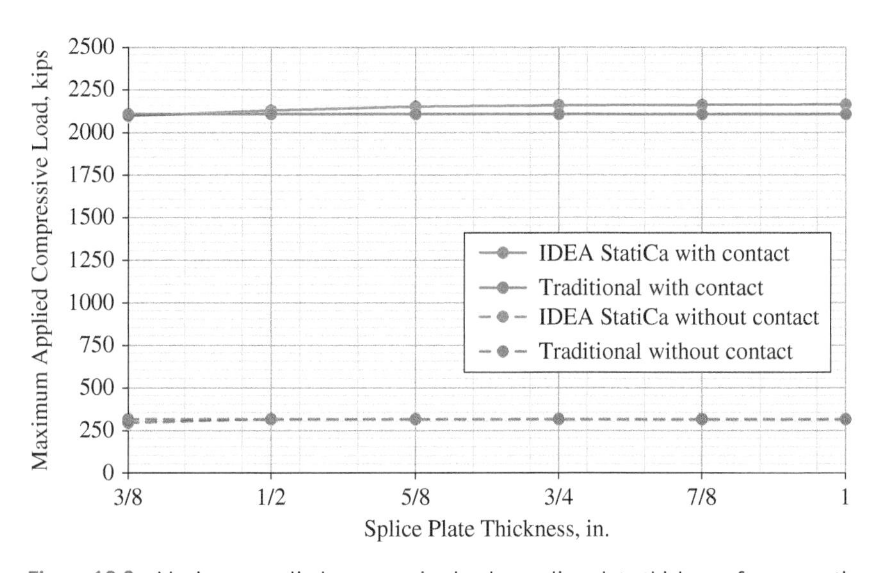

Figure 10.8 Maximum applied compressive load vs. splice plate thickness for connections with filler plates.

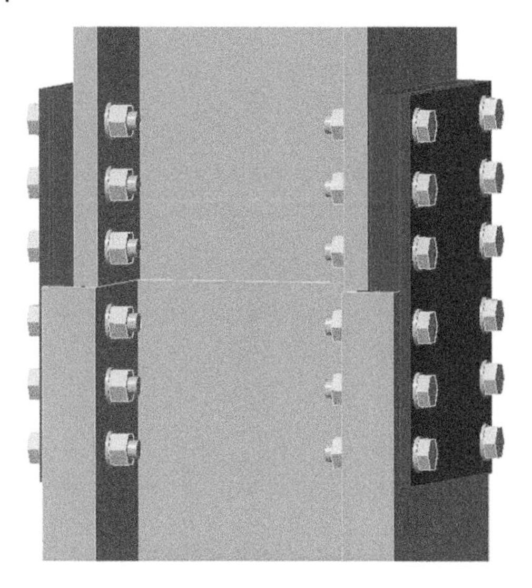

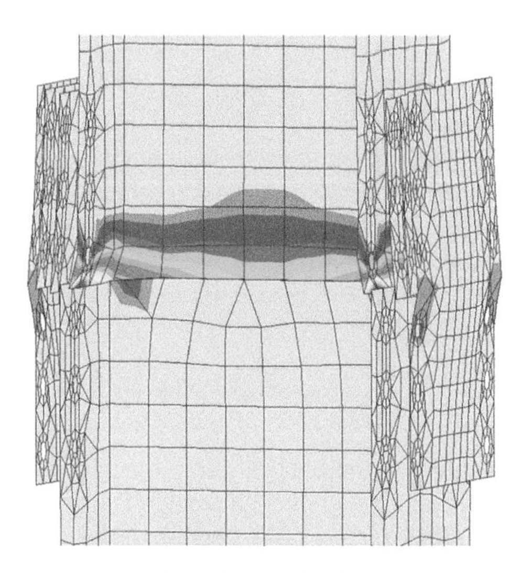

Figure 10.9 Plastic strain results at 1920 kips applied load for connection with 1/2 in. thick splice plate and no contact between the filler plates and the lower wide flange (deformation scale factor = 10).

no fillers. Note, however, that contact was defined both between the upper wide flange and the lower wide flange and between the filler plates and the lower wide flange. If contact was defined only between the two wide flange members, the offset of flange centerlines would result in bending of the flange (Figure 10.9) and somewhat reduced strengths in IDEA StatiCa (1879 kips without contact of the filler plates versus 2121 kips with contact of the filler plates for the connection with 1/2 in. thick splice plates). Full-contact bearing is achieved since the two columns are of the same family (i.e. W14) and the distance between flanges is the same, so the traditional calculations are unaffected.

Without contact bearing, the strength of the splice is much less and IDEA StatiCa shows the exact same strength as the traditional calculations for all but the connection with the thinnest splice plates (i.e. 3/8 in. thick). Note that the reduction in shear strength for fillers defined in Section J5.2 of the AISC *Specification* (AISC, 2016) is applied in both IDEA StatiCa and the traditional calculations. For the connection with the thinnest splice plates, the plastic strain in the splice plates controls in IDEA StatiCa resulting in a lower strength than the traditional calculations.

Variation of the maximum applied axial tension load with splice plate thickness is presented in Figure 10.10. The bolt gage was taken as g = 8 in. for this case to avoid the block shear rupture limit state. As with the compression case, IDEA StatiCa and the traditional calculations give the same strength for all but the connection with the thinnest splice plates. For the connection with the thinnest splice plates, tearout controls some of the bolts and a difference in strength arises because of the different ways that IDEA StatiCa and the traditional calculations handle bolt groups with bolts of different strength.

A strength reduction also applies to the slip limit state for connections with two or more fillers between connected parts. The reduction is defined by h_f, a factor for fillers, in Equation J3-4 of the AISC *Specification* (AISC, 2016); $h_f = 0.85$ for cases with two or more fillers between connected parts and $h_f = 1.0$ otherwise. If the splice connection was slip-critical, the available strength would be 199 kips for the case of no fillers or single-ply fillers and 169 kips for the case of multi-ply fillers. Without contact bearing and defining the connection as slip-critical, the maximum applied axial load in tension per IDEA StatiCa is

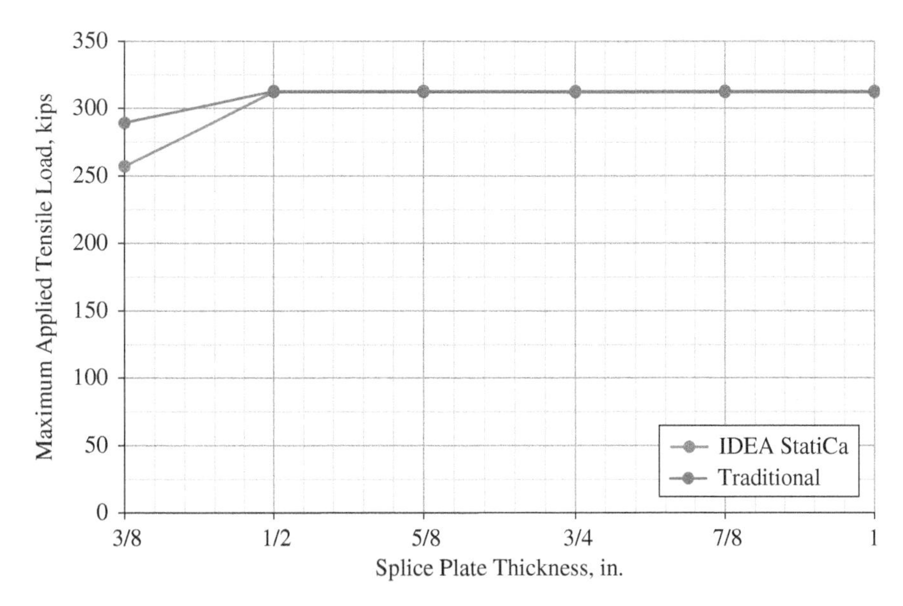

Figure 10.10 Maximum applied tensile load vs. splice plate thickness for connections with filler plates.

152 kips for the connection with fillers and 1/2 in. thick splice plates. IDEA StatiCa detects the multiple fillers and applies the appropriate factor for fillers. The lower strength from IDEA StatiCa is because IDEA StatiCa considers the eccentric loading of the filler plates, which is resisted by a couple formed by contact pressure and bolt tension (Figure 10.11). IDEA StatiCa conservatively ignores the friction due to the contact pressure while accounting for the applied tension in the bolt using the reduction factor k_{sc}. (AISC *Specification*, Section J3.9; AISC, 2016).

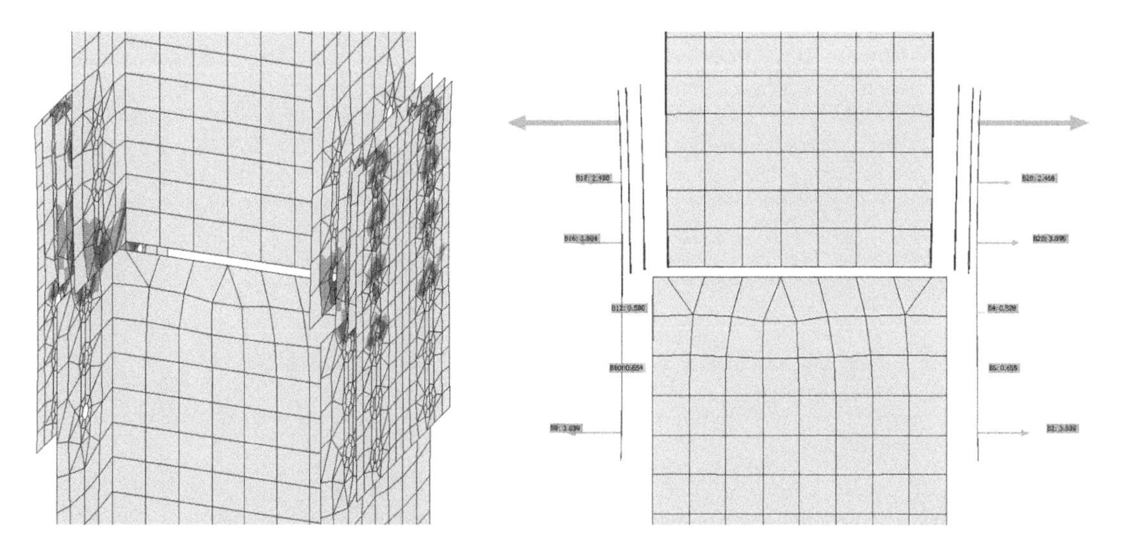

Figure 10.11 Stress in contacts and bolt force results at 152 kips applied tensile load for connection with 1/2 in. thick splice plate and friction (slip-critical) bolts (deformation scale factor = 10).

10.4 Combined Axial and Major-Axis Flexure Loading

Splice connections may need to support more than just axial loads. For the case of 1000 kip-in. of major-axis bending moment applied concurrently with the axial load, the variation of the maximum applied compression load with splice plate thickness is presented in Figure 10.12, and the variation of the maximum applied tensile load with splice plate thickness is presented in Figure 10.13. The bolt gage was taken as $g = 8$ in. for analyses in this section to avoid the block shear rupture limit state.

In compression and with bearing contact, member strength controlled both the IDEA StatiCa analyses and the traditional calculations. Both maximum applied loads are reduced from the case with pure compression (see Figure 10.3) because of the concurrent bending moment. In compression without bearing contact and in tension, IDEA StatiCa gives slightly greater maximum applied loads than the traditional calculations for the connections with thicker splice plates where bolt shear rupture controls. In contrast, the IDEA StatiCa and the traditional calculations gave the same strength under concentric load. For the traditional calculations, the force in each bolt group was determined as $P/2 \pm M/d$ where d is the depth of the wide flange (Tamboli, 2016). This equation assumes that shear in the bolts is the only force at the faying surface between the column flange and the splice plate. With explicit modeling of the connection in IDEA StatiCa, contact stress is observed at the faying surfaces (Figure 10.14) which does not directly increase capacity (since friction at contact surfaces is ignored in IDEA StatiCa) but shifts the lever arm resisting the moment outward and reduces the shear in the bolts.

Variation of maximum applied axial load with applied major-axis moment for the connection with 1/2 in. thick splice plates, no bearing contact, and bolt gage $g = 8$ in. is shown in Figure 10.15. These results confirm that IDEA StatiCa matches well with the traditional calculations over the entire range of applied axial load and bending moment for this connection.

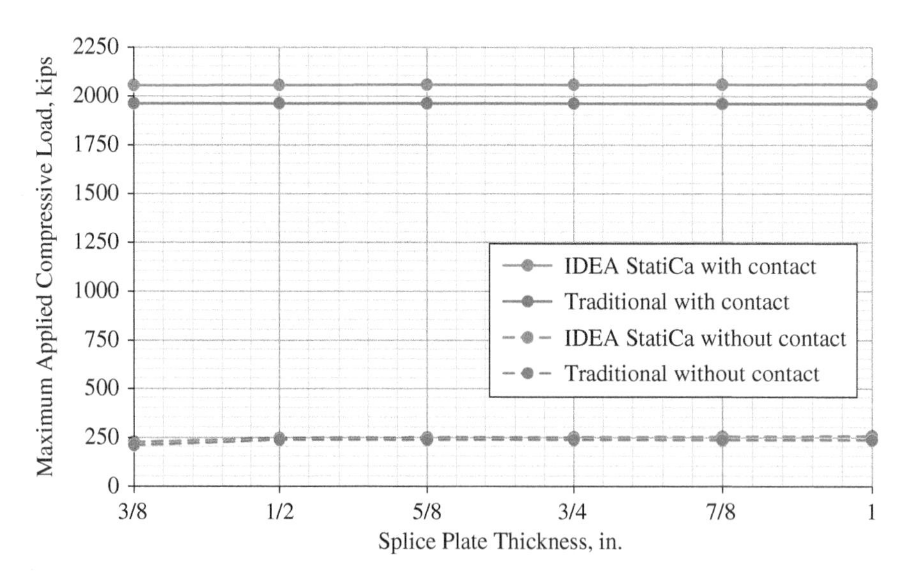

Figure 10.12 Maximum applied compressive load vs. splice plate thickness for connection with concurrent major-axis flexure.

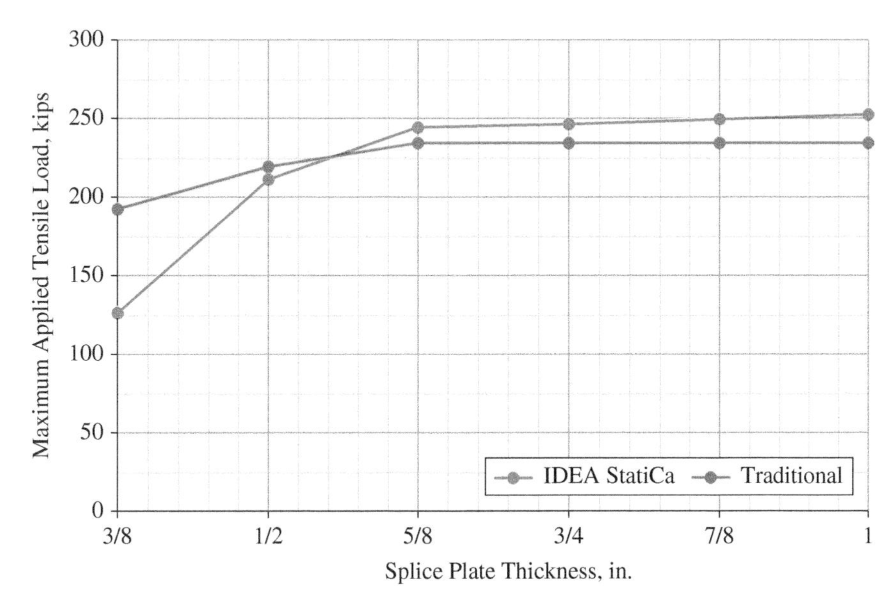

Figure 10.13 Maximum applied tensile load vs. splice plate thickness for connection with concurrent major-axis flexure.

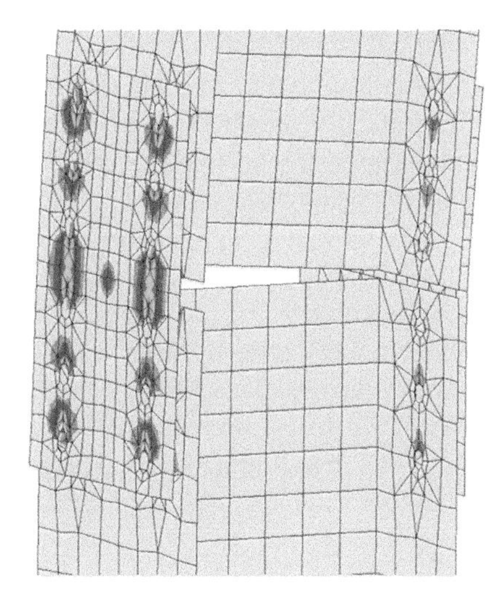

Figure 10.14 Stress in contacts at 212 kips applied tensile load and 1000 kip-in applied major-axis moment for connection with 1/2 in. thick splice plate (deformation scale factor = 10).

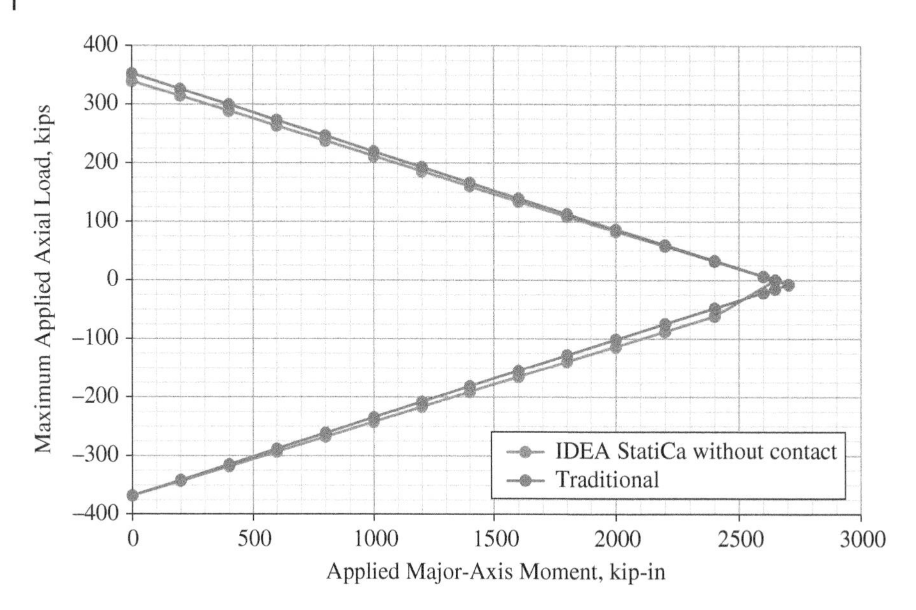

Figure 10.15 Maximum applied axial load vs. applied major-axis moment (compression negative).

10.5 Summary

This study compared the design of bolted wide flange splice connections using traditional calculation methods used in US practice and IDEA StatiCa. Key observations from the study include:

- The available strength obtained from IDEA StatiCa agrees well with the traditional calculations.
- One of the greatest differences in strength were for connections where tearout controlled the strength of some bolts. IDEA StatiCa reached 100% utilization of the bolts controlled by tearout while other bolts did not reach 100% utilization, resulting in conservative comparisons to the traditional calculations, which allow the strength of all bolts in a concentrically loaded bolt group to be obtained simultaneously.
- IDEA StatiCa gives somewhat higher strengths than the traditional calculations when block shear rupture controls.
- IDEA StatiCa correctly identified all connections in this study with undeveloped filler plates and subsequently applied the appropriate bolt shear or slip strength reductions defined in the AISC *Specification* (AISC, 2016). However, the algorithm in IDEA StatiCa for identifying undeveloped filler plates does not cover all cases and engineering judgment is required in non-standard cases to ensure that the strength results are applied when appropriate.

References

AISC (2016). *Specification for Structural Steel Buildings*. American Institute of Steel Construction, Chicago.

AISC (2017). *Steel Construction Manual*, 15th Edition. American Institute of Steel Construction, Chicago.

Muir, L. (2015). Bear It and Grin. *Modern Steel Construction*, December.

Tamboli, A. (2016). *Handbook of Structural Steel Connection Design and Details*, 3rd Edition. McGraw-Hill, New York.

Teh, L.H. and Deierlein, G.G. (2017). Effective Shear Plane Model for Tearout and Block Shear Failure of Bolted Connections. *AISC Engineering Journal*, 54(3), 181–194.

11

Temporary Splice Connection

11.1 Introduction

A comparison between the component-based finite element method (CBFEM) and traditional calculation methods used in US practice for the design of a temporary splice connection (Figure 11.1 and Figure 11.2) is presented in this chapter. The connection is intended to temporarily support an upper column above a lower column while the permanent welded splice connection between the two members is made. The columns are built-up box members with outside dimensions of 32 in. square and 2.5 in. thick walls. Lugs are fillet welded near each corner of both the upper and lower columns, then two strap plates are bolted to each pair (upper and lower) of the lugs. All the plate is ASTM A572 Gr. 50, all bolts are 7/8 in. diameter A325 in standard holes (threads not excluded from the shear plane), and all the weld material is E70XX. Loading on the upper column consists of combined axial compression, shear in two directions, biaxial bending moment, and torsion.

There are no established traditional calculation methods for this connection. The objective of this study is to describe how an engineer might approach the problem using traditional calculations, the limitations they may encounter using the traditional calculations, and how they might use traditional calculations to gain trust in CBFEM results.

The traditional calculations in this work are based upon the requirements for load and resistance factor design (LRFD) in the AISC *Specification* (AISC, 2016). The CBFEM results were obtained from IDEA StatiCa Version 21.1. The model of the connection is shown in Figure 11.3. Contact bearing between the upper and lower columns is ignored and the bevel in the upper column is not modeled in IDEA StatiCa.

The load path for this connection initiates in the upper column. Loads are transferred through the upper fillet welds to the upper lug plates, then through the upper bolt groups to the strap plates, then through the lower bolt groups to the lower lugs, then through the lower fillet welds to the lower column. For the purposes of this study, the columns are assumed to have adequate strength, therefore evaluation of this connection involves a check of each of the following components:

- upper fillet welds
- upper lug plates
- upper bolt groups
- strap plates
- lower bolt groups

Steel Connection Design by Inelastic Analysis: Verification Examples per AISC Specification, First Edition.
Mark D. Denavit, Ali Nassiri, Mustafa Mahamid, Martin Vild, Halil Sezen, and František Wald.

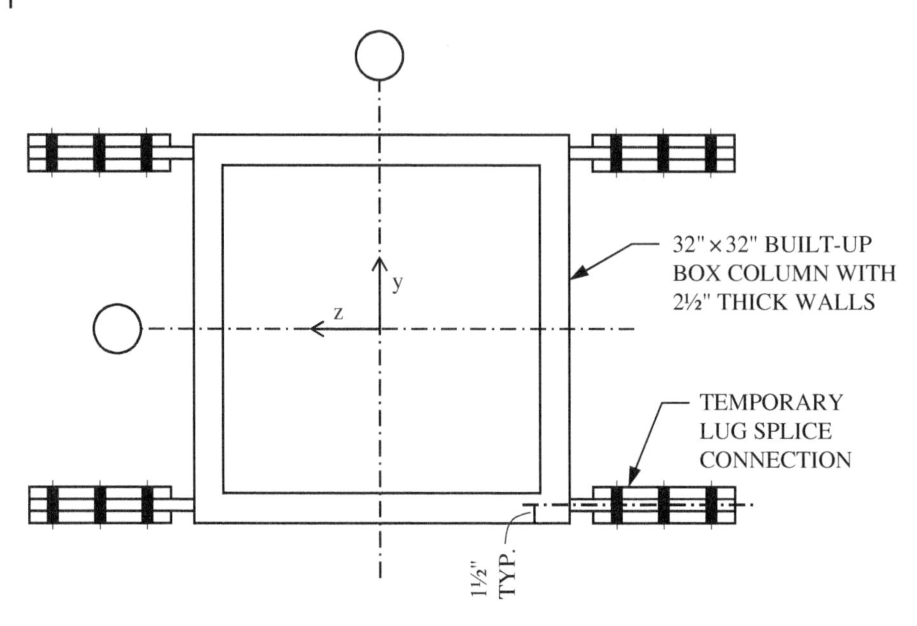

Figure 11.1 Schematic plan view of column and temporary splice connection investigated in this chapter.

- lower lug plates
- lower fillet welds.

The loading condition dictates which limit states apply to each of these components. The complex combined loading applied to the upper column makes evaluation using traditional calculations difficult. While IDEA StatiCa can handle the general loading condition without difficulty, simplified loading conditions will be examined as points of comparison, to better understand the behavior of the connection, and to build trust in the analysis results.

For each type of load, traditional calculations will be evaluated first, essentially forming a hypothesis of the behavior and strength of the connection. Then IDEA StatiCa analyses are performed to test the hypothesis. Agreement between the traditional calculations and IDEA StatiCa results confirms the hypothesis and increases confidence in both methods. Disagreement between the traditional calculations and IDEA StatiCa requires further investigation.

11.2 Axial Load

To approach evaluation of this connection by hand, a simplified model of the connection upon which hand calculations can be performed must be developed. When subjected to axial compression, each lug splice connection can be reasonably simplified to a two-dimensional beam model as shown in Figure 11.4. Hinges are included in the model at a distance "x" from the face of the column to make the model statically determinant.

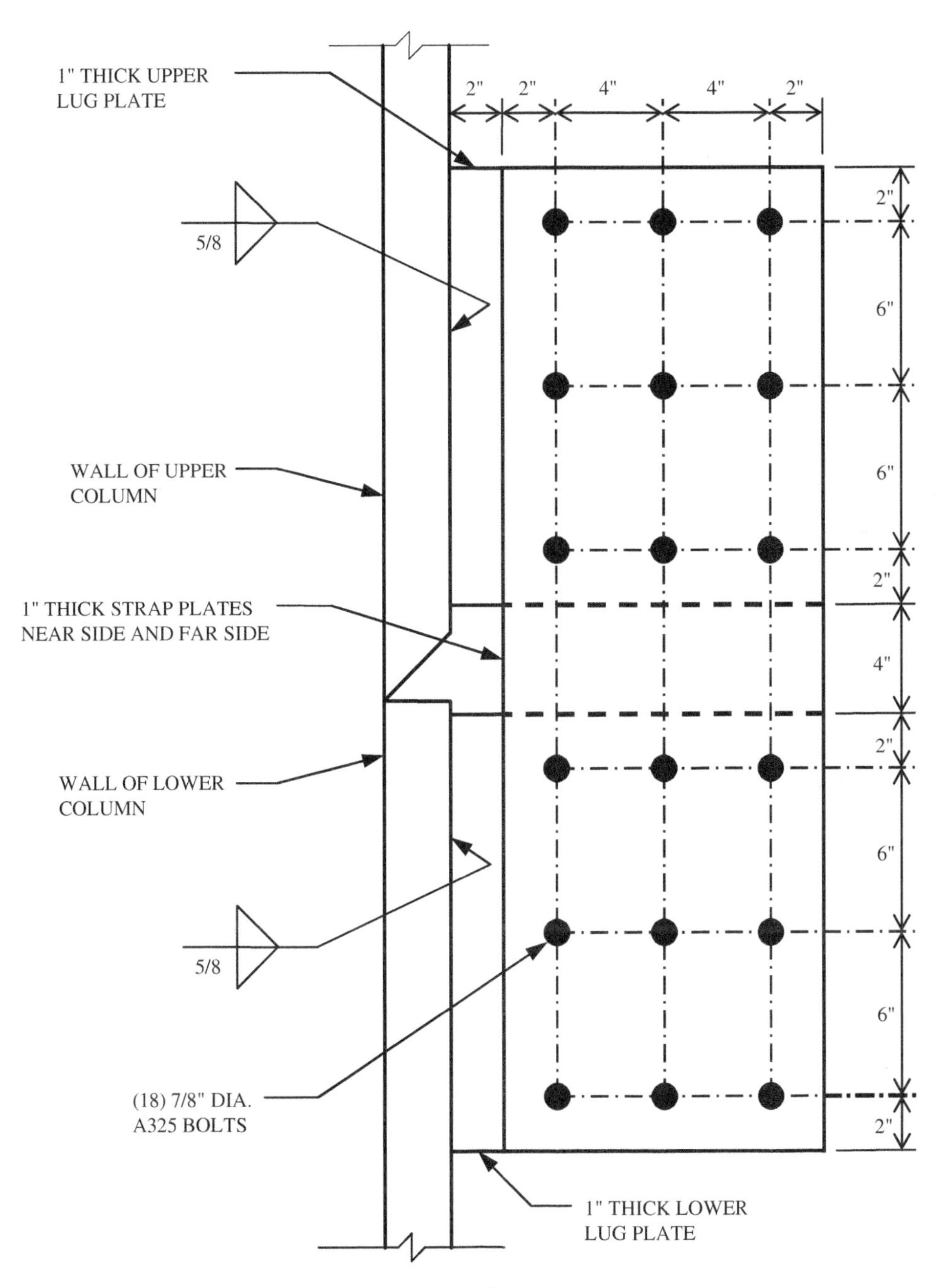

Figure 11.2 Schematic detail of temporary lug splice connection investigated in this chapter.

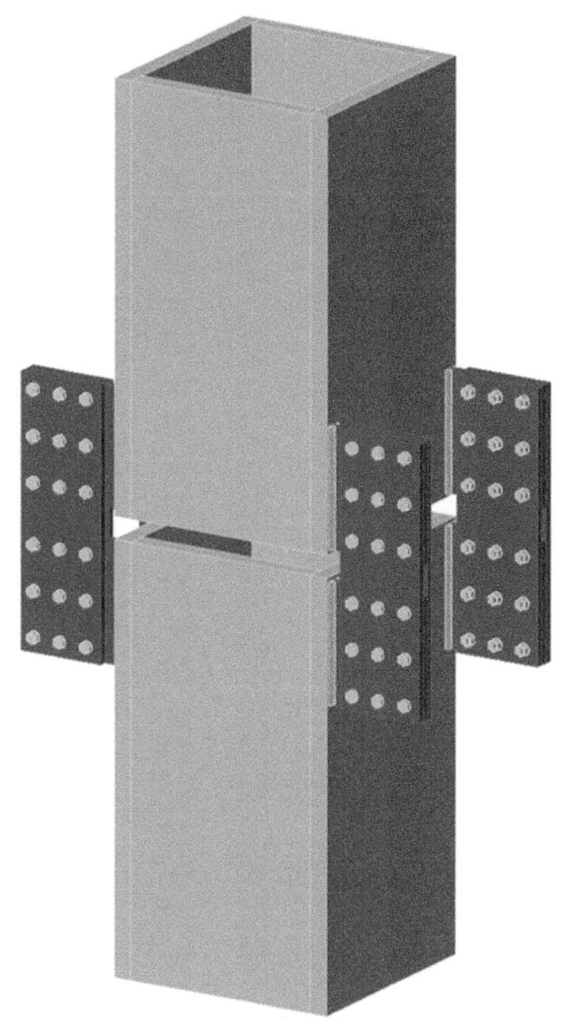

Figure 11.3 Temporary splice connection modeled in IDEA StatiCa.

With this model, the required strength of each component can be computed, and design checks can be performed, starting with the welds, lug plate material adjacent to the welds, and the bolts. Both the welds and the bolt groups are eccentrically loaded. The strength of the welds can be determined as a function of x using Table 8-4 of the AISC *Manual* (AISC, 2017). The strength of the lug plate material adjacent to the weld is controlled by shear and flexural yielding and can be evaluated using the following interaction equation based on Drucker (1956).

$$\left(\frac{V_u}{\phi V_n}\right)^4 + \frac{M_u}{\phi M_n} \leq 1$$

where V_u is the required shear strength of the lug plate, equal to one-fourth of the compression load applied to the upper column; ϕV_n is the design shear strength of the lug plate, equal to 480 kips; M_u is

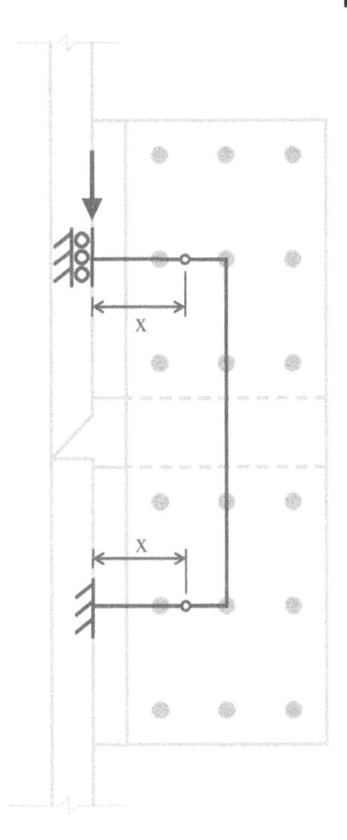

Figure 11.4 Simplified model of lug splice connection for axial loads.

the required flexural strength of the lug plate, equal to $V_u x$; and ϕM_n is the design flexural strength of the lug plate, equal to 2,880 kip-in.

The strength of the bolt group can be determined as a function of x using Tables 7-10 and 7-11 of the AISC *Manual* (AISC, 2017). Interpolation between these tables is necessary since the bolts are spaced at 4 in. horizontally. Note that the design shear strength of an individual bolt in this connection is 48.7 kips for the governing limit state of bolt shear rupture (bearing and tearout do not control for this connection). The maximum permitted vertical load in each lug for each of the limit states is plotted in Figure 11.5.

The "actual" location of the hinges is unknown and must be assumed. By the lower bound theorem of limit analysis, if a distribution of forces within a connection can be found, which is in equilibrium with the external load and which satisfies the limit states, then the externally applied load is less than or at most equal to the load that would cause connection failure (Tamboli, 2016). Therefore, any assumption of the location of the hinge will result in a safe design. The most favorable assumed location is approximately $x = 5$ in., where the strength of the welds and bolt group both equal approximately 360 kips. To complete the design, other limits states, including shear rupture of the lug plate and those associated with the strap plates, need to be evaluated for this load. However, these other limit states do not control, thus the maximum permitted applied compression load on the column is 4×360 kips = 1,440 kips.

With a hypothesis of the behavior and strength of the connection under axial load in place, the connection can be analyzed using IDEA StatiCa to evaluate the hypothesis. The maximum permitted axial compression load per IDEA StatiCa is 1,324 kips. This value was determined iteratively by adjusting the applied load input to a value that the program deems safe but if increased by a small amount

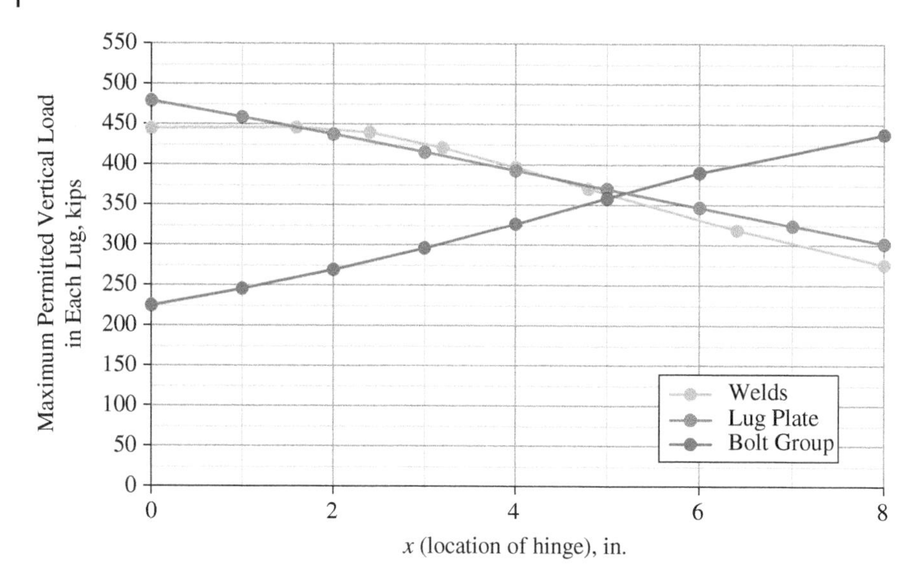

Figure 11.5 Design strength for select limit states as a function of the location of the hinge.

(e.g. 1 kip) the program would deem unsafe. The strength of the welds and bolts controls with both at 100% utilization in IDEA StatiCa.

The behavior of the connection as observed in the IDEA StatiCa results is consistent with the behavior assumed in the traditional calculations. The deformed shape and plastic strain results (Figure 11.6) show in-plane bending of the lug splice connections and weld groups. The bolt forces (Figure 11.7) show in-plane bending of the bolt groups. The strength per IDEA StatiCa is 8% lower than estimated by traditional calculations, a relatively close comparison that is consistent with previous investigations of eccentrically loaded bolt and weld groups described in Chapter 7.

The close agreement between traditional calculations and IDEA StatiCa gives confidence to both results. However, further exploration of the IDEA StatiCa results can bring additional confidence. A buckling analysis can be performed to confirm the appropriateness of ignoring geometric nonlinearity (i.e. P-Δ effects). The buckling factor for this connection at the maximum permitted axial compression load is 19.56. The buckling factor is the ratio of load at which elastic buckling occurs to the applied load, a value this high indicates geometric nonlinearity is negligible. The maximum permitted applied tension load was found to be nearly equal to the compression load, confirming symmetric behavior, as would be expected from the model used in the traditional calculations.

11.3 Bending Moments

When the upper member is subjected to bending moments, it is expected that the behavior and strength of each individual lug splice connection will be similar to the axial load case. Accordingly, for the traditional calculations, the moment strength for bending about the z-axis of the member can be computed as twice the axial strength of an individual lug splice connection times the lever arm between pairs of lugs

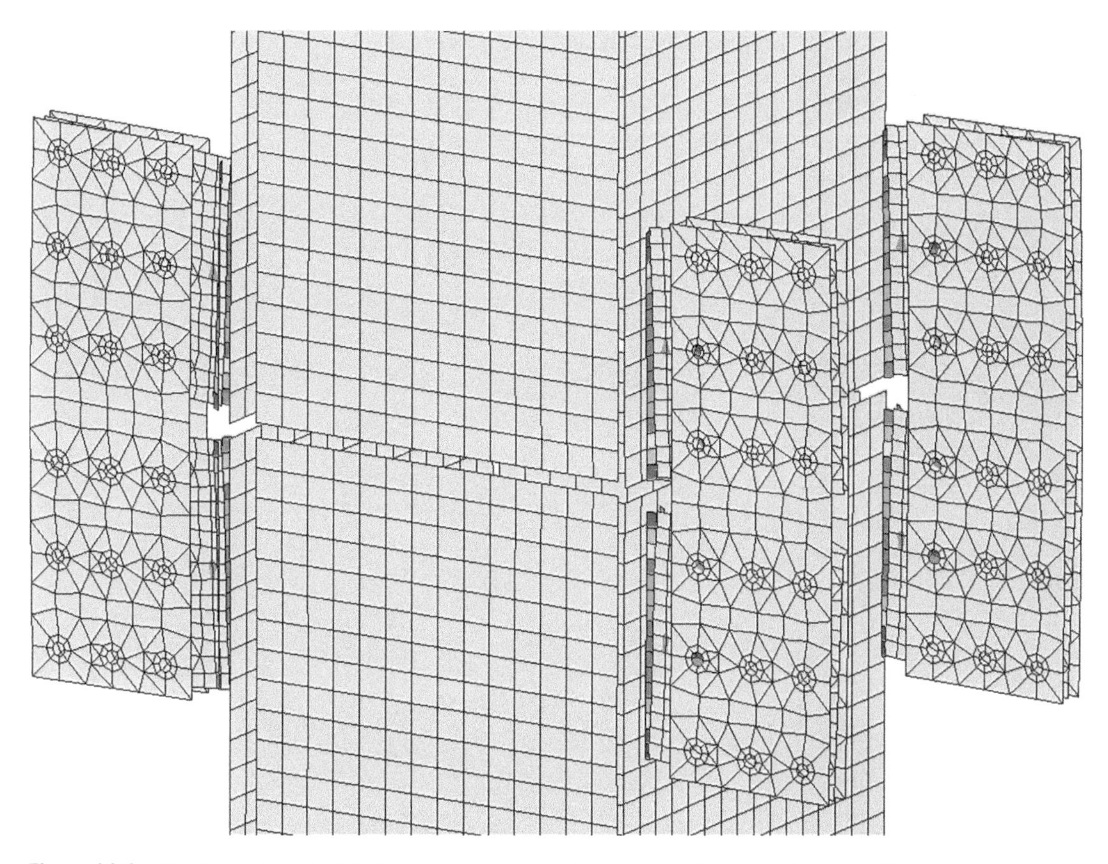

Figure 11.6 Plastic strain at 1,324 kips applied compression (deformation scale factor = 10).

(i.e. 2×360 kips $\times$ 29 in. = 20,880 kip-in.). Similarly, the moment strength for bending about the *y*-axis of the member can be computed as twice the axial strength of an individual lug splice connection times the lever arm between assumed hinge locations (i.e. 2×360 kips $\times$ 39 in. = 28,080 kip-in.).

Using the IDEA StatiCa results for axial compression, the moment strength for bending about the *z*-axis is $2 \times (1{,}324$ kips/4$) \times 29$ in. = 19,200 kip-in. and the moment strength for bending about the *y*-axis is $2 \times (1{,}324$ kips/4$) \times 39$ in. = 25,800 kip-in. The maximum permitted applied bending moments per IDEA StatiCa are 18,810 kip-in. and 25,065 kip-in. for bending about the *z*-axis and *y*-axis, respectively. These values were determined iteratively as described previously. Again, there is close agreement between the traditional calculations and the IDEA StatiCa results, indicating that the assumed behavior is accurate.

To further explore and confirm the assumed relationship between axial load and bending moment, the interaction strength is evaluated using IDEA StatiCa. Based on the assumed behavior, the interaction should be linear with each increment of axial load reducing the moment strength by a constant amount. The interaction strength per IDEA StatiCa is plotted in Figure 11.8. As expected, the interaction relationship between axial load and bending about the *z*-axis is linear. The interaction relationship between axial load and bending about the *y*-axis is nearly linear. The minor deviation from linearity in the interaction for

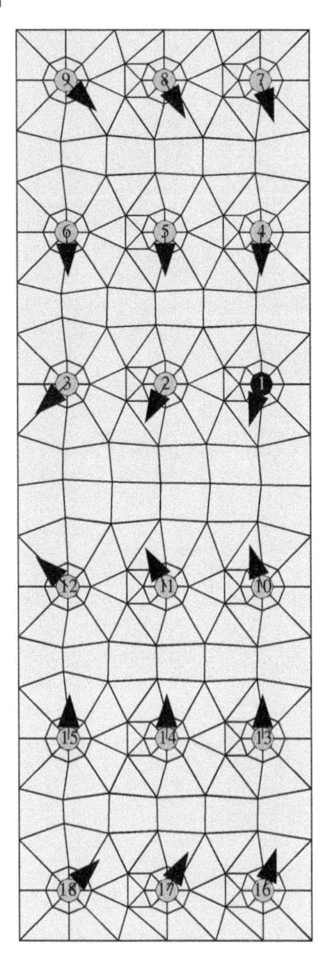

Figure 11.7 Bolt forces in strap plate at 1,324 kips applied compression.

bending about the y-axis could be investigated further, but some differences between simplified assumed behavior and the results of IDEA StatiCa should be expected.

11.4 Shear Along the z-Axis

Evaluation of the connection when subjected to shear along the z-axis requires a different simplified model of behavior. The two-dimensional beam model shown in Figure 11.9 will be used for this evaluation. A hinge, representative of a point of zero moment, is included at mid-height of the strap plates.

As before, when evaluating axial loads, the welds, lug plate material adjacent to the welds, and bolt groups will be evaluated first. The welds can be evaluated using Table 8-5 of the AISC *Manual* (AISC, 2017). Using an interpolated value of C, the maximum shear for an individual lug splice is determined as 218 kips.

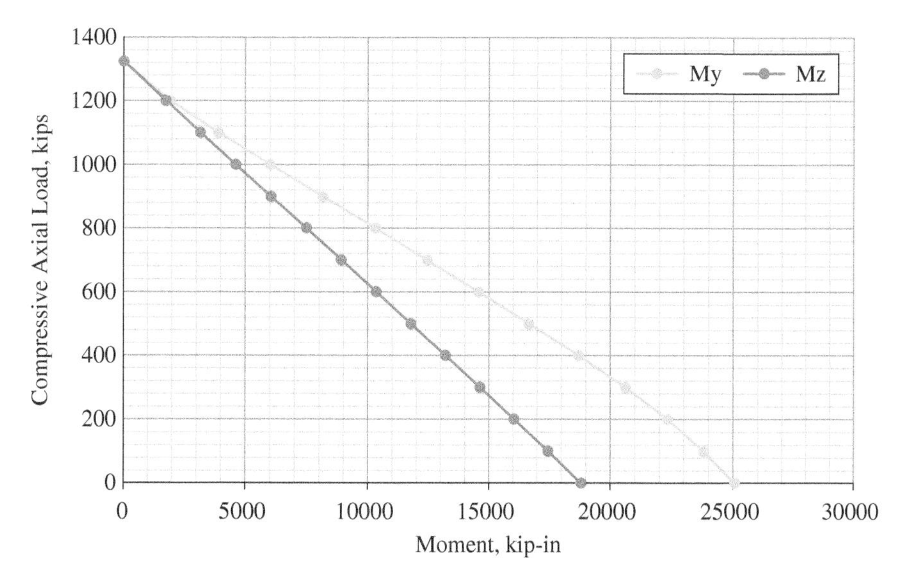

Figure 11.8 Axial compression vs. bending moment interaction strength.

Figure 11.9 Simplified model of lug splice connection for shear along the z-axis.

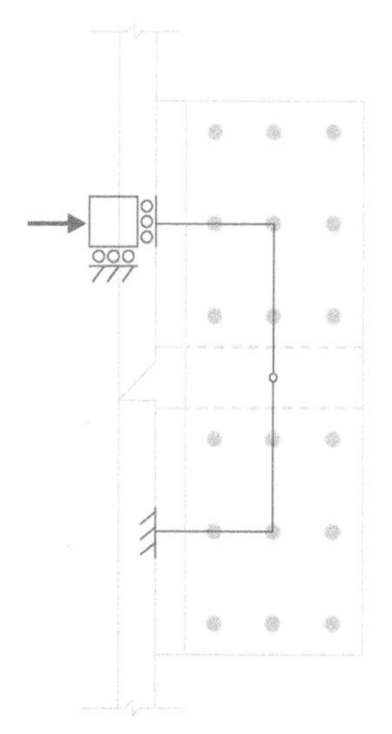

The strength of the lug plate material adjacent to the welds is controlled by axial and flexural yielding and can be evaluated using the following interaction equation based on plastic stress distribution.

$$\left(\frac{P_u}{\phi P_n}\right)^2 + \frac{M_u}{\phi M_n} \leq 1$$

where P_u is the required axial strength of the lug plate, equal to one-fourth of the shear load applied to the upper column; ϕP_n is the design axial strength of the lug plate, equal to 720 kips; M_u is the required flexural strength of the lug plate, equal to $P_u \times (10\text{ in.})$; and ϕM_n is the design flexural strength of the lug plate, equal to 2,880 kip-in. Evaluating the interaction equation for the strength of the weld (i.e. $P_u = 218$ kips) results in a value less than 1, indicating that the strength of the lug plate material adjacent to the welds does not control.

The strength of the bolt group can be determined using Table 7-11 of the AISC *Manual* (AISC, 2017). Using an interpolated value of C, the maximum shear for an individual lug splice is calculated as 186 kips, which controls among the limit states evaluated thus far. To complete the design, other limits states, including tensile rupture of the lug plate and those associated with the strap plates, need to be evaluated for this load. These limit states are found not to control, thus the maximum permitted applied shear along the z-axis of the column is 4×186 kips $= 744$ kips.

The maximum permitted applied shear load along the z-axis per IDEA StatiCa is 694 kips. This value was determined iteratively as described previously. Note that shear was applied, such that the point of zero moment was located between the upper and lower columns. The strength of the bolts controlled in IDEA StatiCa.

As before, the behavior of the connection as observed in the IDEA StatiCa results is consistent with the behavior assumed in the traditional calculations. The deformed shape, plastic strain results, and bolt forces (Figure 11.10 and Figure 11.11) show in-plane bending of the lug splice connections, weld groups,

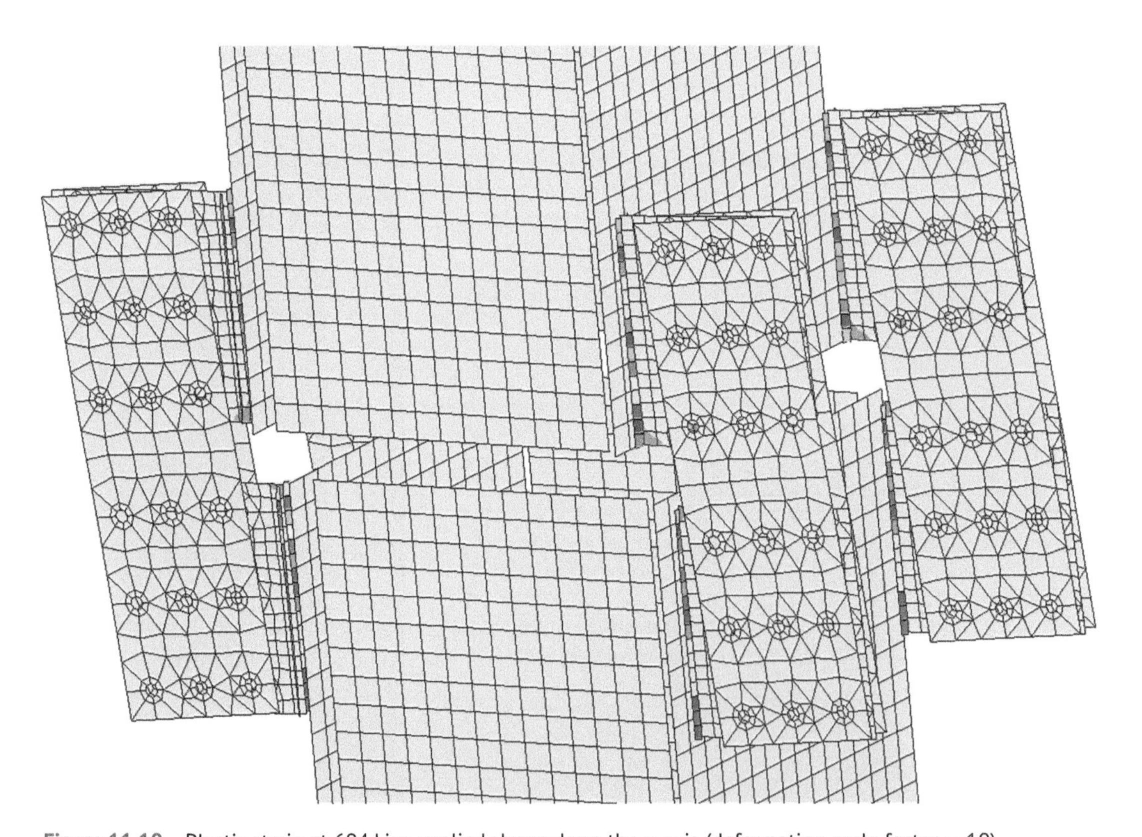

Figure 11.10 Plastic strain at 694 kips applied shear along the z-axis (deformation scale factor = 10).

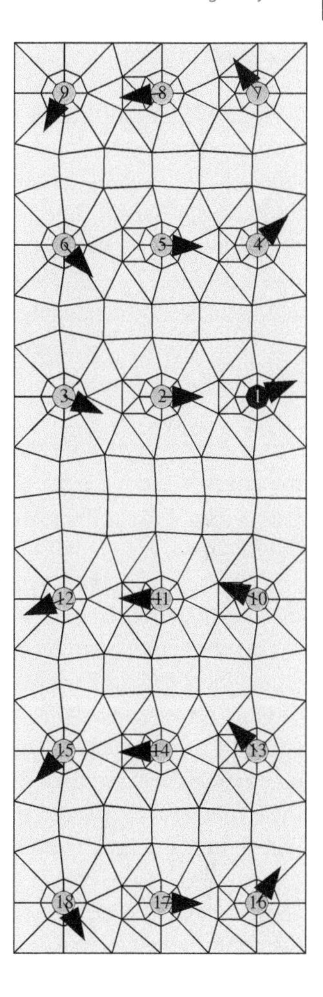

Figure 11.11 Bolt forces in strap plate at 694 kips applied shear along the *z*-axis.

and bolt groups that is consistent with the simplified model of behavior (see Figure 11.9). The strength per IDEA StatiCa is 7% lower than estimated by traditional calculations. These results confirm the hypothesis formed by the traditional calculations.

11.5 Shear Along the *y*-Axis

Evaluation of the connection when subjected to shear along the *y*-axis requires yet another simplified model of behavior. However, this model of behavior is less simple than the others. The beam model shown in Figure 11.9 will be used for this evaluation, but with the load applied perpendicular to the lug connection, resulting in out-of-plane moment, out-of-plane shear, and torsion in the lug plate. The AISC *Specification* (AISC, 2016) has few provisions for this complex loading condition. Recommendations developed by Dowswell (2019) will be used to gain some feel for the strength of the connection.

Dowswell presents the following interaction equation:

$$\left(\frac{T_u}{\phi T_n}\right)^2 + \left(\frac{V_u}{\phi V_n}\right)^4 + \frac{M_u}{\phi M_n} \leq 1$$

where T_u, V_u, and M_u are the required torque, shear, and flexural strengths and ϕT_n, ϕV_n, ϕM_n are the design torque, shear, and flexural strengths. Based on the model presented in Figure 11.9 and assuming no moment in either direction at the hinge, V_u is equal to one-fourth of the shear load applied to the upper column, $T_u = V_u \times (10\text{ in.})$, and $M_u = V_u \times (8\text{ in.})$. Assuming $\phi = 0.9$, ϕT_n can be calculated using equations recommended by Dowswell as

$$\phi T_n = \phi\left(\frac{0.6F_y dt^2}{2}\right)\left(1 + \frac{d}{2.4L}\right)$$

where F_y is the yield strength of the lug plate (50 ksi), d is the depth of the lug plate (16 in.), t is the thickness of the lug plate (1 in.), and L is the length of the lug plate (8 in. per the model presented in Figure 11.9). Using these values, ϕT_n = 396 kip-in. Using standard equations in the AISC *Specification* (AISC, 2016), ϕV_n = 480 kips and ϕM_n = 180 kip-in. With these design strengths, the maximum value of V_u that satisfies the interaction equation is V_u = 17.9 kips. Assuming lug plate yielding controls, the maximum permitted applied shear along the z-axis of the column is 4 × 17.9 kips = 71.6 kips.

This strength is part of the hypothesis which will be evaluated using IDEA StatiCa results. However, an engineer should have less confidence in this expected strength than those for the other loading conditions. Fewer potentially controlling limit states were evaluated, the out-of-plane behavior of the lug splice connection is likely not well approximated by Figure 11.9, and several assumptions were made in computing the strength of the lug plate. Nonetheless, it is helpful to form a hypothesis beforehand. Also, the hypothesis consists of more than just the strength result. The expected behavior, that the lug plate will control and that it will be subjected to combined torsion, out-of-plane shear, and out-of-plane bending moment, is also part of the hypothesis. While explicit modeling of the stiffness and strength of each component will overcome the uncertainties of the traditional calculations and yield a different strength result, the overall behavior should be consistent.

The maximum permitted applied shear load along the y-axis per IDEA StatiCa is 249 kips. This value was determined iteratively as described previously. Note that shear was applied such that the point of zero moment was located between the upper and lower columns. The strength per IDEA StatiCa is significantly greater than that estimated by the traditional calculations. An examination of the deformed shape of the connection (Figure 11.12) reveals the reason for this difference. The strap plates are relatively stiff, meaning that most of the twist and out-of-plane bending of the lug plates occurs over a much shorter length than is assumed in the simplified beam model of the connection (see Figure 11.9). Nonetheless, the strength of the connection is controlled by the plastic strain in the lug plate and the types of demands on the lug are consistent with the assumed behavior.

Recalculating the strength of the lug plate using the traditional calculations and a length, L = 2 in. instead of L = 8 in. produces a maximum permitted applied shear along the z-axis of the column equal to 227 kips, which is closer to the IDEA StatiCa results. However, it would be difficult to arrive at this value, let alone be confident in it, *a priori*.

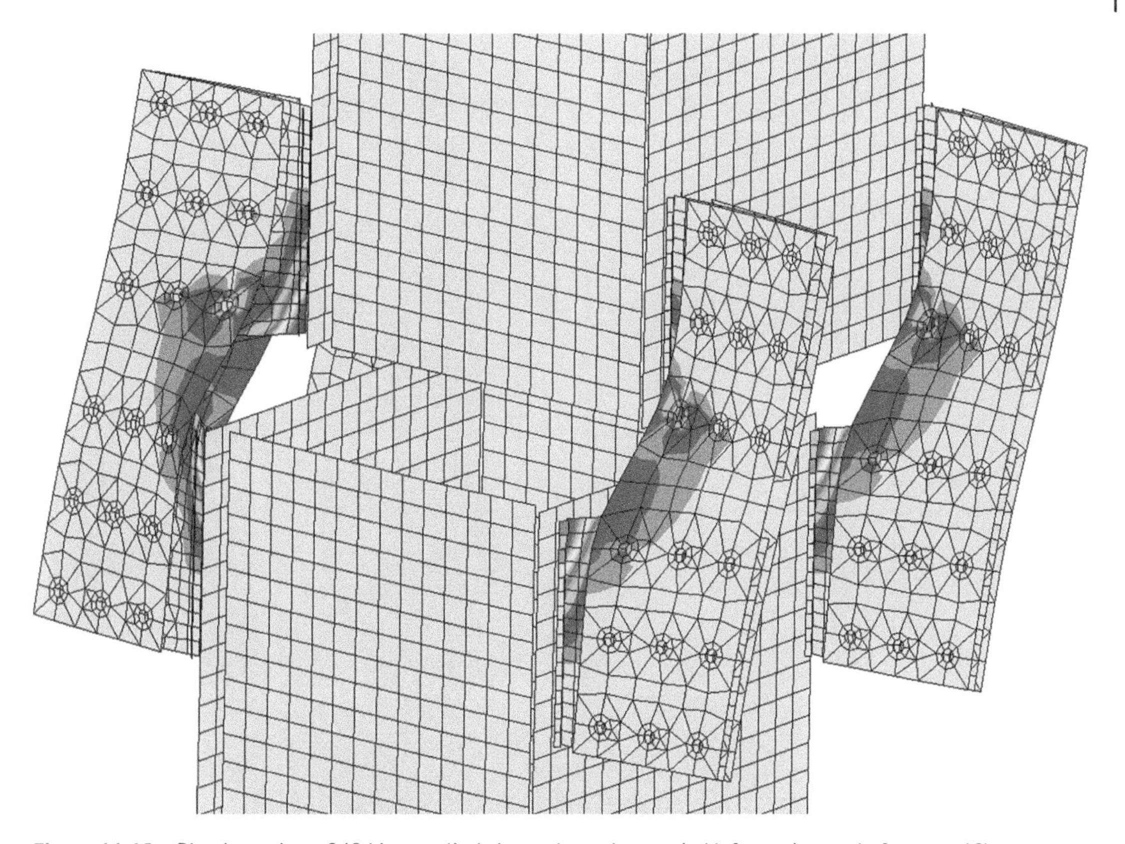

Figure 11.12 Plastic strain at 249 kips applied shear along the y-axis (deformation scale factor = 10).

11.6 Torsion

It is expected that applying torsion to the upper column will place demands on each individual lug splice that are similar to those they experience when the upper column is subjected to shear along the y-axis. Thus, like the flexural strength of the connection, the torsion strength could be estimated from the strength of the individual lugs and the geometry of the cross-section. For example, the torsion strength could be estimated as 4 times the strength of each lug times the distance from the centroid of the column to each lug. However, this may be an overly simplistic approximation. The lugs are near the corners of the column and not centered on the faces, thus twist of the column will impart in-plane demands on the lug in addition to the out-of-plane demands. Also, it is unclear where on the lug each lever arm should be measured. It is likely not possible to come to an accurate and confident result on the torsion strength of this connection without a better understanding and characterization of its behavior from more detailed analysis.

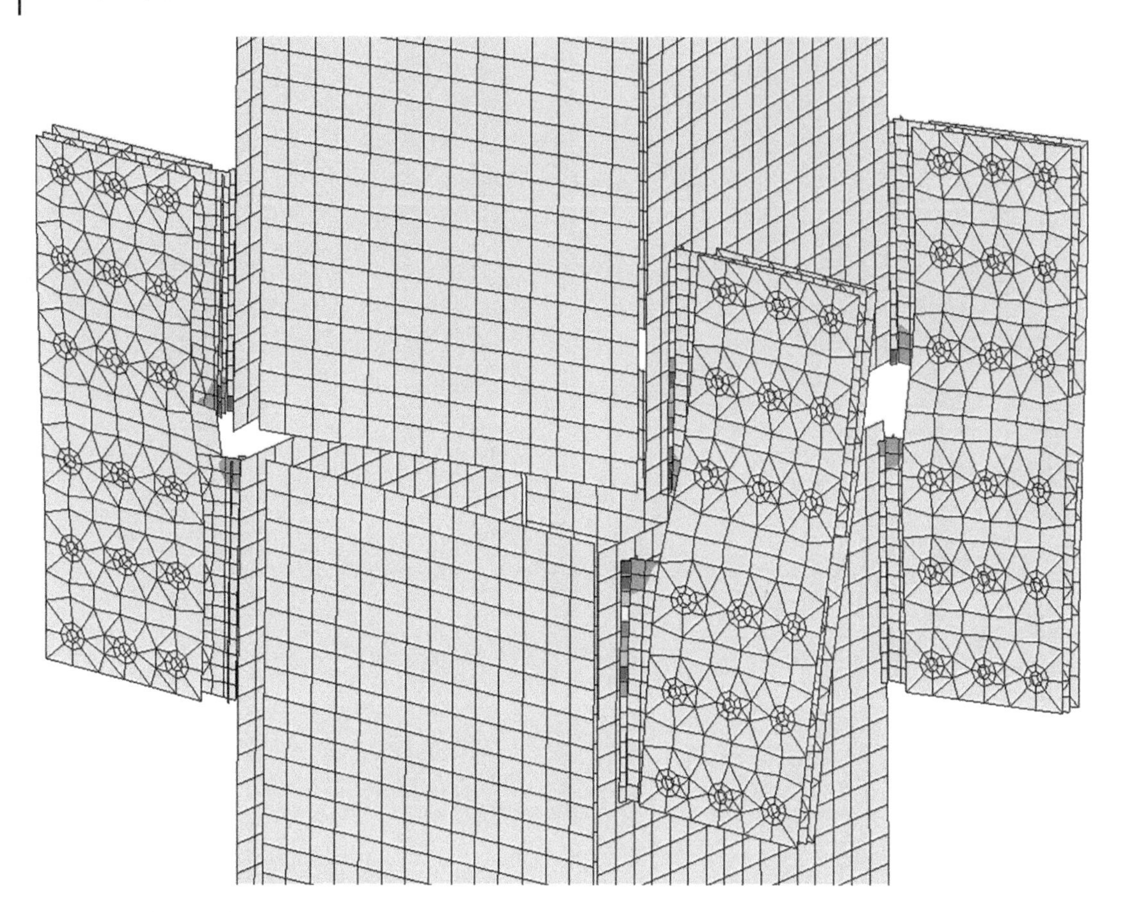

Figure 11.13 Plastic strain at 9,045 kip-in. applied torsion (deformation scale factor = 10).

The maximum permitted applied torsion per IDEA StatiCa is 9,045 kip-in. This value was determined iteratively as described previously. The weld utilization controls the strength. As seen in Figure 11.13, the deformed shape of each lug splice connection is similar when the column is subjected to shear along the *y*-axis (see Figure 11.12). However, there are differences in the behavior, most notably weld utilization controlling for the torsion case in lieu of the plastic strain limit in the lug plate controlling for the shear loaded case. While fewer comparisons can be made for this loading condition, the comparisons to other loading conditions have demonstrated that the model is well defined and capable of providing results that are in line with traditional methods.

11.7 Summary

Design or evaluation of structural connections requires good engineering judgment. Good engineering judgment requires understanding how the connection will behave. Developing this understanding is part of the process of evaluating novel connections that do not have established design procedures. In many cases, logical reasoning can be used to develop simplified models of behavior upon which traditional calculations can be based. However, there are limitations to this approach. More advanced tools, such as the CBFEM, are not subject to the same limitations and can be used to better understand and subsequently design a wide range of connection types. But care must be taken when defining the model and performing the analysis to ensure that the results are meaningful. Comparisons to simplified models of behavior and traditional calculations, such as those presented in this study, can help confirm the model is well defined and the analysis was performed correctly.

References

AISC (2016). *Specification for Structural Steel Buildings*. American Institute of Steel Construction, Chicago, Illinois.

AISC (2017). *Steel Construction Manual*, 15th Edition. American Institute of Steel Construction, Chicago, Illinois.

Dowswell, B. (2019). Torsion of Rectangular Connection Elements. *AISC Engineering Journal*, 56(2), 63–87.

Drucker, D.C. (1956). The Effect of Shear on the Plastic Bending of Beams. *Journal of Applied Mechanics*, 23(4), 509–514.

Tamboli, A. (2016). *Handbook of Structural Steel Connection Design and Details*, 3rd Edition. McGraw-Hill, New York.

12

Vertical Bracing Connections

12.1 Introduction

A braced frame system is a prevalent form of structural steel system, being economical to construct and yet simple to analyze. The economy comes from the inexpensive, nominally pinned connections between beams and columns. Bracing provides stability and resistance to lateral loads, possibly from diagonal steel members or a combination of steel and concrete or grout. In concentrically braced frames, beams and columns are designed under vertical load only, assuming the bracing system carries all the lateral loads. Depending on the applied loads, bracing members are used for gravity loads as well, similar to trusses.

In seismic design, the two major concentrically braced frames are ordinary concentrically braced frames (OCBF) and special concentrically braced frames (SCBF). In contrast, the buckling-restrained braced frame (BRBF) is considered a special type of concentrically braced frame (CBF). As illustrated in Figure 12.1, the braces within a frame are laid out such that the centerlines of the connecting elements (beams, braces, and columns) intersect at their common point – connection node, unlike the eccentrically braced frame in Figure 12.2 (Grusenmeyer, 2012).

Several configurations of braces are possible in CBFs. Commonly used layouts are V, inverted V, X, and K, some are shown in Figure 12.3. Each layout has advantages and disadvantages for the design, structural performance, fabrication, and construction. However, they all function in the same mode.

Ordinary concentrically braced frames (OCBFs) do not have extensive requirements regarding members or connections and are frequently used in low seismic risk areas. OCBF steel frame buildings originated in Chicago, and reinforced concrete frames originated in Germany and France – areas where earthquakes were not an engineering consideration (Reitherman, 2012). Accordingly, special concentrically or eccentrically braced frames were later developed with extensive design requirements and are frequently used in areas of high seismic risk. The major purpose of the concentrically or eccentrically braced design is to ensure adequate ductility in the connecting members.

The gusset plate connections are widely used in steel structures to transfer forces from a bracing member to the framing elements. A typical gusset plate connection in a braced steel frame is shown in Figure 12.4 (Sheng et al., 2002). Depending on the connection details, the gusset plate can be either bolted or welded to the diagonal bracing member and also to the main framing member. If the applied load to the bracing system is such that compression exists in the diagonal member, the compressive strength and stability of the gusset plate connection must be investigated. Due to the complexity of the connections, there are challenges in evaluating the compressive strength of gusset plates

Steel Connection Design by Inelastic Analysis: Verification Examples per AISC Specification, First Edition.
Mark D. Denavit, Ali Nassiri, Mustafa Mahamid, Martin Vild, Halil Sezen, and František Wald.

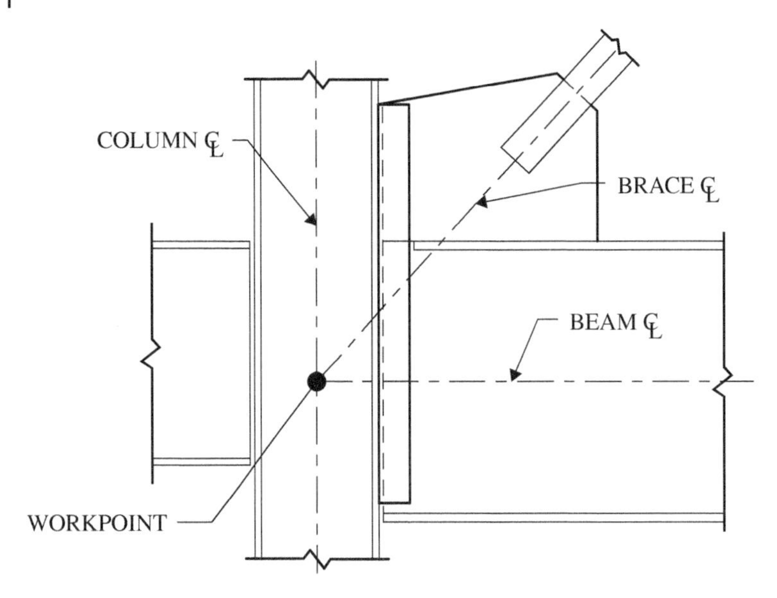

Figure 12.1 Intersection of member centerlines (Grusenmeyer, 2012).

(Bardot et al., 2017). Depending on the connectors, welds or bolts, several limit states must be checked against the applied load and its horizontal and vertical components.

12.2 Verification Examples

The design code guides the analysis and design of steel connections based on manuals such as AISC 360-16 in the United States, CSA S16-14 in Canada, and EN 1993-1-8 in Europe, which are primarily based on an analytical approach. They are easy to apply in practice but time-consuming when complex connection types exist (IDEA StatiCa, n.d.a). Hence, the component-based finite element method (CBFEM) can be used, which validates the results of any topology of connection based on selected design code and parameters by the user and also takes less time to perform an analysis, since it is a synergy of two well-known and established methods, the component method (CM) and the finite element method (IDEA StatiCa, n.d.b; n.d.c).

To evaluate the application of the CBFEM, which is programmed in IDEA StatiCa software, and further validate the results as per AISC *Specification* (AISC 360, AISC, 2016), verification of four vertical braced connections in the non-seismic zone is performed in Chapters 13–16.

Failure in the members and plates due to yielding and rupture limit states is measured based on the 5% plastic strain limit. Typically, the plastic strain starts at the bolt holes, in bolted connections; the stresses are based on von Mises stresses, a combination of normal and shear stresses. The utilization percentage of limit states associated with the welds and bolts in welded connection and bolted connection in CBFEM are based on the AISC *Specification* (AISC 360, AISC, 2016), however, attention should be made to the load distribution to bolts and welds when compared with the AISC 360 provisions.

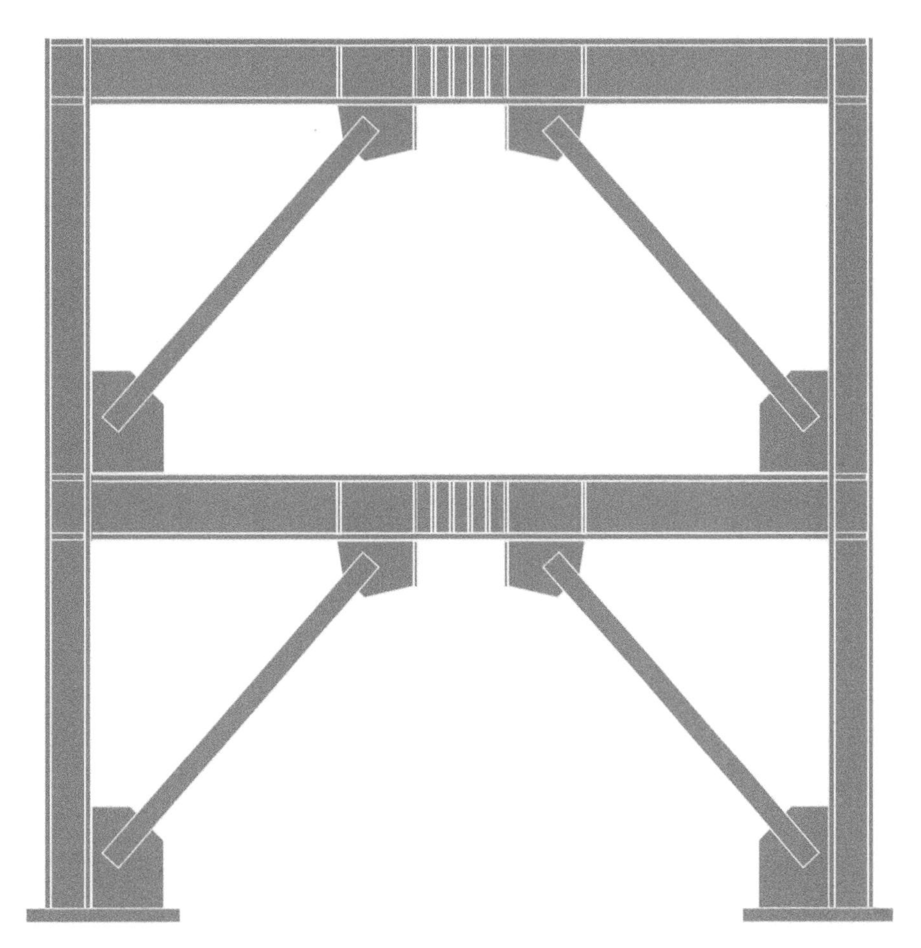

Figure 12.2 Eccentrically braced frame (EBF).

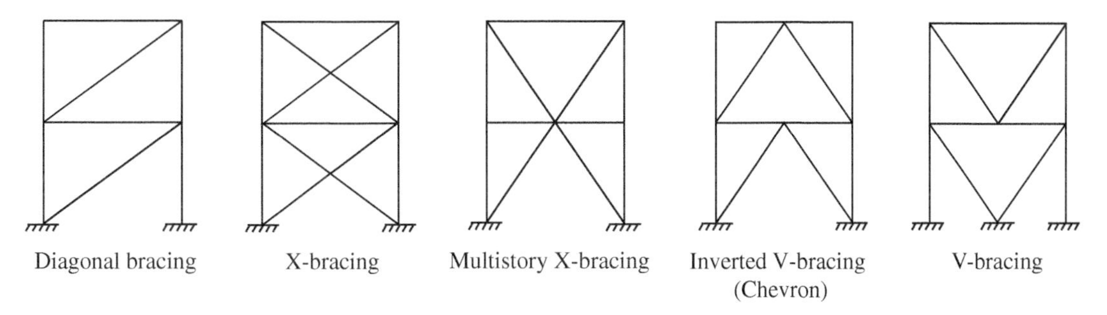

Figure 12.3 Commonly used brace layouts in concentrically braced frames (Sabelli et al., 2013).

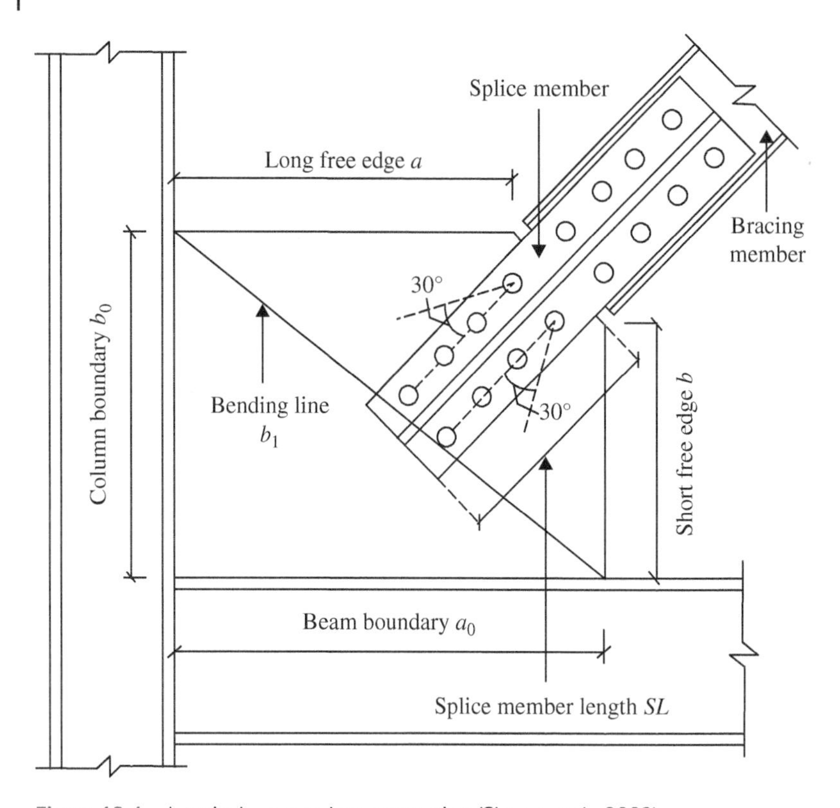

Figure 12.4 A typical gusset plate connection (Sheng et al., 2002).

The calculations performed as per AISC (AISC 360, AISC, 2016) are in accordance with the provisions for load and resistance factor design (LRFD) to obtain the results for the limit states of each connection. The brace limit states, the gusset plate limit states, and the connections limit states (weld and bolt) were found separately. In the case of CBFEM, these limit states were investigated individually in several iterations, and then the capacities were reported according to IDEA StatiCa Version 23.

The maximum permitted loads were determined iteratively by adjusting the applied load input to a value the program deems safe. However, the results would be unsafe if increased by a small amount. All obtained results for the investigated connections in CBFEM are based on EPS type analysis (stress/strain design) in IDEA StatiCa.

Chapters 13–16 present examples of vertical bracing connections, the examples presented are as follows:

1. HSS square braces welded to gusset plates in a concentrically braced frame – Chapter 13.
2. HSS circular braces welded to a gusset plate in a Chevron concentrically braced frame – Chapter 14.
3. Wide flange brace bolted to a gusset plate in a concentrically braced frame – Chapter 15.
4. Double angle brace bolted to a gusset plate in a concentrically braced frame – Chapter 16.

12.3 Connection Design Capabilities of Software for HSS

The selection of an efficient section shape and size for a diagonal brace is one of the crucial aspects in the analysis and design of concentrically braced frames. When used as a brace section, the hollow structural section (HSS) provides the highest brace strength per pound of steel (field welding required for installation) among several section shapes. However, bolting is hard when using an HSS member as the section is closed, especially when a circular HSS is used. Among several shapes of HSS available, square HSS braces are widely preferred. A wide flange section as a brace member is preferred when large axial loads are to be resisted in a braced frame, while the double angles section as a brace member is an efficient connection (double shear bolts). Though single-angle brace members are suitable for small axial tension loads, they are not the preferred section for a compression brace, due to their small buckling capacity.

For the vertical bracing connections investigated, two of the four verification examples presented have a square HSS as their brace member. In contrast, the others have a wide flange and double-angle sections.

Due to this distinction, it is essential to understand which software will properly design connections to the HSS shapes, as this design requires considering many limit states which are unique to HSS. While most connection software packages available can handle many of the typical wide-flange connection cases, they may be more limited when designing HSS connections. Figure 12.5 provides a summary of the capabilities of each software grouped by connection type (Manor, 2021).

As per observations from the tabulated data in Figure 12.5, among the nine leading software presented, IDEA StatiCa stands out as having all the capabilities regarding HSS connection design.

Software HSS Connection Capabilities									
	DESCON	GIZA	HSS CONNEX	IDEA STATICA	QNECT	RAM CONNECTION	RISA CONNECTION	SDS/2	SKYCIV
Round HSS	-	✓	✓	✓	✓	✓	✓	✓	✓
Rectangular HSS	✓	✓	✓	✓	✓	✓	✓	✓	✓
WF Beam at HSS Column - Shear[3] (single/double angle, WT, end plate, seat)	✓	✓[6]	-	✓	✓	✓	✓[11]	✓	✓
WF Beam at HSS Column - Moment	✓[2]	✓[2]	-	✓	-	✓[2]	✓[2]	✓[2]	✓
Longitudinal/Through Plate – Shear	✓[5]	✓	✓[1]	✓	✓[10]	✓	✓	✓	✓
Longitudinal Plate – Axial	✓	✓	✓[1]	✓	✓	✓	✓	✓	✓
Longitudinal Plate – Moment[8]	-	-	-	✓	-	-	-	-	-
Transverse Plate – Shear	-	-	✓[1]	✓	-	✓	✓	-	-
Transverse Plate – Axial	-	-	✓[1]	✓	-	-	-	-	-
Transverse Plate – Moment[8]	-	-		✓	-	-	-	-	-
HSS to HSS – Truss (T-, Y-, X- and K-connection)	-	-	✓	✓	-	✓	✓[13]	✓	-
HSS to HSS – Moment (T-, Y- and X-connection)	-	-	✓	✓	-	✓	✓[12]	-	-
Cap Plate -Axial	-		✓[1]	✓	-	✓	-	✓	-
Base Plate	✓[4]	✓[4]		✓	-	✓	✓	-	✓[14]
HSS Brace Connections		✓[7]	-	✓	✓[9]	✓[7]	✓[7]	✓[7]	-

1. Partial check of HSS wall only
2. Direct weld or flange plate only
3. Programs may not include all shear connection types listed
4. Only with brace connections
5. Through plate not included
6. Angles only
7. HSS brace and/or column; HSS is not an option for the beam
8. AISC includes design option for round HSS only
9. HSS for brace only
10. Through plate is not an option
11. Single angle and end plate only
12. In-plane moment only
13. T-connection only for rectangular HSS
14. Free module on website without login

Figure 12.5 Software HSS connection capabilities (Manor, 2021).

12.4 Summary

For the four presented non-seismic verification examples related to concentrically braced frames, the results of the limit states obtained for AISC approximately agree with the CBFEM results. Hence, the CBFEM can predict the actual behavior and several failure modes of the four braced frame connections. Also, for examples in Chapters 13 and 14, parametric studies were carried out not only to understand the governing parameters or limit states in detail, but also to get an idea of the capabilities of CBFEM in predicting the actual behavior of the connection subjected to the action of the design loads as per LRFD AISC 360-16 *Specification*.

As calculated for several locations, the weld capacity agrees in both AISC and CBFEM. Also, the bolt limit states, including bolt shear and bearing per AISC 360-16, agree with CBFEM. The plate limit states, including yielding, rupture in tension, and shear, are based on a 5% plastic strain limit and are found to agree for the case of presented connections.

Tension yielding and tension rupture in the brace agree as per AISC and CBFEM, with approximately a 10% average difference in the capacities. The block shear limit state can be observed in the gusset plate, the end plate, and the connecting brace section. It is not in other places, such as the brace angles, as the shear and tension rupture limit states of the angles precede block shear rupture.

AISC specifications require the limit state of prying action, for the example in Chapter 16, which is considered in CBFEM by the additional tension forces applied to the bolts. Beam web buckling, web crippling, and shear yielding would occur at high loads, and the model would not converge at such high loads; all other limit states would occur before these limit states. These limit states can be checked as per the AISC *Specification*, as shown in the Appendices of Chapters 13, 14, 15, and 16. The buckling limit state of the gusset plate was not observed as a limit state in AISC and CBFEM in the case of hot-rolled sections in steel connections, buckling is critical in cold-formed sections (IDEA StatiCa, n.d.a). According to a recent study by the Steel Tube Institute (STI), among nine leading software presented, IDEA StatiCa stands out as having most of the capabilities when it comes to HSS connection design.

References

AISC 360 (2016). *Specification for Structural Steel Buildings*. American Institute of Steel Construction, Chicago.

Dowswell, B. (2021). Analysis of the Shear Lag Factor for Slotted Rectangular HSS Members. *AISC Engineering Journal*, 58(3), 77–89.

Grusenmeyer, E. (2012). Design Comparison of Ordinary Concentric Brace Frames and Special Concentric Brace Frames for Seismic Lateral Force Resistance for Low Rise Buildings. Kansas State University. https://krex.k-state.edu/handle/2097/14986

IDEA StatiCa (n.d.a). IDEA StatiCa Support Center: FAQ. https://www.ideastatica.com/support-center-faq

IDEA StatiCa (n.d.b). Structural Design of Steel Connection and Joints. Steel. https://www.ideastatica.com/

IDEA StatiCa (n.d.c). Theoretical Background of Steel Connections in IDEA StatiCa. https://www.ideastatica.com/support-center-theoretical-backgrounds?product=steel&label=connection

IDEA StatiCa (n.d.d). Verifications in IDEA StatiCa. https://www.ideastatica.com/support-center-verifications

Manor, M. (2021, October). HSS Connection Design Software. Steel Tube Institute. https://steeltubeinstitute.org/resources/hss-connection-design-software/

Reitherman, R.K. (2012). *Earthquakes and Engineers: An International History*. American Society of Civil Engineers, Reston, VA.

Sabelli, R., Roeder, C.W., and Hajjar, J.F. (2013). Seismic Design of Steel Special Concentrically Braced Frame Systems: A Guide for Practicing Engineers. NEHRP Seismic Design Technical Brief No. 8, NIST GCR 13-917-24.

Sheng, N., Yam, C.H., and Iu, V.P. (2002). Analytical Investigation and the Design of the Compressive Strength of Steel Gusset Plate Connections. *Journal of Constructional Steel Research*, 58(11), 1473–1493. https://doi.org/10.1016/S0143-974X(01)00076-1

13

HSS Square Braces Welded to Gusset Plates in a Concentrically Braced Frame

13.1 Problem Description

This example consists of a connection at the beam-column connection in a concentrically braced frame such that square hollow structural section (HSS) braces are welded to a gusset plate. The study is prepared for the size of the brace, beam, column, geometry and thickness of gusset plates, bolts, and welds as obtained from the design. All components are designed according to AISC 360-16 *Specification* (AISC, 2016) and CBFEM analysis is performed using IDEA StatiCa Version 23.

13.2 Verification of Resistance as per AISC

This example uses the sections and dimensions shown in Figure 13.1, which are as follows. The brace is HSS 10×10×5/8 (ASTM A500 Gr. B), beam W21 × 83 (ASTM A992), column W12 × 170 (ASTM A992), ½ in. thick gusset plate (ASTM A572 Gr. 50), 7/8 in. ASTM A325-N bolts and ASTM E70XX weld. All section and material properties are obtained from the *Steel Construction Manual* (AISC, 2017).

The results of the analytical solution are represented by Table 13.1 for the different limit states shown. The limit states that should be considered for this connection and the capacities of the different limit states are shown in Table 13.1.

Based on obtained results, the governing limit state of this connection is weld shear strength at the brace-gusset plate followed by block shear rupture on the gusset plate, and then followed by tension rupture on the brace net section. For the given connection, the brace below the beam is subjected to an axial compression force, while the brace above the beam is subjected to an axial tension force, such that both load cases converged to 100% in CBFEM analysis.

13.3 Resistance by CBFEM

The details of the given connection as modeled in IDEA StatiCa are shown in Figure 13.2, which represents three different types of models that can be observed: solid model, transparent model, and wireframe model.

Steel Connection Design by Inelastic Analysis: Verification Examples per AISC Specification, First Edition.
Mark D. Denavit, Ali Nassiri, Mustafa Mahamid, Martin Vild, Halil Sezen, and František Wald.

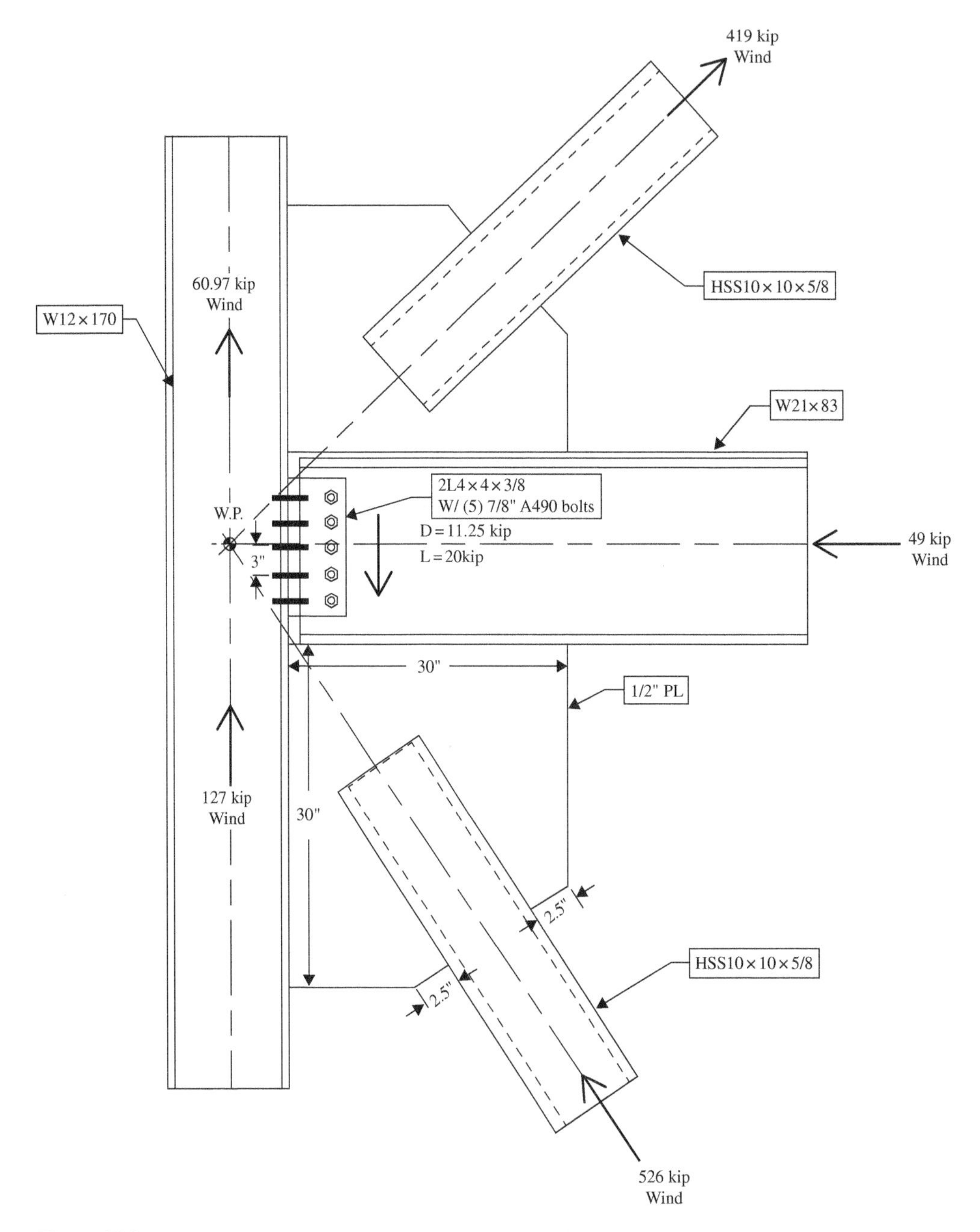

Figure 13.1 Brace connection at beam-column connection in a braced frame – geometry and design.

Table 13.1 Limit states checked using the AISC 360 provisions.

Limit state	AISC
Tensile yielding on the brace gross section	ϕR_n =*869.4* kips
Shear rupture of the brace wall	Required weld length based on this limit state = 8.7 in. provided length = 20 in. ϕR_n =1213 kips
Weld shear strength at the brace-gusset plate	Required weld length based on this limit state = 18.9 in. provided length = 20 in. ϕR_n =557 kips
Tensile rupture on the brace net section	ϕR_n =*716.6* kips
Block shear rupture on the gusset plate	ϕR_n =*693.8* kips
Tension yielding of gusset plate at Whitmore section	ϕR_n =*746* kips
Buckling of gusset plate at Whitmore section	ϕR_n =*746* kips
Gusset-to-Beam Connection – tension yielding and shear yielding of gusset plate and along the beam flange	Required weld size based on this limit state = 3.8/16 in. provided weld size = 5/16 in.
Gusset-to-Column Connection – tension yielding and shear yielding of gusset plate and along the beam flange	Required weld size based on this limit state = 2.4/16 in. provided weld size = 5/16 in.
Beam-to-Column Connection	ϕR_n =185 kips compared to beam reaction
Beam web yielding	ϕR_n =*859* kips
Beam web crippling	ϕR_n =*441.5* kips

The overall check of the connection is verified in CBFEM, as shown in Figure 13.3. The check shows that the connection works well according to the CBFEM for the applied load which conforms to the AISC provisions. It can be concluded that the CBFEM can predict the actual behavior and failure modes of the braced frame connections presented herein.

Failure in members and plates due to yielding and rupture limit states are measured based on the 5% plastic strain limit. Figure 13.3 shows that the plastic strain is 0.7%, which is less than the 5% plastic strain limit. The connection presented includes elements that are welded at braces and gusset plates, and others that are bolted, beam-to-column connection. It can be observed that the maximum weld check utilization is 84.9% (based on the AISC 360-16 *Specification*; AISC, 2016) which is less than 100% and therefore is found to be safe for the application of the calculated loads. The maximum weld check utilization is observed at welds connecting the bottom axial loaded brace to the gusset plate below the beam.

At the beam-to-column connection, all shear connection limit states, including bolt shear, bolt bearing, double angles limit states, such as shear yielding, shear rupture and block shear, and limit states in the beam web and the column flange agree in both AISC 360-16 *Specification* and CBFEM.

It is worth mentioning that CBFEM checks the bolts individually for bolt shear and bearing; the utilization ratio is based on a single bolt, whereas the utilization ratio as per AISC is based on the summation of the bearing capacities of all the bolts. Hence, this would result in the CBFEM values being safer and slightly more conservative than AISC values.

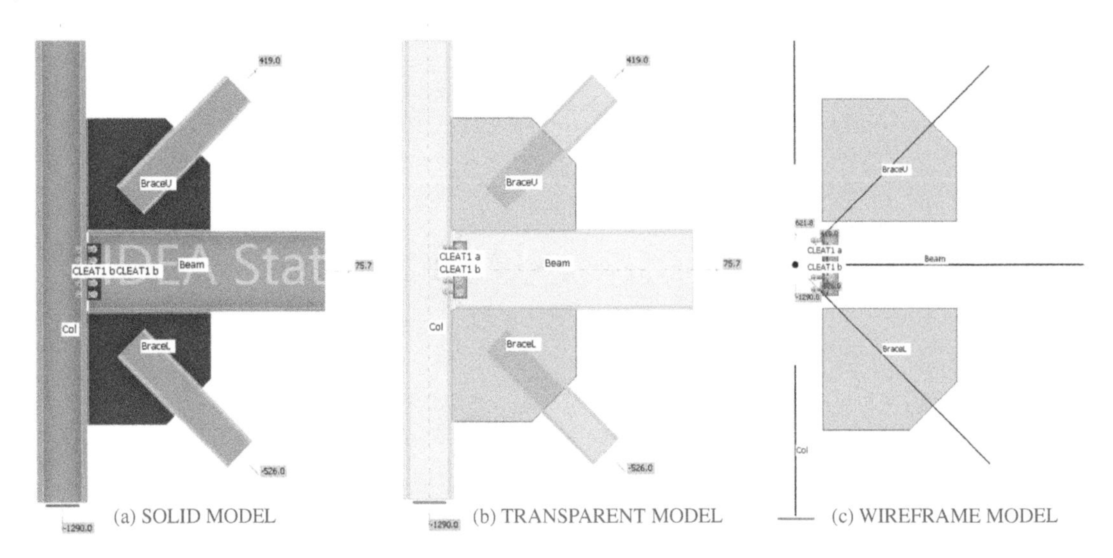

Figure 13.2 Models of given connection in IDEA StatiCa: (a) solid; (b) transparent; and (c) wireframe.

13.3.1 Limit States (AISC and CBFEM)

The results for limit states were obtained using the various limit states per the AISC procedures. These limit states were investigated individually as per CBFEM iteratively by adjusting the applied axial load input to a value that the program deems safe but if increased by a small amount (1 kip) until the results would deem it unsafe, and the capacities were then reported accordingly. All the results for this example are based on EPS type analysis (stress/strain design) in CBFEM. The brace limit states the gusset plate limit states, and the weld limit states were found separately.

The plastic strain for the overall connection, as shown in Figure 13.4, starts at the gusset-column connection; these stresses are based on von Mises stresses which are a combination of normal and shear stresses. Figure 13.5 shows the von Mises stress distribution for overall connection as per FEA – Equivalent stresses in CBFEM.

In CBFEM, the load distribution in the weld is derived from multi-point constraint (MPC), as shown in Figure 13.6, such that the stresses are calculated in the throat section.

Welds are modeled as elastoplastic elements in CBFEM to allow for plastic redistribution at the joints. The computation of the design strength of the weld in CBFEM is based on the equations provided in AISC 360-16-J2.4, but the weld angle, if required, is calculated between the longitudinal axis of the weld and the resultant force direction in the most stressed finite element of the weld. It is not calculated as a simple 2D resultant, rather, it is based on the FEA relations as mentioned in Figure 13.7. Therefore, the weld capacity computed by CBFEM is accurate when compared with AISC values.

The value of the weld capacity as per CBFEM is based on the critical element among several elements in the weld, as shown in Figure 13.8.

The first limit state that was investigated is the weld shear strength at the brace-gusset plate since it is found to be governing in both AISC and CBFEM. Von Mises stresses in the bottom gusset plate along with the equivalent stress values as computed by FEA are shown in Figure 13.9. Weld capacity computed

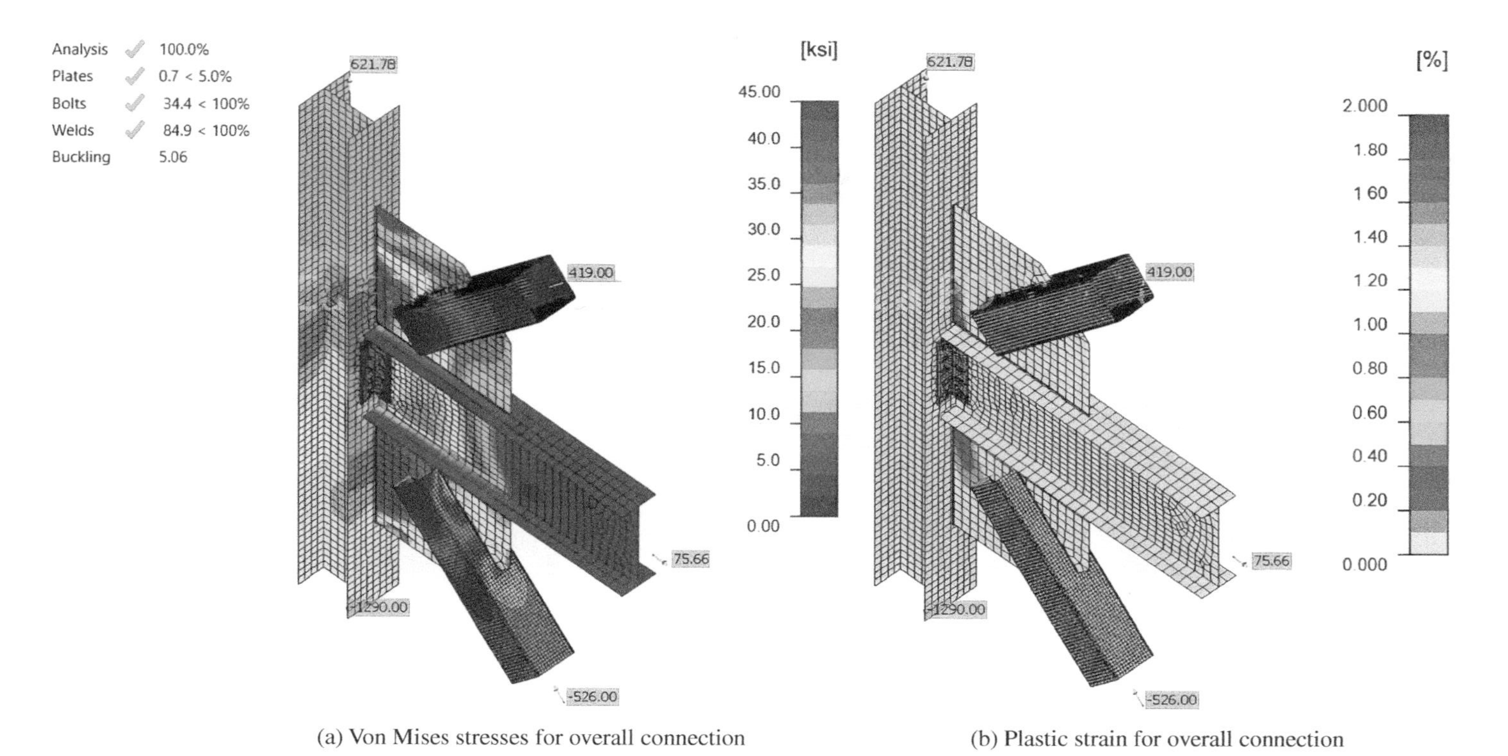

Figure 13.3 Overall solution of the connection in CBFEM: (a) Von Mises stresses; (b) plastic strains.

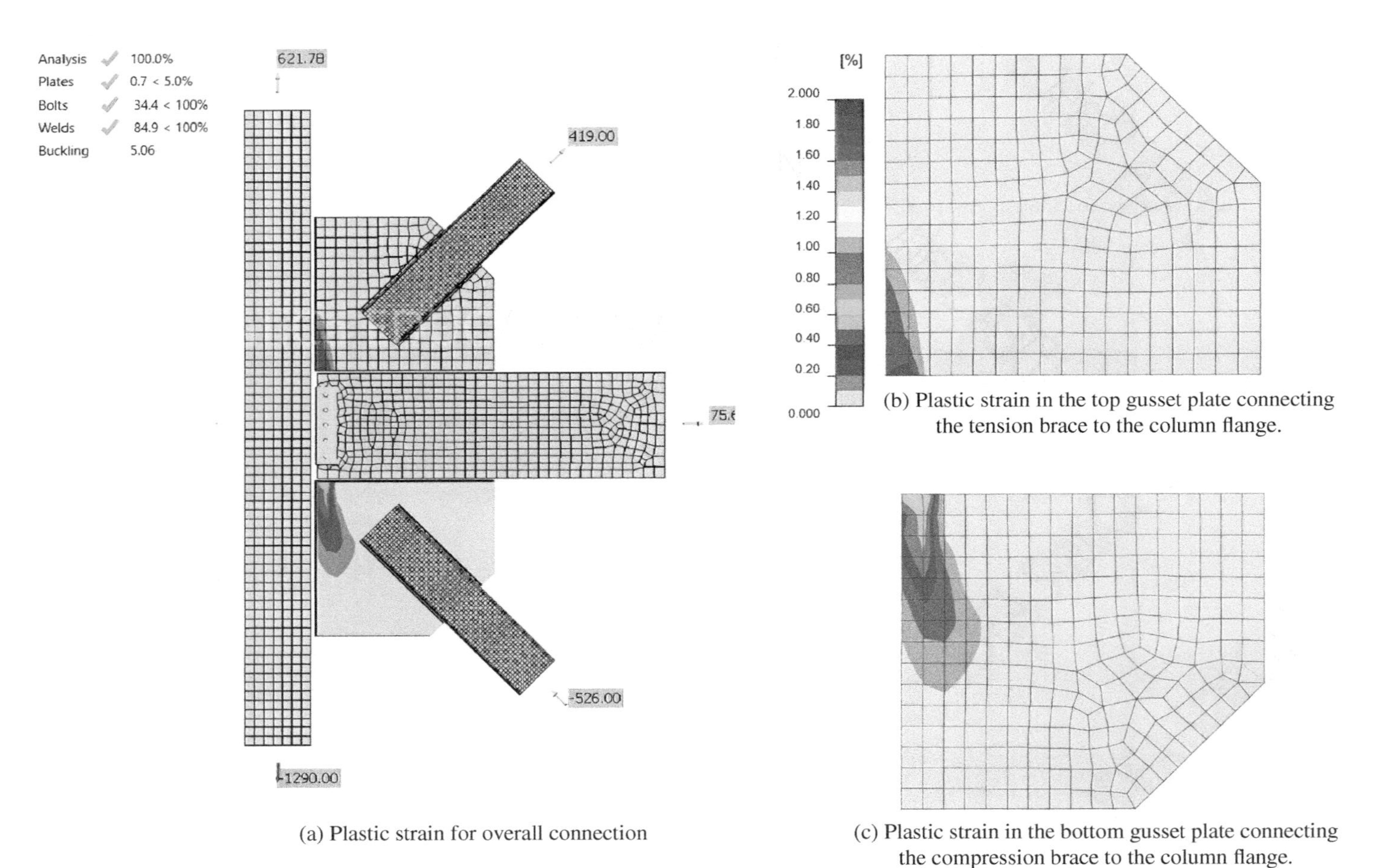

(a) Plastic strain for overall connection

(b) Plastic strain in the top gusset plate connecting the tension brace to the column flange.

(c) Plastic strain in the bottom gusset plate connecting the compression brace to the column flange.

Figure 13.4 Plastic strains for: (a) overall connection; (b) top gusset plate; and (c) bottom gusset plate.

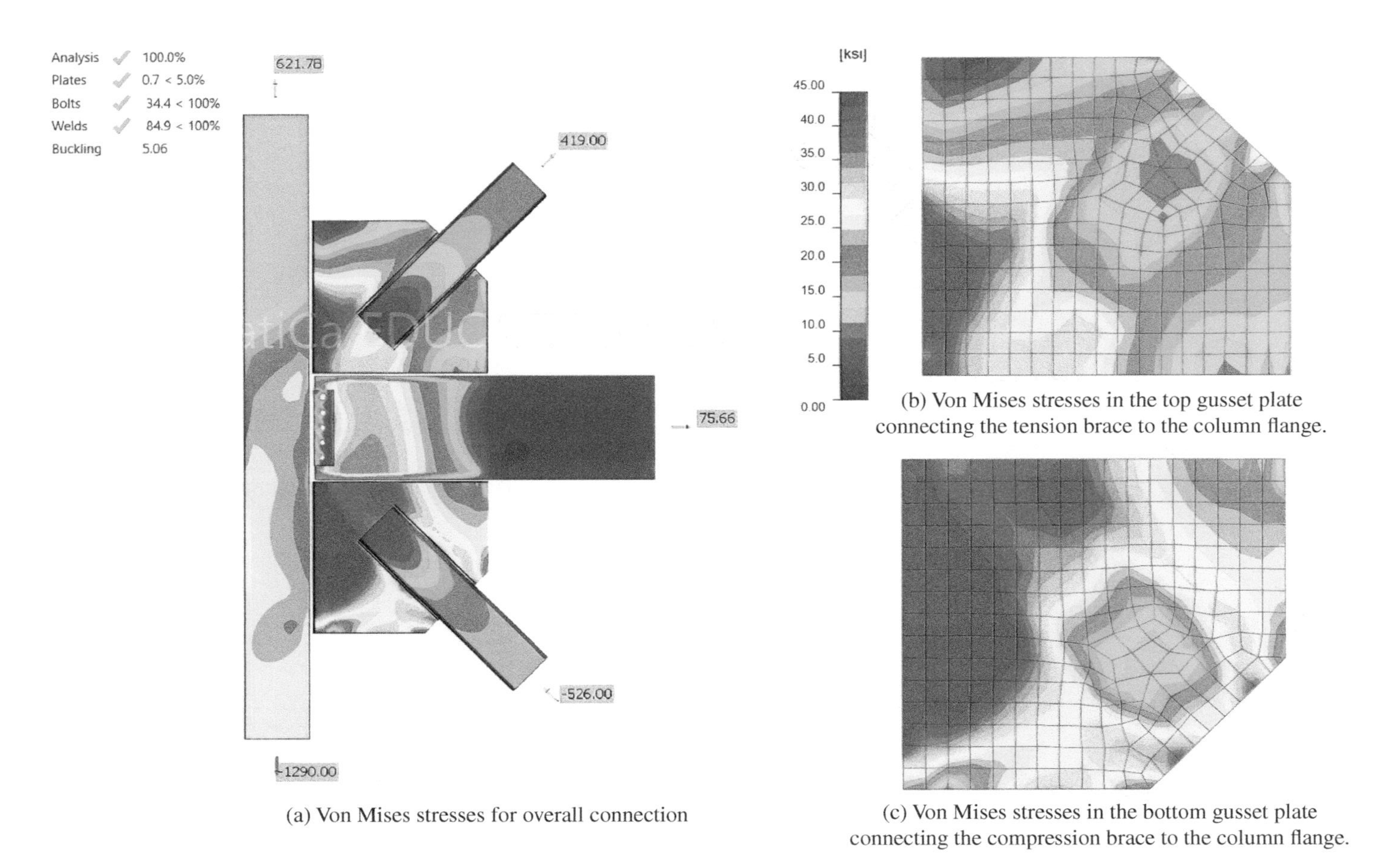

(a) Von Mises stresses for overall connection

(b) Von Mises stresses in the top gusset plate connecting the tension brace to the column flange.

(c) Von Mises stresses in the bottom gusset plate connecting the compression brace to the column flange.

Figure 13.5 Von Mises stresses for: (a) overall connection; (b) top gusset plate; and (c) bottom gusset plate.

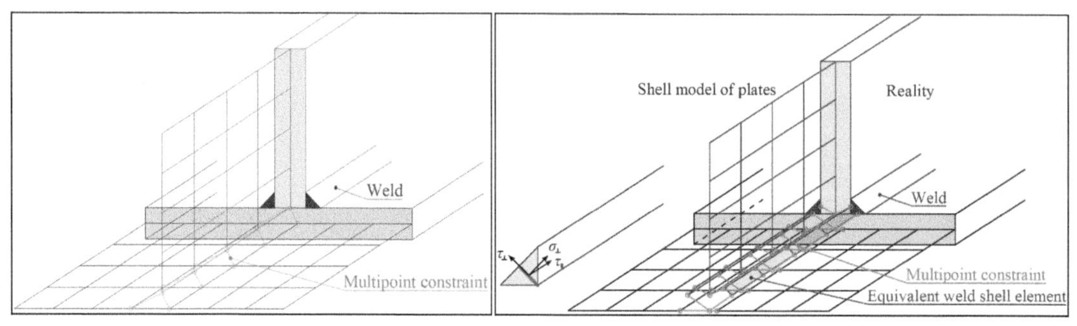

Figure 13.6 Multi-point constraint (MPC) in CBFEM (IDEA StatiCa Support Center).

The force, F_n, and weld angle, θ, are derived from stresses $\sigma_\perp$, $\tau_\perp$, $\tau_\parallel$, length and effective area of weld finite element. These stresses are the basic output of finite element solver.

The weld diagrams show stress according to the following formulas:

If base metal is deactivated (matching electrode is used):

$$\sigma = \frac{\sqrt{\sigma_\perp^2 + \tau_\perp^2 + \tau_\parallel^2}}{1 + 0.5\sin^{1.5}\theta}$$

If base metal is activated (matching electrode is not used):

$$\sigma = \max\left\{\frac{\sqrt{\sigma_\perp^2 + \tau_\perp^2 + \tau_\parallel^2}}{1 + 0.5\sin^{1.5}\theta}, \frac{\sqrt{\sigma_\perp^2 + \tau_\perp^2 + \tau_\parallel^2}}{\sqrt{2}F_u / F_{EXX}}\right\}$$

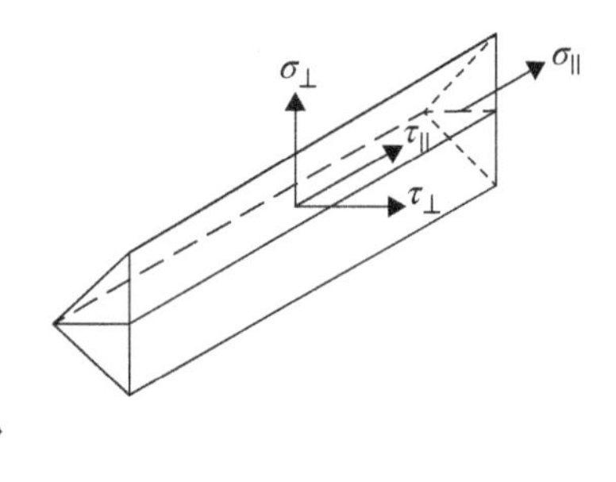

Figure 13.7 FEA relations for weld angle computation (IDEA StatiCa Support Center).

from the obtained CBFEM (556.8 kips) is found to be similar when compared with AISC (556.8 kips). One important point to note here is that the nominal stress of weld material (F_{nw}) is independent of the weld angle for the case of weld connecting brace to gusset plate, the directional strength increase is not used for the HSS weld, which gives us approximately the same values for a rectangular HSS brace section with a single concentric gusset plate with respect to weld capacity.

Similarly, the weld capacity between the column and brace member and also between the beam and brace members as per AISC and CBFEM is found to be approximately the same, assuming the total weld length as shown in Table 13.2.

The tension rupture of the brace capacity per AISC is 869.4 kips, and the tensile yielding on the brace net section per AISC is 716.6 kips, as per Table 13.1, which are greater than the applied loads. It is obvious that rupture would occur before yielding, which is also observed from the stress distribution of CBFEM. As per CBFEM, the tensile rupture capacity on the brace net section is 820 kips, which is 14.43% larger

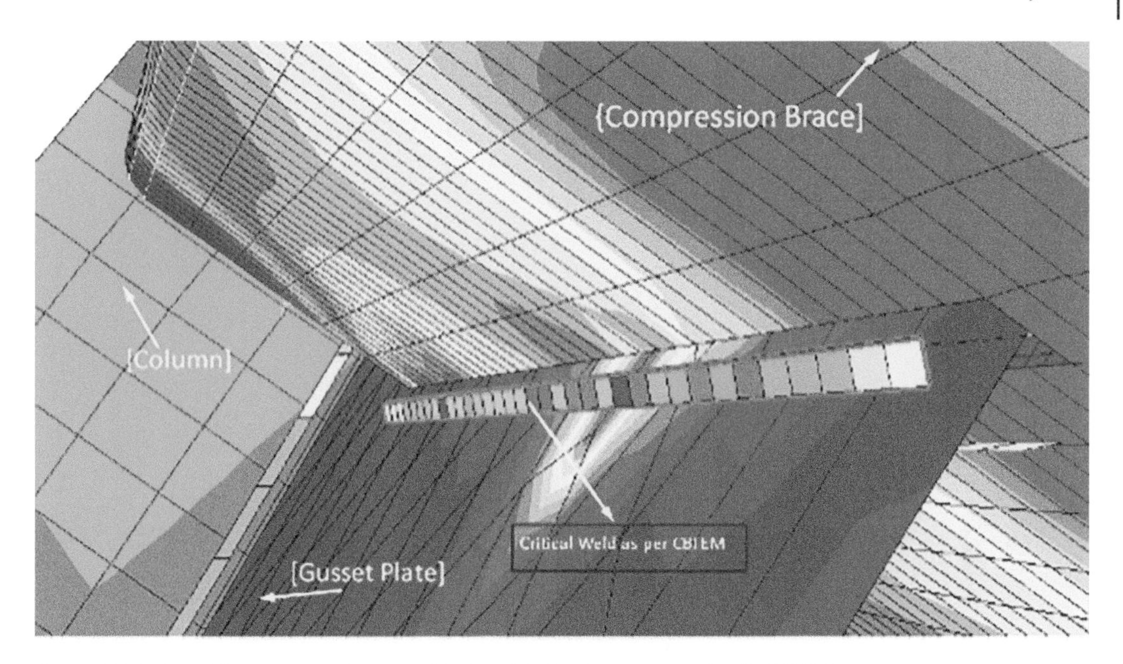

Figure 13.8 Von Mises stresses for the critical weld of a given connection in CBFEM.

Table 13.2

Sr No.	Weld between	AISC		CBFEM	
		Weld capacity (Kips)	θ (Angle of loading measured from the weld longitudinal axis)	Weld capacity (Kips)	θ (Angle of loading measured from the weld longitudinal axis)
1	Brace and Gusset Plate	556.8	Not applicable	556.8	Not applicable
2	Column and Gusset Plate	690.8	56.3 degree	690.8	56.3 degree
3	Beam and Gusset Plate	723	67.3 degree	723	67.3 degree

when compared with the value as per AISC (716.6 kips). Similarly, the tension yielding of a brace as computed from CBFEM is 880 kips, which is approximately the same when compared with the value as per AISC (869.4 kips). The plastic strain distribution for the tension rupture of the brace is shown in Figure 13.10 and for tension yielding of the brace in Figure 13.11.

Tension yielding of the gusset plate above the beam connected to the brace subjected to axial tensile load is based on the Whitmore section. As per CBFEM, the value obtained for tension yielding of the gusset plate limit state is 800 kips, which is 13.94% higher than the value obtained from AISC (746 kips).

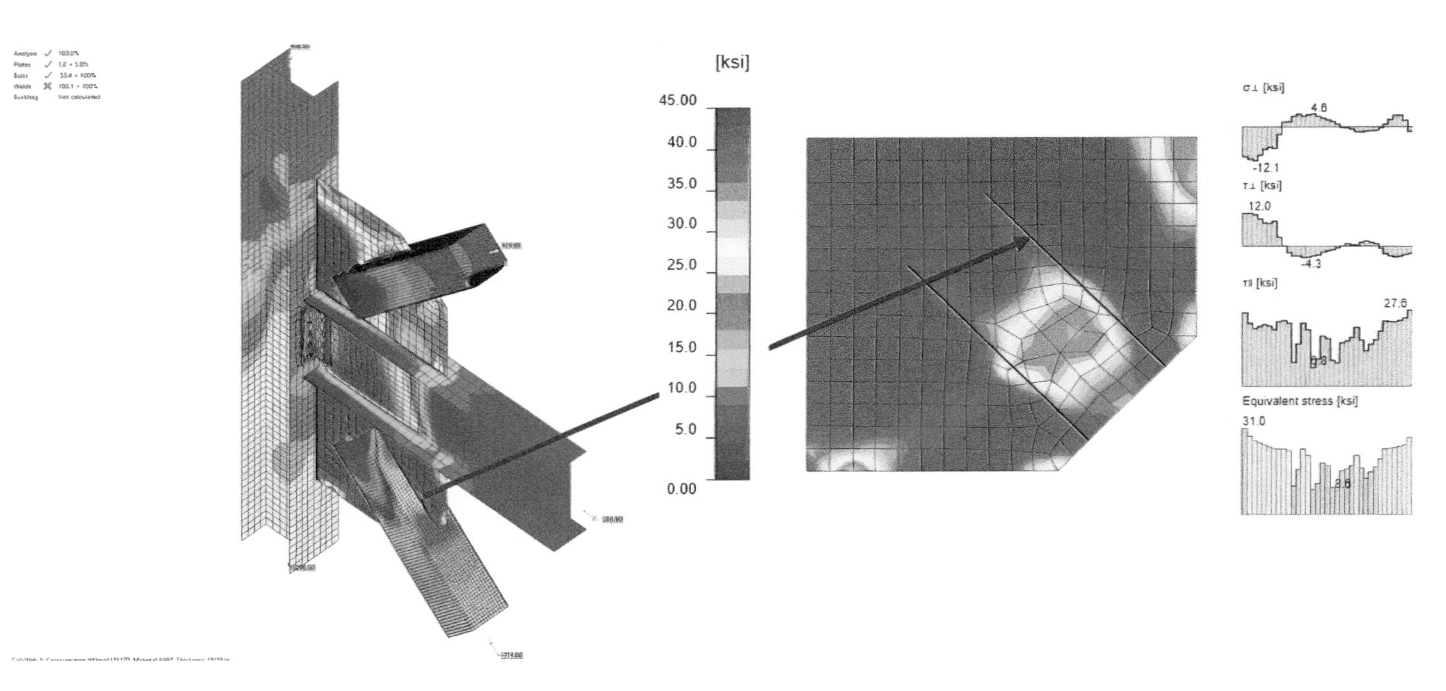

Figure 13.9 Von Mises stresses in the bottom gusset plate for weld capacity computation in CBFEM.

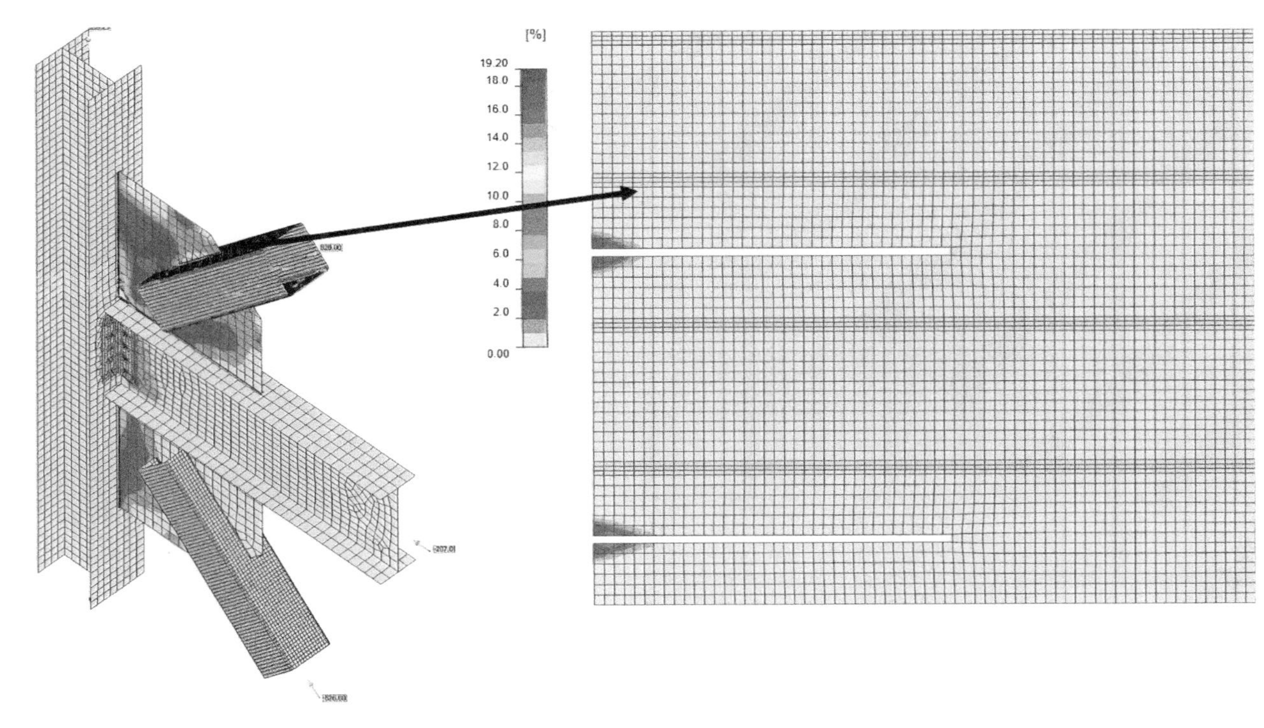

Figure 13.10 Plastic strains in CBFEM for tension rupture of brace above beam that is loaded with an axial tensile force.

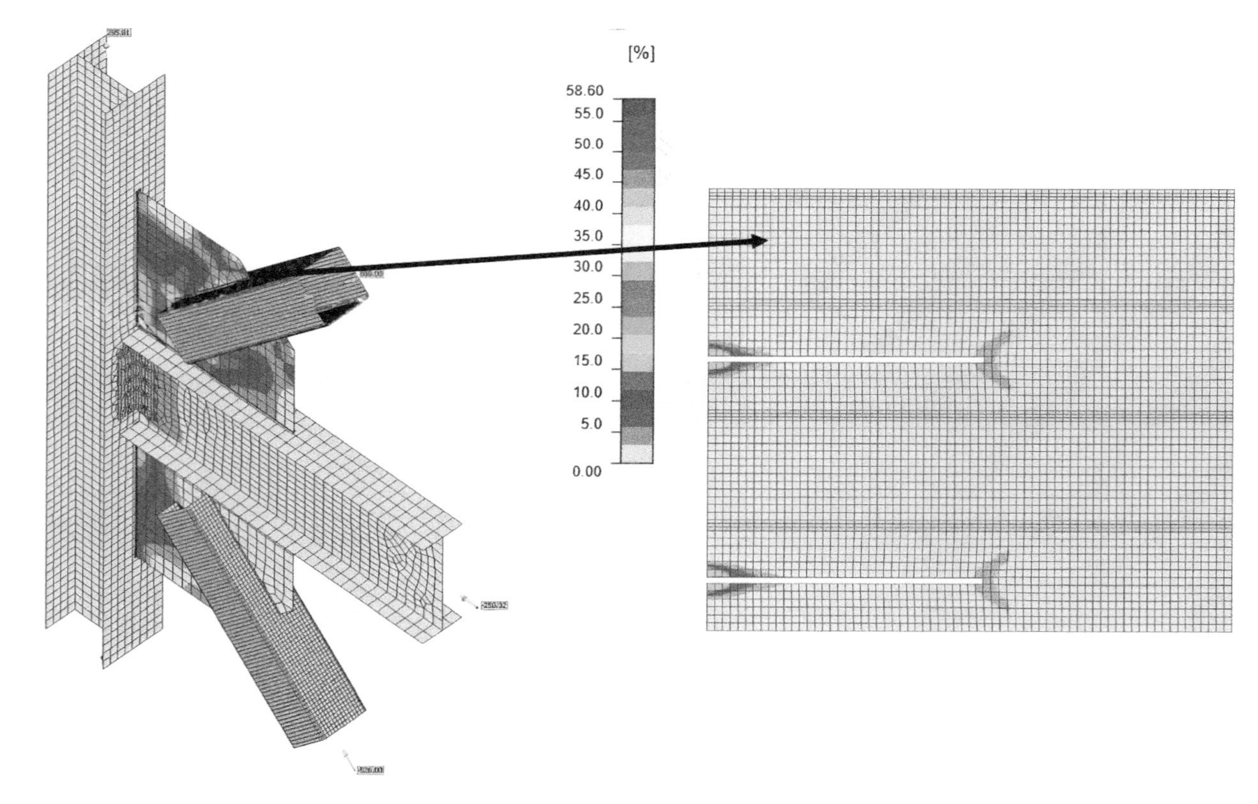

Figure 13.11 Plastic strains in CBFEM for tension yielding of brace above beam that is loaded with an axial tensile force.

The plastic strain distribution and the von Mises stress distribution in the gusset plate above the beam for the limit state of tension yielding of the gusset plate are as shown in Figure 13.12.

Block shear on the gusset plate above the beam was investigated as per CBFEM, the value obtained as 850 kips was found to be 22.51% higher than the value obtained from AISC. The plastic strain distribution and von Mises stress distribution as observed in the gusset plate above the beam for the block shear limit state are as shown in Figure 13.13.

The buckling of the gusset plate is the function of the axial compression load in the connecting braces and the requirement for the same in CBFEM can be checked by a buckling multiplier factor obtained using analysis. Currently, it is the only measure, and it is hard to differentiate between the buckling resistance of various connecting parts, e.g. the buckling of the gusset plate on the Whitmore section or the gusset plate sideways buckling. Mode 1 is the governing mode among several modes in CBFEM and for the given connection, mode 1 buckling is found to exist in the bottom gusset plate as it is subjected to the action of the compression load. The buckling factor obtained for this connection by CBFEM is 3.66 which is for the critical mode (mode 1), observed for an axial compressive load of 796.70 kips in the bottom brace as shown in Figure 13.14. The buckling factor here is obtained for a plastic strain of equal to or just greater than 5% in the gusset plate. For local buckling, the preferred value for the buckling factor is greater than 9.16 for 50 ksi steel per AISC 360-16 Section J.4, therefore, it is expected that the gusset would buckle at this load, according to CBFEM, however, buckling according to CBFEM still requires more research to allow accurate prediction. The buckling capacity obtained as per CBFEM after multiplying the axial compressive load and the buckling factor, which is elastic buckling capacity, is higher than the inelastic buckling capacity. Hence, the buckling limit state is not a controlling limit state per both AISC and CBFEM.

The forces on the beam obtained from the AISC uniform force method are shear and axial forces only. The beam must be checked to ensure that the web strength and stability are maintained. For tension loads applied to the beam web, local yielding is a critical failure mode. For beam web local yielding, the most critical location is near the end of the beam at the connection location where loads are transferred from the braces. However, web local yielding must be checked regardless of where the tensile load is applied.

When the applied brace force is in compression, web crippling is a critical limit state. Section J10.3, AISC *Specification* (AISC, 2016) provides equations to determine the web crippling strength. If the beam does not have enough capacity in either web local yielding or web crippling, then the transverse stiffener plates must be added to the beam web.

Beam web local and shear yielding would occur at larger loads when compared to the applied loads. Almost all the limit states in this connection would occur before these two limit states which typically do not control the design. If required, these limit states can be checked using the AISC *Specification* (AISC, 2016), using the procedure presented in the Appendix in this chapter for the beam web local and shear yielding.

Beam web crippling would occur after yielding and at higher loads, therefore the model may not converge under such a high load and would not be able to capture this failure mode. If the crippling capacity is needed, it can be calculated as per the AISC specifications using the procedure presented in the Appendix in this chapter.

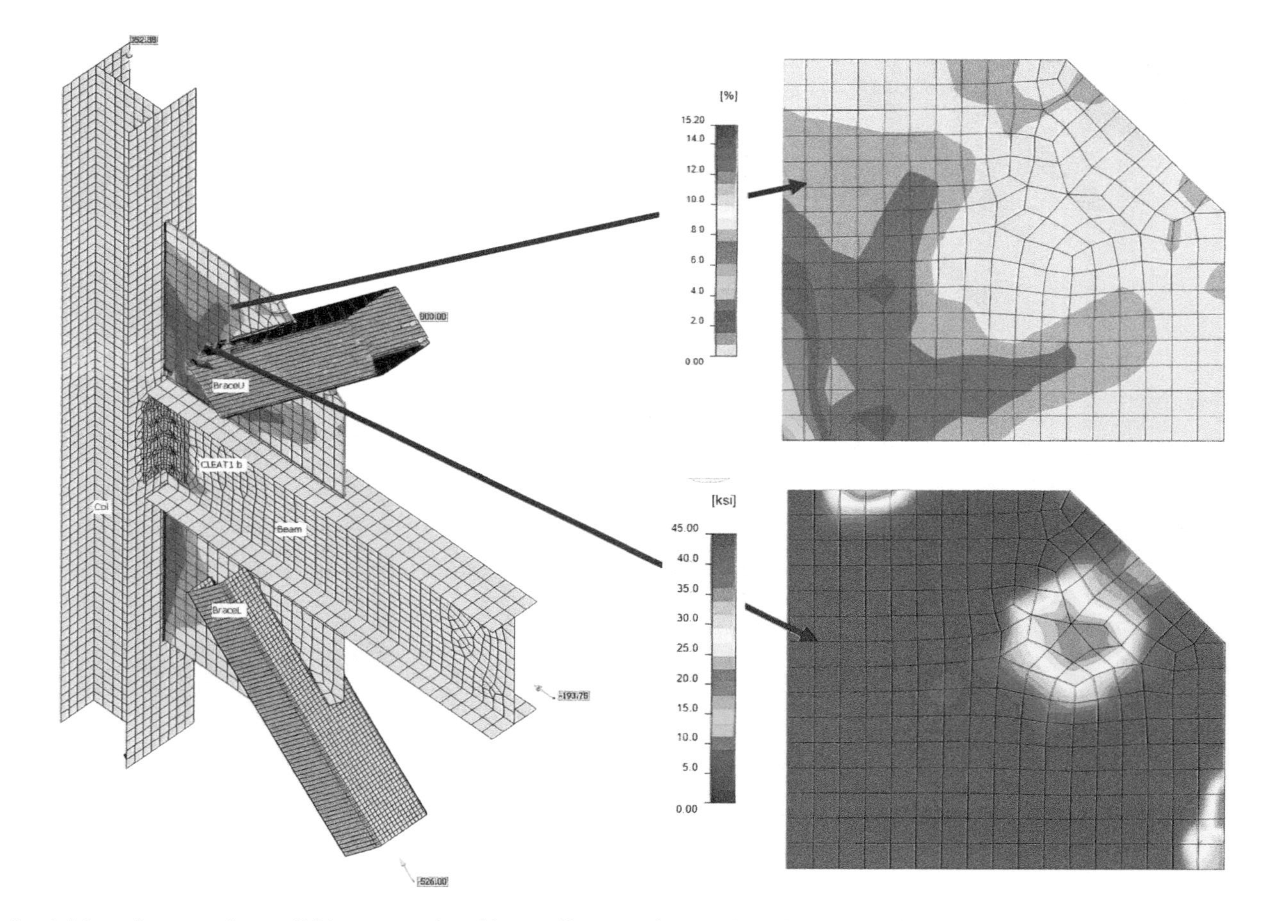

Figure 13.12 Tensile yielding of gusset plate at Whitmore section: (a) overall connection results; (b) plastic strain; (c) equivalent stresses.

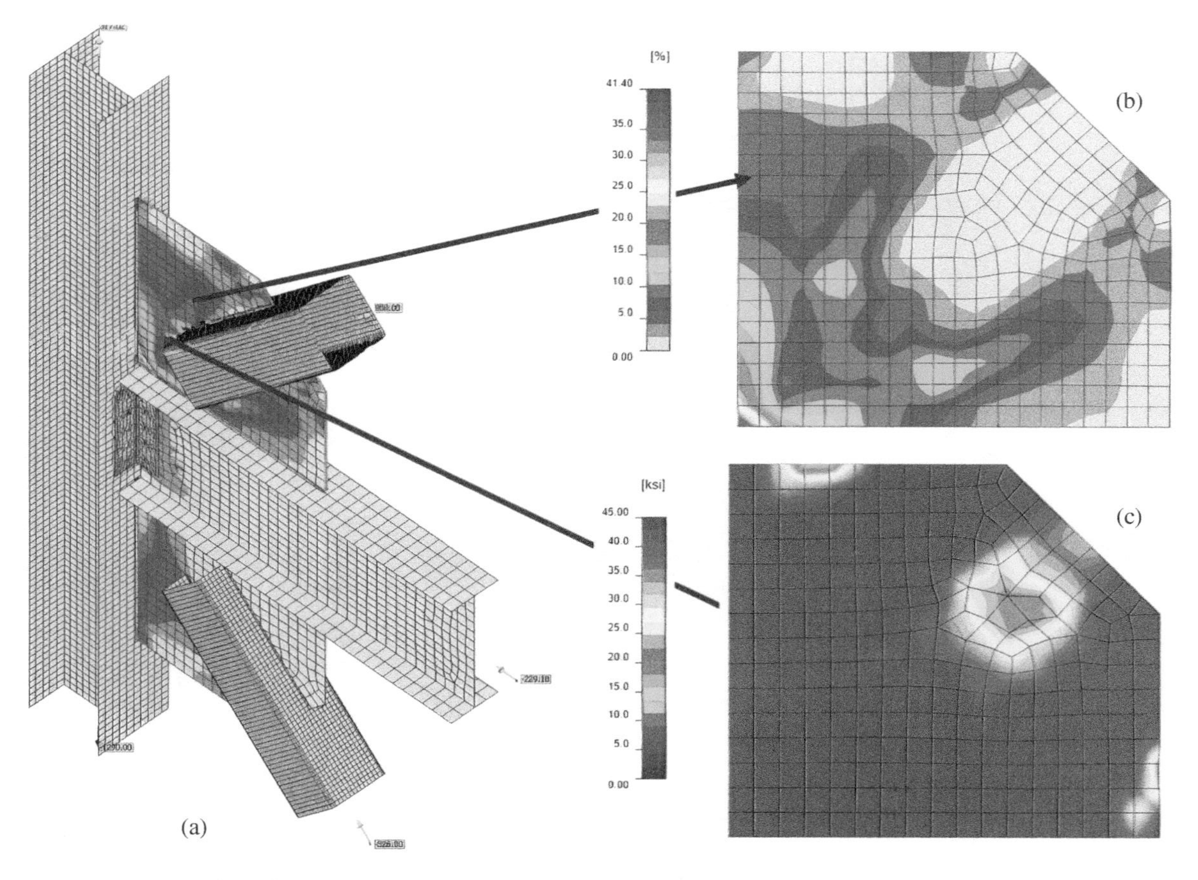

Figure 13.13 Block shear rupture on gusset plate above beam: (a) overall connection results; (b) plastic strain; (c) equivalent stresses.

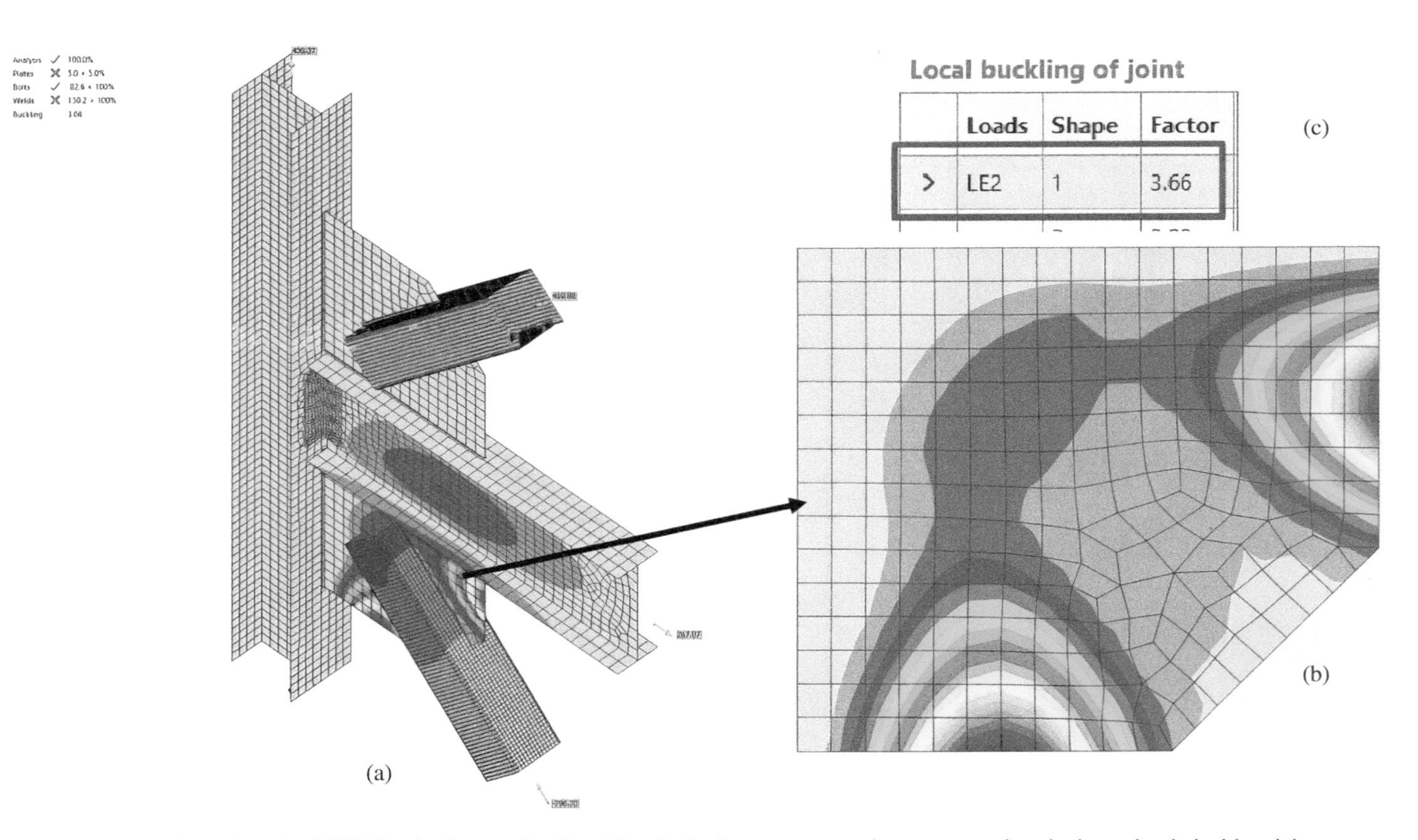

Figure 13.14 Deformations in CBFEM for the first mode of buckling in the bottom gusset plate connected to the brace loaded with axial compressive force.

13.3.2 Parametric Study

To verify the resistance of several components in detail and to understand the ability of CBFEM in capturing the various failure modes, the parametric study is performed as follows:

Modification 1: Fillet welds at all braces changed to butt welds

The governing limit state as per AISC and CBFEM for the given connection is the weld shear strength between the brace which is loaded with an axial compression load and the gusset plate below the beam. Therefore, the fillet welds at all braces to the connecting gusset plate are changed to full penetration butt weld in CBFEM.

As per the design, for the connection with fillet welds between the brace and gusset plate in CBFEM, for an axial load of 685 kips in both braces, the limit state of the fillet weld was found to be governing. But when the fillet welds between the brace and gusset plate are modified to the butt weld in CBFEM, the axial load can be increased up to 691.66 kips, which is in close agreement with the block shear rupture of the gusset plate as given in Table 13.1. At this load, the welds are 100% utilized and the overall resistance is increased by 1%. Based on the investigation from CBFEM results, the critical weld was observed between the flange of the column and the connecting bottom gusset plate as shown in Figure 13.15 for modification 1. In CBFEM, butt welds are not checked and their resistance is assumed to be higher than the welded member. Plastic strain in the cleat angle is shown in Figure 13.16, wherein the critical bolts observed are bolt numbers 8 and 13.

Modification 2: Change in maximum element size of mesh in CBFEM

To evaluate the impact on the CBFEM results of the percentage utilization in the connections due to the variation of the maximum element size in the mesh, a mesh sensitivity analysis was performed. Six specific maximum element sizes: 0.35 in., 0.5 in., 1 in., 1.5 in., 2 in., and 2.5 in. are considered and the utilization of bolts and welds is evaluated for calculated standard loads. As per the "default" setting for the model and mesh in IDEA StatiCa, the minimum element size was kept as 0.3 in. (the default value), while the maximum element size was varied for the six specific element sizes mentioned above. The obtained results are plotted as shown in Figure 13.17. Based on the observations, the variation between the utilization percentage of bolts and welds is not significant. For smaller maximum element sizes, the values are slightly conservative, but the overall analysis time for the same is higher when compared to the higher values for the maximum element size of the mesh. As per CBFEM, the preferred default maximum element mesh size is 1.97 in. for IDEA StatiCa Version 23.0.1.1401 software.

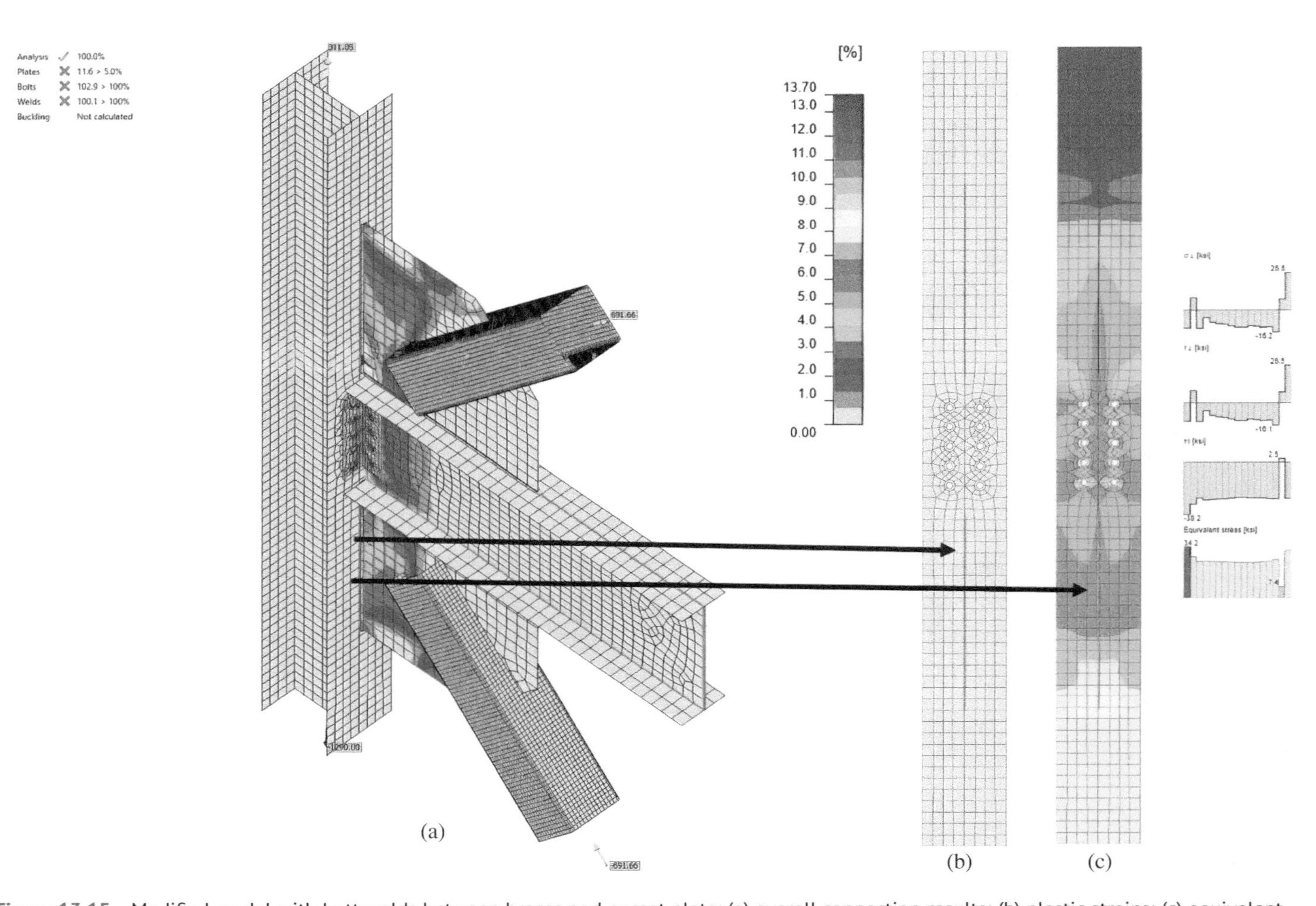

Figure 13.15 Modified model with butt welds between braces and gusset plate: (a) overall connection results; (b) plastic strains; (c) equivalent stresses.

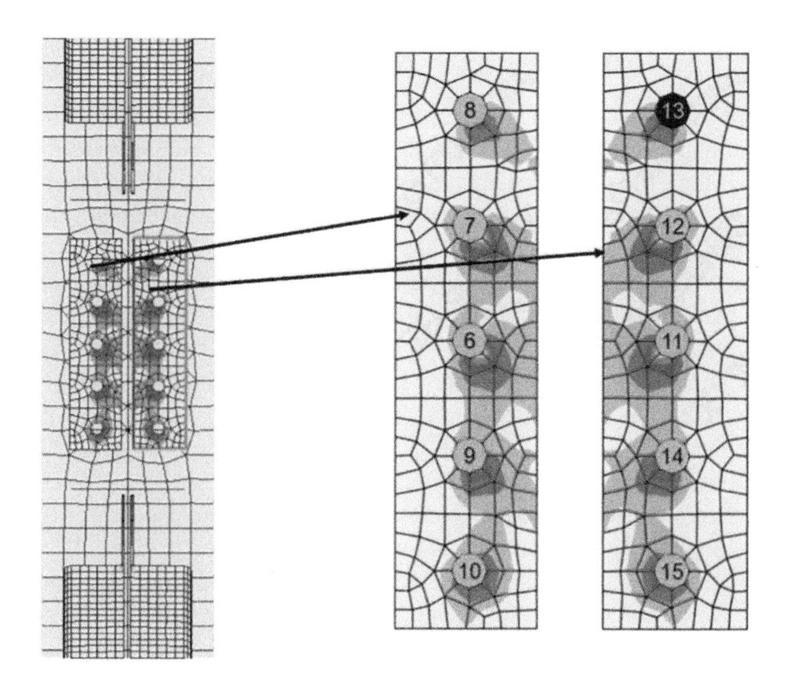

Figure 13.16 Plastic strain in the cleat angle connecting the beam web to the column bottom flange.

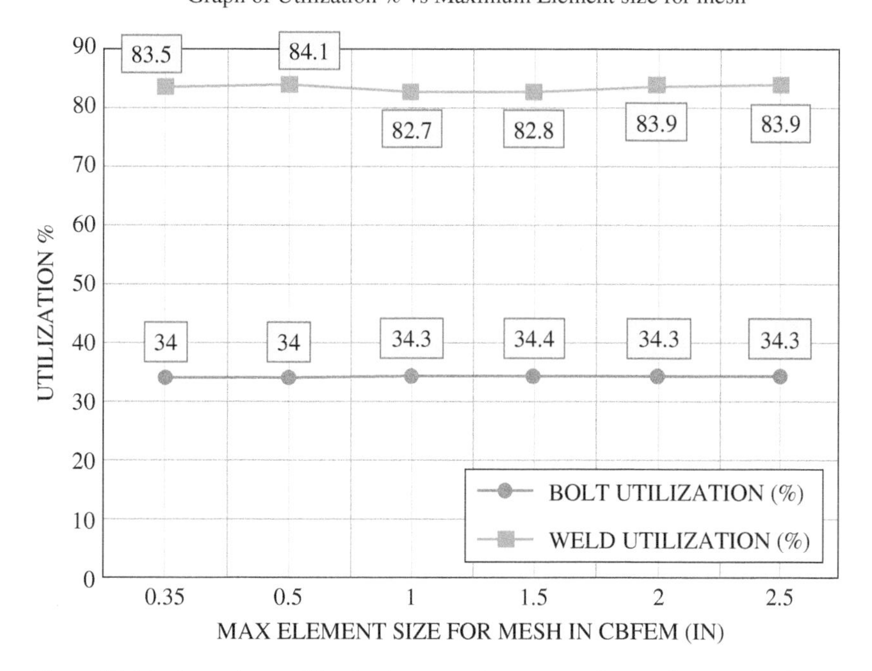

Figure 13.17 Percentage utilization in connections vs. maximum element size for mesh in IDEA StatiCa.

13.4 Summary

The connection presented herein has two braces, one is subjected to tension, and another is subjected to compression. The connection controlling limit state, per AISC, is the weld limit state, which agrees with the CBFEM. It can be concluded that the CBFEM can predict the actual behavior and the failure modes of the braced frame connections presented herein. The various limit states were investigated carefully by performing a parametric study to compare the AISC vs. the CBFEM capacities.

The weld capacity for the weld between the gusset and the brace, between the gusset plate and beam top flange, and between the gusset plate and the column flange, agrees in both AISC and CBFEM. Bolt limit states, including bolt shear and bolt bearing in AISC agree with CBFEM. The plates limit states, including yielding, rupture in tension and in shear, are based on the 5% plastic strain limit, according to CBFEM.

Tension rupture in the net section of the brace as per CBFEM is 14.43% higher than the value from AISC, while the tensile yielding in the gross section of the brace value as per CBFEM is approximately similar to the value from AISC. Similarly, the value for block shear rupture in the gusset plate is 22.51% higher, while the tension yielding of the gusset plate is 13.94% higher when compared with the AISC value. Beam web buckling, web crippling, and shear yielding would occur at high loads and the model would not converge at such high loads; all other limit states would occur before these limit states. The buckling limit state of the gusset plate was not observed as a limit state in AISC and CBFEM.

Benchmark Case

Input

Beam cross-section

- W21X83
- Steel ASTM A992

Braces cross-section

- HSS 10X10X5/8
- Steel ASTM A500 Gr. B

Column cross-section

- W12X170
- Steel ASTM A992

Gusset plates

- Thickness 1/2 in.
- Steel ASTM A572 Gr. 50

Beam-to-column connecting angles

- 2-L4 × 4 × 3/8
- Steel ASTM A572 Gr. 50

Loading

- Axial force in braces $P = 419$ kips in tension and $P = -526$ kips compression
- Weld
- Gusset-to-beam flanges 5/16” ASTM E70XX
- Gusset-to-column flange 5/16” ASTM E70XX
- Brace-to-gusset plate 5/16” ASTM E70XX

Output

- Weld 84.9%
- Bolts 34.4%
- Plastic strain 0.7% < 5%
- Buckling factor 5.06

Appendix

The AISC Solution

ORIGIN:=1

Refer to AISC 14th Edition shapes database:

T1 :=READEXCEL(“.\AISC Shapes Database v14.1.xlsx” , “Database v14.1!A1:CA1998”)

```
Row(shape) := ‖ for i ∈ 1..rows(T1) − 1
              ‖   ‖ if ⟨T1^⟨2⟩⟩_i = shape
              ‖   ‖   ‖ R ← i
              ‖ R
```

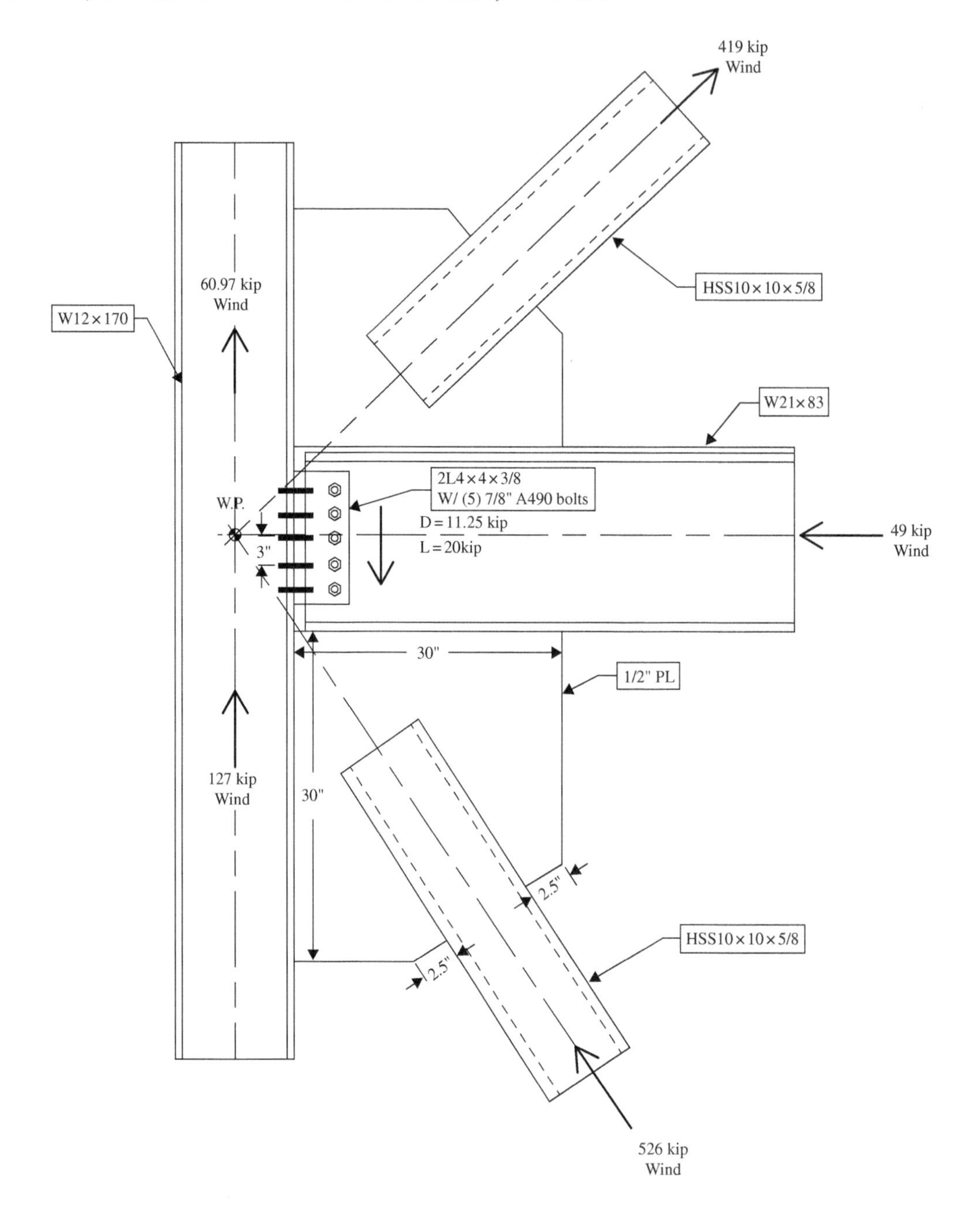
419 kip
Wind
HSS10 × 10 × 5/8
60.97 kip
Wind
W12 × 170
W21 × 83
2L4 × 4 × 3/8
W/ (5) 7/8" A490 bolts
W.P.
D = 11.25 kip
L = 20kip
3"
49 kip
Wind
30"
1/2" PL
127 kip
Wind
30"
2.5"
HSS10 × 10 × 5/8
2.5"
526 kip
Wind

Member Properties

Beam/Column Properties

Beam section	$Beam := \text{"W21} \times \text{83"}$	$Column := \text{"W12} \times \text{170"}$
Depth of beam	$d_{beam} := T1_{Row(Beam),7} \cdot in = 21.4\ in$	$d_{col} := T1_{Row(column),7} \cdot in = 14\ in$
Width of flange	$bf_{beam} := T1_{Row(Beam),12} \cdot in = 8.4\ in$	$bf_{col} := T1_{Row(column),12} \cdot in = 12.6\ in$
Thickness of flange	$tf_{beam} := T1_{Row(Beam),20} \cdot in = 0.835\ in$	$tf_{col} := T1_{Row(column),20} \cdot in = 1.56\ in$
Thickness of web	$tw_{beam} := T1_{Row(Beam),17} \cdot in = 0.515\ in$	$tw_{col} := T1_{Row(column),17} \cdot in = 0.96\ in$
Design weld depth	$k\_des_{beam} := T1_{Row(Beam),25} \cdot in = 1.34\ in$	$k\_des_{col} := T1_{Row(Column),25} \cdot in = 2.16\ in$

Brace Properties

Brace section	$Brace := \text{"HSS10} \times 10 \times .625\text{"}$	
Cross-sectional area	$Ag_{brace} := T1_{Row(Brace),6} \cdot in^2 = 21\ in^2$	this is the r.x value
Radius of gyration	$r_{brace} := T1_{Row(Brace),42} \cdot in = 3.8\ in$	(sim to r.y for square shapes)
Depth to end of weld	$t\_des_{brace} := T1_{Row(Brace),24} \cdot in = 0.581\ in$	
	$B := T1_{Row(Brace),9} \cdot in = 10\ in$	
	$H := T1_{Row(Brace),14} \cdot in = 10\ in$	
Slope of brace	$slope := \text{atan}\left(\frac{12}{12}\right) = 45\,deg$	slope of brace is assumed

Gusset Plate Properties

Thickness	$t_{PL} := 0.5 \cdot in$	Distance brace to W.P.	$e_{WP} := 27.375 \cdot in$
Length	$L := 30 \cdot in$	Eccentricity of brace CL to W.P.	$e := 13.65 \cdot in$
Height	$h := 30 \cdot in$		
Cut length	$aa := 15 \cdot in$	End of PL to W.P.	$L_1 := 32 \cdot in$
		Length (centered)	$L_2 := L_1$

Material Properties

Beams grade A992	$Fy_{beam} := 50 \cdot ksi$	$Fu_{beam} := 65 \cdot ksi$	material properties are assumed for beam. column, HSS, PL, and welds
Brace grade A500 GrB	$Fy_{brace} := 46 \cdot ksi$	$Fu_{brace} := 58 \cdot ksi$	
Plate grade A572 Gr. 50	$Fy_{PL} := 50 \cdot ksi$	$Fu_{PL} := 65 \cdot ksi$	
Modulus of Elasticity	$E := 29000 \cdot ksi$		
Weld strength	$F_{EXX} := 70 \cdot ksi$	$C_1 := 1.0$	

Loads

Beam reactions (no factors)	$R_{ub} := 11.25 \cdot kip + 20 \cdot kip$	$A_{ub} := 49 \cdot kip$
Column reactions (no factors)	$A_{uc\_top} := 60.79 \cdot kip$	$A_{uc\_bott} := 127.0 \cdot kip$

Brace axial load

1 denotes bottom brace $\quad P_{u1} := -526 \cdot kip \quad P_{u2} := 419 \cdot kip$

2 denotes top brace

$e_{beam} := d_{beam} \div 2 = 10.7\,in$

$e_{col} := d_{col} \div 2 = 7\,in$

$\theta := slope = 45\ deg$

$\alpha - \beta \cdot \tan(\theta) = e_{beam} \cdot \tan(\theta) - e_{col}$

take $\beta := 12\ in$

$\alpha := e_{beam} \cdot \tan(\theta) - e_{col} + \beta \cdot \tan(\theta) = 15.7\ in$

$\alpha\_bar = \alpha = \frac{l_h}{2} + t_{PL}$

$l_h := 2 \cdot (\alpha - t_{PL}) = 30.4\ in$

$r := \sqrt{(\alpha + e_{col})^2 + (\beta + e_{beam})^2} = 32.1\ in$

recall
$d_{beam} = 21.4\ in$
$d_{col} = 14\ in$
$slope = 45\ deg$
$t_{PL} = 0.5\ in$

$V_{col} := \frac{\beta}{r} \cdot P_{u1} = -196.6\ kip \quad H_{col} := \frac{e_{col}}{r} \cdot P_{u1} = -114.7\ kip$

$V_{beam} := \frac{e_{beam}}{r} \cdot P_{u1} = -175.3\ kip \quad H_{beam} := \frac{\alpha}{r} \cdot P_{u1} = -257.2\ kip$

$P_{u1} \cdot \cos(\theta) - V_{col} - V_{beam} = 0\ kip \quad P_{u1} \cdot \sin(\theta) - H_{col} - H_{beam} = 0\ kip$

Check tension yielding on the brace (J4.1a)

Reduction factor $\quad \phi := 0.90$

Tensile yielding $\quad \phi R_n := \phi \cdot Fy_{brace} \cdot Ag_{brace} = 869.4\,kip$

if $(P_{u2} < \phi R_n, \text{"OK"}, \text{"NOT OK"}) = \text{"OK"}$

recall
$Fy_{brace} = 46\ ksi$
$Ag_{brace} = 21\ in^2$
$P_{u2} = 419\ kip$

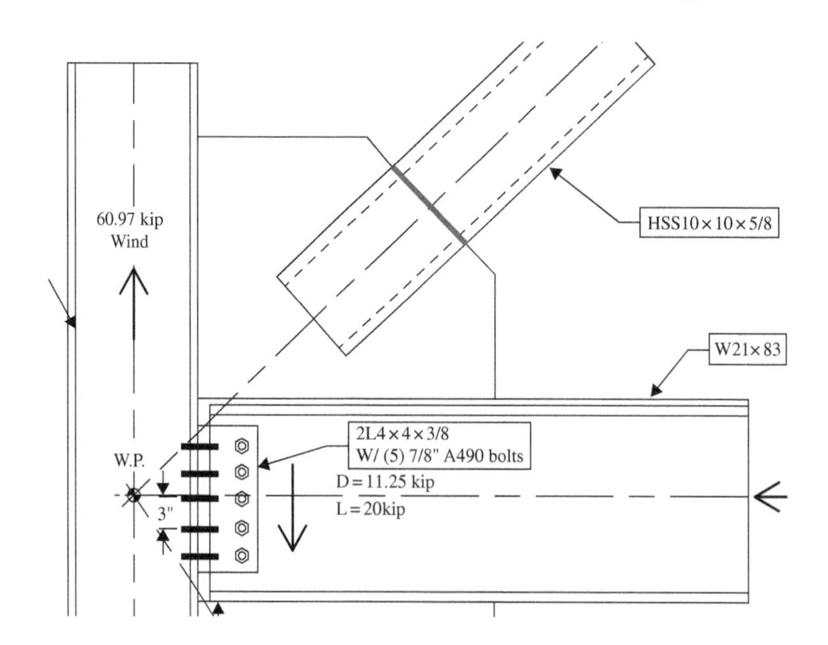

Shear rupture in the brace wall (J4.2b)

Reduction factor $\phi := 0.75$

recall

$P_{u1} = -526$ kip

$Fu_{brace} = 58$ ksi

$t\_des_{brace} = 0.6$ in

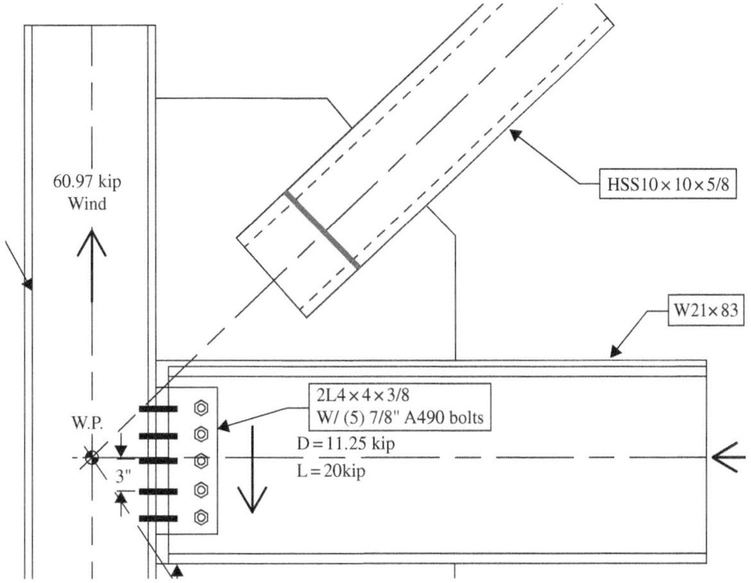

$$P_u \le \phi \cdot 0.6 \cdot F_u \cdot A_{nv} = \phi \cdot 0.6 \cdot F_u \cdot t\_des_{brace} \cdot (4) \cdot l_{weld}$$

Calculated weld length (4 lengths of weld)

$$l_{weld} := \frac{-P_{u1}}{\left(0.6 \cdot \left(\phi \cdot Fu_{bruce} \cdot t\_des_{brace} \cdot (4)\right)\right)} = 8.7 \text{ in}$$

Check weld strength (Eq. 8-1)

Reduction factor $\phi := 0.75$

Try weld size 5/16 inch $w_{weld} := 5 \cdot \text{in}$ 16 th of an inch

Weld strength

$$P_u \le \phi \cdot R_n = \phi \cdot 0.60 \cdot F_{EXX} \cdot \left(\frac{1}{\sqrt{2}}\right) \cdot w_{weld} \cdot (4) \cdot l_{weld}$$

Minimum weld length

$$l_{weld} \ge \frac{|P_{u1}|}{\left(\phi \cdot 0.60 \cdot F_{EXX} \cdot \left(\frac{1}{\sqrt{2}}\right) \cdot w_{weld} \cdot (4)\right)}$$

$$l_{weld} := \frac{|P_{u1}|}{\left(\phi \cdot 0.60 \cdot F_{EXX} \cdot \left(\frac{1}{\sqrt{2}}\right) \cdot \frac{w_{weld}}{16} \cdot (4)\right)} = 18.9 \text{in}$$

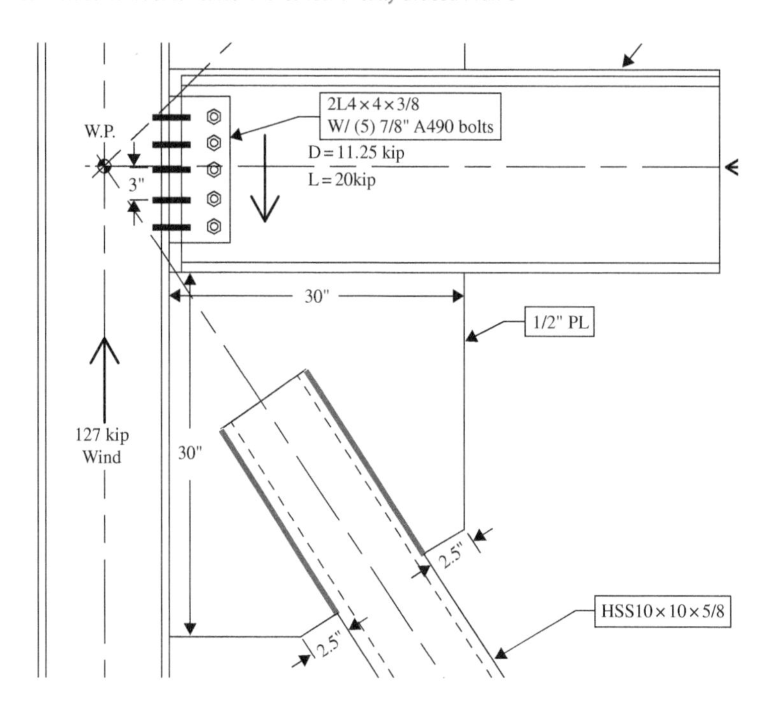

Check weld strength (Eq. 8-2)

Weld strength $\phi \cdot P_n = 1.392 \cdot D \cdot l_{weld} \cdot (4)$

Minimum weld length $l_{weld} := \dfrac{|\,P_{u1}\,|}{1.392 \cdot ksi \cdot w_{weld} \cdot (4)} = 18.9\ in$

Use $l_{weld} = 20^{\varepsilon}$ $l_{weld} := 20 \cdot in$

recall
$P_{u1} = -526\ kip$
$w_{weld} = 5\ in$

Check tensile rupture on the brace (J4.1b)

Slot width $d_{slot} := t_{PL} + \dfrac{2}{16} \cdot in = 0.6 in$

Net area $An_{brace} := Ag_{brace} - 2 \cdot t\_des_{brace} \cdot d_{slot} = 20.27\ in^2$

Eccentricity of connection $x\_bar := \dfrac{B^2 + 2 \cdot B \cdot H}{4 \cdot (B + H)} = 3.75\ in$

recall
$t_{PL} = 0.5 in$
$Ag_{brace} = 21\ in^2$
$t\_des_{brace} = 0.581\ in$
$B = 10\ in$
$H = 10\ in$
$l_{weld} = 20\ in$
$Fu_{brace} = 58\ ksi$
$P_{u2} = 419\ kip$

Table D3.1, Case 6

6	Rectangular HSS	with a single concentric gusset plate	$l \geq H \ldots U = 1 - \overline{x}/l$ $\overline{X} = \dfrac{B^2 + 2BH}{4(B+H)}$

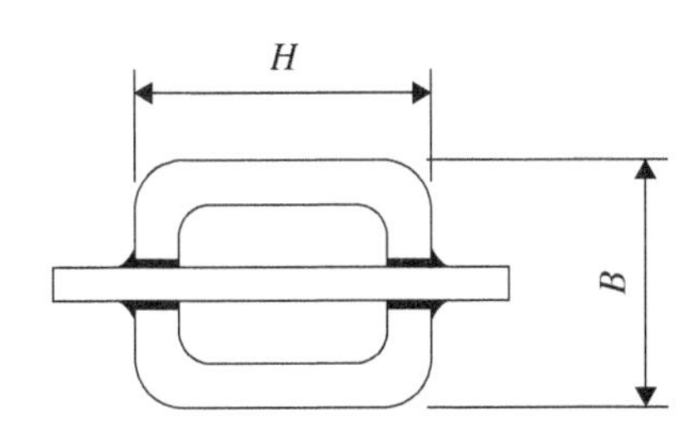

Shear lag factor

Table D3.1, Case 6 $\quad U := 1 - \left(\dfrac{X\_bar}{l_{weld}}\right) = 0.81$

Effective net area $\quad Ae_{brace} := An_{brace} \cdot U = 16.47\,\text{in}^2$

$\phi := 0.75$

Reduction factor $\quad \phi R_n := \phi \cdot Fu_{brace} \cdot Ae_{brace} = 716.6\ \text{kip}$

Tensile rupture strength $\quad \textbf{if}\left(P_{u2} < \phi R_n, \text{"OK"}, \text{"NOT OK"}\right) = \text{"OK"}$

Check block shear rupture on the gusset plate (J4.3)

Block shear strength $\quad R_n = 0.60 \cdot F_u \cdot A_{nv} + U_{bs} \cdot F_u \cdot A_{nt} \leq 0.60 \cdot F_y \cdot A_{gv} + U_{bs} \cdot F_u \cdot A_{nt}$

Gross shear area $\quad Agv_{PL} := 2 \cdot t_{PL} \cdot l_{weld} = 20\ \text{in}^2$

Net shear area $\quad Anv_{PL} := Agv_{PL} = 20\ \text{in}^2$

$0.60 \cdot Fy_{PL} \cdot Agv_{PL} = 600\ \text{kip}$

$0.60 \cdot Fu_{PL} \cdot Anv_{PL} = 780\ \text{kip}$

$U_{bs} := 1$

Gross tensile area $\quad Agt_{PL} := t_{PL} \cdot B = 5\ \text{in}^2$

Net tensile area $\quad Ant_{PL} := Agt_{PL} = 5\ \text{in}^2$

$U_{bs} \cdot Fu_{PL} \cdot Ant_{PL} = 325\ \text{kip}$

Reduction factor $\quad \phi := 0.75$

recall

$t_{PL} = 0.5\ \text{in}$

$l_{weld} = 20\ \text{in}$

$Fy_{PL} = 50\ \text{ksi}$

$Fu_{PL} = 65\ \text{ksi}$

$B = 10\ \text{in}$

$P_{u2} = 419\ \text{kip}$

Block shear strength $\quad \phi R_n := \phi \cdot \left(min\left(0.60 \cdot Fy_{PL} \cdot Agv_{PL}, 0.60 \cdot Fu_{PL} \cdot Anv_{PL}\right) + U_{bs} \cdot Fu_{PL} \cdot Ant_{PL}\right) = 693.8\ \text{kip}$

$\textbf{if}\left(P_{u2} < \phi R_n, \text{"OK"}, \text{"NOT OK"}\right) = \text{"OK"}$

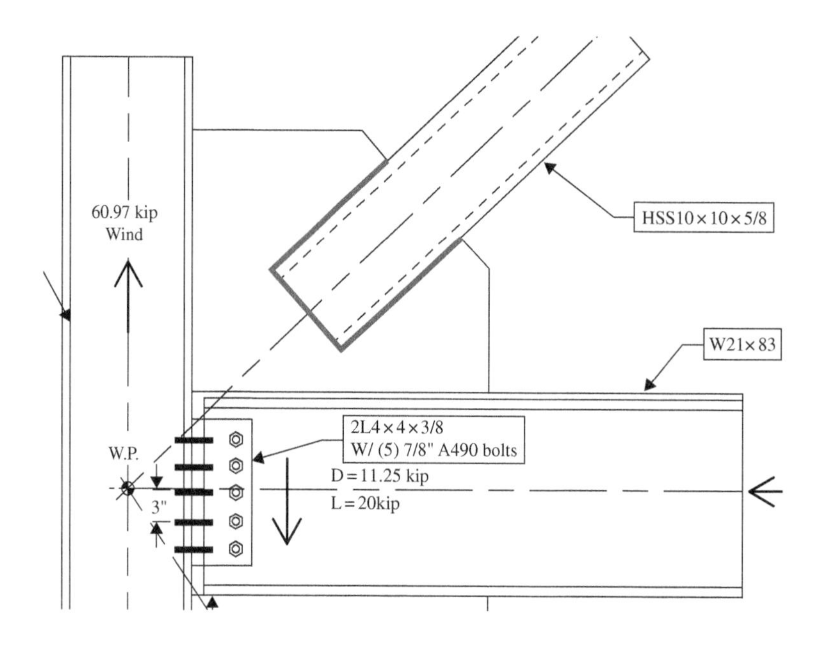

Check the gusset plate for tensile yielding on the Whitmore section (J4.1a)

Width of Whitmore section (p. 9-3)
See Appendix figure above

$$I_w := B + 2 \cdot I_{weld} \cdot \tan(30 \cdot \text{deg}) = 33.1 \text{ in}$$

$$A_w := (I_w - 2 \cdot \text{in}) \cdot t_{PL} + 2 \cdot \text{in} \cdot (tw_{beam}) = 16.6 \text{ in}$$

Gross tensile area
Reduction factor

$$\phi := 0.90$$

Tensile yielding strength

$$\phi R_n := \phi \cdot Fy_{PL} \cdot A_w = 746 \text{ kip}$$

$$\textbf{if} \langle P_{u2} < \phi R_n, \text{"OK"}, \text{"NOT OK"} \rangle = \text{"OK"}$$

recall
$B = 10$ in
$I_{weld} = 20$ in
$t_{PL} = 0.5$ in
$tw_{beam} = 0.515$ in
$P_{u2} = 419$ kip

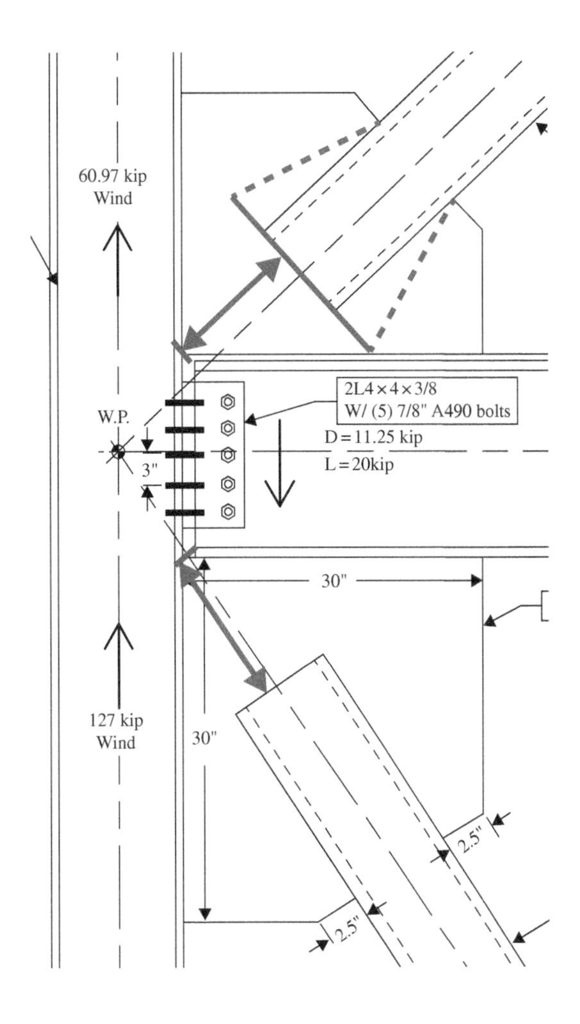

Check the gusset plate for buckling on the Whitmore section

Effective length factor $K_{PL} := 0.65$

Length of plate to buckle radius of gyration $L_{PL} := 5.5 \cdot in$

$\dfrac{K \cdot L}{r}$ $r_{PL} := \sqrt{\dfrac{L_{PL} \cdot t_{PL}^{\ 3}}{12 \cdot t_{PL} \cdot L_{PL}}} = 0.144\ in$

$\dfrac{K_{PL} \cdot L_{PL}}{r_{PL}} = 24.8$ < 25 yielding controls

Reduction factor $\phi := 0.90$

Yielding strength $\phi P_n := \phi \cdot FY_{PL} \cdot A_W = 746\ kip$

$\mathbf{if}\left(P_{u2} < \phi P_n, \text{"OK"}, \text{"NOT OK"}\right) = \text{"OK"}$

Gusset-to-Beam Connection

Check the gusset plate for tensile yielding and shear yielding along the beam flange (J4.1a and J4.2a)

$ductility := 1.25$

$$f_{a.beam} := \frac{V_{beam}}{L} = -5.8\ \frac{kip}{in}$$

$$f_{v.beam} := \frac{H_{beam}}{L} = -8.6\ \frac{kip}{in}$$

$$f_{a.beam} := 0\ \frac{kip}{in}$$

$$\theta_{load.beam} := \operatorname{atan}\left(\frac{f_{a.beam}}{f_{v.beam}}\right) = 34.3\ deg$$

$$f_{peak.beam} := \sqrt{\left(f_{a.beam} + f_{b.beam}\right)^2 + {f_{v.beam}}^2} = 10.4\ \frac{kip}{in}$$

$$f_{avg.beam} := \frac{1}{2}\left(\sqrt{\left(f_{a.beam} + f_{b.beam}\right)^2 + {f_{v.beam}}^2} + \sqrt{\left(f_{a.beam} + f_{b.beam}\right)^2 + {f_{v.beam}}^2}\right) = 10.4\ \frac{kip}{in}$$

$$f_{weld.beam} := \max\left\langle ductility \cdot f_{avg.beam}, f_{peak.beam}\right\rangle = 13\ \frac{kip}{in}$$

$$D.beam := \frac{f_{weld.beam}}{2 \cdot \left(1.392\ \frac{kip}{in}\right) \cdot \left(1 + 0.5 \cdot \sin\left\langle \theta_{load.beam}\right\rangle^{1.5}\right)} = 3.8$$

use 5/16th

$D := 5 \cdot in$

16th of an inch

Gusset-to-Column Connection

Check the gusset plate for tensile yielding and shear yielding along the column flange (J4.1a and J4.2a)

$$f_{a.col} := \frac{V_{col}}{L} = -6.6\ \frac{kip}{in}$$

$$f_{v.col} := \frac{H_{col}}{L} = -3.8\ \frac{kip}{in}$$

$$f_{b.col} := 0\ \frac{kip}{in}$$

$$\theta_{load.col} := \operatorname{atan}\left(\frac{f_{a.col}}{f_{v.col}}\right) = 59.7\ deg$$

$$f_{peak.col} := \sqrt{\left\langle f_{a.col} + f_{b.col}\right\rangle^2 + {f_{v.col}}^2} = 7.6\ \frac{kip}{in}$$

$$f_{avg.col} := \frac{1}{2}\left(\sqrt{\left\langle f_{a.col} - f_{b.col}\right\rangle^2 + {f_{v.col}}^2} + \sqrt{\left\langle f_{a.col} + f_{b.col}\right\rangle^2 + {f_{v.col}}^2}\right) = 7.6\ \frac{kip}{in}$$

$$f_{weld.col} := \max\left\langle ductility \cdot f_{avg.col}, f_{peak.col}\right\rangle = 9.5\ \frac{kip}{in}$$

$$D.col := \frac{f_{weld.col}}{2. \left(1.392\ \frac{kip}{in}\right) \cdot \left(1 + 0.5 \cdot \sin\left\langle \theta_{load.col}\right\rangle^{1.5}\right)} = 2.4$$

use 5/16th

$D = 5\ in$

16th of an inch

Beam-to-Column Connection

For an all bolted double angle connection with 1.5ε edge distance using (5) 7/8ε A490 (assume threads included)

Nominal 2L strength $\phi P_n := min\left(185 \cdot kip, 512 \cdot \frac{kip}{in} \cdot tw_{beam}\right) = 185\ kip$ AISC Table 10-1

$R_{ub} = 31.3\ kip$

$\mathbf{if}\left\langle R_{ub} < \phi P_n, \text{"OK"}, \text{"NOT OK"}\right\rangle = \text{"OK"}$

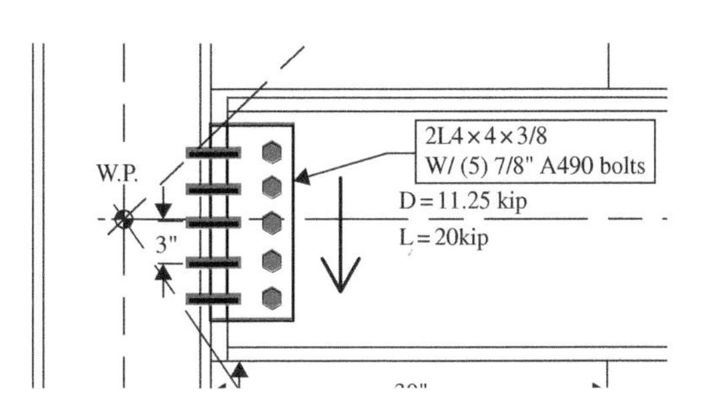

Beam Web Yielding (J10.2.b)

$\phi := 1.00$

$\phi R_n := \phi \cdot Fy_{beam} \cdot tw_{beam} \cdot \left\langle 2.5 \cdot k\_des_{beam} + L\right\rangle = 858.8\ kip$

$\mathbf{if}\left\langle |V_{beam}| < \phi R_n, \text{"OK"}, \text{"NOT OK"}\right\rangle = \text{"OK"}$

recall

$k\_des_{beam} = 1.3\ in$

$L = 30\ in$

$tw_{beam} = 0.5\ in$

$tf_{beam} = 0.8\ in$

$V_{beam} = -175.3\ kip$

$d_{beam} = 21.4\ in$

Beam Web Crippling (J10.3b)

$\phi := 0.75$

$$\phi R_n := \phi \cdot 0.40 \cdot tw_{beam}^{2} \cdot \sqrt{\frac{E \cdot Fy_{beam} \cdot tf_{beam}}{tw_{beam}}} \cdot \begin{cases} 1 + \left(\frac{4 \cdot L}{d_{beam}} - 0.2\right) \cdot \left(\frac{tw_{beam}}{tf_{beam}}\right)^{1.5} & \text{if } \frac{L}{d_{beam}} > 0.2 \\ 1 + 3 \cdot \left(\frac{L}{d_{beam}}\right) \cdot \left(\frac{tw_{beam}}{tf_{beam}}\right)^{1.5} & \text{else} \end{cases} = 441.5\ kip$$

$\mathbf{if}\left\langle |V_{beam}| < \phi R_n, \text{"OK"}, \text{"NOT OK"}\right\rangle = \text{"OK"}$

References

AISC (2016). *Specification for Structural Steel Buildings*. American Institute of Steel Construction, Chicago.

AISC (2017). *Steel Construction Manual*, 15th Edition. American Institute of Steel Construction, Chicago.

IDEA StatiCa Support Center. https://www.ideastatica.com/support-center

14

HSS Circular Braces Welded to a Gusset Plate in a Chevron Concentrically Braced Frame

14.1 Problem Description

This example consists of a chevron-braced frame connection in a concentrically braced frame with square hollow structural section (HSS) braces welded to a gusset plate. The study is prepared for the size of the braces, beam, geometry, and thickness of the plate and welds as obtained from the design. All components are designed according to AISC 360-16 *Specification* (AISC 360) (AISC, 2016) and CBFEM analysis is performed in IDEA StatiCa Version 23. The connection presented is taken from Design Guide 29 (AISC, 2015).

14.2 Verification of Resistance as Per AISC

The example uses the sections and dimensions shown in Figure 14.1 and as follows. Braces are HSS8x8x1/2 (ASTM A500 Gr. C), beam W27 × 114 (ASTM A992), gusset plate ¾ in. (ASTM A572, Gr. 50) and ASTM E70XX weld. All section and material properties are obtained from the *Steel Construction Manual* (AISC, 2017).

The results of the analytical solution are represented by the comparison table for the different limit states shown in Table 14.1. The limit states that should be considered for these connections are as follows and the analysis results of these limit states are presented in Table 14.1.

1. Weld between gusset plate and brace
2. Weld between gusset plate and bottom flange of beam
3. Tension yielding of brace
4. Tension rupture of brace
5. Shear rupture in the brace wall
6. Block shear rupture
7. Gusset plate for tensile yielding and shear yielding along the beam flange
8. Gusset plate tension yielding on the Whitmore section
9. Gusset plate for buckling on Whitmore section
10. Yielding of the beam web
11. Crippling of the beam web.

Steel Connection Design by Inelastic Analysis: Verification Examples per AISC Specification, First Edition.
Mark D. Denavit, Ali Nassiri, Mustafa Mahamid, Martin Vild, Halil Sezen, and František Wald.

W-shape: A992
HSS: A500 Gr. B
Plates: A572 Gr. 50

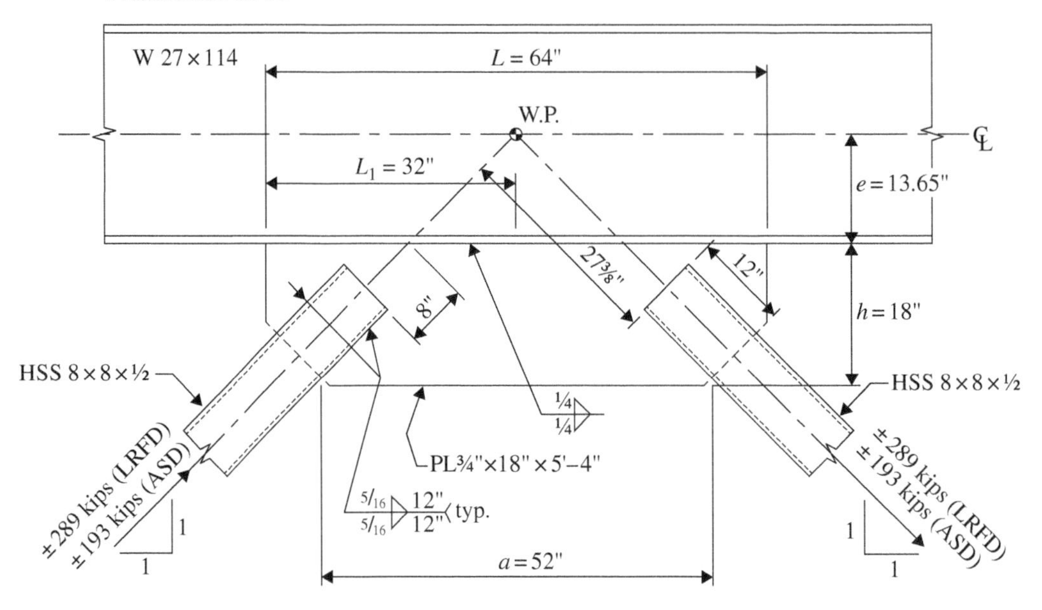

Figure 14.1 Chevron braced frame connection.

Table 14.1 Limit states checked and compared with CBFEM.

Limit state	AISC
Weld between gusset plate and brace	$\phi R_n = 333$ kips
Weld between gusset plate and beam bottom flange	$\phi R_n = 385$ kips
Tension yielding of brace	$\phi R_n = 559$ kips
Shear rupture of brace wall	$\phi R_n = 583$ kips
Brace tension rupture	$\phi R_n = 414$ kips
Block shear rupture of gusset plate	$\phi R_n = 697$ kips
Gusset plate tensile yielding on Whitmore section	$\phi R_n = 721$ kips
Gusset plate tensile yielding and shear yielding along the beam flange	$\phi R_n = 45$ ksi
	$f_{un} = 15.8$ ksi
Gusset plate for buckling on the Whitmore section	$\phi R_n = 671$ kips
Check the gusset plate for sidesway buckling	$\phi R_n = 2009$ kips
Beam web local buckling	N/A
Beam local web yielding	$\phi R_n = 2042$ kips
Beam web shear yielding	$\phi R_n = 1094$ kips
Beam web crippling	$\phi R_n = 1311$ kips

The governing component of this connection is the weld shear between the gusset plate and the brace (ϕR_n =333 kips > P_u = 289 kips). The utilization of this weld is 87%. The next critical check is the brace tension rupture with the load resistance of ϕR_n =414 kips > Pu = 289 kips (utilization 70%).

14.3 Resistance by CBFEM

The overall check of the connection is verified as shown in Figures 14.2–14.4. The check shows that the connection works according to the CBFEM. Failure in members and plates due to yielding and rupture limit states is measured based on the 5% plastic strain limit. Figure 14.3 shows that the plastic strain is 0.1% which is well below the 5% plastic strain limit. The connection presented is a welded connection. The weld shear limit state is usually accurate when compared with the AISC *Specification* procedure. CBFEM uses the AISC 360-16 provisions of Chapter J to check the weld strength. It can be seen that the weld check utilization is 86.6%. The analysis is materially nonlinear, and it should not be relied only upon utilization. By overloading the basic model from 333 kips to 334 kips in each brace, the load resistance is revealed – the weld just barely holds at 333 kips and fails at 334 kips. Both AISC and CBFEM give the weld as the governing component and provide the same load resistance.

For the gusset plate tensile yielding and shear yielding along the beam flange, the AISC 360-16 procedure requires comparing the combined yielding and shear stresses to the permitted stress (ϕR_n = ϕF_y=0.9(50 ksi)=45 ksi). The results of the comparison are shown in Table 14.1 and are in agreement. Figure 14.5 shows the stress distribution in the overall connection and in the gusset plate.

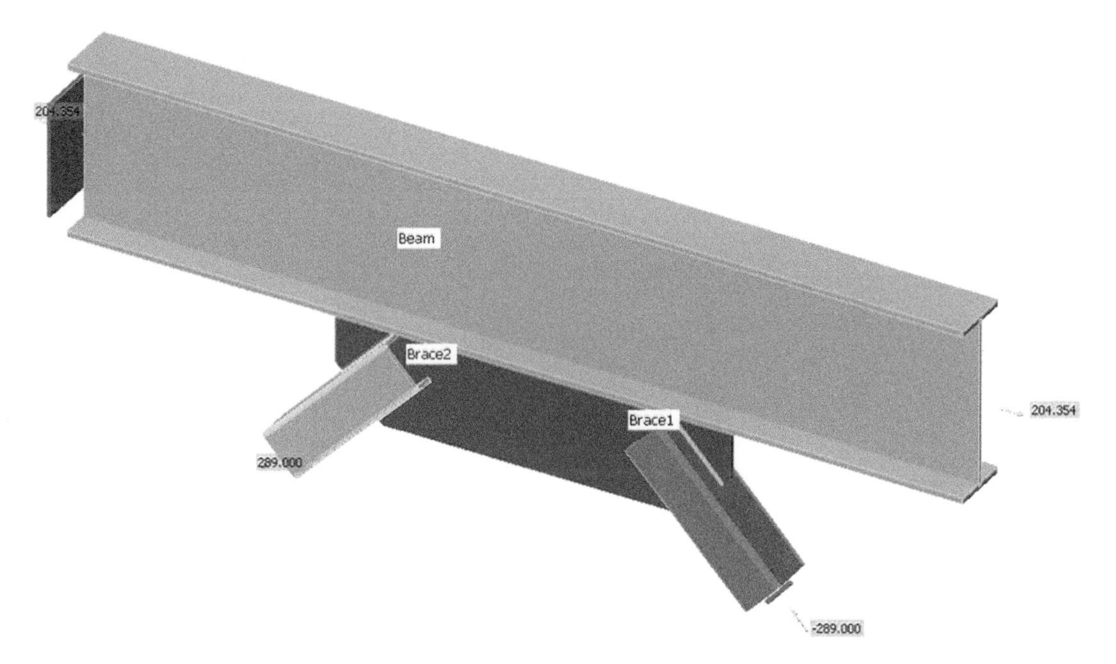

Figure 14.2 Design model.

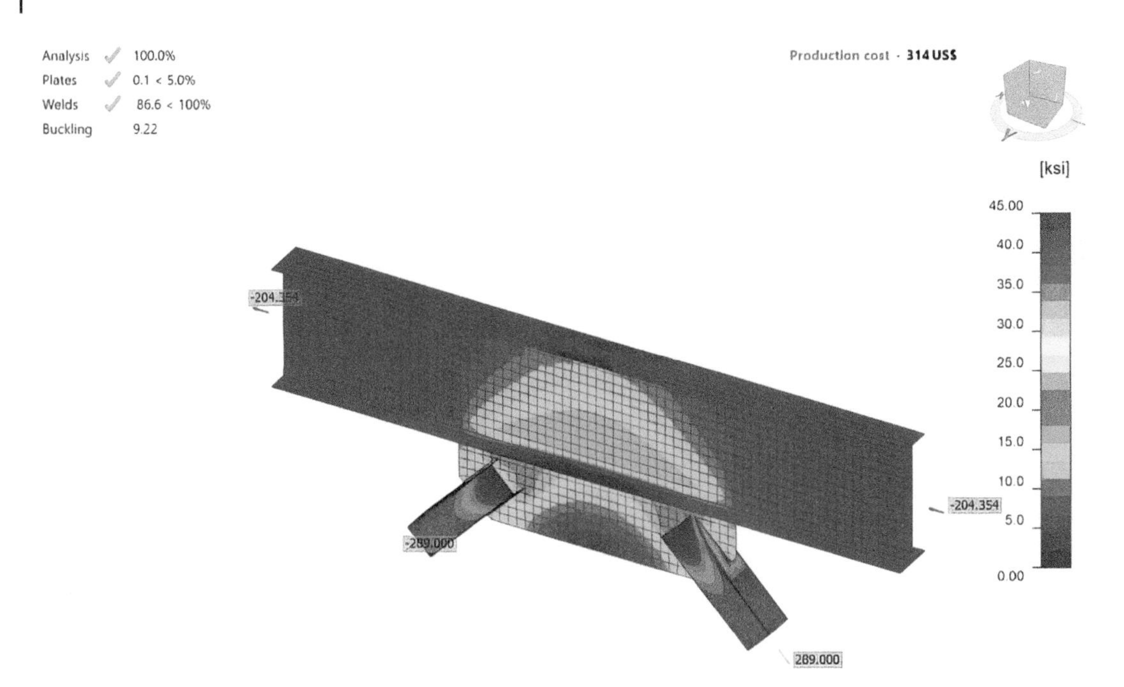

Figure 14.3 Overall solution of the connection – plastic strains.

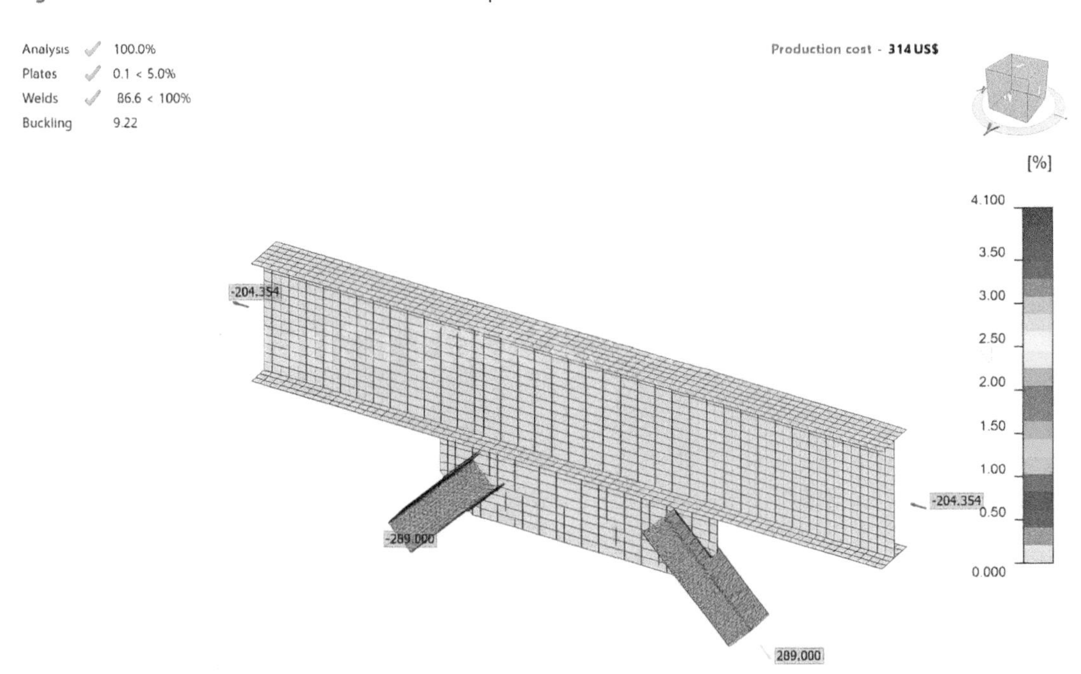

Figure 14.4 Overall solution of the connection – stresses.

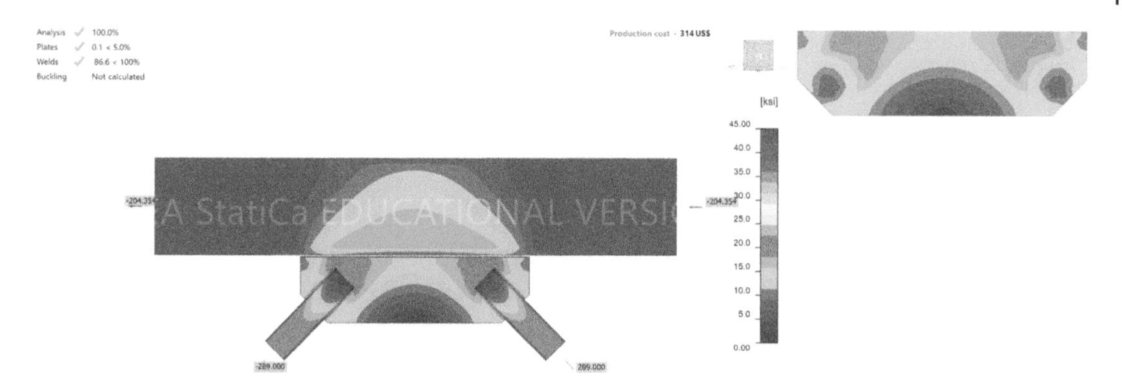

Figure 14.5 Gusset plate for tensile yielding and shear yielding along the beam flange.

The buckling of the gusset plate is the function of the axial compression load in the connecting braces, and the requirement for the same in CBFEM can be checked by a buckling multiplier factor obtained using analysis. The buckling capacity is obtained as per CBFEM after multiplying the axial compressive load and the buckling factor, which is the elastic buckling capacity. The elastic buckling capacity is higher than the inelastic buckling capacity. Currently, it is the only measure, and it is hard to differentiate between the buckling resistance of various connecting parts, e.g. the buckling of the gusset plate on the Whitmore section or the gusset plate sideways buckling. Mode 1 is the governing mode among several modes in CBFEM and for the given connection, mode 1 buckling is found to exist in the bottom gusset plate, as it is subjected to the action of the compression load. The buckling factor obtained for this connection with ¾ in. gusset plate by CBFEM is 7.86, which is for the critical mode (mode 1), observed for an axial compressive load of 289 kips in the bottom brace as shown in Figure 14.6. For local buckling, the preferred value for the buckling factor is greater than 9.16 for 50 ksi steel per AISC 360-16 Section J.4,

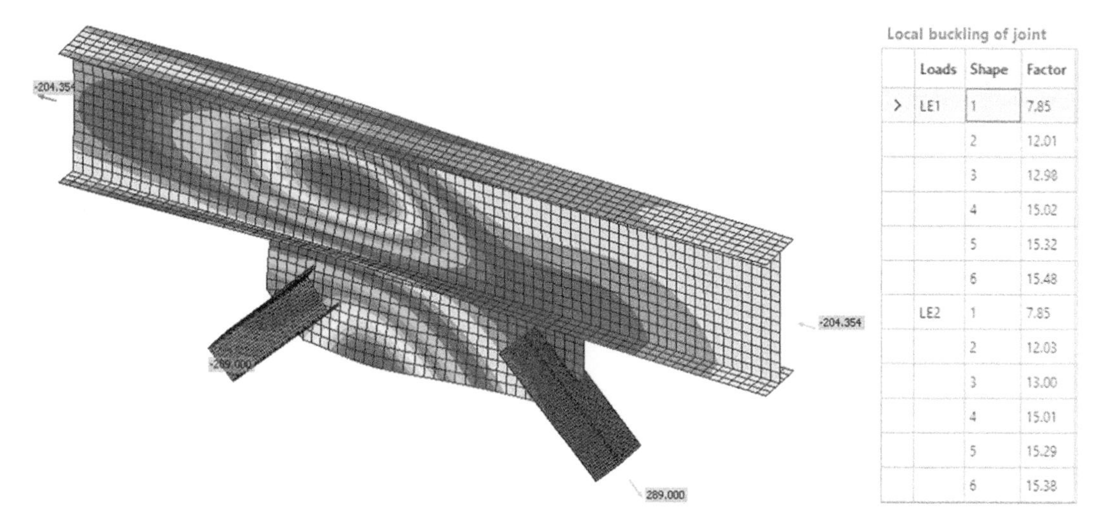

Local buckling of joint

	Loads	Shape	Factor
>	LE1	1	7.85
		2	12.01
		3	12.98
		4	15.02
		5	15.32
		6	15.48
	LE2	1	7.85
		2	12.03
		3	13.00
		4	15.01
		5	15.29
		6	15.38

Figure 14.6 First buckling mode shape with the factor of 7.85.

therefore, it is expected that the gusset would buckle at this load according to CBFEM, however, buckling according to CBFEM still requires more research to allow accurate prediction. The buckling limit state is not a controlling limit state per both AISC (Table 14.1) and CBFEM.

14.4 Parametric Study

To verify the resistance of other components and the ability of CBFEM to capture all the failure modes, the parametric study is prepared by varying plate thicknesses and weld size.

Modification 1 – fillet welds at braces changed to butt welds

The governing failure mode of the basic model is the failure of fillet welds at braces. Therefore, these fillet welds are changed in the model to full-penetration butt welds. The load at braces may be increased to 479 kips. At this load, the fillet welds between the gusset plate and the beam are utilized at 100%; see Figure 14.7. Hand calculation provides resistance 430 kips. CBFEM provides higher resistance by 10%.

Modification 2: all fillet welds changed to butt welds

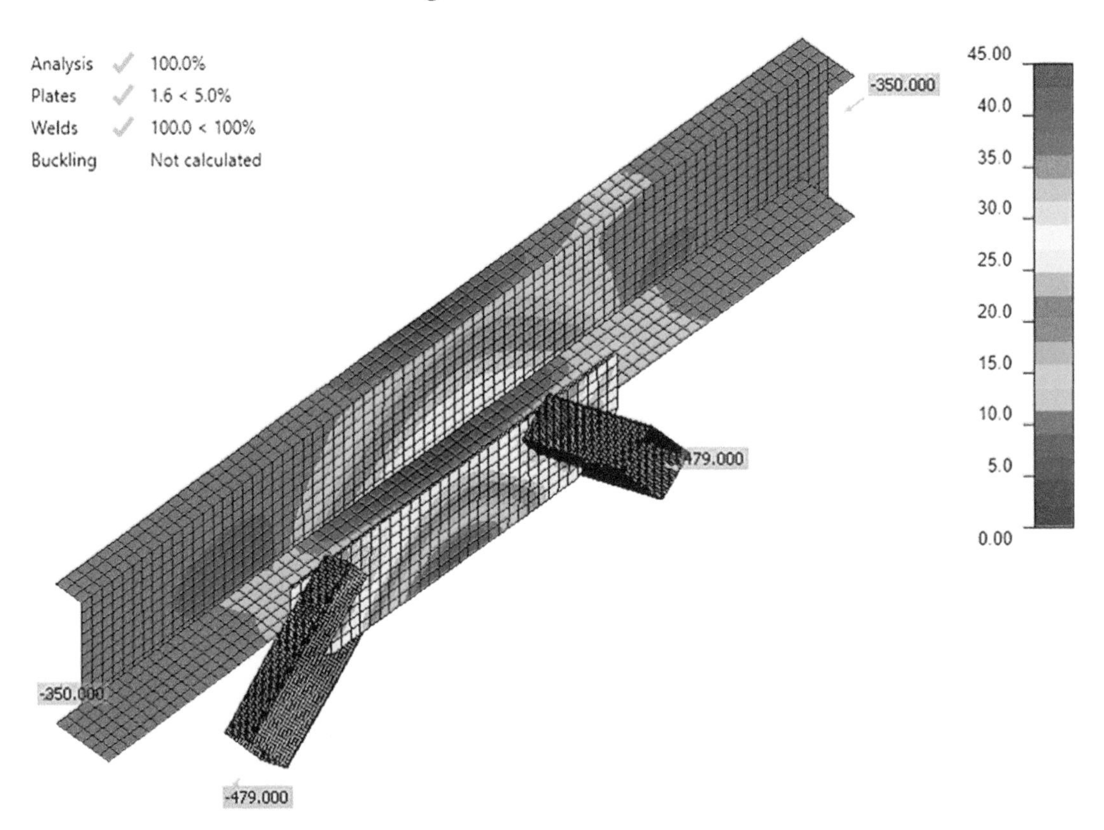

Figure 14.7 Modified model with butt welds between braces and gusset plate.

The second modification avoids the failure mode of fillet welds between the gusset plate and the brace. A plastic limit strain limit check is used to simulate the following checks in hand calculation: Tension yielding of the brace: $\phi R_n = 559$ kips, shear rupture of brace wall: $\phi R_n = 583$ kips, and brace tension rupture: $\phi R_n = 414$ kips. The plastic strains start at the net section of the brace and propagate to the gross area as the load increases. The load can be increased to 540 kips, when the plates of both braces just barely satisfy the plastic strain limit check. This load is in agreement with the AISC capacities shown in Table 14.1 for tension yielding and shear rupture. The tension rupture per AISC 360 specifications is less than what was obtained from CBFEM, and this is due to the shear lag factor, U, which is equal to 0.75 in this case and as required by Table D3.1 case 6 (AISC 360-16) (AISC, 2016); the shear lag factor is multiplied by the net area of the brace. The effect of shear lag is apparent in Figure 14.8. Without the shear lag factor, the tension rupture capacity of the section is 552 kips per AISC which is more in line with CBFEM capacity.

Modification 3: all butt welds and gusset plate thickness lowered to 3/8 in.

This modification is used to investigate failure modes connected to the gusset plate. The plastic strain limit is exceeded at load 400 kips in each brace, as shown in Figure 14.9. This check simulates the block shear rupture of the gusset plate, gusset plate tensile yielding on Whitmore section, gusset plate tensile yielding, and shear yielding along the beam flange. According to CBFEM, gusset plate tensile yielding and shear yielding along the beam flange are the governing failure mode, and the block shear rupture of gusset plate will soon follow because the significant plastic strain is along the full length of the braces.

The AISC procedure predicts tensile yielding of the gusset plate at the Whitmore section followed by block shear in the gusset plate. Since CBFEM uses von Mises stresses which include both normal and

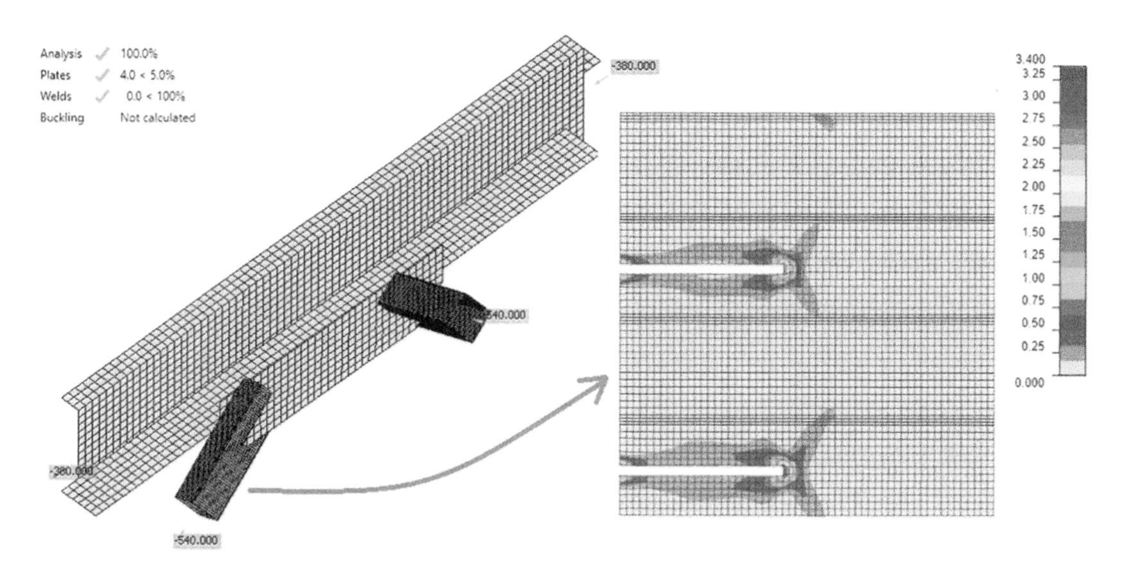

Figure 14.8 Plastic strain in the connection with butt welds only.

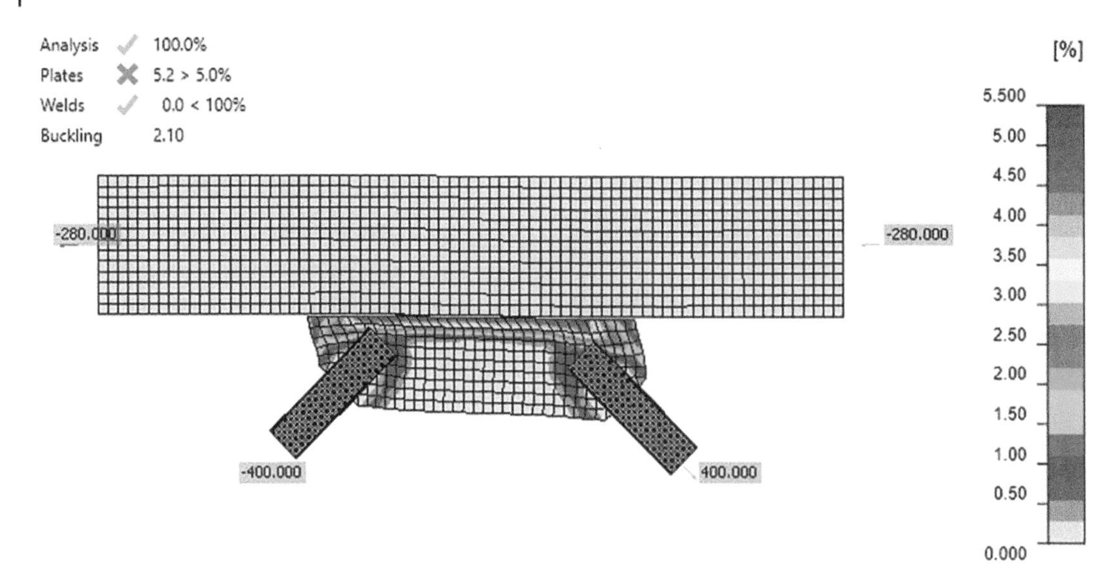

Figure 14.9 Plastic strain at the model with thin gusset plate.

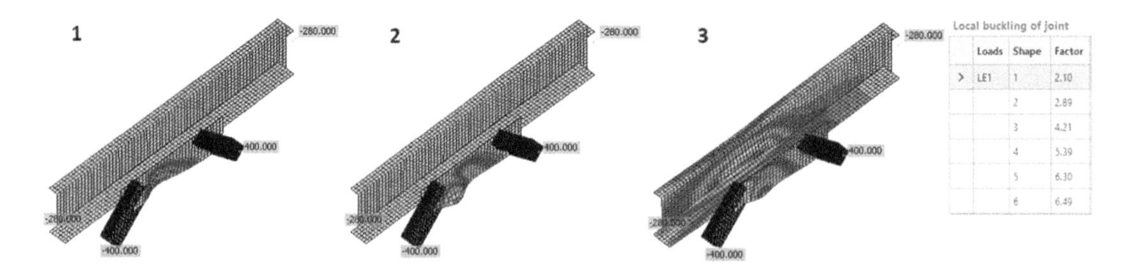

Figure 14.10 First three buckling mode shapes of model with thin gusset plate.

shear stresses, the prediction by CBFEM is accurate. For buckling analysis of the gusset plate, both AISC and CBFEM predicted the buckling in the 3/8 in. plate. The AISC buckling capacity for the gusset plate is 359 kips, where the applied load is 400 kips. Figure 14.10 shows the first three buckling modes from CBFEM. Figures 14.9 and 14.10 show the buckling factor at a load equals to 400 kips which produces an equivalent plastic strain equals to 5.2%, which is just above the 5% plastic strain limit. The buckling factor for the first mode is 2.10 which is associated with buckling of the gusset plate, as shown in Figure 14.10. The buckling factor is less than the 9.16 limit per AISC 360-16 Section J.4 (AISC, 2016), which indicates that the gusset plate would buckle at this load.

Beam web local and shear yielding have a very large load resistance compared to the applied load. Almost all limit states in this connection would occur before these two limit states, which typically do not control the design. These limit states are checked by a 5% limit strain in the beam.

Beam web crippling is a buckling state that would occur after yielding; therefore, linear buckling analysis is not perfectly ideal. In CBFEM, using geometrically linear analysis without imperfections, the buckling factor limit is the only way to capture this failure mode. A separate model was not created specifically for these failure modes to govern.

14.5 Summary

It can be concluded that the CBFEM is capable of predicting the actual behavior and failure mode of chevron braced frame connections similar to the one presented herein.

The various limit states were investigated carefully by performing a parametric study which resulted in obtaining the capacity for each limit state using CBFEM. The weld capacity between the braces and the gusset based on AISC 360 *Specification* (AISC, 2016) matches what is obtained by CBFEM, where the weld between the gusset plate and the beam has the capacity per AISC is less than the capacity per CBFEM by 10%. The plate limit states, including yielding, rupture, are based on a 5% plastic strain limit in CBFEM; for these limit states, the difference between the AISC and CBFEM is within 10%. The buckling limit state was investigated per AISC and per CBFEM; in the connection investigated, buckling was not a governing limit state. To investigate buckling, a 3/8 in. plate was investigated and in both the AISC procedure and CBFEM, buckling of the plate was observed in both methods.

Benchmark Case

Input

Beam cross-section

- W27X114
- Steel ASTM A992

Braces cross-section

- HSS 8X8X1/2
- Steel ASTM A500 Gr. C

Gusset plate

- Thickness 3/4 in.
- Steel ASTM A572 Gr. 50

Loading

- Axial force $N = \pm 289$ kips

Output

- Weld 86.6%
- Plastic strain 0.1% < 5%
- Buckling factor 7.85

Appendix

The AISC Solution

ORIGIN := 1
Refer to AISC 14th Edition shapes database:

$$T1 := \quad Row(shape) := \left| \begin{array}{l} \text{for } i \in 1..rows(T1) - 1 \\ R \leftarrow 1 \text{ if } \left(T1^{\langle 2 \rangle}\right)_i = shape \\ R \end{array} \right.$$

Solve Example 5.9 Chevron Brace Connection

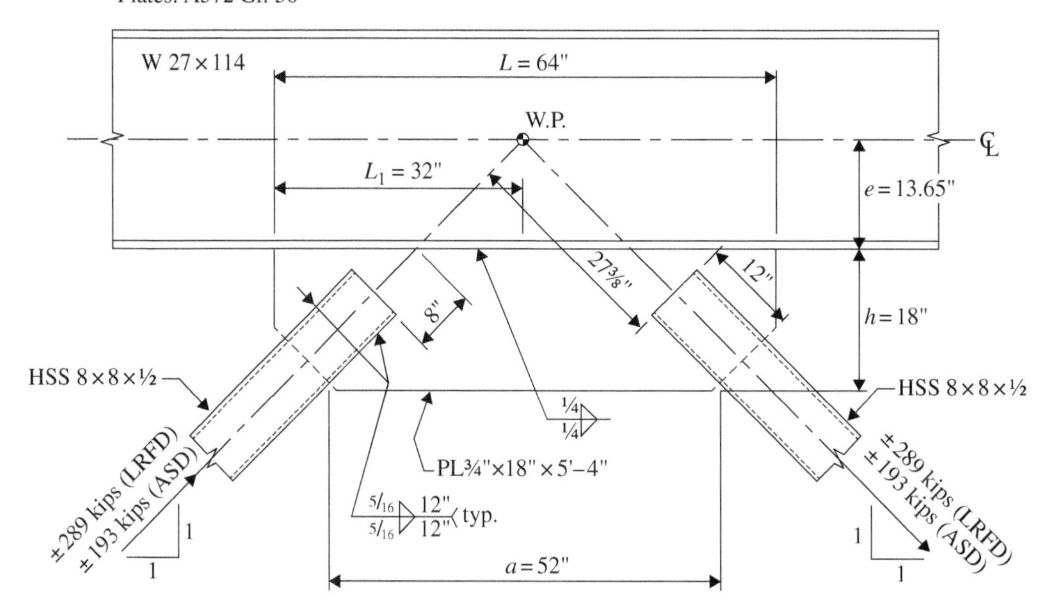

Loads

LRFD Brace axial force

Note : similar in both braces $\quad P_u := 289\,kip$

Material Properties

Beams steel grade A992	$Fy_{beam} := 50 \cdot ksi$	$Fu_{beam} := 65 \cdot ksi$
Brace steel grade A500 GrB	$Fy_{brace} := 46 \cdot ksi$	$Fu_{brace} := 58 \cdot ksi$
Plate steel grade A572 Gr. 50	$Fy_{PL} := 50\,ksi$	$Fu_{PL} := 65\,ksi$
Modulus of Elasticity	$E := 29000\,ksi$	
Weld strength	$F_{EXX} := 70\,ksi$	$C_1 := 1.0$

Member Properties

Beam Properties

Beam section $\boxed{Beam := \text{“W27} \times \text{114”}}$

Depth of beam $d_{beam} := T1_{(Row(Beam),7)} \cdot in = 27.3 \cdot in$

Width of flange $df_{beam} := T1_{(Row(Beam),12)} \cdot in = 10.1 \cdot in$

Thickness of flange $tf_{beam} := T1_{(Row(Beam),20)} \cdot in = 0.93 \cdot in$

Thickness of web $tw_{beam} := T1_{(Row(Beam),17)} \cdot in = 0.57 \cdot in$

Design weld depth $k\_des_{brace} := T1_{(Row(Beam),25)} \cdot in = 1.53 \cdot in$

Brace Properties

Brace section $\boxed{Brace := \text{“HSS8} \times 8 \times .500\text{”}}$

Cross-sectional area $Ag_{brace} := T1_{(Row(Brace),6)} \cdot in^2 = 13.5 \cdot in^2$

Radius of gyration $r_{brace} := T1_{(Row(Brace),42)} \cdot in = 3 \cdot in$

Depth to end of weld $t\_des_{brace} := T1_{(Row(Beam),24)} \cdot in = 0.465 \cdot in$ this is the r.x value (sim to r.y for square shapes)

$B := T1_{(Row(Beam),9)} \cdot in = 8\,in$

$H := T1_{(Row(Beam),14)} \cdot in = 8\,in$

Slope of brace $slope := atan\left(\frac{12}{12}\right) = 45 \cdot deg$

Plate Properties

Thickness	$t_{PL} := 0.75\,in$
Length	$L := 64\,in$
Height	$h := 18\,in$
Cut length	$a := 52\,in$
Distance brace to W.P.	$e_{WP} := 27.375\,in$
Eccentricity of brace CL to W.P.	$e := 13.65\,in$
End of PL to W.P.	$L_1 := 32\,in$
Length (centered)	$L_2 := L_1$

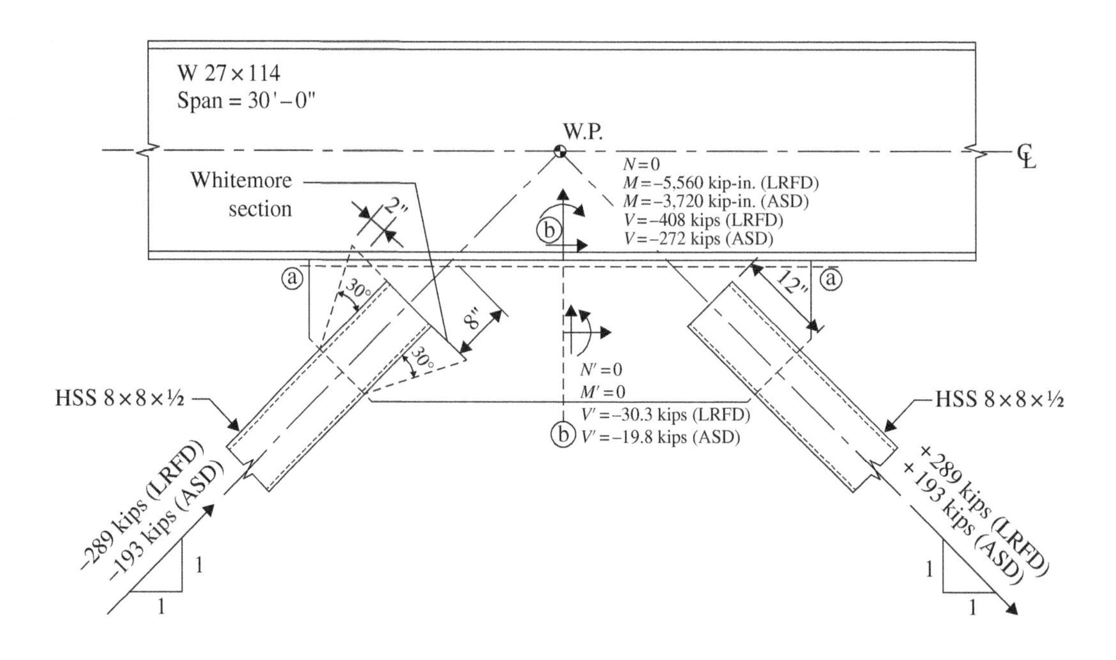

Subscript "1" denotes Brace 1, on the left side of Figure 5-19, and the subscript "2" denotes Brace 2 on the right side

$$\Delta := \frac{L_1 - L_2}{2} = \frac{32 \cdot in - 32 \cdot in}{2} = 0\,in$$

$P_{u1} := -289kip$

$H_{u1} := P_{u1} \cdot \cos(slope) = -204.35 \cdot kip$

$V_{u1} := P_{u1} \cdot \sin(slope) = -204.4 \cdot kip$

$M_{u1} : H_{u1} \cdot e + V_{u1} \cdot \Delta = -2789.4 \cdot kip \cdot in$

$M'_{u1} : \frac{1}{8} \cdot V_{u1} L - \frac{1}{4} \cdot H_{u1} \cdot h - \frac{1}{2} \cdot M_{u1} = 679 \cdot kip \cdot in$

$P_{u2} := -289kip$

$H_{u2} := P_{u2} \cdot \cos(slope) = -204.4 \cdot kip$

$V_{u2} := P_{u2} \cdot \sin(slope) = -204.4 \cdot kip$

$M_{u2} : H_{u2} \cdot e + V_{u2} \cdot \Delta = -2789.4 \cdot kip \cdot in$

$M'_{u2} : \frac{1}{8} \cdot V_{u2} L - \frac{1}{4} \cdot H_{u2} \cdot h - \frac{1}{2} \cdot M_{u2} = -679 \cdot kip \cdot in$

Axial $N_u := V_{u1} + V_{u2} = 0 \cdot kip$

Shear $V_u := H_{u1} + H_{u2} = -408.7 \cdot kip$

Moment $M_u := M_{u1} + M_{u2} = -5578.9 \cdot kip \cdot in$

Axial $N'_u := \frac{1}{2} \cdot (H_{u1} + H_{u2}) = 0 \cdot kip$

Shear $V'_u := \frac{1}{2} \cdot (V_{u1} - V_{u2}) - \frac{2M_u}{L} = -30 \cdot kip$

Moment $M'_u := M'_{u1} + M'_{u2} = 0 \cdot kip \cdot in$

recall

$V_{u1} = -204.4\,kip$

$V_{u2} = 204.4\,kip$

$H_{u1} = -204.4\,kip$

$H_{u2} = 204.4\,kip$

$L = 64\,in$

$M'_{u1} = 679.5\,in \cdot kip$

$M'_{u2} = -679.5\,in \cdot kip$

Check tension yielding on the brace (J4.1a)

Reduction factor	$\phi := 0.90$	recall
Tensile yielding	$\phi R_n := \phi \cdot Fy_{brace} \cdot Ag_{brace} = 558.9 \cdot kip$	$Fy_{brace} = 46 \cdot ksi$
	$if\left(P_u < \phi R_n, \text{"OK"}, \text{"NOT OK"}\right) = \text{"OK"}$	$Ag_{brace} = 13.5\ in^2$
		$P_u = 289\, kip$

Shear rupture in the brace wall (J4.2b)

Reduction factor	$\phi := 0.75$	recall
	$P_u \le \phi \cdot 0.6 \cdot F_u \cdot A_{nv} = \phi \cdot 0.6 \cdot F_u \cdot t\_des_{brace} \cdot (4) \cdot I_{weld}$	$P_u = 289\ kip$
Calculated weld length (4 lengths of weld)	$I_{weld} := \dfrac{P_u}{\left[0.6 \cdot \left[\phi \cdot Fu_{brace} \cdot t\_des_{brace} \cdot (4)\right]\right]} = 6\ in$	$Fu_{brace} = 58 \cdot ksi$
		$t\_des_{brace} = 0.5\ in$

Check weld strength (Eq. 8-1)

Reduction factor $\qquad \phi := 0.75$

Try weld size 5/16 inch $\qquad W_{weld} := 5in \qquad$ 16th of an inch

Weld strength

$$P_u \le \phi \cdot R_n = \phi \cdot 0.60 \cdot F_{EXX} \cdot \left(\frac{1}{\sqrt{2}}\right) \cdot W_{weld} \cdot (4) \cdot I_{weld}$$

recall
$P_u = 289\ kip$
$F_{EXX} = 70\ ksi$

Minimum weld length

$$I_{weld} \ge \frac{P_u}{\left[\phi \cdot 0.60 \cdot F_{EXX} \cdot \left(\frac{1}{\sqrt{2}}\right) \cdot W_{weld} \cdot (4)\right]}$$

$$I_{weld} := \frac{P_u}{\left[\phi \cdot 0.60 \cdot F_{EXX} \cdot \left(\frac{1}{\sqrt{2}}\right) \cdot \frac{W_{weld}}{16} \cdot (4)\right]} = 10.4\ in$$

Check weld strength (Eq.: 8-2)

Weld strength	$\phi \cdot P_n = 1.392D \cdot I_{weld} \cdot (4)$	recall
Minimum weld length	$I_{weld} := \dfrac{P_u}{1.392ksi \cdot w_{weld} \cdot (4)} = 10.4\ in$	$P_u = 289\ kip$
		$w_{weld} = 5\ in$
Use $I_{weld} = 12''$	$I_{weld} := 12\ in$	

Check tensile rupture on the brace (J4.1b)

		recall
Slot width	$d_{slot} := t_{PL} + \frac{2}{16} in = 0.9 in$	$t_{PL} = 0.75$ in
Net area	$An_{brace} := Ag_{brace} - 2 \cdot t\_des_{brace} \cdot d_{slot} = 12.69\ in^2$	$Ag_{brace} = 13.5\ in^2$
Eccentricity of connection	$x\_bar := \frac{B^2 + 2 \cdot B \cdot H}{4(B + H)} = 3 in$	$t\_des_{brace} = 0.465$ in
Table D3.1, Case 6		$B = 8 in$
Shear lag factor		$H = 8 in$
Table D3.1, Case 6	$U := 1 - \left(\frac{x\_bar}{I_{weld}}\right) = 0.75$	$I_{weld} = 12 in$
Effective net area	$Ae_{brace} := An_{brace} \cdot U = 9.51\ in^2$	$Fu_{brace} = 58$ ksi
Reduction factor	$\phi := 0.75$	$P_u = 289$ kip
Tensile rupture strength	$\phi R_n := \phi \cdot Fu_{brace} \cdot Ae_{brace} = 413.9 \cdot kip$	
	if $(P_u < \phi R_n$, "OK", "NOT OK"$)$ = "OK"	

Check block shear rupture on the gusset plate (J4.3)

		recall
Block shear strength	$R_n = 0.60F_u \cdot A_{nv} + U_{bs} \cdot F_u \cdot A_{nt} \leq 0.60F_y \cdot A_{gv} + U_{bs} \cdot F_u \cdot A_{nt}$	$t_{PL} = 0.75$ in
Gross shear area	$Agv_{PL} := 2 \cdot t_{PL} \cdot I_{weld} = 18 in^2$	$I_{weld} = 12$ in
Net shear area	$Anv_{PL} := Agv_{PL} = 18 in^2$	$Fy_{PL} = 50$ ksi
	$0.60 \cdot Fy_{PL} \cdot Agv_{PL} = 540 \cdot kip$	$Fu_{PL} = 65$ ksi
	$0.60 \cdot Fu_{PL} \cdot Agv_{PL} = 702 \cdot kip$	$B = 8$ in
	$U_{bs} := 1$	$P_u = 289$ kip
gross tensile area	$Agt_{PL} := t_{PL} \cdot B = 6\ in^2$	
net tensile area	$Ant_{PL} := Agt_{PL} = 6\ in^2$	
	$U_{bs} \cdot Fu_{PL} \cdot Ant_{PL} = 390 \cdot kip$	
Reduction factor	$\phi := 0.75$	
Block shear strength	$\phi R_n := \phi \cdot (\min(0.60 \cdot Fy_{PL} \cdot Agv_{PL}, 0.60 \cdot Fu_{PL} \cdot Anv_{PL}) + U_{bs} \cdot Fu_{PL} \cdot Ant_{PL}) = 697.5 \cdot kip$ if $(P_u < \phi R_n$, "OK", "NOT OK"$)$ = "OK"	

Check the gusset plate for tensile yielding on the Whitmore section (J4.1a)

		recall
Width of Whitmore section (pg: 9-3) see Appendix figure above	$I_w := B + 2 \cdot I_{weld} \cdot \tan(30 deg) = 21.9$ in	$B = 8$ in
		$I_{weld} = 12$ in
Gross tensile area	$A_w := (I_w - 2 in) \cdot t_{PL} + 2 in \cdot (tw_{beam}) = 16\ in^2$	$t_{PL} = 0.75$ in
Reduction factor	$\phi := 0.90$	$tw_{beam} = 0.57$ in
Tensile yielding strength	$\phi R_n := \phi Fy_{PL} \cdot A_w = 721.5 \cdot kip$	$P_u = 289$ kip
	if $(P_u < \phi R_n$, "OK", "NOT OK"$)$ = "OK"	

Check the gusset plate for buckling on the Whitmore section

Effective length factor	$K_{PL} := 0.65$	recall
Length of plate to buckle	$L_{PL} := 8\ in$	$B = 8\ in$
Radius of gyration	$r_{PL} := \sqrt{\dfrac{L_{PL} \cdot t_{PL}^3}{12 \cdot t_{PL} \cdot L_{PL}}} = 0.217\ in$	$t_{PL} = 0.8\ in$
$\dfrac{K \cdot L}{r}$	$\dfrac{K_{PL} \cdot L_{PL}}{r_{PL}} = 24$ <25	

Gusset-to-Beam Connection

Check the gusset plate for tensile yielding and shear yielding along the beam flange (J4.1a and J4.2a)

Gross area	$Ag_{PL} := t_{PL} \cdot L = 48\ in^2$	recall
Shear yielding stress	$f_{uv} := \dfrac{\mid V_u \mid}{Ag_{PL}} = 8.5 \cdot ksi$	$t_{PL} = 0.75\,in$
		$L = 64\,in$
Reduction factor	$\phi := 1.00$	$V_u = -408.7 \cdot kip$
	$\phi R_n := \phi \cdot 0.60 \cdot Fy_{PL} = 30\,ksi$	$N_u = 0 \cdot kip$
	$if\left(f_{uv} < \phi R_n, \text{“OK”}, \text{“NOT OK”}\right) = \text{“OK”}$	$M_u = -5578.9 \cdot kip \cdot in$
Tensile yielding stress (due to tension)	$f_{ua} := \dfrac{N_u}{Ag_{PL}} = 0 \cdot ksi$	$Fy_{PL} = 50\,ksi$
Plastic section modulus	$Z_{PL} := \dfrac{t_{PL} \cdot L^2}{4} = 768\,in^3$	
Tensile yielding stress (due to moment)	$f_{ub} := \dfrac{\mid M_u \mid}{Z_{PL}} = 7.3 \cdot ksi$	
Total tensile yielding stress	$f_{un} := f_{uv} + f_{ub} = 15.8 \cdot ksi$	
Reduction factor	$\phi := 0.90$	
	$\phi R_n := \phi \cdot Fy_{PL} = 45\,ksi$	
	$if\left(f_{un} < \phi R_n, \text{“OK”}, \text{“NOT OK”}\right) = \text{“OK”}$	

Design weld at gusset-to-beam flange connection (Eq. 8-13)

Effective eccentricity of the shear force	$e_{effective} := \dfrac{M_u}{V_u} = 13.65\,in$	recall
		$M_u = -5578.9\,in \cdot kip$
Value in Table 8-4	$a_{weld} := \dfrac{e_{effective}}{L} = 0.213$	$V_u = -408.7\,kip$
Table 8-4	$C := 3.458$	$L = 64\,in$
Additional ductility (p: 13-11)	$ductility := 1.25$	$C_1 = 1$
Reduction factor	$\phi := 0.75$	
Weld strength	$V_u < \dfrac{\phi R_n}{ductility} = \dfrac{(\phi \cdot C \cdot C_1 \cdot D \cdot L)}{ductility}$ $D_{min} > \dfrac{V_u \cdot ductility}{(\phi \cdot C \cdot C_1 \cdot L)}$	
Minimum weld size	$D_{min} := \dfrac{(\lvert V_u \rvert \div kip) \cdot ductility}{\phi \cdot C \cdot C_1 \cdot (L \div in)} = 3.1$	
Use D = 4 /16th	$D_{weld} := 4\ in$ 16th of an inch	

Alternative method to determine the required weld size (DG29 Appendix B), (Eq. 8-2 and p. 8-9)

Maximum equivalent normal force $N_{max} := |N_u| + \left|\frac{2 \cdot M_u}{L \div 2}\right| = 348.7 \cdot kip$

Minimum equivalent normal force $N_{min} := \left||N_u| - \left|\frac{4M_u}{L}\right|\right| = 348.7 \cdot kip$

Peak weld resultant force $R_{peak} := \sqrt{{V_u}^2 + {N_{max}}^2} = 537.2 \cdot kip$

Average weld resultant force $R_{avg} := \sqrt{{V_u}^2 + \left(\frac{N_{max} + N_{min}}{2}\right)} = 537.2 \cdot kip$

Angle of weld loading $\theta := atan\left(\frac{N_{max}}{|V_u|}\right) = 40.5 \cdot deg$

Weld strength $R_{avg} < \frac{\phi R_n}{ductility} = \frac{\phi \cdot 1.392 \cdot \left(1 + 0.5 \cdot \sin(\theta)^{1.5}\right) \cdot D \cdot L \cdot (2)}{ductility}$

Minimum weld size $D_{min.2} := \frac{ductility \cdot \left(R_{avg} \div kip\right)}{1.392 \cdot \left(1 + 0.5 \cdot \sin(\theta)^{1.5}\right) \cdot (2) \cdot (L \div in)} = 2.99$

% difference in calculated weld size $\frac{D_{min} - D_{min.2}}{D_{min}} = 2.93 \cdot \%$

recall
$M_u = -5578.9 in \cdot kip$
$V_u = -408.7\ kip$
$N_u = 0\ kip$
$L = 64\ in$
$D_{min} = 3.1$
$ductility = 1.25$

$1 + 0.5 \cdot \sin(\theta)^{1.5} = 1.26$
% strength increase

Check gusset internal strength

Check the gusset plate for shear yielding (J4.2a)

$\phi := 1.00$
$\phi R_n := \phi \cdot 0.60 \cdot Fy_{PL} \cdot Agv_{PL} = 540 \cdot kip$
$if\left(V'_u < \phi R_n, \text{"OK"}, \text{"NOT OK"}\right) = \text{"OK"}$

recall
$V'_u = -30.01 \cdot kip$
$Agv_{PL} = 18\ in^2$
$Fy_{PL} = 50\ ksi$

Check gusset stresses on Section b-b for a hypothetical load case (P.1 = P.2 = compression)

$M_{u1\_hypothetical} := H_{u1} \cdot e = -2789.4 \cdot kip \cdot in$
$M_{u2\_hypothetical} := M_{u1\_hypothetical} = -2789.4 \cdot kip \cdot in$
$M'_{u1\_hypothetical} := \left(\frac{1}{8} \cdot H_{u1} \cdot L - \frac{1}{4} \cdot V_{u1} \cdot h\right) - \frac{1}{2} \cdot M_{u1\_hypothetical}$
$M'_{u1\_hypothetical} = 679.5 in \cdot kip$
$M_{u2\_hypothetical} := M_{u1\_hypothetical} = 679.5 \cdot kip \cdot in$
$N_{u\_hypothetical} := H_{u1} + V_{u1} = -408.7 \cdot kip$
$V_{u\_hypothetical} := H_{u1} - V_{u1} = -0 \cdot kip$
$M_{u\_hypothetical} := M'_{u1\_hypothetical} - M_{u2\_hypothetical} = 0 \cdot kip \cdot in$
$N'_{\_hypothetical} := \frac{1}{2} \cdot \left(H_{u1} + V_{u1}\right) = -204.4 \cdot kip$
$V'_{\_hypothetical} := \frac{1}{2} \cdot \left(H_{u1} - V_{u1}\right) + \frac{2in}{L}\left(V_{u\_hypothetical}\right) = -0 \cdot kip$
$M'_{hypothetical} := M'_{u1\_hypothetical} + M'_{u2\_hypothetical} = 1359 \cdot kip \cdot in$

recall
$N'_u = 0 \cdot kip$
$M'_u = 0 \cdot kip \cdot in$
$M_{u1} = -2789.4 \cdot kip \cdot in$
$H_{u1} = -204.4 kip$
$e = 13.7 in$
$L = 64 in$
$V_{u1} = -204.4 kip$
$h = 18 in$
$a = 52 in$
$t_{PL} = 0.75 in$
$Fy_{PL} = 50 ksi$

Worst case equivalent normal force (compression)

DG 29 Appendix B $N_{ue\_hypothetical} := \left|N'_{\_hypothetical}\right| + \frac{M'_{hypothetical}}{h \div 2} \cdot (2) = 506.3 \cdot kip$

DG 29 Appendix C.4 $b := h = 18\,in$

$$\lambda := \frac{(b \div t_{PL}) \cdot \sqrt{Fy_{PL} \div ksi}}{5 \cdot \sqrt{475 + \frac{1120}{(a \div b)^2}}} = 1.4$$

$Q := 1.34 - 0.486 \cdot \lambda = 0.7$

$\phi := 0.9$

Critical buckling stress $\phi F_{cr} := \phi \cdot Q \cdot Fy_{PL} = 30.2 \cdot ksi$

DG 29 Eq. C-6

Actual (worst case)

Compressive stress $f_{ua\_theoretical} := \frac{N_{ue\_hypothetical}}{t_{PL} \cdot h} = 37.5 \cdot ksi$

$if\left(f_{ua\_theoretical} < \phi F_{cr}, \text{"OK"}, \text{"NOT OK"}\right) = \text{"NOT OK"}$

Using 7/8' plate $t_{PL} := \frac{7}{8} in$

$$\lambda := \frac{(b \div t_{PL}) \cdot \sqrt{Fy_{PL} \div ksi}}{5 \cdot \sqrt{475 + \frac{1120}{(a \div b)^2}}} = 1.2$$

$Q := 1.34 - 0.486 \cdot \lambda = 0.8$

Critical buckling stress $\phi F_{cr} := \phi \cdot Q \cdot Fy_{PL} = 34.5 \cdot ksi$

DG 29 Eq.:C-6

Actual (worst case)

Compressive stress $f_{ua\_theoretical} := \frac{N_{ue\_hypotheyical}}{t_{PL} \cdot h} = 32.1 \cdot ksi$

$if\left(f_{ua\_theoretical} < \phi F_{cr}, \text{"OK"}, \text{"NOT OK"}\right) = \text{"OK"}$

Plate thickness $t_{PL} : \frac{3}{4}\ in$

$b := 13.6\ in$ see Tamboli (2010)

$f_u := 29.9\ ksi$

$$\lambda := \frac{(b \div t_{PL}) \cdot \sqrt{Fy_{PL} \div ksi}}{5 \cdot \sqrt{475 + \frac{1120}{(a \div b)^2}}} = 1.1$$

$Q := 1.34 - 0.486 \cdot \lambda = 0.8$

Critical buckling stress $\phi F_{cr} := \phi \cdot Q \cdot Fy_{PL} = 36.4 \cdot ksi$

DG 29 Eq. C-6 $if\left(f_u < \phi F_{cr}, \text{"OK"}, \text{"NOT OK"}\right) = \text{"OK"}$

Check the gusset plate for buckling on the Whitmore section

Effective length factor (conservative) $K_{buckle} := 1.2$

Length of buckling member $L_{buckle} := \frac{L_{PL}}{\sqrt{2}} = 5.66\ in$

$\frac{K \cdot L}{r}$ $\frac{K_{buckle} \cdot L_{buckle}}{r_{PL}} = 31.4 > 25$

recall

$L_{PL} = 8\ in$

$r_{PL} = 0.217\ in$

$A_w = 16.03\ in^2$

$P_u = 289\ kip$

Available critical stress (Table 4-22) $\phi F_{cr} := 41.86\,ksi$

Compressive strength (E.3) $\phi P_n := \phi F_{cr} \cdot A_w = 671.1 \cdot kip$

$if\left(P_u < \phi P_n, \text{"OK"}, \text{"NOT OK"}\right) = \text{"OK"}$

Check the gusset plate for sidesway buckling

Compressive strength (E.3) $\phi P_n : \phi F_{cr} \cdot t_{PL} \cdot L = 2009.3 \cdot kip$

$if\left(\left|N_{u\_hypothetical}\right| < \phi P_n, \text{"OK"}, \text{"NOT OK"}\right) = \text{"OK"}$

recall
$\phi F_{cr} = 41.9\,ksi$
$t_{PL} = 0.75\,in$
$L = 64\,in$
$\left|N_{u\_hypothetical}\right| = 408.7 \cdot kip$

Check the beam for web local yielding and local crippling

Check beam web local yielding (J10.2)

Length of bearing

Strength reduction factor $\phi := 1.0$

Weld local yielding strength $\phi R_n := \phi \cdot Fy_{beam} \cdot tw_{beam} \cdot \left(5k\_des_{brace} + L\right) = 2042 \cdot kip$

$if\left(N_{max} < \phi R_n, \text{"OK"}, \text{"NOT OK"}\right) = \text{"OK"}$

recall
$N_{max} = 348.7 \cdot kip$
$k\_des_{brace} = 1.53\,in$
$tw_{beam} = 0.57\,in$
$Fy_{beam} = 50\,ksi$
$L = 64\,in$

Check beam web shear yielding (J4.2a)

Strength reduction factor $\phi := 1.0$

$\phi R_n := \phi \cdot 0.60 \cdot Fy_{beam} \cdot Agv_{beam}$

Shear yielding strength $\phi R_n := \phi \cdot 0.60 \cdot Fy_{beam} \cdot tw_{beam} \cdot L = 1094.4 \cdot kip$

$if\left(\left|N_{u\_hypothetical}\right| < \phi R_n, \text{"OK"}, \text{"NOT OK"}\right) = \text{"OK"}$

recall
$Fy_{beam} = 50\,ksi$
$tw_{beam} = 0.57\,in$
$L = 64\,in$
$N_{u\_hypothetical} = -408.7\,kip$

Check beam web local crippling (J10.3)

Strength reduction factor $\phi := 0.75$

Web local crippling strength

$$\phi R_n := \phi \cdot 0.80 \cdot tw_{beam}^{\ 2} \cdot \left[1 + 3 \cdot \left(\frac{L}{d_{beam}}\right)\left(\frac{tw_{beam}}{tf_{beam}}\right)^{1.5}\right] \cdot \sqrt{\frac{E \cdot Fy_{beam} \cdot tf_{beam}}{tw_{beam}}} = 1311.7 \cdot kip$$

$if\left(N_{max} < \phi R_n, \text{"OK"}, \text{"NOT OK"}\right) = \text{"OK"}$

recall
$E = 29000\,ksi$
$tf_{beam} = 0.93\,in$
$d_{beam} = 27.3\,in$

Check transverse section web yielding (G2.1)

		recall
	$V_u : H_{u2} + V'_u = 174.3\,kip$	$H_{u2} = 204.4 \cdot kip$
Area of beam web	$Aw_{beam} := tw_{beam} \cdot d_{beam} = 15.6\ in^2$	$V'_u = -30 \cdot kip$
Strength reduction factor	$\phi := 1.0 \qquad C_v := 1.0$	$Fy_{beam} = 50\ ksi$
Shear yielding strength	$\phi V_n := \phi \cdot 0.6 \cdot Fy_{beam} \cdot Aw_{beam} \cdot C_v = 466.8 \cdot kip$	$tw_{beam} = 0.57\ in$
	$\text{if}\left(V_u < \phi V_n, \text{"OK"}, \text{"NOT OK"}\right) = \text{"OK"}$	$d_{beam} = 27.3\ in$

References

AISC (2015). *Vertical Bracing Connections: Analysis and Design, Design Guide 29*. American Institute of Steel Construction, Chicago.

AISC (2016). *Specification for Structural Steel Buildings*. American Institute of Steel Construction, Chicago.

AISC (2017). *Steel Construction Manual*, 15th Edition. American Institute of Steel Construction, Chicago.

15

Wide Flange Brace Bolted to a Gusset Plate in a Concentrically Braced Frame

15.1 Problem Description

The aim of this example is a comparison of the component-based finite element method (CBFEM) of a brace connection at beam-column connection in a braced frame with the AISC 360-16 specifications. The verification is prepared for the size of the brace, beam, column, connecting angles, geometry, the thickness of the plate, bolts, and welds. In this study, ten components are examined: brace, beam flange and web, column flange and web, connecting angles, gusset plate, splice plates between brace and gusset plate, connecting angles to the column, connecting angles to beam, bolts, and welds. All components are designed according to AISC 360-16 *Specification* (AISC, 2016).

15.2 Verification of Resistance as Per AISC

The connection presented is taken from AISC Design Guide 29 (AISC, 2015), the details are as shown in Figures 15.1 and 15.2. The brace is W12x87 (ASTM A992), beam W18x106 (ASTM A992), column W14x605, gusset plate ¾ in. (ASTM A36), connecting angles L4x4x3/4 between brace and gusset plate (ASTM A36), connecting angles to column L5x3 ½ x 5/8, 3/8 in. splice plates (ASTM A36), connecting angles to beam L8x6x7/8 (ASTM A36), 7/8 in. ASTM A325 bolts, and ASTM E70XX weld. All section and material properties are obtained from the *Steel Construction Manual* (AISC, 2017).

Based on the results obtained for an analytical solution, the governing component of this connection is the bolt shear between the gusset plate and the brace followed by tension yielding of the connecting angles between the brace flanges and the gusset plate and then followed by the tension rupture of the angles. For the given connection, the brace below the beam is subjected to an axial tension force. The load cases converged to 100% in the CBFEM analysis.

15.3 Verification of Resistance as Per CBFEM

After obtaining values for several limit states of the presented connection as per the AISC360-16 *Specification* (AISC, 2016), the CBFEM analysis was then performed in IDEA StatiCa. The overall check of the

Steel Connection Design by Inelastic Analysis: Verification Examples per AISC Specification, First Edition.
Mark D. Denavit, Ali Nassiri, Mustafa Mahamid, Martin Vild, Halil Sezen, and František Wald.

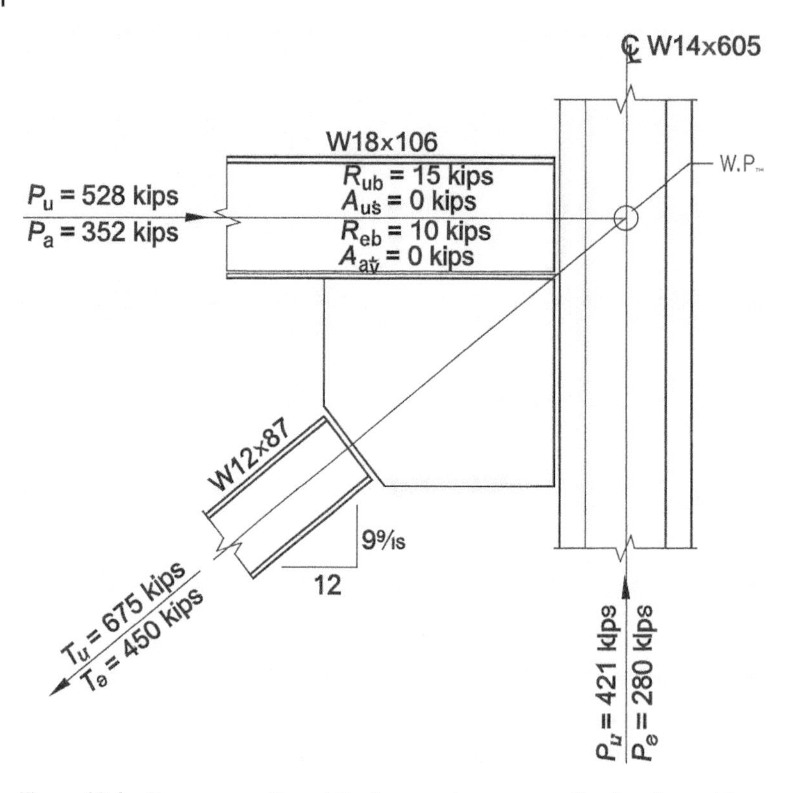

Figure 15.1 Brace connection at the beam-column connection in a braced frame – geometry.

connection is verified in CBFEM as shown in Figures 15.3 and 15.4. The check shows that the connection works well according to the CBFEM for the applied load which conforms to AISC provisions. The connection presented includes elements that are welded and others that are bolted.

The results of the analytical solution are represented by the comparison table for the different limit states shown below. The limit states that should be considered for this connection are as follows and the comparison of the capacities of the different limit states is shown in Table 15.1.

1. Bolt shear at brace to gusset connection
2. Tension yielding of angles
3. Tension rupture of angles
4. Block shear rupture of angles
5. Yielding of splice connecting brace to gusset plate
6. Rupture of splice connecting brace to gusset plate
7. Block shear of splice connecting brace to gusset plate
8. Yielding of brace
9. Rupture of brace
10. Block shear rupture of gusset plate
11. Tension yielding of Whitmore section

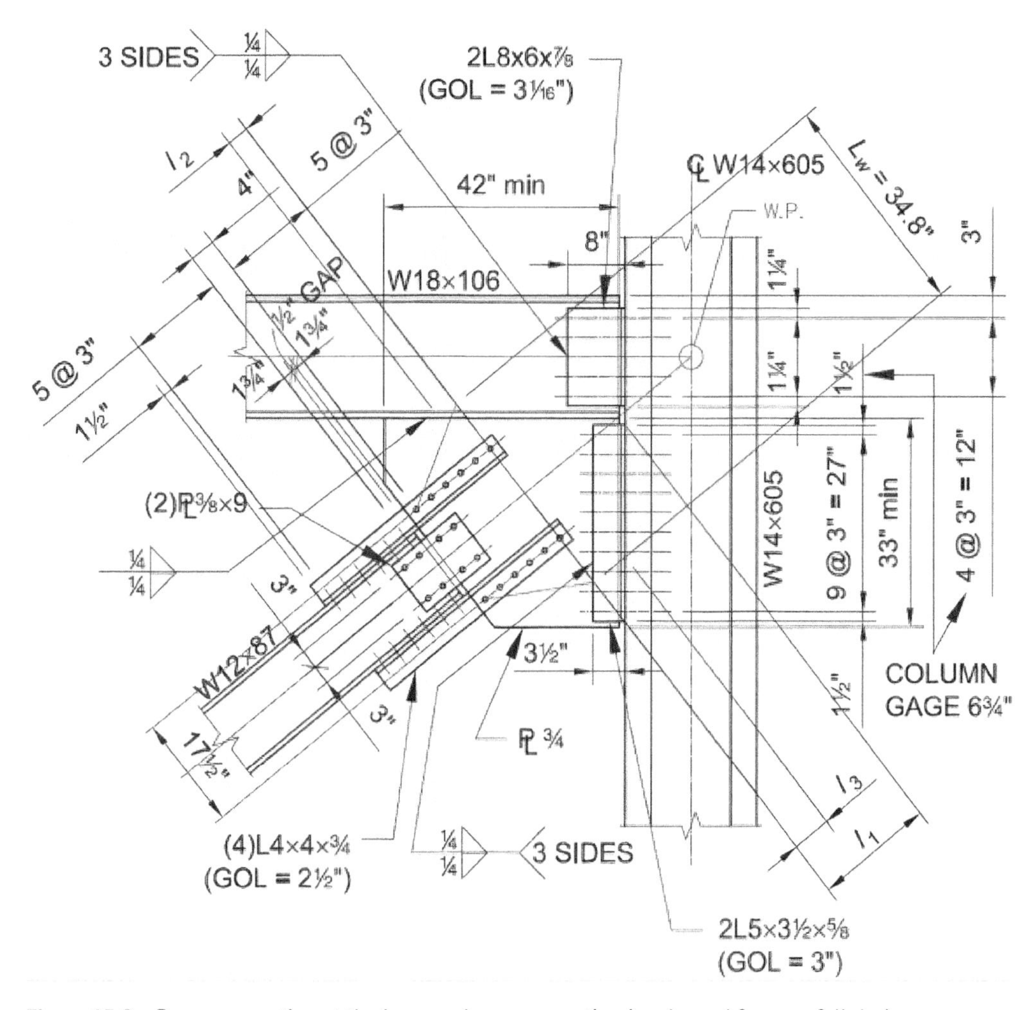

Figure 15.2 Brace connection at the beam-column connection in a braced frame – full design.

12. Bolt capacity at gusset-column connection – shear and tension
13. Bolt capacity at gusset-column connection – bolt bearing
14. Prying action on double angle
15. Shear yielding of angles at gusset-column connection
16. Shear rupture of angles at gusset-column connection
17. Block shear strength at angles at the gusset-column connection
18. Tension yielding and shear yielding of plate at gusset-beam connection
19. Weld between gusset plate and beam bottom flange
20. Web local yielding and crippling of beam
21. Beam-to-column connection
22. Beam-to-column connection, bolt strength, and weld

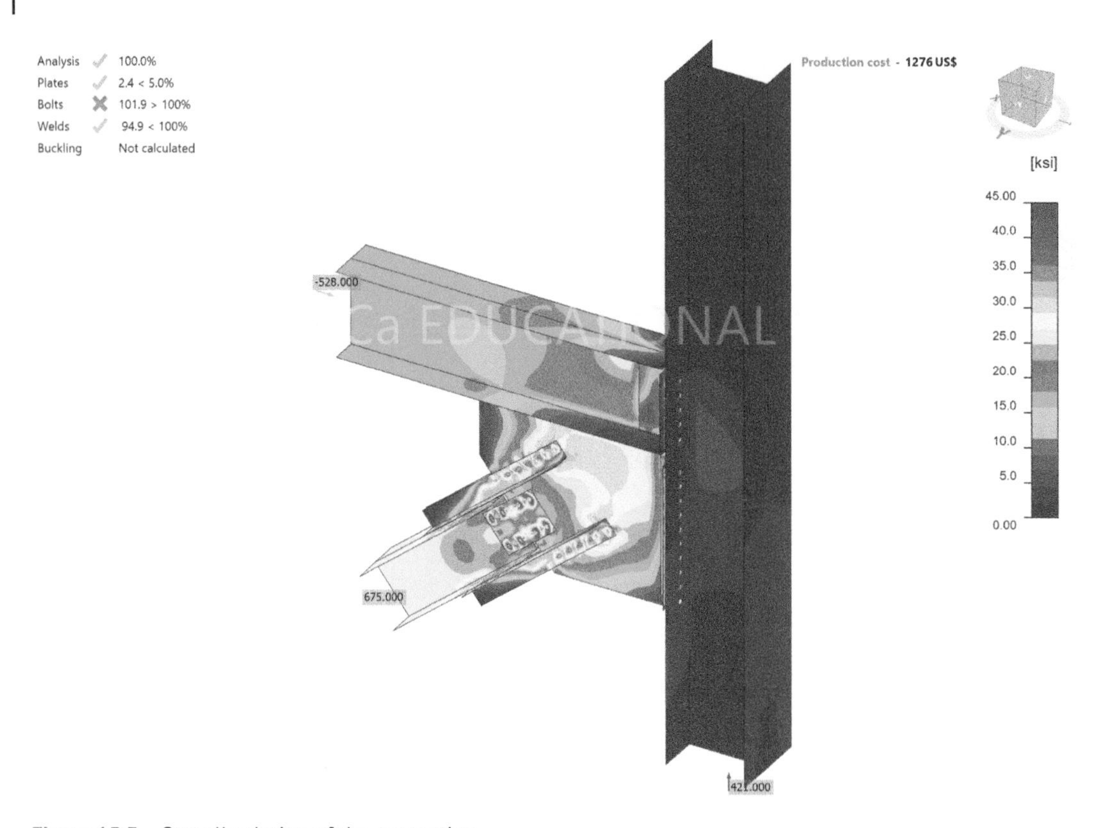

Figure 15.3 Overall solution of the connection.

The governing component of this connection is bolt shear between the gusset plate and the brace with load resistance ϕR_n =681 kips > P_u = 675 kips (utilization 99%). The next critical is tension yielding of the connecting angles between the brace flange and the gusset plate with the load resistance of ϕR_n =705 kips > P_u = 675 kips (utilization 96%) and the tension rupture of the angles with ϕR_n =746 kips > P_u = 675 kips (utilization 90%).

For the gusset plate buckling, the preferred value for the buckling factor is greater than 9.16 for 50 ksi steel per AISC 360-16 Section J.4 (AISC, 2016). Figure 15.8 below shows that the buckling factor is 10.48, therefore, it is expected that the gusset would not buckle at this load according to CBFEM, however, buckling according to CBFEM still requires more research to allow predicting it accurately.

15.4 Parametric Study

The results were obtained using the various limit states per the AISC procedure. These limit states were investigated individually per CBFEM and the capacities were reported accordingly. Bolt limit states including bolt shear, bolt tension, combined bolt shear and tension. and bolt bearing are accurate. For the tension yielding, tension rupture, shear yielding. and shear rupture limit states, they are found

Table 15.1 Limit states checked by AISC

Limit state	AISC
Bolt shear at brace to gusset connection	$\phi R_n = 40.59$ kips $\phi r_{nv} = 24.35$ kips
Tension yielding of angles	$\phi R_n = 705$ kips
Tension rupture of angles	$\phi R_n = 746$ kips
Block shear rupture of angles	$\phi R_n = 932$ kips
Yielding of splice connecting brace to gusset plate	$\phi R_n = 219$ kips
Rupture of splice connecting brace to gusset plate	$\phi R_n = 228$ kips
Block shear of splice connecting brace to gusset plate	$\phi R_n = 175$ kips
Block shear of brace web	$\phi R_n = 216$ kips
Yielding of brace	$\phi R_n = 1152$ kips
Rupture of brace	$\phi R_n = 1040$ kips
Block shear rupture of gusset plate	$\phi R_n = 945$ kips
Tension yielding of Whitmore section	$\phi R_n = 855$ kips
Bolt capacity at gusset-column connection – shear and tension	$\phi R_n = 30.39$ kips
Bolt capacity at gusset-column connection – bolt bearing	$\phi R_n = 33.64$ kips
Prying action on double angle	See Appendix in this chapter for calculations
Shear yielding of angles at gusset-column connection	$\phi R_n = 810$ kips
Shear rupture of angles at gusset-column connection	$\phi R_n = 652$ kips
Block shear strength at angles at the gusset-column connection	$\phi R_n = 658$ kips
Tension yielding and shear yielding of plate at gusset-beam connection	$\phi R_n = 21.6$ ksi
Weld between gusset plate and beam bottom flange	$\phi R_n =12.024$ kips
Web local yielding of beam	$\phi R_n =1338$ kips Compared to force in beam equals 152 kips
Web local crippling of beam	$\phi R_n =852$ kips *Compared to force in beam equals 152 kips*
Beam to column connection bolt shear	$\phi r_{nv} =24.33$ kips
Beam to column connection, weld strength	$\phi R_n =8.32$ kips

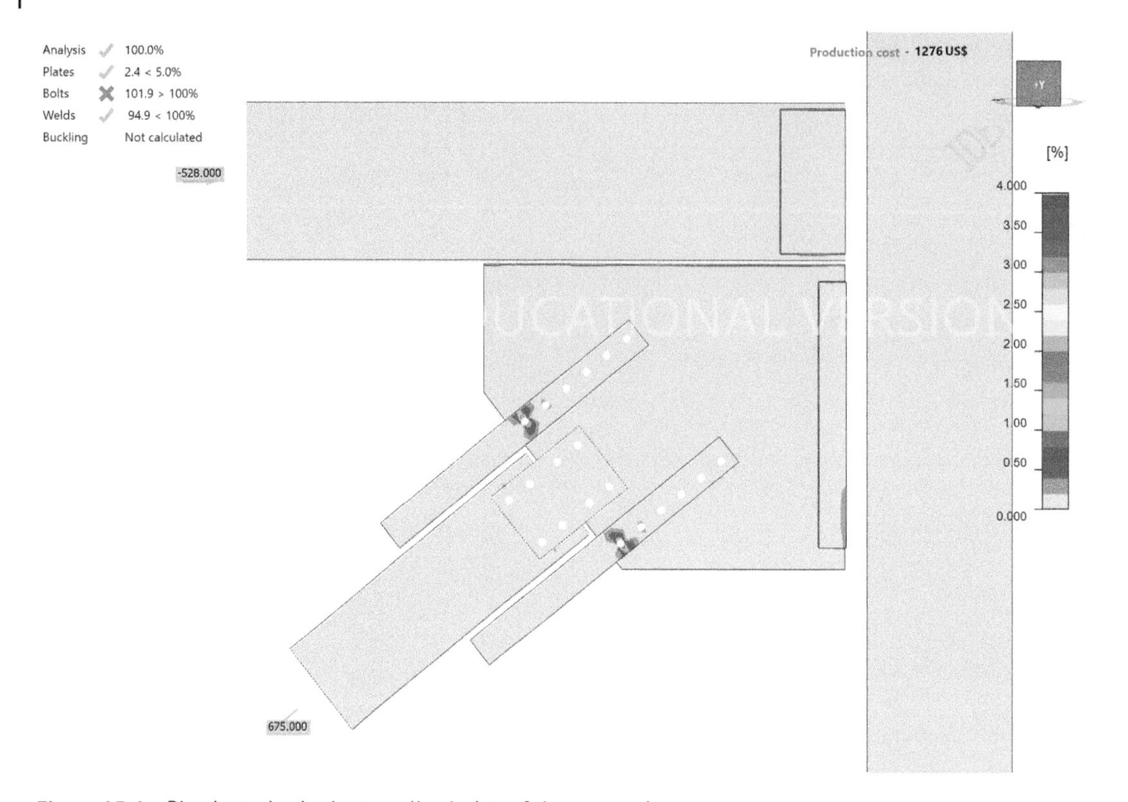

Figure 15.4 Plastic strains in the overall solution of the connection.

separately. The plastic strain starts at the bolt holes; these stresses are based on von Mises stresses which is a combination of normal and shear stresses. Figure 15.5 shows the stress distribution in the angles connecting the brace to the gusset plate. CBFEM results show that the plastic strain in the angles is exceeded at a load (780 kips) higher than the originally applied (675 kips) and noted as the failure load for the limit states in the angles. This load is in agreement with the AISC 360-16 (AISC, 2016) requirements as shown in Table 15.1 for tension rupture of the angles.

The block shear limit state can be observed in some members and not in some others. Examples of these two cases are shown in Figures 15.6, 15.7, and 15.8. Figure 15.6 shows that the stresses are increased around the hole without extending to the adjacent holes. This is in accordance with AISC 360-16 where the governing failure mode of the angles is the tensile rupture. Figure 15.7 shows that block shear can be observed accurately in the gusset plate which is also in agreement with AISC 360-16, as shown in Table 15.1. Figure 15.8 shows the block shear rupture for the splice plate connecting the brace web to the gusset plate; which is in agreement with the AISC 360-16 *Specification* and the capacity shown in Table 15.1.

The brace rupture failure mode occurs in the web and in the flange as shown in Figures 15.9 and 15.10. The brace failure loads are in agreement with AISC 360-16, as presented in Table 15.1.

The AISC specifications require checking yielding at the Whitmore section on the gusset plate. Figure 15.11 shows the stress distribution in the gusset plate at the failure load for yielding at the

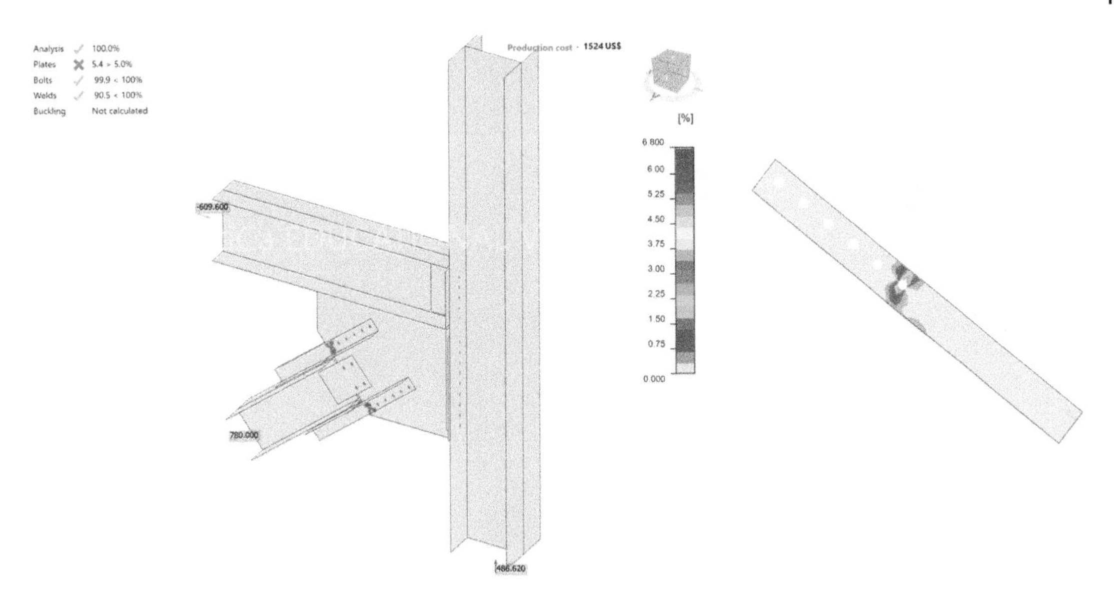

Figure 15.5 Plastic strains in angles connecting the brace to the gusset plate.

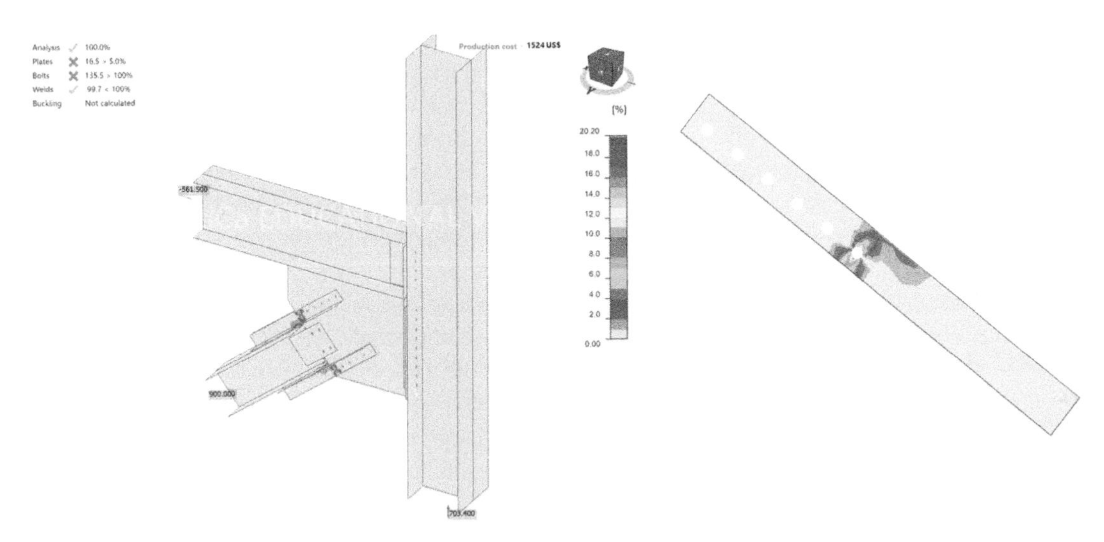

Figure 15.6 Plastic strains in angles connecting the brace to the gusset plate at high loads to investigate block shear limit state in the angles.

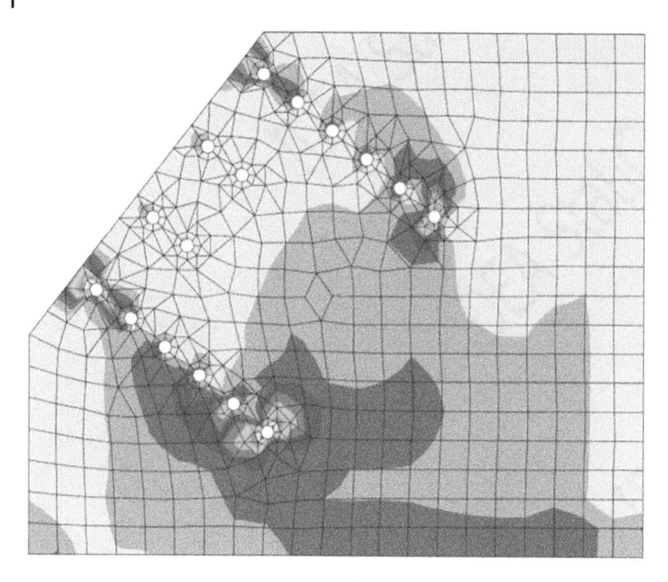

Figure 15.7 Plastic strains in the gusset plate to investigate the block shear limit state.

Whitmore section per AISC specifications. It is obvious that rupture along the bolt lines would occur before yielding of the gusset plate and as observed in the yielding and rupture capacities in Table 15.1.

Prying action is another limit state that is required per the AISC specifications; the prying action limit states is taken into consideration in CBFEM by the additional tension forces applied to the bolts.

In investigating the limit states in the angles connecting the gusset plates to the column flange, the capacities for shear yielding, shear rupture in combination with tension rupture and tension yielding are shown in Figure 15.12. As discussed above, rupture along the bolt line was observed and as the load is increased, the stresses increase along the bolt line without clear observation of the block shear in the angles; this is expected as shear rupture along the bolt line is expected to happen before block shear rupture. Figure 15.12 also shows yielding in the gross part of the angle.

Beam web local and shear yielding would occur at a large load compared to the applied load. Almost all limit states in this connection would occur before these two limit states which typically do not control the design. If needed, these limit states can be checked using the AISC specifications using the procedure presented in the Appendix in this chapter for beam web local and shear yielding.

Beam web crippling would occur after yielding and at high loads, therefore, the model may not converge under such high loads and would not be able to capture this failure mode. If the crippling capacity is needed, it can be calculated per the AISC specifications using the procedure presented in the Appendix in this chapter.

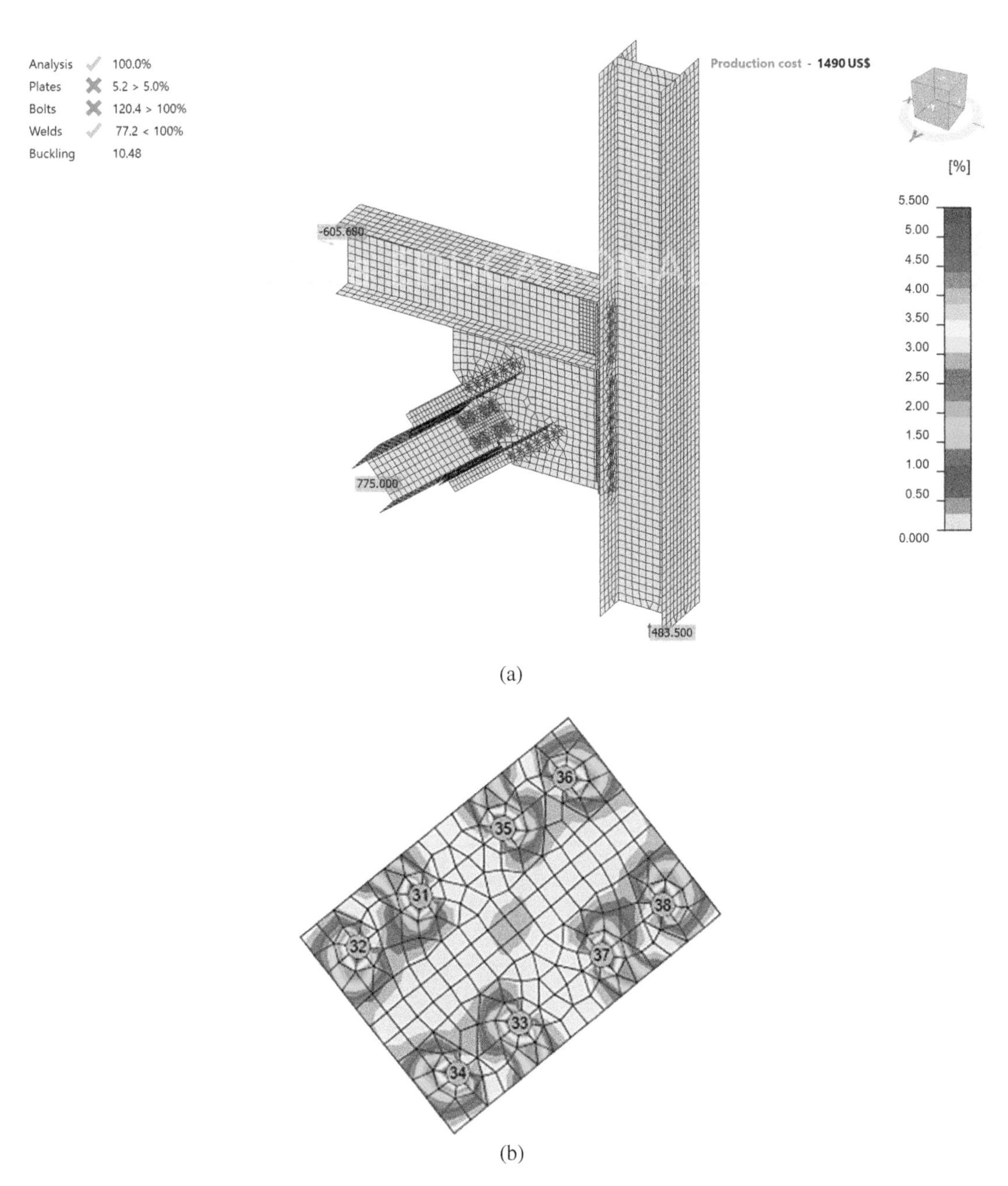

Figure 15.8 Plastic strains in the splice plate to investigate the block shear limit state. (a) Plastic strains in the whole connection; (b) plastic strains in the splice plate.

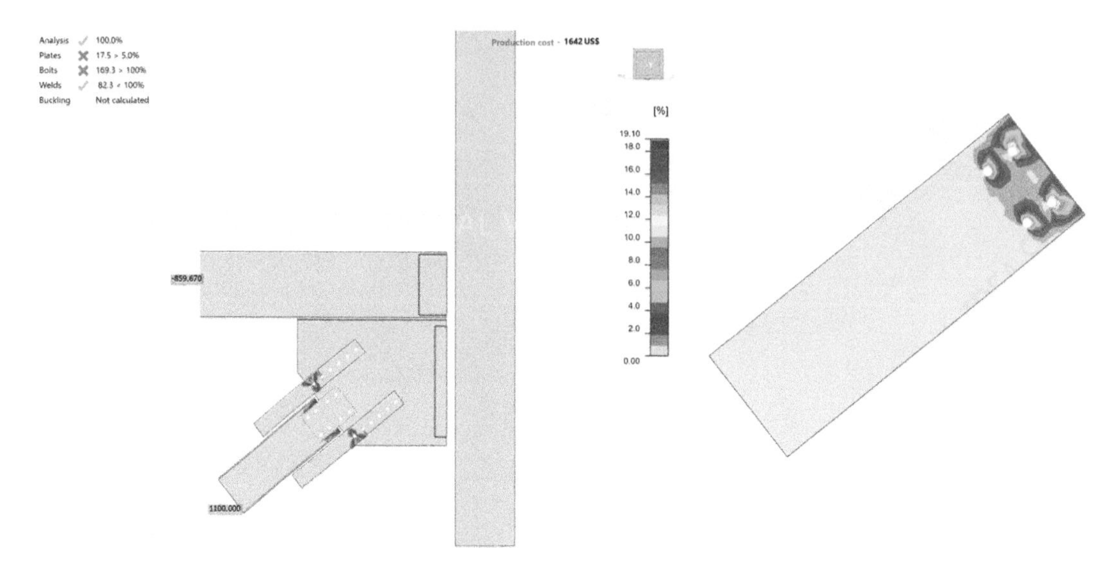

Figure 15.9 Plastic strains in the brace web.

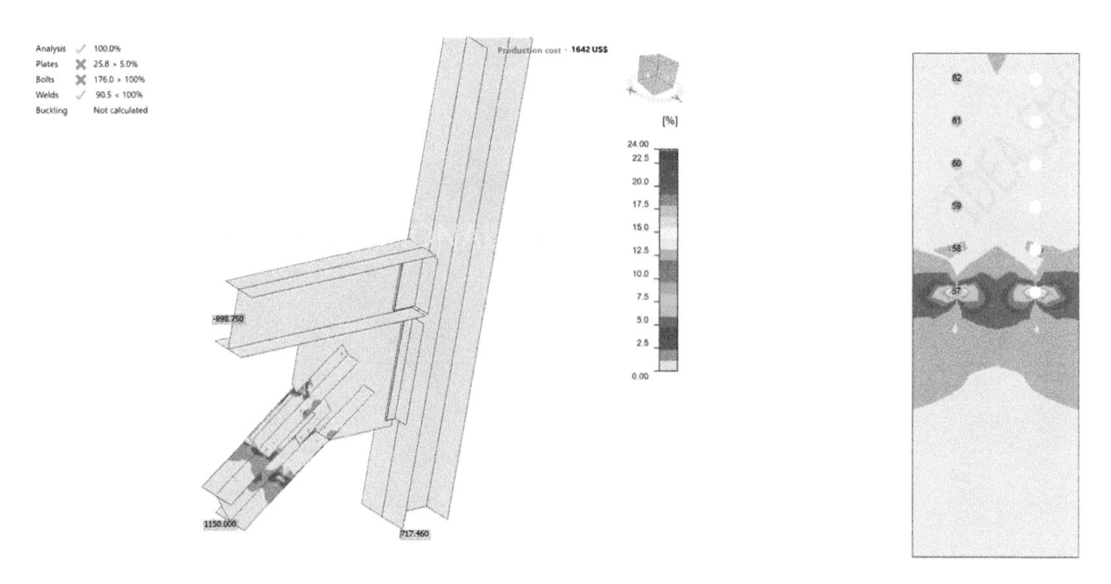

Figure 15.10 Plastic strains in the brace flange.

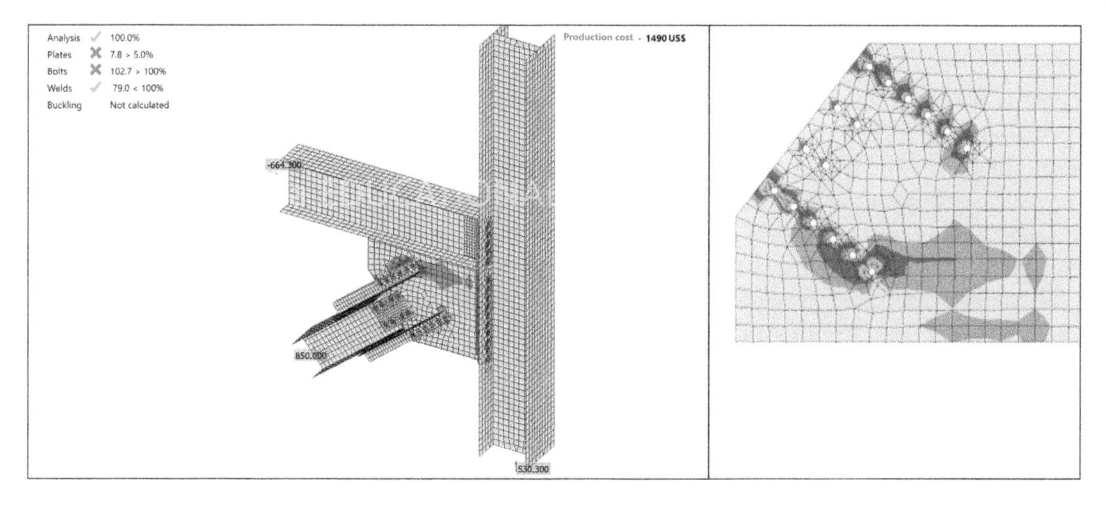

Figure 15.11 Plastic strains in the gusset plate at 850 kip load.

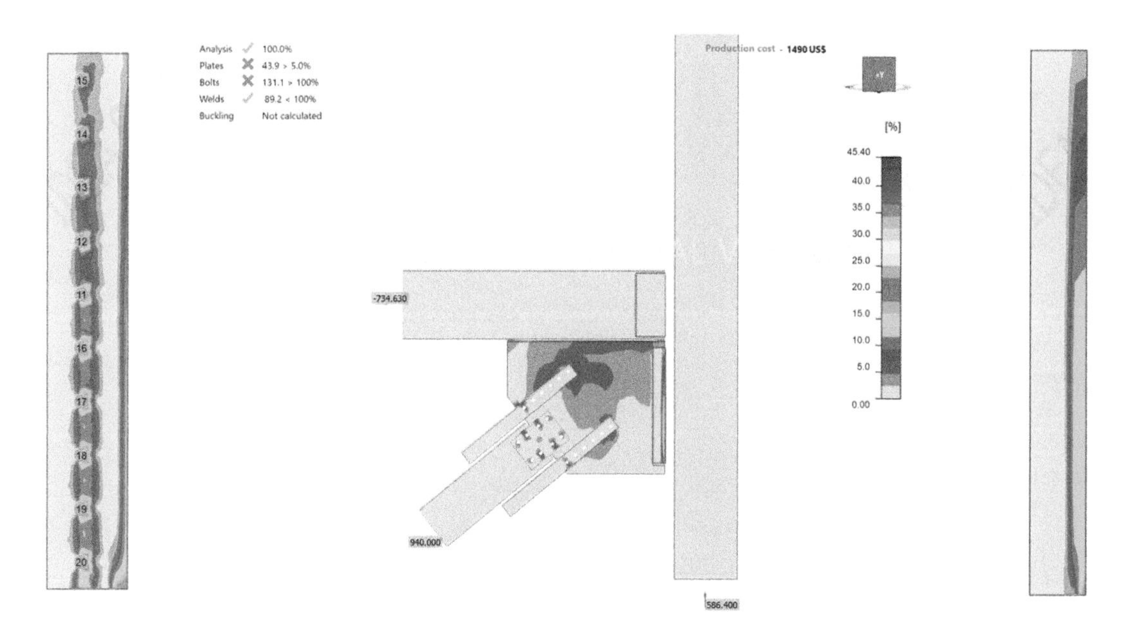

Figure 15.12 Plastic strains in the angles connecting the gusset plate to the column flange.

15.5 Summary

It can be concluded that the CBFEM is capable of predicting the actual behavior and failure mode of the braced frame connections presented herein. The various limit states were investigated carefully by performing a parametric study which resulted in obtaining the capacity for each limit state using CBFEM. The weld capacity for the weld between the gusset and the beam bottom flange and between the beam and the column are in agreement in both AISC and CBFEM. Bolt limit states, including bolt shear, bolt tension, combined bolt shear and tension and bolt bearing in AISC, are in agreement with CBFEM. The plates' limit states including yielding, rupture in tension, and in shear are based on 5% plastic strain limit according to CBFEM. The tension rupture in the angles is in agreement per AISC and CBFEM with less than 10% difference in the capacities. For the block shear limit state, it can be observed in the gusset plate and in the web connecting plate, but not in other plates, such as the angles connecting the gusset plate to the column; this is because shear and tension rupture of the angles precede the block shear rupture. The prying action limit state, required by the AISC *Specification*, is taken into consideration in CBFEM by the additional tension forces applied to the bolts. Beam web buckling, web crippling, and shear yielding would occur at high loads and the model would not converge at such high loads; all other limit states would occur before these limit states. If necessary, these limit states can be checked per the AISC specifications as shown in the Appendix in this chapter. The buckling limit state of the gusset plate was not observed as a limit state in AISC and in CBFEM.

Benchmark Case

Input

Beam cross-section

- W18X106
- Steel ASTM A992

Braces cross-section

- W27X84
- Steel ASTM A992

Column cross-section

- W14X605
- Steel ASTM A992

Gusset plate

- Thickness 3/4 in.
- Steel ASTM A36

Splice plate connecting beam web to gusset plate

- 2-3/8"x9" plates
- Steel ASTM A36

Angles connecting brace to gusset plate

- 4-L4x4x3/4
- Steel ASTM A36

Angles connecting gusset to column

- 2-L5x3½x5/8
- Steel ASTM A36

Angles connecting beam to column

- 2-L8x6x7/8
- Steel ASTM A36

Loading

- Axial force $N = 675$ kips in tension

Output

- Weld 94.9%
- Bolts 101.9%
- Plastic strain 2.4% < 5%
- Buckling factor 10.48

Appendix

The AISC Solution

ORIGIN := 1

Refer to AISC 14th Edition shapes database:

$$T1 := \mathrm{Row}(\mathrm{shape}) := \left| \begin{array}{l} \text{for } i \in 1..\mathrm{rows}(T1) - 1 \\ R \leftarrow 1 \text{ if } \left(T1^{\langle 2 \rangle}\right)_i = \mathrm{shape} \\ R \end{array} \right.$$

Solve Example 3 Diagonal Bracing Connection

Loads (LRFD)

Brace axial force	$T_u := 675\text{kip}$
Beam end reaction	$R_{u\_beam} := 15 \text{ kip}$
Beam axial force	$P_{u\_beam} := 528 \text{ kip}$
Column axial force	$P_{u\_column} := 421 \text{ kip}$

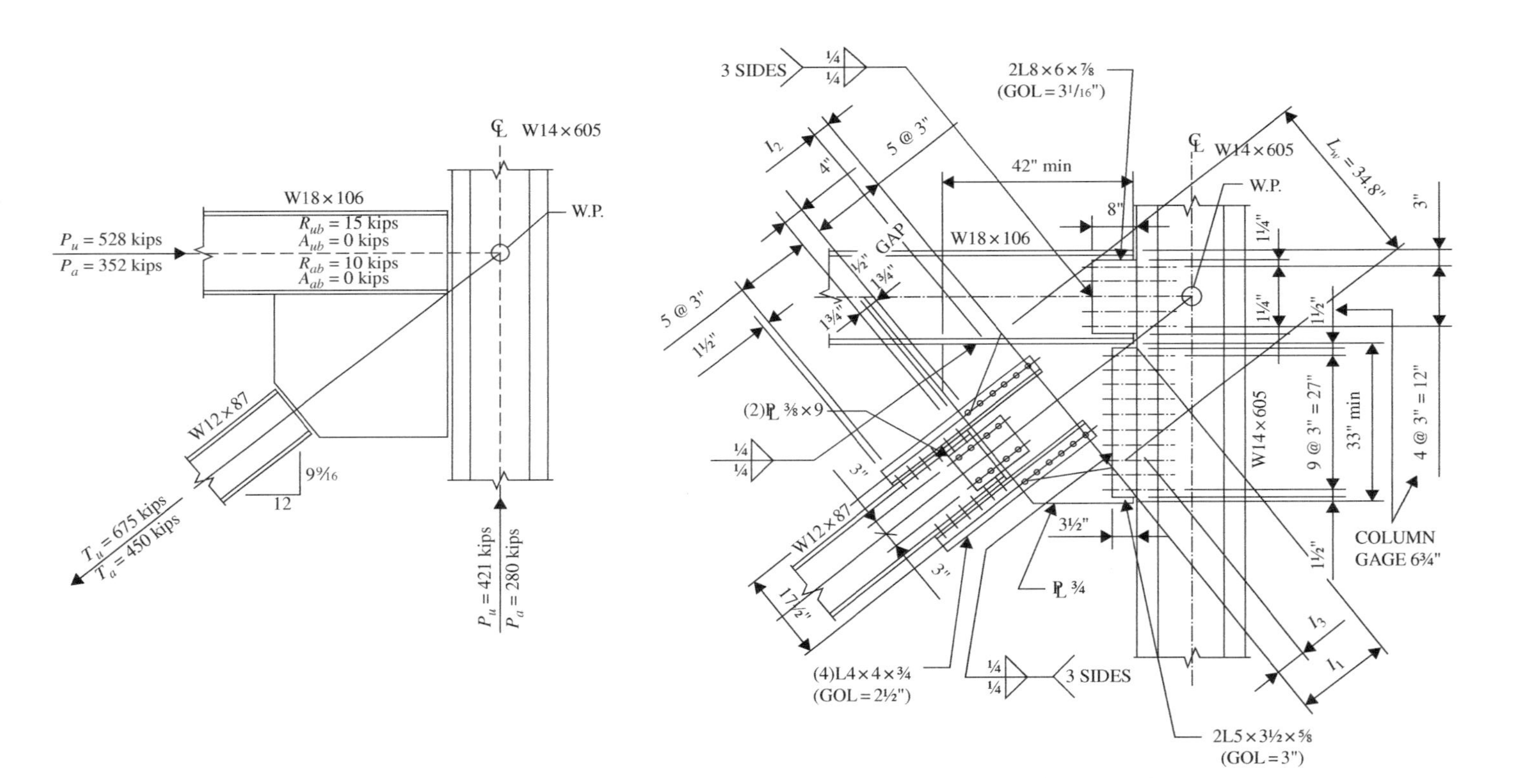

(a) members and forces

(b) connection

Material Properties

Beams, Column and Brace A992	$Fy_{A992} := 50 \cdot ksi$	$Fu_{A992} := 65 \cdot ksi$
Plate and Angle A36	$Fy_{A36} := 36\ ksi$	$Fu_{A36} := 58\ ksi$
Modulus of Elasticity	$E := 2900\ ksi$	
Weld strength	$F_{EXX} := 70\ ksi$	$C_1 := 1.0$
Bolt nominal stress (A325)	$F_{nt} := 90\ ksi$	$F_{nv} := 54\ ksi$

Member Properties

Brace Properties

Brace section	Brace := "W12X87"
Cross-sectional area	$Ag_{brace} := T1_{(Row(Brace),6)} \cdot in^2 = 25.6 \cdot in^2$
Depth	$d_{brace} := T1_{(Row(Brace),7)} \cdot in = 12.5 \cdot in$
Web thickness	$tw_{brace} := T1_{(Row(Brace),17)} \cdot in = 0.515 \cdot in$
Flange width	$bf_{brace} := T1_{(Row(Brace),12)} \cdot in = 12.1 \cdot in$
Flange thickness	$tf_{brace} := T1_{(Row(Brace),20)} \cdot in = 0.81\ in$

Beam Properties

Beam section	Beam := "W18X35"
Depth of beam	$d_{beam} := T1_{(Row(Beam),7)} \cdot in = 17.7 \cdot in$
Width of flange	$bf_{beam} := T1_{(Row(Beam),12)} \cdot in = 6.0 \cdot in$
Thickness of flange	$tf_{beam} := T1_{(Row(Beam),20)} \cdot in = 0.425 \cdot in$
Thickness of web	$tw_{beam} := T1_{(Row(Beam),17)} \cdot in = 0.3 \cdot in$
Design weld depth	$k\_des_{beam} := T1_{(Row(Beam),25)} \cdot in = 0.827 \cdot in$

Column Properties

	Column := "W14X605"
Depth	$d_{column} := T1_{(Row(column),7)} \cdot in = 20.9 \cdot in$
Flange width	$bf_{column} := T1_{(Row(column),12)} \cdot in = 17.4 \cdot in$
Flange thickness	$tf_{column} := T1_{(Row(column),20)} \cdot in = 4.16 \cdot in$
Web thickness	$tw_{column} := T1_{(Row(column),17)} \cdot in = 2.6 \cdot in$
Strong moment of inertia	$Ix_{column} := T1_{(Row(column),39)} \cdot in^4 = 10800\ in^4$

Gusset Plate Properties

Thickness	$t_{gusset} := 0.75\,in$		
Length	$L_{gusset} := 42\,in$		
Height	$h_{gusset} := 33\,in$		
Bolt Properties (A325)			
Bolt diameter	$d_{bolt} := \frac{7}{8}in$	$d'_{bolt} := d_{bolt} + \frac{1}{16}in$ $d''_{bolt} := d_{bolt} + \frac{2}{16}in$	
Area of bolt	$A_{bolt} := \pi \cdot d_{bolt}{}^2 \div 4 = 0.6\ in^2$		
Strength reduction factor	$\phi := 0.75$		
Bolt shear strength (single shear)	$\phi r_{nv} := \phi F_{nv} \cdot A_{bolt} = 24.35\ kip$	EQ: J3-1	
Bolt tensile strength	$\phi r_{nt} := \phi \cdot F_{nt} \cdot A_{bolt} = 40.59\ kip$	EQ: J3-1	
Pitch (bolt spacing)	$p := 3in$		recall

Brace to gusset connection

recall: $T_u = 675\ kip$, $bf_{brace} = 12.1\ in$, $tf_{brace} = 0.81\ in$, $Ag_{brace} = 25.6\ in^2$, $\phi r_{nv} = 24.35\ kip$

Force in one flange $\quad P_{uf} := T_u \cdot \frac{bf_{brace} \cdot tf_{brace}}{Ag_{brace}} = 258.42\ kip$

Force in web $\quad P_{uw} := T_u - 2 \cdot P_{uf} = 158.15\ kip$

Brace Flange to Gusset Connection

Number of bolts in brace flange to gusset plate rounded up to nearest even number $\quad n_{bolts\_brcflng2gusset} := \frac{P_{uf}}{\phi r_{nv}} = 10.61$

$n_{bolts\_brcflng2gusset} := Ceil\left(n_{bolts\_brcflng2gusset}, 2\right) = 12$

six rows of 2 bolts in flange and 6 bolts total in gusset (double shear)

Angle property

Section $\quad$ $\boxed{L_1 := \text{"L4X4X3/4"}}$ $\quad$ double angle (properties are doubled in definition)

Gross area $\quad Ag_{L1} := 2T1_{(Row(L_1),6)} \cdot in^2 = 10.88\ in \cdot in$

Thickness $\quad t_{L1} := T1_{(Row(L_1),22)} \cdot in = 0.75\ in$

Leg length (same both legs) $\quad L_{L1} := T1_{(Row(L_1),15)} \cdot in = 4\ in$

Centroid $\quad x\_bar_{L1} := T1_{(Row(L_1),28)} \cdot in = 1.27\,in$

Tensile yielding of angles (J4.1a)

recall: $Ag_{L1} = 10.88\ in^2$, $p = 3\ in$, $x\_bar_{L1} := 1.27\ in$, $d''_{bolt} = 1\ in$, $t_{L1} = 0.75\ in$, $Fy_{A36} = 36\ ksi$, $Fu_{A36} = 58\ ksi$, $P_{uf} = 258.42\ kip$

Strength reduction factor $\quad \phi := 0.90$

Tensile yielding strength $\quad \phi R_n := \phi \cdot Fy_{A36} \cdot Ag_{L1} = 352.51\ kip$

Check capacity $\quad if\left(P_{uf} < \phi R_n, \text{"OK"}, \text{"NOT OK"}\right) = \text{"OK"}$

Tensile rupture of angles (J4.1b)

Table D3.1 (2)

$$U := 1 - \frac{x\_bar_{L1}}{p \cdot \left(\frac{n_{bolts\_brcflng2gusset}}{2} - 1\right)} = 0.915$$

Net area $\quad An_{L1} := Ag_{L1} - 2t_{L1} \cdot \left(d''_{bolt}\right) = 9.38\ in^2$

Effective area $\quad Ae_{L1} := An_{L1} \cdot U = 8.59\,in^2$

Strength reduction factor $\quad \phi := 0.75$

Tensile rupture strength $\quad \phi R_{nv} := \phi Fu_{A36} \cdot Ae_{L1} = 373.48\ kip$

Check capacity $\quad if\left(P_{uf} < \phi R_n, \text{"OK"}, \text{"NOT OK"}\right) = \text{"OK"}$

Block shear strength (J4.3)

Block shear strength	$R_n = 0.60F_u \cdot A_{nv} + U_{bs} \cdot F_u \cdot A_{nt} \le 0.60F_y \cdot A_{gv} + U_{bs} \cdot F_u \cdot A_{nt}$		
Edge distance of bolt centers	$L_{ev} := 1.5\,in$	$L_{eh} := 1.5\,in$	$U_{bs} := 1.0$
Net tensile area	$Ant_{L1} := 2t_{L1} \cdot (L_{eh} - 0.5 \cdot d''_{bolt}) = 1.5\ in^2$		
Gross shear area	$Agv_{L1} := 2t_{L1} \cdot \left[p \cdot \left(\frac{n_{bolts\_brcflng2gusset}}{2} - 1\right) + L_{ev}\right] = 24.75\ in^2$		
Net shear area	$Anv_{L1} := 2Agv_{L1} - t_{L1} \cdot \left(\frac{n_{bolts\_brcflng2gusset}}{2} - 0.5\right) \cdot d''_{bolt} = 45.37\ in^2$		
Strength reduction factor	$\phi := 0.75$	$\phi \cdot 0.60Fu_{A36} \cdot Anv_{L1} = 1184.29\ kip$ $\phi \cdot U_{bs} \cdot Fu_{A36} \cdot Ant_{L1} = 65.25\ kip$ $\phi \cdot 0.60Fy_{A36} \cdot Agv_{L1} = 400.95\ kip$	

Block shear strength: $\phi R_n : \phi \cdot U_{bs} \cdot Fu_{A36} \cdot Ant_{L1} + \min\left(\phi \cdot 0.60Fy_{A36} \cdot Agv_{L1}, \phi \cdot 0.60Fu_{A36} \cdot Anv_{L1}\right)$

$\phi R_n = 466.2\,kip$

Check capacity: $if\left(P_{uf} < \phi R_n, \text{"OK"}, \text{"NOT OK"}\right) = \text{"OK"}$

recall
$n_{bolts\_brcflng2gusset} = 12$
$t_{L1} = 0.75\ in$
$p = 3\ in$
$d''_{bolt} = 1\ in$
$P_{uf} = 258.42\ kip$

Brace Web to Gusset Connection

number of bolts in brace web to gusset plate rounded up to nearest even number (double shear): $n_{bolts\_brcweb2gusset} := \frac{P_{uw}}{(2)\,\phi r_{nv}} = 3.25$

recall
$P_{uw} = 158.15\ kip$
$\phi r_{nv} = 24.35\ kip$

2 rows of 2 bolts: $n_{bolts\_brcweb2gusset} := Ceil\left(n_{bolts\_brcweb2gusset}, 2\right) = 4$

Web plate dimensions: $t_{PL1} := 0.375\,in$ $\quad h_{PL1} := 9\,in$

Tensile yielding of plate (J4.1a)

Gross tensile area	$Ag_{PL1} := (2) \cdot t_{PL1} \cdot h_{PL1} = 6.75\,in^2$
Strength reduction factor	$\phi := 0.90$
Tensile yielding strength	$\phi R_n := \phi \cdot Fy_{A36} \cdot Ag_{PL1} = 218.7\,kip$
Check capacity	$if\left(P_{uw} < \phi R_n, \text{"OK"}, \text{"NOT OK"}\right) = \text{"OK"}$

Tensile Rupture of Plates (J4.1b)

Net area	$An_{PL1} := Ag_{PL1} - (2)\,t_{PL1} \cdot (2)\,d''_{bolt} = 5.25\,in^2$	recall
Effective area	$Ae_{PL1} := \min\left(An_{PL1}, 0.85 \cdot Ag_{PL1}\right) = 5.25\,in^2$	$t_{PL1} = 0.38\,in$
Strength reduction factor	$\phi := 0.75$	$d''_{bolt} = 1\,in$
Tensile rupture strength	$\phi R_n := \phi \cdot Fu_{A36} \cdot Ae_{PL1} = 228.38\,kip$	$P_{uw} = 158.15\,kip$
Check capacity	$if\left(P_{uw} < \phi R_n, \text{"OK"}, \text{"NOT OK"}\right) = \text{"OK"}$	

Block shear strength of PL-1 (J4.3)

Block shear strength $R_n = 0.60F_u \cdot A_{nv} + U_{bs} \cdot F_u \cdot A_{nt} \le 0.60F_y \cdot A_{gv} + U_{bs} \cdot F_u \cdot A_{nt}$

Edge distance of bolt centers $L_{ev} := 1.5\,in$ $\quad L_{eh} := 1.5\,in$ $\quad U_{bs} := 1.0$

Net tensile area $Ant_{PL1} := (4)\,t_{PL1}\left(L_{eh} - 0.5 \cdot d''_{bolt}\right) = 1.5\,in^2$

Gross shear area $$Agv_{PL1} := (4)\,t_{PL1} \cdot \left[p \cdot \left(\frac{n_{bolts\_brcweb2gusset}}{2} - 1\right) + L_{ev}\right] = 6.75\,in^2$$

Net shear area $$Anv_{PL1} := (4)\,Agv_{PL1} - t_{PL1} \cdot \left(\frac{n_{bolts\_brcweb2gusset}}{2} - 0.5\right) \cdot d''_{bolt} = 26.44\,in^2$$

Strength reduction factor $\phi := 0.75$

$\phi \cdot 0.60Fu_{A36} \cdot Anv_{PL1} = 690.02\,kip$

$\phi \cdot U_{bs} \cdot Fu_{A36} \cdot Ant_{PL1} = 65.25\,kip$

$\phi \cdot 0.60Fy_{A36} \cdot Agv_{PL1} = 109.35\,kip$

Block shear strength $\phi R_n : \phi \cdot U_{bs} \cdot Fu_{A36} \cdot Ant_{PL1} + \min\left(\phi \cdot 0.60Fy_{A36} \cdot Agv_{PL1}, \phi \cdot 0.60Fu_{A36} \cdot Anv_{PL1}\right)$

$\phi R_n = 174.6\,kip$

Check capacity $if\left(P_{uw} < \phi R_n, \text{"OK"}, \text{"NOT OK"}\right) = \text{"OK"}$

Block shear strength of Brace (J4.3)

Block shear strength $R_n = 0.60F_u \cdot A_{nv} + U_{bs} \cdot F_u \cdot A_{nt} \le 0.60F_y \cdot A_{gv} + U_{bs} \cdot F_u \cdot A_{nt}$

Edge distance of bolt centers $L_{ev} := 1.5\,in \quad L_{eh} := 3\,in \quad U_{bs} := 1.0$

Net tensile area $Ant_{brace1} := tw_{brace}\left(6in - d''_{bolt}\right) = 2.57\,in^2$

Gross shear area $Agv_{brace\ 1} := (2)\,tw_{brace} \cdot \left[p \cdot \left(\frac{n_{bolts\_brcweb2gusset}}{2} - 1\right) + L_{ev}\right] = 4.63\,in^2$

Net shear area $Anv_{brace\ 1} := Agv_{brace\ 1} - (2)\,tw_{brace} \cdot \left(\frac{n_{bolts\_brcweb2gusset}}{2} - 0.5\right) \cdot d''_{bolt} = 3.09\,in^2$

recall
$tw_{brace} = 0.52\,in$
$n_{bolts\_brcweb2gusset} = 4$
$P_{uw} = 158.15\,kip$
$T_u = 675\,kip$

Strength reduction factor $\phi := 0.75$

$\phi \cdot 0.60Fu_{A992} \cdot Anv_{brace1} = 90.38\,kip$

$\phi \cdot U_{bs} \cdot Fu_{A992} \cdot Ant_{brace1} = 125.53\,kip$

$\phi \cdot 0.60Fy_{A992} \cdot Agv_{brace1} = 104.29\,kip$

Block shear strength $\phi R_n := \phi \cdot U_{bs} \cdot Fu_{A992} \cdot Ant_{brace1} + \min\left(\phi \cdot 0.60Fy_{A992} \cdot Agv_{brace1}, \phi \cdot 0.60Fu_{A992} \cdot Anv_{brace1}\right)$

$\phi R_n = 215.9\,kip$

Check capacity if $\left(P_{uw} < \phi R_n, \text{"OK"}, \text{"NOT OK"}\right) = \text{"OK"}$

Tensile yielding of brace (J4.1a)

Strength reduction factor $\phi := 0.90$

Tensile yielding strength $\phi R_n := \phi \cdot Fy_{A992} \cdot Ag_{brace} = 1152\,kip$

Check capacity if $\left(T_u < \phi R_n, \text{"OK"}, \text{"NOT OK"}\right) = \text{"OK"}$

Tensile rupture of brace (J4.1b)

Net area $An_{brace1} := Ag_{brace} - \left[tw_{brace} \cdot (2) + tf_{brace} \cdot (4)\right] d''_{bolt}$

$An_{brace1} = 21.33\,in^2$

Effective area $Ae_{brace1} := An_{brace1}$

Strength reduction factor $\phi := 0.75$

Tensile rupture strength $\phi R_n := \phi \cdot Fu_{A992} \cdot Ae_{brace1} = 1039.84\,kip$

Check capacity if $\left(T_u < \phi R_n, \text{"OK"}, \text{"NOT OK"}\right) = \text{"OK"}$

Gusset plate $t_{gusset} = 0.75\,in = 2 \cdot t_{PL1} = 0.75\,in$

recall
$Ag_{brace} = 25.6\,in^2$
$tw_{brace} = 0.52\,in$
$tf_{brace} = 0.81\,in$
$d''_{bolt} = 1\,in$
$T_u = 675\,kip$
$t_{PL1} = 0.375\,in$
$t_{gusset} = 0.75\,in$

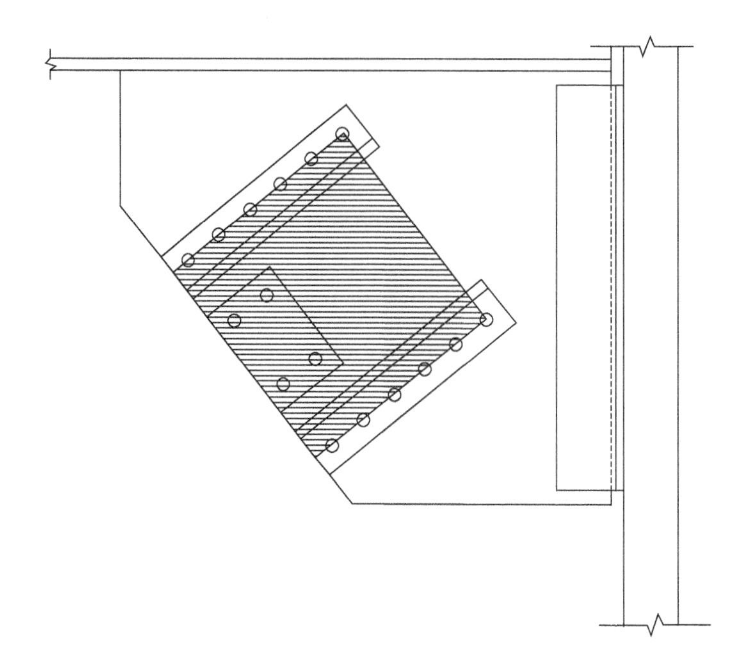

block shear strength of gusset (J4.3)

Block shear strength $R_n = 0.60\ F_u \cdot A_{nv} + U_{bs} \cdot F_u \cdot A_{nt} \leq 0.60\ F_y \cdot A_{gv} + U_{bs} \cdot F_u \cdot A_{nt}$

$U_{bs} = 1.0$

Net tensile area $Ant_{gusset1} := t_{gusset} \left[d_{brace} + (2) \cdot L_{L1} - 2 \cdot L_{ev} - d''_{bolt}\right] = 12.37\ in^2$

Gross shear area $Agv_{gusset\ 1} := (2)\ t_{gusset} \cdot \left[p \cdot \left(\frac{n_{bolts\_brcflng2gusset}}{2} - 1\right) + 1.75\ in\right] =$ 1.75″ edge distance

$25.12\ in^2$

Net shear area $Anv_{gusset1} := Agv_{gusset1} - (2)\ t_{gusset} \cdot \left(\frac{n_{bolts\_brcflng2gusset}}{2} - 0.5\right) \cdot$

$d''_{bolt} = 16.87 in^2$

Strength reduction factor $\phi := 0.75$

$\phi \cdot 0.60Fu_{A36} \cdot Anv_{gusset1} = 440.44\ kip$

$\phi \cdot U_{bs} \cdot Fu_{A36} \cdot Ant_{gusset\ 1} = 538.31\ kip$

$\phi \cdot 0.60Fy_{A36} \cdot Agv_{gusset\ 1} = 407.02\ kip$

Block shear strength $\phi R_n := \phi \cdot U_{bs} \cdot Fu_{A36} \cdot Ant_{gusset1} +$

$\min\left(\phi \cdot 0.60Fy_{A36} \cdot Agv_{gusset1}, \phi \cdot 0.60Fu_{A36} \cdot Anv_{gusset1}\right.$

$\phi R_n = 945.3 kip$

Check capacity $\text{if}\left(T_u < \phi R_n, \text{“OK”}, \text{“NOT OK”}\right) = \text{“OK”}$

Tensile yielding of Whitmore section (J3.1a)

Total length of Whitmore section

$$l_w := d_{brace} + (2) \cdot L_{L1} - 2 \cdot L_{ev} + 2 \cdot \tan(30\,deg) \cdot \left[p \cdot \left(\frac{n_{bolts\_brcflng2gusset}}{2} - 1\right)\right] = 34.82\ in$$

Length of Whitmore section in gusset

$lw_{gusset} := 30.9\ in$ $\quad lw_{web} := l_w - lw_{gusset} = 3.92\ in$

Strength reduction factor

$\phi := 0.90$

Tensile yielding strength (gusset portion)

$\phi R_{n1} := \phi \cdot Fy_{A36} \cdot lw_{gusset} \cdot t_{gusset} = 750.87\ kip$

(web portion)

$\phi R_{n2} := \phi \cdot Fy_{A992} \cdot lw_{web} \cdot tw_{beam} = 52.93\ kip$

Tensile yielding strength

$\phi R_n := \phi R_{n1} + \phi R_{n2} = 803.8\ kip$

Check capacity

$if(T_u < \phi R_n, \text{"OK"}, \text{"NOT OK"}) = \text{"OK"}$

recall

$d_{brace} = 12.5\ in$

$L_{L1} = 4\ in$

$L_{ev} = 1.5\ in$

$p = 3\ in$

$n_{bolts\_brcflng2gusset} = 12\ in$

$t_{gusset} = 0.75\ in$

$tw_{beam} = 0.3\ in$

$T_u = 675\ kip$

$2\phi r_{nv} \div \phi = 54.12\ kip$

Bolt bearing strength (angles, brace flange, and gusset) (J3.10)

by inspection - gusset controls

Edge bolt clear distance

$l_{c1\_edge} := 1.75\ in - 0.5 \cdot d'_{bolt} = 1.28\ in$ $\quad$ input 1.75″

Interior bolt clear distance

$l_{c1\_interior} := p - d'_{bolt} = 2.06\ in$

Strength reduction factor

$\phi := 0.75$

Interior bolt bearing strength

$$\phi r_{n\_interior} := \phi \cdot \min\left(1.2 \cdot l_{c1\_interior} \cdot t_{gusset} \cdot Fu_{A36}, 2.4 \cdot d_{bolt} \cdot t_{gusset} \cdot Fu_{A36}, 2\phi r_{nv} \div \phi\right)$$

Edge bolt bearing strength

$$\phi r_{n\_edge} := \phi \cdot \min\left(1.2 \cdot l_{c1\_edge} \cdot t_{gusset} \cdot Fu_{A36}, 2.4 \cdot d_{bolt} \cdot t_{gusset} \cdot Fu_{A36}, 2\phi r_{nv} \div \phi\right)$$

$\phi r_{n\_interior} = 48.71\ kip$ $\quad \phi r_{n\_edge} = 48.71\ kip$

Total bolt bearing strength

$$\phi R_n := (1) \cdot \phi r_{n\_edge} + \left(\frac{n_{bolts\_brcflng2gusset}}{2} - 1\right) \cdot \phi r_{n\_interior} = 292.24\ kip$$

Check capacity

$if(P_{uf} < \phi R_n, \text{"OK"}, \text{"NOT OK"}) = \text{"OK"}$

Bolt bearing strength (brace web) (J3.10)

Edge bolt clear distance $l_{c1\_edge} := 1.5\ in - 0.5 \cdot d'_{bolt} = 1.03\ in$

Interior bolt clear distance $l_{c1\_interior} := p - d'_{bolt} = 2.06\ in$

Strength reduction factor $\phi := 0.75$

Interior bolt bearing strength $\phi r_{n\_interior} := \phi \cdot \min\left(1.2 \cdot l_{c1\_interior} \cdot tw_{brace} \cdot Fu_{A992}, 2.4 \cdot d_{bolt} \cdot tw_{brace} \cdot Fu_{A992}, 2\phi r_{nv} \div \phi\right)$

Edge bolt bearing strength $\phi r_{n\_edge} := \phi \cdot \min\left(1.2 \cdot l_{c1\_edge} \cdot tw_{brace} \cdot Fu_{A992}, 2.4 \cdot d_{bolt} \cdot tw_{brace} \cdot Fu_{A992}, 2\phi r_{nv} \div \phi\right)$

$\phi r_{n\_interior} = 48.71\ kip$ $\phi r_{n\_edge} = 31.07\ kip$

recall
$P_{uw} = 158.15\ kip$

Total bolt bearing strength $\phi R_n := (2) \cdot \phi r_{n\_edge} + \left(n_{bolts\_brcweb2gusset} - 2\right) \cdot \phi r_{n\_interior}$

$\phi R_n = 159.55\ kip$

Check capacity if $(P_{uw} < \phi R_n$, "OK", "NOT OK") = "OK"

Distribution of brace forces to beam and column (AISC Chapter 13)

recall
$d_{beam} = 17.8\ in$
$d_{column} = 20.9\ in$
$L_{gusset} = 42\ in$
$h_{gusset} = 33\ in$
$T_u = 675\ kip$

Half beam depth $e_{beam} := d_{beam} \div 2 = 8.85\ in$

Half column depth $e_{column} := d_{column} \div 2 = 10.45\ in$

Slope of brace

Used input values $\theta_{brace} : \operatorname{atan}\left(\dfrac{12}{9 + \dfrac{9}{16}}\right) = 51.45 \cdot deg$ $\qquad \dfrac{12}{9 + \dfrac{9}{16}} = 12.5$

$e_{beam} \cdot \tan\left(\theta_{brace}\right) - e_{column} = 0.66\ in$

$$\alpha\_bar := \frac{L_{gusset}}{2} + 0.5\ in = 21.5\ in$$

$$\beta\_bar := \frac{h_{gusset}}{2} = 16.5\ in$$

$\beta := \beta\_bar$

$\alpha - \beta \cdot \tan(\theta) = e_{beam} \cdot \tan(\theta) - e_{column}$ Eq. 13-1

$\alpha = e_{beam} \cdot \tan(\theta) - e_{column} + \beta \cdot \tan(\theta)$

$\alpha := e_{beam} \cdot \tan\left(\theta_{brace}\right) - e_{column} + \beta\_bar \cdot \tan\left(\theta_{brace}\right) = 21.36\ in$

$e_{additional} := \alpha - \alpha\_bar = -0.14\ in$

$r := \sqrt{\left(\alpha + e_{column}\right)^2 + \left(\beta + e_{column}\right)^2} = 40.68\ in$ Eq. 13-6

Required axial force on gusset to column connection	$H_{column} := \frac{e_{column}}{r} \cdot T_u = 173.41\ kip$	Eq. 13-3
Required shear force on gusset to column connection	$V_{column} := \frac{\beta}{r} \cdot T_u = 273.8\ kip$	Eq. 13-2
Required axial force on gusset to beam connection	$H_{beam} := \frac{\alpha}{r} \cdot T_u = 354.48\ kip$	Eq. 13-5
Required shear force on gusset to beam connection	$V_{beam} := \frac{e_{beam}}{r} \cdot T_u = 146.86\ kip$	Eq. 13-4
Check balance of forces	$T_u \cdot \sin(\theta_{brace}) - (H_{column} + H_{beam}) = -0\ kip$	
section	$\boxed{L_2 := \text{"L5X3} - 1/2\text{X5/8"}}$	double angle (properties are doubled in definition)
Gross area	$Ag_{L2} := 2T1_{(Row(L_2),6)} \cdot in^2 = 9.86 in \cdot in$	
Thickness	$t_{L2} := T1_{(Row(L_2),22)} \cdot in = 0.625\ in$	
Leg length long	$L_{L2} := T1_{(Row(L_2),15)} \cdot in = 5\ in$	
Leg length short	$h_{L2} := T1_{(Row(L_2),7)} \cdot in = 3.5\ in$	
Centroid	$x\_bar_{L2} : T1_{(Row(L_2),28)} \cdot in = 0.95\ in$	
Gage of angle	$GOL_2 := 3 in$	
Number of bolts in gusset to column connection	$n_{bolts\_gusset2column} := 20$	

Check bolt capacity (J3.7)

Ultimate tensile force per bolt	$r_{ut} := \frac{H_{column}}{n_{bolts\_gusset2column}} = 8.67 kip$	recall
		$H_{column} = 173.41\ kip$
Check capacity	$if(r_{ut} < \phi r_{nt}, \text{"OK"}, \text{"NOT OK"}) = \text{"OK"}$	$V_{column} = 273.8\ kip$
Ultimate tensile force per bolt	$r_{uv} := \frac{V_{column}}{n_{bolts\_gusset2column}} = 13.69\ kip$	$\phi r_{nt} = 40.59\ kip$
		$A_{bolt} = 0.6\ in^2$
Check capacity	$if(r_{uv} < \phi r_{nv}, \text{"OK"}, \text{"NOT OK"}) = \text{"OK"}$	$F_{nt} = 90\ ksi$
	$f_{uv} := \frac{r_{uv}}{A_{bolt}} = 22.77\ ksi$	$F_{nv} = 54\ ksi$
Strength reduction factor	$\phi := 0.75$	
Modified nominal tensile Stress factored to include Shear stress effects	$F'_{nt} = 1.3 \cdot F_{nt} - \frac{F_{nt}}{\phi \cdot F_{nv}} \cdot f_{rv} < F_{nt}$	
	$F'nt_{bolt} := \min\left(1.3 \cdot F_{nt} - \frac{F_{nt}}{\phi \cdot F_{nv}} \cdot f_{uv}, F_{nt}\right) = 66.41\ ksi$	
Combined loading strength	$\phi R_n := \phi \cdot F'nt_{bolt} \cdot A_{bolt} = 29.95\ kip$	
Check capacity	$if(r_{ut} < \phi R_n, \text{"OK"}, \text{"NOTOK"}) = \text{"OK"}$	

Bolt bearing on double angle (J3.10)

Bolt edge distance $l_{c1\_edge} = 1.03\ in$

Strength reduction factor $\phi := 0.75$

Edge bolt bearing strength $\phi r_{n\_edge} := \phi \cdot \min\left(1.2 \cdot l_{c1\_edge} \cdot t_{L2} \cdot Fu_{A36}, 2.4 \cdot d_{bolt} \cdot t_{L2} \cdot Fu_{A36}\right)$

$\phi r_{n\_edge} = 33.64\ kip > \phi r_{nv} = 24.35\ kip$

$> r_{nv} = 13.369\ kip$

$if\left(r_{uv} < \phi r_{n\_edge}, \text{“OK”}, \text{“NOT OK”}\right) = \text{“OK”}$

Since edge bolt bearing > bolt shear, and bolt shear force, no need to check

Prying action on double angles (Chapter 9)

recall $t_{L2} = 0.63\ in$, $GOL_2 = 3\ in$, $L_{L2} = 5\ in$, $d_{bolt} = 0.88\ in$, $p = 3\ in$, $r_{ut} = 8.67\ kip$

$b_2 := GOL_2 - \frac{t_{L2}}{2} = 2.69\ in$

$a_2 := L_{L2} - GOL_2 = 2\ in$

$b_2' : b_2 - \frac{d_{bolt}}{2} = 2.25\ in$

$a_2' := \min\left(a_2 + \frac{d_{bolt}}{2}, 1.25b_2 + \frac{d_{bolt}}{2}\right) = 2.44\ in$ Eq. 9-27

$\rho_2 : \frac{b_2'}{a_2'} = 0.923$ Eq. 9-26

$B := \phi r_{nt} = 40.59\ kip$

Available tension per bolt $B := 40.59\ kip$

$\beta := \frac{1}{\rho 2}\left(\frac{B}{r_{ut}} - 1\right) = 3.99$ Eq. 9-25

$\alpha_2' := 1.0$ for $\beta > 1.0$

$\delta_2 : 1 - \frac{d_{bolt}'}{p} = 0.688$ Eq. 9-24

Strength reduction factor $\phi := 0.90$

Minimum thickness to neglect prying $t_{req \cdot 2} := \sqrt{\frac{4 \cdot r_{ut} \cdot b_2'}{\phi \cdot p \cdot Fu_{A36} \cdot (1 + \delta_2 \cdot \alpha_2')}} = 0.54\ in$ Eq. 9-23

Check thickness requirements $if\left(t_{req.2} < t_{L2}, \text{“OK”}, \text{“NOT OK”}\right) = \text{“OK”}$

Weld design

Force on weld $Pu_{column} := \sqrt{H_{column}^{\ 2} + V_{column}^{\ 2}} = 324.1\ kip$

Load angle on weld $\theta_{load.2} := atan\left(\frac{H_{column}}{V_{column}}\right) = 32.35 \cdot deg$

recall $H_{column} = 173.41\ kip$, $V_{column} = 273.8\ kip$, $n_{bolts\_gusset2column} = 20L_{ev} = 1.5\ in$, $p = 3\ in$, $h_{L2} = 3.5\ in$, $t_{gusset} = 0.75\ in$

Side length of weld $l_{weld.2} := p \cdot \left(\frac{n_{bolts\_gusset2column}}{2} - 1\right) + (2) \cdot L_{ev} = 30\ in$

Bottom length of weld $kl_{weld.2} := h_{L2} - 0.5\ in = 3\ in$

$k_{weld.2} := \frac{kl_{weld.2}}{l_{weld.2}} = 0.1$

$xl_2 := \frac{(2)\frac{1}{2}kl_{weld.2}^{\ 2}}{(2)\,kl_{weld.2} + l_{weld.2}} = 0.25\ in$

$al_{weld.2} := h_{L2} - xl_2 = 3.25\ in$

$a_{weld.2} := \frac{al_{weld.2}}{l_{weld.2}} = 0.108$

Table 8-8

Eccentric weld factor $C_{weld.2} := 2.55\frac{kip}{in}$

Strength reduction factor $\phi := 0.75$

Minimum weld size $D_2 := \frac{Pu_{column}}{\phi \cdot C_{weld.2} \cdot C_1 \cdot (2)\,l_{weld.2}} = 2.82$

Minimum thickness of gusset plate $t_{gusset.min} := \frac{6.19\frac{kip}{in} \cdot D_2}{Fu_{A36}} = 0.301 \cdot in$ Eq. 9-3

Check thickness requirements $\text{if}\left(t_{gusset.min} < t_{gusset}, \text{“OK”}, \text{“NOT OK”}\right) = \text{“OK”}$

Weld size $D_2 := \max\left(\text{ceil}\left(D_2\right), 4\right) = 4$ use 4/16 as minimum

Shear yielding of angles (J4.2a)

Gross shear area $Agv_{L2} := (2) \cdot t_{L2} \cdot l_{weld.2} = 37.5\ in^2$

Strength reduction factor $\phi := 1.00$

Shear yielding strength $\phi R_n := \phi \cdot 0.60 \cdot Fy_{A36} \cdot Agv_{L2} = 810\ kip$

Check capacity $\text{if}\left(V_{column} < \phi R_n, \text{“OK”}, \text{“NOT OK”}\right) = \text{“OK”}$

recall
$t_{L2} = 0.63\ in$
$l_{weld.2} = 30\ in$
$V_{column} = 273.8\ kip$
$d''_{bolt} = 1\ in$

Shear rupture of angles (J4.2b)

Net shear area $Anv_{L2} := t_{L2} \cdot \left(2 \cdot l_{weld.2} - n_{bolts\_gusset2column} \cdot d''_{bolt}\right) = 25\ in^2$

Strength reduction factor $\phi := 0.75$

Shear rupture strength $\phi R_n := \phi \cdot 0.60 \cdot Fu_{A36} \cdot Anv_{L2} = 652.5\ kip$

Check capacity $\text{if}\left(V_{column} < \phi R_n, \text{“OK”}, \text{“NOT OK”}\right) = \text{“OK”}$

Block shear strength of angles (J4.3)

Block shear strength	$R_n = 0.60F_u \cdot A_{nv} + U_{bs} \cdot F_u \cdot A_{nt} \leq 0.60F_y \cdot A_{gv} + U_{bs} \cdot F_u \cdot A_{nt}$
Edge distance of bolt centers	$L_{ev\_L2} := 1.5\ in \quad L_{eh\_L2} := 2\ in \quad U_{bs} := 1.0\ in$
Net tensile area	$Ant_{L2} := (2)\, t_{L2} \left(L_{eh\_L2} - 0.5\, d''_{bolt} \right) = 1.88\ in^2$
Gross shear area	$Agv_{L2} := (2)\, t_{L2} \cdot \left[p \cdot \left(\frac{n_{bolts\_gusset2column}}{2} - 1 \right) + L_{ev\_L2} \right] = 35.63\ in^2$
Net shear area	$Anv_{L2} := Agv_{L2} - (2)\, t_{L2} \cdot \left(\frac{n_{bolts\_gusset2column}}{2} - 0.5 \right) \cdot d''_{bolt} = 23.75\ in^2$
Strength reduction factor	$\phi = 0.75$ $\quad \phi \cdot 0.60Fu_{A36} \cdot Anv_{L2} = 619.87\ kip$
	$\phi \cdot U_{bs} \cdot Fu_{A36} \cdot Ant_{L2} = 81.56\ kip$
	$\phi \cdot 0.60Fy_{A36} \cdot Agv_{L2} = 577.13\ kip$
Block shear strength	$\phi R_n := \phi \cdot U_{bs} \cdot Fu_{A36} \cdot Ant_{L2} + \min \left(\phi \cdot 0.60Fy_{A36} \cdot Agv_{L2}, \phi \cdot 0.60Fu_{A36} \cdot Anv_{L2} \right)$
	$\phi R_n = 658.7\ kip$
Check capacity	$\text{if}\left(V_{column} < \phi R_n, \text{"OK"}, \text{"NOT OK"}\right) = \text{"OK"}$

Gusset plate to beam connection

Additional ductility factor (p. 13-11) $\quad ductility := 1.25$

recall
$V_{beam} = 146.86\ kip$
$H_{beam} = 354.48\ kip$
$t_{gusset} = 0.75\ in$
$L_{gusset} = 42\ in$

Tensile yielding strength (J4.1a)

Tensile stress	$f_{ua\_gusset2beam} := \frac{V_{beam}}{t_{gusset} \cdot L_{gusset}} = 4.66\ ksi$
Strength reduction factor	$\phi := 0.9 \quad\quad \phi \cdot Fy_{A36} = 32.4\ ksi$
Check capacity	$\text{if}\left(f_{ua\_gusset2beam} < \phi \cdot Fy_{A36}, \text{"OK"}, \text{"NOT OK"}\right) = \text{"OK"}$

Shear yielding strength (J4.2a)

Shear stress $f_{uv\_gusset2beam} := \frac{H_{beam}}{t_{gusset} \cdot L_{gusset}} = 11.25\ ksi$

Strength reduction factor $\phi := 1.00$ $\phi \cdot 0.6Fy_{A36} = 21.6\ ksi$

Check capacity $\text{if}\left(f_{uv\_gusset2beam} < \phi \cdot 0.6Fy_{A36}, \text{"OK"}, \text{"NOT OK"}\right) = \text{"OK"}$

Load angle $\theta_{gusset2beam} := \text{atan}\left(\frac{V_{beam}}{H_{beam}}\right) = 22.5 \cdot deg$

Eq J2-5

Effect of load angle $\mu := 1.0 + 0.5 \cdot \sin\left(\theta_{gusset2beam}\right)^{1.5} = 1.12$

Weld strength (Eq 8-2) $\phi rw := \mu \cdot 1.392 \frac{kip}{in} = 1.56 \cdot \frac{kip}{in}$ per 1/16 weld size

Peak stress $fu_{peak\_gusset2beam} := \left(\frac{t_{gusset}}{2}\right) \cdot \sqrt{f_{ua\_gusset2beam}{}^2 + f_{uv\_gusset2beam}{}^2} = 4.57 \cdot \frac{kip}{in}$

Average stress $fu_{average\_gusset2beam} = 4.57 \cdot \frac{kip}{in}$

$$fu_{average\_gusset2beam} := \frac{\left(\frac{t_{gusset}}{2}\right)\left(\sqrt{f_{ua\_gusset2beam}{}^2 + f_{uv\_gusset2beam}{}^2} + \sqrt{f_{ua\_gusset2beam}{}^2 + f_{uv\_gusset2beam}{}^2}\right)}{2}$$

Stress on weld $fu_{weld} := \max\left(fu_{peak\_gusset2beam}, ductility \cdot fu_{average\_gusset2beam}\right) = 5.71 \cdot \frac{kip}{in}$

Minimum required weld size $D_{gusset2beam\ ceil\ Dgusset2beam} := \frac{fu_{weld}}{\phi rw} = 3.67$

Actual weld size $D_{average\_gusset2beam} := \text{ceil}\left(D_{gusset2beam}\right) = 4$

Web local yielding of beam (J10.2.b)

Strength reduction factor $\phi := 1.00$

Web local yielding strength $\phi R_n := \phi \cdot Fy_{A992} \cdot tw_{beam} \cdot \left(2.5 \cdot k\_des_{beam} + L_{gusset}\right)$

$\phi R_n = 661.01\ kip$

Check capacity $\text{if}\left(V_{beam} < \phi R_n, \text{"OK"}, \text{"NOT OK"}\right) = \text{"OK"}$

recall

$tw_{beam} = 0.3\ in$

$tf_{beam} = 0.43\ in$

$d_{beam} = 17.7\ in$

$k\_des_{beam} = 0.83\ in$

$L_{gusset} = 42\ in$

$V_{beam} = 146.86\ kip$

Web local crippling of beam (J10.3b)

Strength reduction factor $\phi := 0.75$

Web local crippling strength

$$\phi R_n := \begin{cases} \phi \cdot 0.40 \cdot tw_{beam}^{\ 2} \cdot \left[1 + \left(\dfrac{4L_{gusset}}{d_{beam}} - 0.2\right) \cdot \left(\dfrac{tw_{beam}}{tf_{beam}}\right)^{1.5}\right] \cdot \sqrt{\dfrac{E \cdot Fy_{A992} \cdot tf_{beam}}{tw_{beam}}} & \text{if } \dfrac{L_{gusset}}{d_{beam}} > 0.2 \\ \phi \cdot 0.40 \cdot tw_{beam}^{\ 2} \cdot \left[1 + \left(\dfrac{L_{gusset}}{d_{beam}}\right) \cdot \left(\dfrac{tw_{beam}}{tf_{beam}}\right)^{1.5}\right] \cdot \sqrt{\dfrac{E \cdot Fy_{A992} \cdot tf_{beam}}{tw_{beam}}} & \text{otherwise} \end{cases}$$

$\phi R_n = 251.94 \text{ kip}$

Check capacity $\text{if}\left(V_{beam} < \phi R_n, \text{"OK"}, \text{"NOT OK"}\right) = \text{"OK"}$

Beam-to-column connection

Net shear in beam $Vu_{beam} := R_{u\_beam} + V_{beam} = 161.86 \text{ kip}$

Net axial force in beam $Hu_{beam} := P_{u\_beam} - H_{beam} = 173.52 \text{ kip}$

Section $L_3 := \text{"L8X6X7/8"}$ double angle (properties are doubled in definition)

recall
$R_{u\_beam} = 15 \text{ kip}$
$V_{beam} = 146.86 \text{ kip}$
$P_{u\_beam} = 528 \text{ kip}$
H beam =354.48 kip

Gross area $Ag_{L3} := 2\ T1_{(Row(L_3),6)} \cdot in^2 = 23 \text{ in} \cdot \text{in}$

Thickness $t_{L3} := T1_{(Row(L_3),22)} \cdot in = 0.875 \text{ in}$

Leg length long $L_{L3} := T1_{(Row(L_3),15)} \cdot in = 8 \text{ in}$

Leg length short $h_{L3} := T1_{(Row(L_3),7)} \cdot in = 6 \text{ in}$

Centroid $x\_bar_{L3} := T1_{(Row(L_3),28)} \cdot in = 1.6 \text{ in}$

Gage of angle $GOL_3 := 3 \text{ in} + \dfrac{1}{16} \text{ in}$

Number of bolts in gusset to column connection $n_{bolts\_beamclip} := 10$

Length of angle $Length_{L3} := 1 \text{ ft} + (2)\ 1.25 \text{ in}$

Check bolt shear capacity (J3.6)

Ultimate tensile force per bolt $r_{uv\_beamclip} := \dfrac{Vu_{beam}}{n_{bolts\_beamclip}} = 16.19\ kip$

Check capacity $\text{if}\left(r_{uv\_beamclip} < \phi r_{nv}, \text{"OK"}, \text{"NOT OK"}\right) = \text{"OK"}$

recall
$Vu_{beam} = 161.86\ kip$
$n_{bolts\_beamclip} = 10$
$\phi r_{nv} = 24.35\ kip$

Bolt bearing on double angle (J3.10)

Bolt edge clear distance $l_{c3\_edge} := 1.25\ in - 0.5 \cdot d''_{bolt} = 0.78\ in$ — user input the edge distance (1.25″)

Strength reduction factor $\phi := 0.75$

Bolt edge bearing strength $\phi r_{n3\_edge} := \phi \cdot \min\left(1.2 \cdot l_{c3\_edge} \cdot t_{L3} \cdot Fu_{A36}, 2.4 \cdot d_{bolt} \cdot t_{L3} \cdot Fu_{A36}\right)$

$\phi r_{n3\_edge} = 35.68\ kip > \phi r_{nv} = 24.35\ kip$ — bolt shear strength

Check capacity $\text{if}\left(\phi r_{nv} < \phi r_{n3\_edge}, \text{"OK"}, \text{"NOT OK"}\right) = \text{"OK"}$

Since edge bolt bearing > bolt shear strength, no need to check bearing

Weld design Table 8-8

Force on weld $Pu_{clip} := \sqrt{Hu_{beam}^2 + Vu_{beam}^2} = 237.29\ kip$

Force angle on weld $\theta_{clip} := \text{atan}\left(\dfrac{Vu_{beam}}{Hu}\right) = 43.01 \cdot deg$

Length of weld side $l_{eccweld} := Length_{L3} = 14.5\ in$

recall
$Hu_{beam} = 173.52\ kip$
$Vu_{beam} = 161.86\ kip$
$Length_{L3} = 14.5\ in$
$tw_{beam} = 0.3\ in$

Length of weld bottom and top $kl_{eccweld} := L_{L3} - 0.5in = 7.5\ in$

$k_{eccweld} := \dfrac{kl_{eccweld}}{l_{eccweld}} = 0.52$

$xl_{eccweld} := \dfrac{kl_{eccweld}^2}{l_{eccweld} + 2 \cdot kl_{eccweld}} = 1.91\ in$

$al_{eccweld} := L_{L3} - xl_{eccweld} = 6.09\ in$

$a_{eccweld} := \dfrac{al_{eccweld}}{l_{eccweld}} = 0.42$

Table 8-8

Eccentric weld factor	$C_{clip} := 3.55kip \div in$
Strength reduction factor	$\phi := 0.75$
Minimum weld size	$D_{clip} := \dfrac{Pu_{clip}}{\phi \cdot C_{clip} \cdot C_1 \cdot 2\ Length_{L3}} = 3.07$
Minimum thickness of beam web	$tw_{beam\_min} := 6.19\dfrac{kip}{in} \cdot D_{clip} \div Fu_{A992} = 0.29\ in$
Check thickness requirements	$if\left(tw_{beam\_min} < tw_{beam}, \text{"OK"}, \text{"NOT OK"}\right) = \text{"OK"}$
Weld size	$D_{clip:} := ceil\left(D_{clip}\right) = 4$

Shear yielding of angles (J4.2a)

Gross shear area	$Agv_{L3} := (2) \cdot Length_{L3} \cdot t_{L3} = 25.38\ in^2$	recall
Strength reduction factor	$\phi := 1.00$	$Length_{L3} = 14.5in$
		$t_{L3} = 0.88\ in$
Shear yielding strength	$\phi R_n := \phi \cdot 0.6 \cdot Fy_{A36} \cdot Agv_{L3} = 548.1\ kip$	$n_{bolts\_beamclip} = 10$
Check capacity	$if\left(Vu_{beam} < \phi R_n, \text{"OK"}, \text{"NOT OK"}\right) = \text{"OK"}$	$d''_{bolt} = 1\ in$
		$Vu_{beam} = 161.86\ kip$

Shear rupture of angles (J4.2b)

Net shear area	$Anv_{L3} := Agv_{L3} - t_{L3} \cdot n_{bolts\_beamclip} \cdot d''_{bolt} = 16.63\ in^2$
Strength reduction factor	$\phi := 0.75$
Shear rupture strength	$\phi R_n := \phi \cdot 0.6 \cdot Fy_{A36} \cdot Anv_{L3} = 433.91\ kip$
Check capacity	$if\left(Vu_{beam} < \phi R_n, \text{"OK"}, \text{"NOT OK"}\right) = \text{"OK"}$

Block shear strength of angles (J4.3)

Block shear strength	$R_n = 0.60F_u \cdot A_{nv} + U_{bs} \cdot F_u \cdot A_{nt} \leq 0.60F_y \cdot A_{gv} + U_{bs} \cdot F_u \cdot A_{nt}$
Bolt centerline edge distance	$L_{ev\_L3} := 1.25in$ $L_{eh\_L3} := 2.94in$ $U_{bs} := 1.0$

Net tensile area $Ant_{L3} := (2)\, t_{L3} \left(L_{eh\_L3} - 0.5\, d''_{bolt} \right) = 4.27\ in^2$

Gross shear area $Agv_{L3} := (2)\, t_{L3} \cdot \left[p \cdot \left(\frac{n_{bolts\_beamclip}}{2} - 1 \right) + L_{ev\_L3} \right] = 23.19\ in^2$

Net shear area $Anv_{L3} := Agv_{L3} - (2)\, t_{L3} \cdot \left(\frac{n_{bolts\_beamclip}}{2} - 0.5 \right) \cdot d''_{bolt} = 15.31\ in^2$

Strength reduction factor $\phi = 0.75$

$\phi \cdot 0.60 Fu_{A36} \cdot Anv_{L3} = 399.66\ kip$

$\phi \cdot U_{bs} \cdot Fu_{A36} \cdot Ant_{L3} = 185.74\ kip$

$\phi \cdot 0.60 Fy_{A36} \cdot Agv_{L3} = 375.64\ kip$

recall
$t_{L3} = 0.88\ in$
$d''_{bolt} = 1\ in$
$n_{bolts\_beamclip} = 10$
$Vu_{beam} = 161.86\ kip$

Block shear strength $\phi R_n := \phi \cdot U_{bs} \cdot Fu_{A36} \cdot Ant_{L3} + \min \left(\phi \cdot 0.60 Fy_{A36} \cdot Agv_{L3}, \phi \cdot 0.60 Fu_{A36} \cdot Anv_{L3} \right)$

$\phi R_n = 561.4\, kip$

Check capacity if $\left(Vu_{beam} < \phi R_n, \text{"OK"}, \text{"NOT OK"} \right) = \text{"OK"}$

Check gusset buckling (J4.4 and E3)

Effective length factor $K := 0.5$

Whitmore cross-sectional area $A_w := t_{gusset} \cdot lw_{gusset} + lw_{web} \cdot tw_{beam} \cdot \frac{Fy_{A992}}{Fy_{A36}} = 24.81\ in^2$

$l_1 := 17 in$ user input guess l1 length

recall
$t_{gusset} = 0.75\ in$
$lw_{gusset} = 30.9\ in$
$lw_{web} = 3.92\ in$
$tw_{beam} = 0.36\ in$
$T_u = 675\ kip$

$\frac{KL}{r}$ $\frac{K \cdot l_1}{\sqrt{\frac{t_{gusset}^2}{12}}} = 39.26 > 25 \ll 4.71 \cdot \sqrt{\frac{E}{Fy_{A36}}} = 133.68$

Elastic buckling stress	$F_e := \dfrac{\pi^2 \cdot E}{\left(\dfrac{K \cdot l_1}{\sqrt{t_{gusset}^2 \div 12}}\right)^2} = 185.7\ \text{ksi}$	Eq. E3-4
Critical stress	$F_{cr} := Fy_{A36} \cdot 0.658^{\frac{Fy_{A36}}{F_e}} = 33.19\ \text{ksi}$	Eq. E3-2
Strength reduction factor	$\phi = 0.90$	
Buckling/compressive strength	$\phi R_n := \phi \cdot F_{cr} \cdot A_w = 741.15\ \text{kip}$	Eq. E3-1
Check capacity	$\text{if}\left(Tu < \phi R_n, \text{“OK”}, \text{“NOT OK”}\right) = \text{“OK”}$	

References

AISC (2015). *Vertical Bracing Connections: Analysis and Design*, Design Guide 29. American Institute of Steel Construction, Chicago.

AISC (2016). *Specification for Structural Steel Buildings*. American Institute of Steel Construction, Chicago, Illinois.

AISC (2017). *Steel Construction Manual*, 15th Edition. American Institute of Steel Construction, Chicago, Illinois.

16

Double Angle Brace Bolted to a Gusset Plate in a Concentrically Braced Frame

16.1 Problem Description

The objective of this example is verification of the component-based finite element method (CBFEM) of a brace connection at beam-column connection in a braced frame with the AISC design procedure. The study is prepared for size of the brace, beam, column, connecting angles, geometry, thickness of the plate, bolts, and welds. In this study, ten components are examined: brace, beam flange and web, column flange and web, connecting angles, gusset plate, splice plates between brace and gusset plate, connecting angles to column, connecting angles to beam, bolts, and welds. All components are designed according to AISC-360-16 *Specification* (AISC, 2016). The connection presented is taken from AISC Design Guide 29 (AISC, 2015).

16.2 Verification of Resistance as Per AISC

The example uses the sections and dimensions shown in Figure 16.1 and are as follows. The brace is 2L8×6×l LLBB (ASTM A36), beam W21×83 (ASTM A992), column W14×90 (ASTM A992), 1 in. thick gusset plate (ASTM A36), 3/4 in. thick end plate connecting gusset plate to column flange (ASTM A572 Gr. 50), 7/8 in. ASTM A490-X bolts and ASTM E70XX weld. All section and material properties are obtained from the *Steel Construction Manual* (AISC, 2017).

16.3 Verification of Resistance as Per CBFEM

The results of the analytical solution are represented by the comparison table for the different limit states shown below. The limit states that should be considered for this connection are as follows and the comparison of the capacities of the different limit states is shown in Table 16.1.

1. Bolts at brace to gusset plate
2. Tensile yielding on the brace gross section

Steel Connection Design by Inelastic Analysis: Verification Examples per AISC Specification, First Edition.
Mark D. Denavit, Ali Nassiri, Mustafa Mahamid, Martin Vild, Halil Sezen, and František Wald.

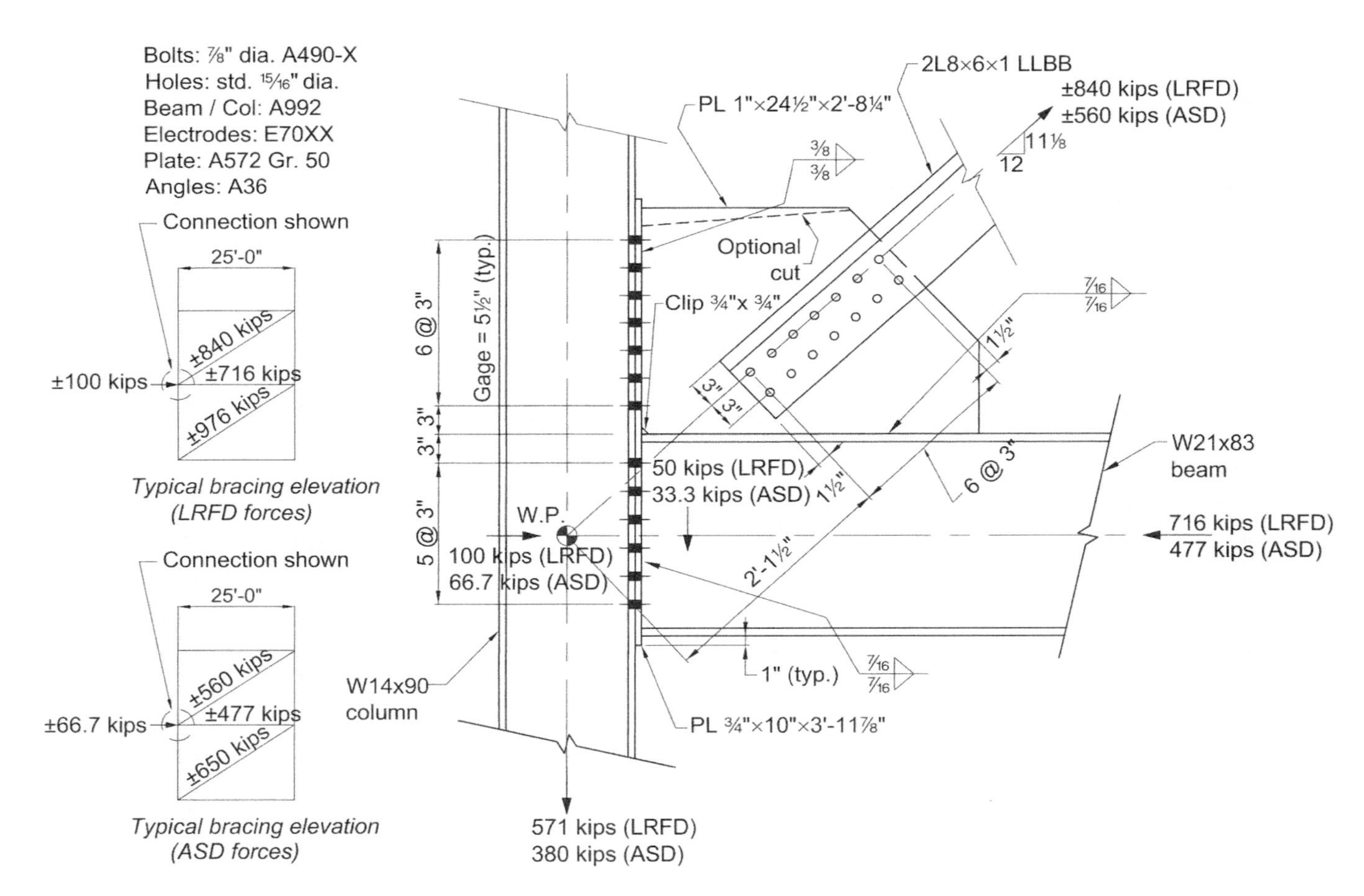

Figure 16.1 Brace connection at beam-column connection in a braced frame – geometry and full design.

Table 16.1 Limit states checked by AISC

Limit state	AISC
Bolts at brace to gusset plate	$\phi R_{nt} = 51$ kips $\phi R_{nv} = 37.9$ kips
Tensile yielding on the brace gross section	$\phi R_n = 849$ kips
Tensile rupture on the brace net section	$\phi R_n = 877$ kips
Block shear rupture on the brace	$\phi R_n = 936$ kips
Block shear rupture on the gusset plate	$\phi R_n = 855$ kips
Bolt bearing on the gusset plate	*Single bolt:* $\phi R_{n\_edge} = 60.3$ kips $\phi R_{n\_edge} = 75.8$ kips *Connection:* $\phi R_n = 1029.9$ kips
Tensile yielding on the Whitmore section of the gusset plate tensile yielding strength of the gusset plate	$\phi R_n = 968$ kips
Compression buckling on the Whitmore section of the gusset	$\phi R_n = 940.5$ kips
Gusset plate for shear yielding and tensile yielding along the beam flange	*Shear yielding* $\phi R_n = 940$ kips *Tension yielding* $\phi R_n = 1416$ kips
Weld at gusset-to-beam flange connection	7/16" weld, required 6.2/16" weld
Beam web local yielding	$\phi R_n = 896.6$ kips *Compared to* $V_{beam} = 269.2$ kips
Beam web local crippling	$\phi R_n = 765.4$ kips *Compared to* $V_{beam} = 269.2$ kips
Bolts at gusset-to-column connection	$\phi R_n = 37.2$ kips *combined shear and tension*
Gusset-to-end plate weld	6/16" weld, required 5.1/16"
Gusset plate tensile and shear yielding at the gusset-to-end-plate interface	*Tensile yielding:* $\phi R_n = 1070.3$ kips *Compared to* $H_{column} = 176.1$ kips *Shear yielding:* $\phi R_n = 713.5$ kips *Compared to* $V_{column} = 301.9$ kips

(Continued)

Table 16.1 (Continued)

Limit state	AISC
Prying action on bolts at the end plate	$\phi R_n = 24.2$ kips
Bolt bearing at bolt holes on end plate	$\phi R_n = 37.9$ kips *Bolt shear governs*
Block shear rupture of the end plate	$\phi R_n = 591$ kips *Compared to* $V_{column} = 301.9$ kips
Prying action on column flange	$\phi R_n = 17.8$ kips
Bearing on column flange	$t_f = 0.7$ in. $> t_{PL} = 0.625$ in. *does not govern*
Bolts at beam-to-column connection	$\phi R_n = 30.5$ kips
Beam web-to-end plate weld	7/16" weld, required 6.4/16"
Prying action on bolts and end plate	$\phi R_n = 20.3$ kips
Prying action on column flange	$\phi R_n = 17.8$ kips
Beam shear strength	
Beam shear strength	$\phi R_n = 330.6$ kips *Compared to* $H_{ucolumn} = 319.2$ kips
Column shear strength	$\phi R_n = 184.8$ kips *Compared to* $H_{ucolumn} = 176.1$ kips

3. Tensile rupture on the brace net section
4. Block shear rupture on the brace
5. Block shear rupture on the gusset plate
6. Bolt bearing on the gusset plate
7. Tensile yielding on the Whitmore section of the gusset plate, tensile yielding strength of the gusset plate
8. Compression buckling on the Whitmore section of the gusset
9. Gusset plate for shear yielding and tensile yielding along the beam flange
10. Weld at gusset-to-beam flange connection
11. Beam web local yielding
12. Beam web local crippling
13. Bolts at gusset-to-column connection
14. Gusset-to-end plate weld
15. Gusset plate tensile and shear yielding at the gusset-to-end-plate interface
16. Prying action on bolts at the end plate

17. Bolt bearing at bolt holes on end plate
18. Block shear rupture of the end plate
19. Prying action on column flange
20. Bearing on column flange
21. Bolts at beam-to-column connection
22. Beam web-to-end plate weld
23. Prying action on bolts and end plate
24. Prying action on column flange
25. Beam shear strength
26. Column shear strength

The governing component of this connection is the tension yielding of the brace, followed by the tension rupture of the brace. Detailed calculation of this connection is shown in the Appendix in this chapter.

16.4 Resistance by CBFEM

The overall check of the connection is verified as shown in Figures 16.2 and 16.3. In this connection, there are two load cases, one in compression and one in tension. The compression load case converged completely to 100% of the applied load, while the tension load case converged to 91% of the load which gives conservative results compared to AISC. It can be concluded that the CBFEM is capable of predicting the actual behavior and failure modes of the braced frame connections presented herein. Failure in members and plates due to yielding and rupture limit states is measured based on a 5% plastic strain limit. Figure 16.3 shows that the plastic strain is 3.6% at 91% of the load which is less than the 5% plastic strain limit. The connection presented includes elements that are welded and others that are bolted. It can be seen that the weld check utilization is 94.9% and is based on the AISC 360-16 *Specification* (AISC, 2016). Both AISC and CBFEM give the same results for the weld check. The bolt shear check is in agreement with both AISC 360-16 specification and CBFEM, which is based on the compression load, which converged to 100%. Similarly, the bolt bearing check in CBFEM and AISC are in agreement per a single bolt check; it is worth mentioning that CBFEM checks the bolts individually for bearing and the utilization ratio is based on that, where the utilization ratio per AISC is based on the summation of the bearing capacities of all the bolts; this would result in CBFEM being safer and showing slightly more conservative results than AISC.

The results were obtained using the various limit states per the AISC procedure. These limit states were investigated individually per CBFEM and the capacities were reported accordingly. Bolt limit states including bolt shear, bolt tension, combined bolt shear and tension, and bolt bearing are accurate. For the tension yielding, tension rupture, shear yielding, and shear rupture limit states, they are found separately. The plastic strain starts at the bolt holes; these stresses are based on von Mises stresses which is a combination of normal and shear stresses. Figure 16.2 shows the stress distribution in the angles connecting the brace to the gusset plate. CBFEM results show tension yielding and tension rupture would occur at the first row of bolts, which is in agreement with AISC solution. The capacity in these limit states per AISC (Table 16.1) is within 3%, and the CBFEM results are within 9% (91% convergence) and provide safer and more conservative results than AISC.

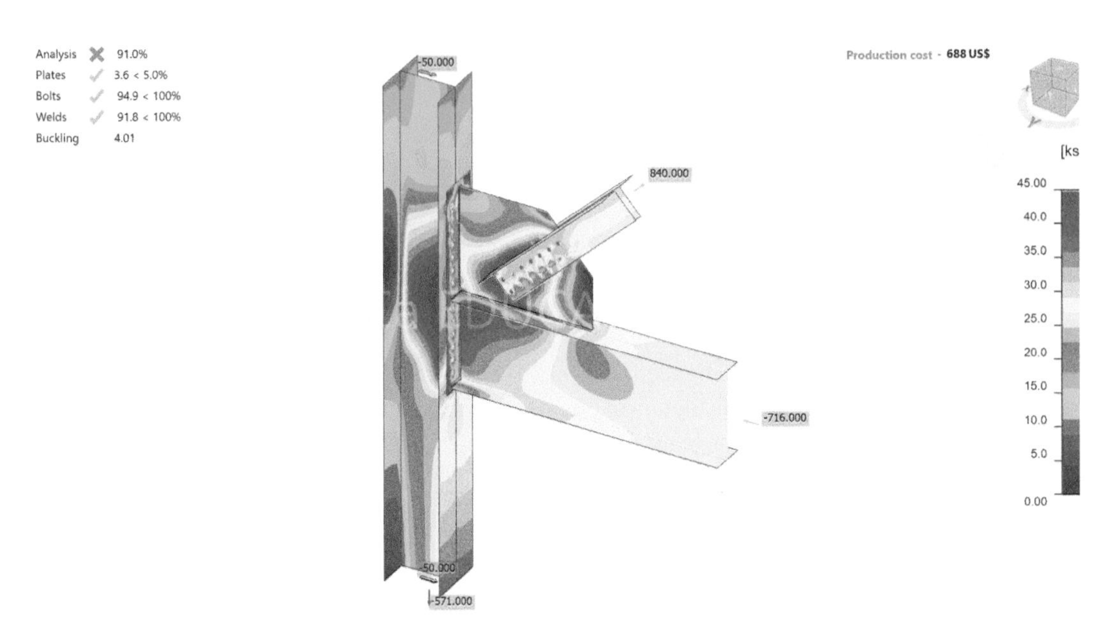

Figure 16.2 Overall solution of the connection.

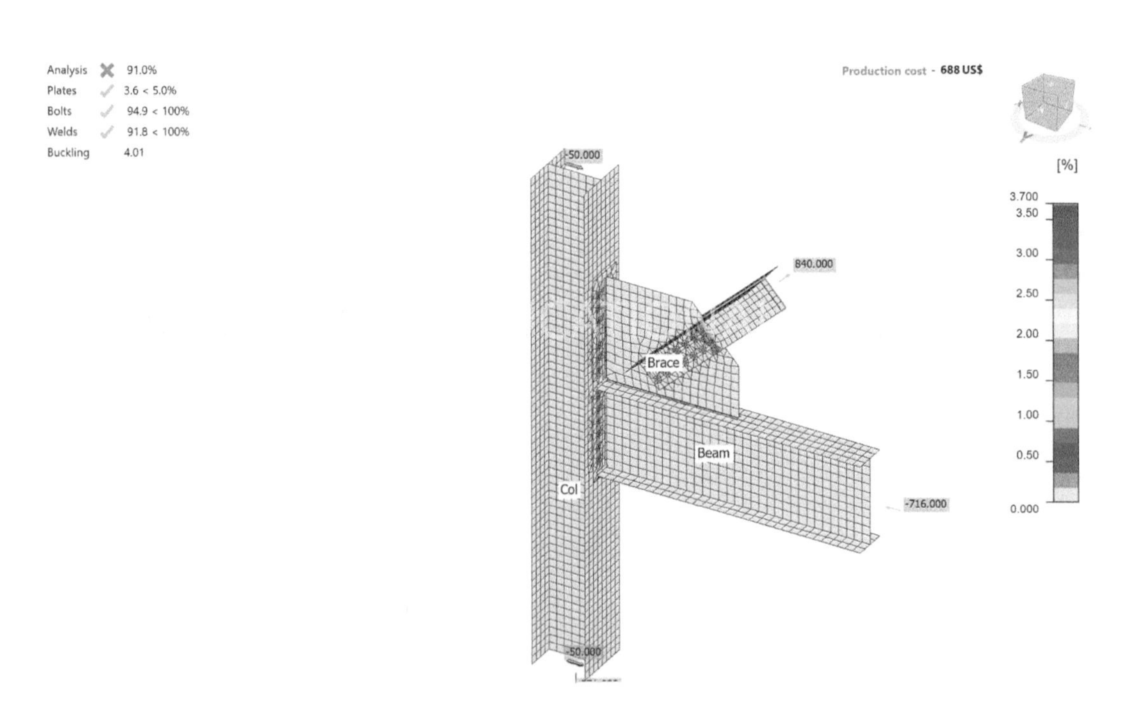

Figure 16.3 Plastic strains in the overall solution of the connection.

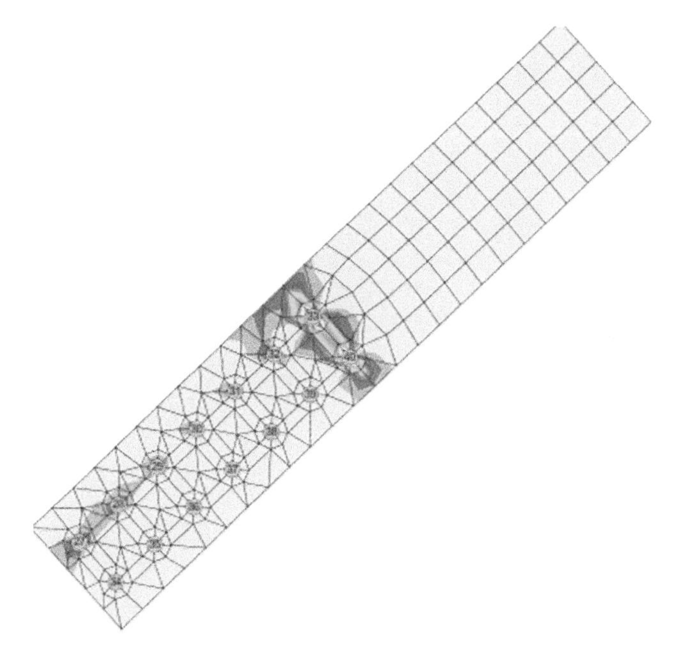

Figure 16.4 Plastic strains in angles connecting the brace to the gusset plate in an investigation of tension yielding, tension rupture, and block shear.

The block shear capacity for the brace per AISC occurs at 936 kips per Table 16.1, which is greater than the capacity of the brace in tension yielding and tension rupture. It has been observed that with increasing load, the plastic strain increases at the first bolt line, where failure would initially occur. The block shear capacity in the gusset plate is 855 kips, which is close to the tension yielding and tension rupture of the brace that controls the design of the connection; as mentioned above, with load increase, the plastic strain increases at the first bolt line. Figures 16.4 and 16.5 show the concentration of plastic strains at the first bolt line and the block shear path. This is in accordance with AISC 360-16 where the governing failure mode of the angles is tensile yielding with a capacity of 849 kips as shown in Table 16.1.

The AISC specifications require checking yielding at Whitmore section on the gusset plate. The AISC capacity for tension yielding at the Whitmore section is 968 kips, which is higher than the governing failure modes. It is obvious that rupture along the bolt lines would occur before yielding of the gusset plate and as observed in the yielding and rupture capacities in Table 16.1.

Prying action is another limit state that is required per the AISC specifications; the prying action limit states is taken into consideration in CBFEM by the additional tension forces applied to the bolts.

The compression buckling capacity of thc gusset plate per AISC 360-16 *Specification* (AISC, 2016) is 940.5 kips, which is higher than the controlling limit states. The buckling factor obtained by CBFEM is 4.10 for the compression load case. For the gusset plate buckling, the preferred value for the buckling factor is greater than 9.16 for 50 ksi steel per AISC 360-16 Section J.4. Figure 16.3 shows that the buckling factor is 4.01 at 91% of the applied load (840 kips), therefore, it is expected that the gusset would buckle at this load, according to CBFEM, however, buckling according to CBFEM still requires more research to allow accurate prediction.

The first mode buckling shape is shown in Figure 16.6. Both AISC and CBFEM are in agreement in checking the buckling failure mode of the gusset plate.

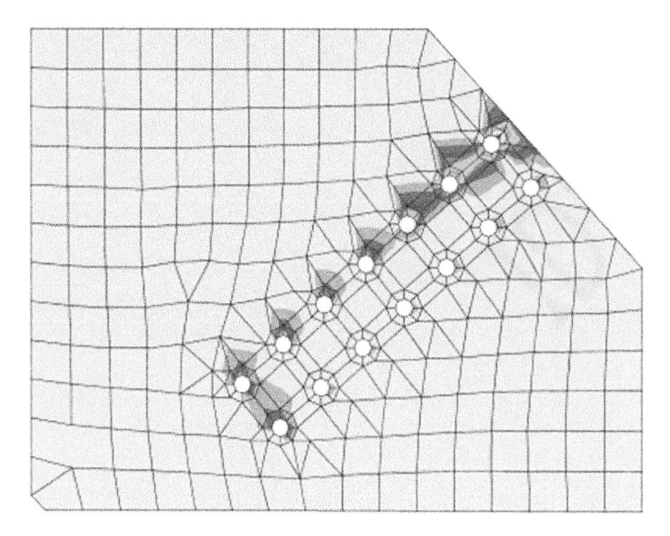

Figure 16.5 Plastic strains in the gusset plate to investigate the block shear limit state.

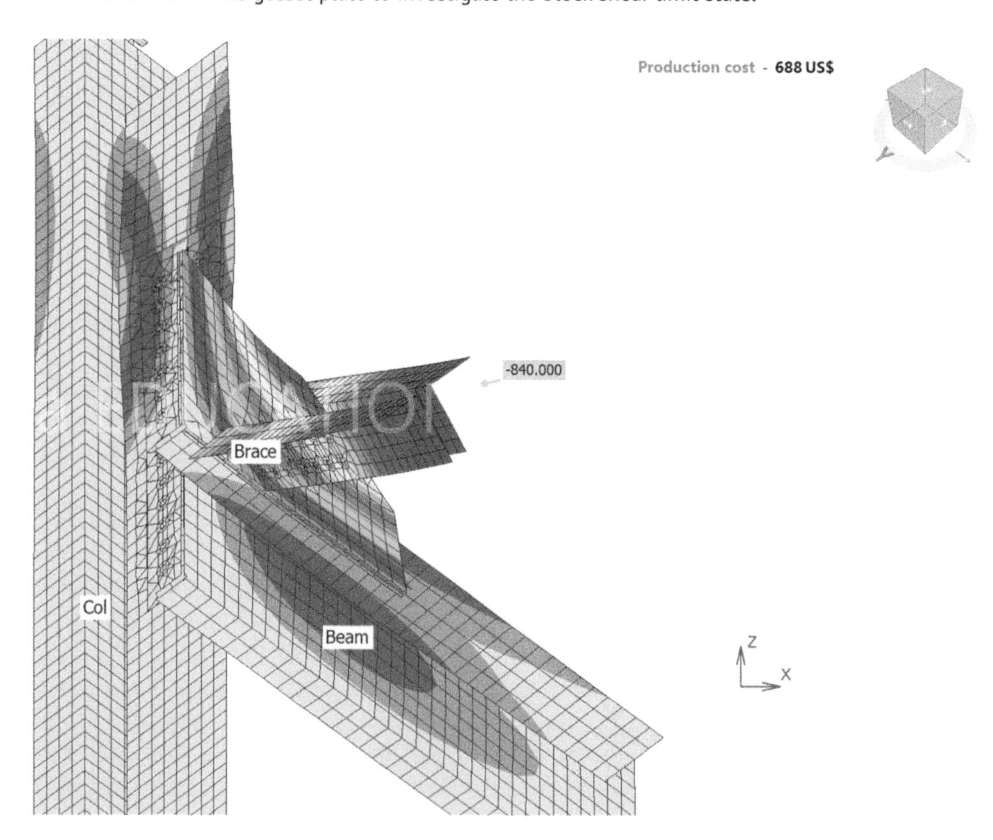

Figure 16.6 First mode buckling shape.

For the combination of shear yielding and tensile yielding along the gusset plate-beam top flange, AISC gives a very small interaction, CBFEM low stresses and zero plastic strains, see Figure 16.7.

For the gusset plate tensile and shear yielding at the gusset-to-end-plate interface, the AISC capacity for tension yielding capacity is 1,073 kips, compared to applied horizontal force on the column, H_{column} = 175 kips, and for shear yielding capacity is 713 kips, compared to applied vertical force, V_{column} = 302 kips. CBFEM gives combined stresses from tension and shear, as shown in Figure 16.8; it is also obvious that there are no plastic strains in the end plate. To investigate this failure mode, a much higher force needs to be applied, at which the model will not converge. The controlling limit states indicated above would occur at a much lower load.

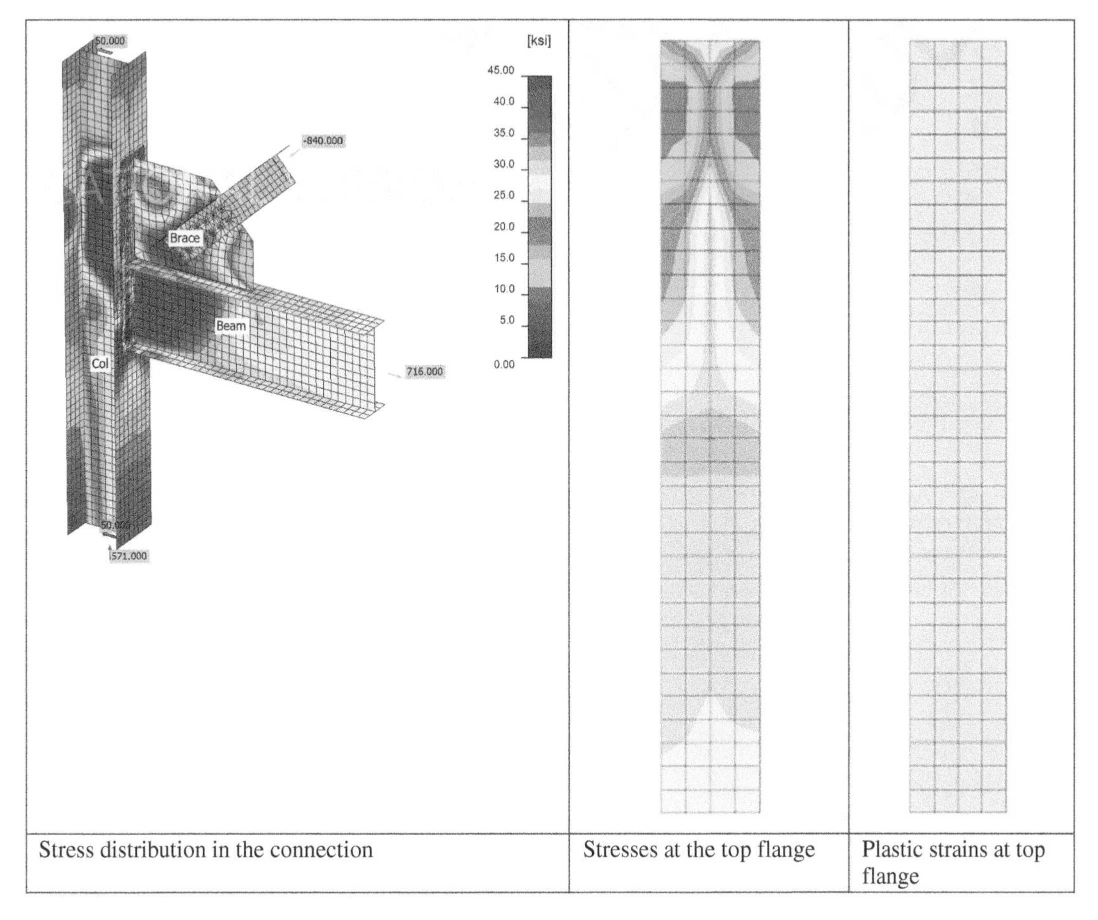

Figure 16.7 Stress distribution in the connection and top flange and plastic strains at the top flange.

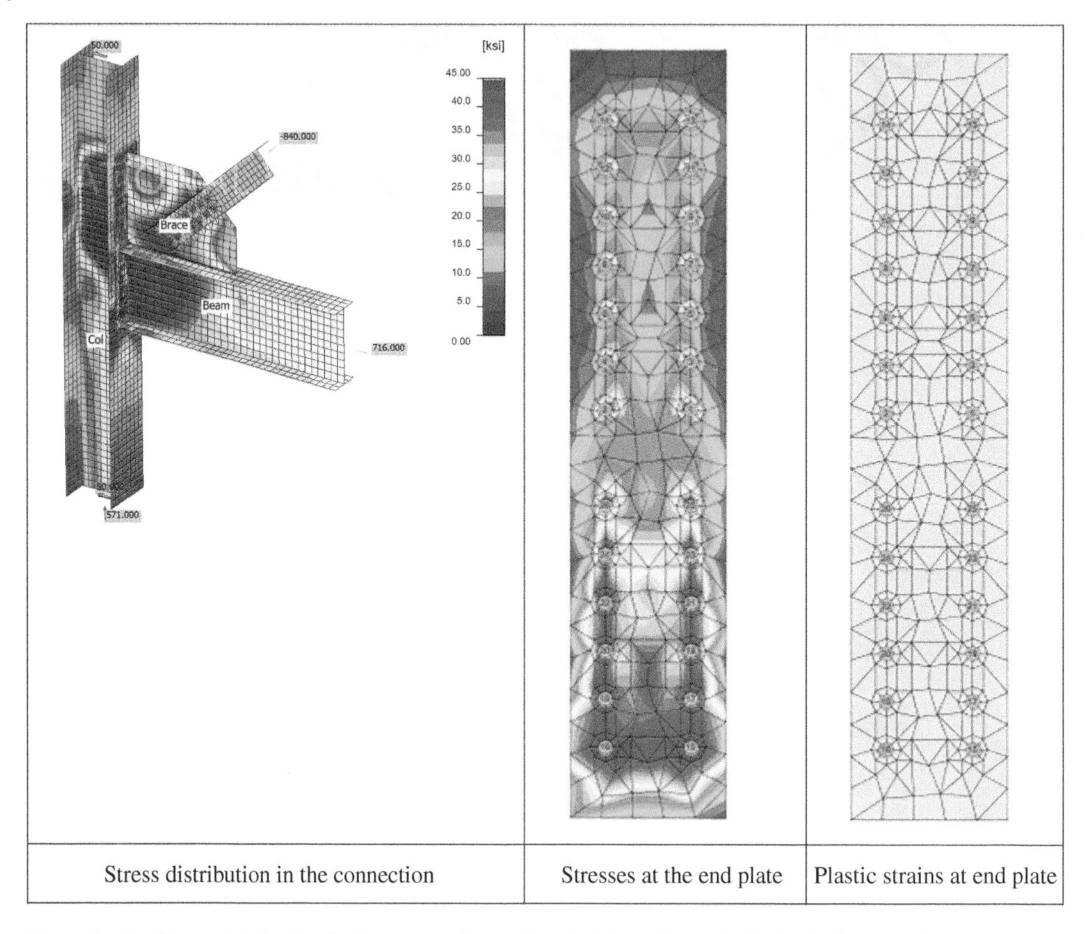

Figure 16.8 Stress distribution in the connection and end plate and plastic strains in the end plate.

The bolt shear and bolt bearing capacity in the end plate and column flange per AISC and CBFEM are in agreement. The block shear rupture capacity for the end plate is 591 kips, compared to the applied force, V_{column} = 302 kips. Again, to reach the capacity of block shear at the end plate, a much higher force should be applied at which the model will not converge. The controlling limit states occur at a much smaller load than the load that would cause block shear failure at the end plate.

Beam web local buckling and web crippling would occur at a large load compared to the applied load. The beam web local buckling capacity shown in Table 16.1 is compared to V_{beam} =269 kips and the web crippling capacity shown in Table 16.1 is compared to V_{beam} =269 kips. Almost all limit states in this connection would occur before these two limit states which typically do not control the design. If needed, these limit states can be checked using the AISC specifications, using the procedure presented in the attachment for beam web local and shear yielding.

Beam web crippling would occur after yielding and at high loads, therefore, the model may not converge under such high loads and would not be able to capture this failure mode. If the crippling capacity is needed, it can be calculated per the AISC specifications, using the procedure presented in the Appendix in this chapter.

16.5 Summary

The connection presented herein has two load cases: tension in the brace and compression in the brace. The load case with compression force in the brace converged to 100% while the load case with the tension force converged to 91%. The connection controlling limit state, per AISC, is tension yielding, with a capacity of 849 kips, compared to applied load equals to 840 kips. This means that CBFEM is safer and more conservative by approximately 10% for the tension load case. It can be concluded that the CBFEM is capable of predicting the actual behavior and failure mode of the braced frame connections presented herein. The various limit states were investigated carefully by checking all the relevant limit states and comparing the AISC vs. the CBFEM capacities. The weld capacity for the weld between the gusset and the beam top flange, between gusset plate and end plate are in agreement in both AISC and CBFEM. Bolt limit states including bolt shear, bolt tension, combined bolt shear and tension and bolt bearing in AISC are in agreement with CBFEM. The plates' limit states including yielding, rupture in tension and in shear are based on a 5% plastic strain limit, according to CBFEM.

Tension yielding and tension rupture in the brace are in agreement in AISC and CBFEM with approximately 10% difference in the capacities. For the block shear limit state, it can be observed in the gusset plate and in the end plate, but not in other plates, such as the brace angles; this is because the shear and tension rupture of the angles precede the block shear rupture. The prying action limit state, required by AISC specifications, is taken into consideration in CBFEM by the additional tension forces applied to the bolts. Beam web buckling, web crippling, and shear yielding would occur at high loads and the model would not converge at such high loads; all other limit states would occur before these limit states. If necessary, these limit states can be checked per the AISC specifications, as shown in the Appendix in this chapter. The buckling limit state of the gusset plate was not observed as a limit state in AISC and in CBFEM.

Benchmark Case

Input

Beam cross-section

- W21X83
- Steel ASTM A992

Braces cross-section

- 2L8X6X1 LLBB
- Steel ASTM A36

Column cross-section

- W14X90
- Steel ASTM A992

Gusset plate

- Thickness 1 in.
- Steel ASTM A572 Gr. 50

End plate connecting gusset to column

- Thickness 3/4 in.
- Steel ASTM A572 Gr. 50

Loading

- Axial force N = 840 kips in tension and compression
- Weld
- Gusset to end plate 3/8" ASTM E70
- Gusset to beam to flange 7/16" ASTM E70
- Beam to end plate 7/16" ASTM E70

Output

- Weld 91.8%
- Bolts 94.9%
- Plastic strain 3.6% < 5%
- Buckling factor 4.01

Appendix

The AISC Solution

ORIGIN := 1

Refer to AISC 14th Edition shapes database:

$$T1 := Row(shape) := \left| \begin{array}{l} \text{for } i \in 1..rows(T1) - 1 \\ R \leftarrow 1 \text{ if } \left(T1^{\langle 2 \rangle}\right)_i = shape \\ R \end{array} \right.$$

Solve Example 5.1 Corner tensile Flange: General Uniform Force Method

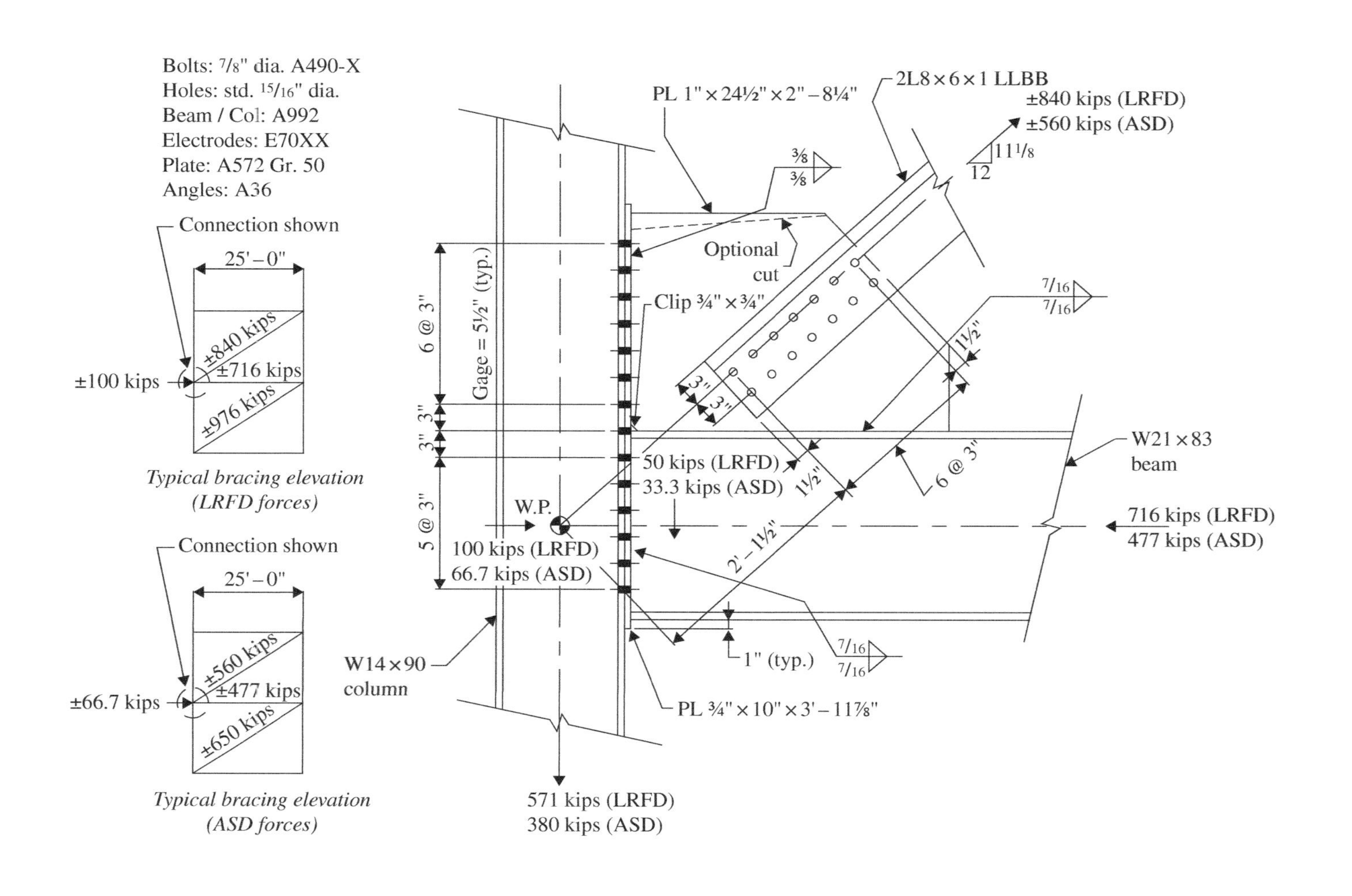

Bolts: 7/8" dia. A490-X
Holes: std. 15/16" dia.
Beam / Col: A992
Electrodes: E70XX
Plate: A572 Gr. 50
Angles: A36
Connection shown
25'–0"
±840 kips
±716 kips
±976 kips
±100 kips
Typical bracing elevation
(LRFD forces)
Connection shown
25'–0"
±560 kips
±477 kips
±650 kips
±66.7 kips
Typical bracing elevation
(ASD forces)
PL 1" × 24½" × 2" – 8¼"
2L8 × 6 × 1 LLBB
±840 kips (LRFD)
±560 kips (ASD)
11 1/8
12
3/8
3/8
Optional
cut
Clip ¾" × ¾"
7/16
7/16
1½"
3"
3"
6 @ 3"
Gage = 5½" (typ.)
3"
3"
5 @ 3"
W.P.
50 kips (LRFD)
33.3 kips (ASD)
1½"
6 @ 3"
W21 × 83
beam
716 kips (LRFD)
477 kips (ASD)
100 kips (LRFD)
66.7 kips (ASD)
2' – 1½"
7/16
7/16
1" (typ.)
W14 × 90
column
PL ¾" × 10" × 3' – 11⅞"
571 kips (LRFD)
380 kips (ASD)

Loads

LRFD Brace axial force	$P_u := 840\ kip$
Horizontal nodal force	$Hu_{node} := 100\ kip$
Beam axial reaction	$Au_{beam} := 716\ kip$
Beam shear reaction	$Vu_{beam} := 50\ kip$

Material Properties

Beams steel grade A992	$Fy_{beam} := 50\ ksi$	$Fu_{beam} := 65\ ksi$
Brace steel grade A36	$Fy_{brace} := 36\ ksi$	$Fu_{brace} := 58\ ksi$
Plate steel grade A572 Gr. 50	$Fy_{PL} := 50\ ksi$	$Fu_{PL} := 65\ ksi$
Modulus of Elasticity	$E := 29000\ ksi$	
Weld strength	$F_{EXX} := 70\ ksi$	

Member Properties

Beam Properties

Beam section $\boxed{Beam := \text{“W21X83”}}$ $\quad Span := 25ft$

Depth of beam $d_{beam} := T1_{(Row(Beam),7)} \cdot in = 21.4 \cdot in$

Width of flange $bf_{beam} := T1_{(Row(Beam),12)} \cdot in = 8.4 \cdot in$

Thickness of flange $tf_{beam} := T1_{(Row(Beam),20)}\ in = 0.835\ in$

Thickness of web $tw_{beam} := T1_{(Row(Beam),17)} \cdot in = 0.515 \cdot in$

Design weld depth $k\_des_{brace} := T1_{(Row(Beam),25)} \cdot in = 1.34 \cdot in$

Center of web to flange toe of fillet $k\_1_{beam} := T1_{(Row(Beam),27)} \cdot in = 0.875 in$

Strong moment of inertia $Ix_{beam} := T1_{(Row(Beam),39)} \cdot in^4 = 1830\ in^4$

Column Properties

$\boxed{Column := \text{“W14X90”}}$

Depth $d_{column} := T1_{(Row(Column),7)} \cdot in = 14 \cdot in$

Flange width $bf_{column} := T1_{(Row(Column),12)} \cdot in = 14.5 \cdot in$

Flange thickness $tf_{column} := T1_{(Row(Column),20)}\ in = 0.71\ in$

Web thickness $tw_{column} := T1_{(Row(Column),17)} \cdot in = 0.44 \cdot in$

Strong moment of inertia $Ix_{column} := T1_{(Row(Column),39)} \cdot in^4 = 999\ in^4$

Brace Properties

Brace section $\boxed{Brace := \text{“2L8X6X1LLBB”}}$

Cross-sectional area $Ag_{brace} := T1_{(Row(Brace),6)} \cdot in^2 = 26.2 \cdot in^2$

x distance to centroid $x\_bar_{brace} := T1_{(Row(L1),28)} \cdot in = 1.65 \cdot in$ $\quad L1 := \text{“L8X6X1”}$ value for single angle

$t_{brace} := T1_{(Row(Brace),22)} \cdot in = 1\ in$

Plate Properties

Thickness	$t_{PL} := 1\,in$
Clip thickness	$t_{clip} := \frac{3}{4}\,in$
End PL width	$h_{endPL} := 10in$

Member Design

Brace-to-Gusset Connection

Determine required number of bolts (J3.6)

Bolt pitch (spacing)	$p := 3\ in$		
Bolt diameter	$d_{bolt} := \frac{7}{8}in$		
Area of bolt	$A_{bolt} := {d_{bolt}}^2 \cdot \frac{\pi}{4} = 0.6in^2$		
Bolt shear strength	$Fnv_{bolt} := 84\ ksi$	Table J3.2 Group A threads X	
Bolt tensile strength	$Fnt_{bolt} := 113\ ksi$		
Strength reduction factor	$\phi := 0.75$		
Bolt shear strength (single shear)	$\phi r_{nv} := \phi Fnv_{bolt} \cdot A_{bolt} = 37.9\ kip$	Eq. J3-1	
Bolt tensile strength	$\phi r_{nt} := \phi \cdot Fnt_{bolt} \cdot A_{bolt} = 51\ kip$	Eq. J3-1	
Number of bolts Double shear	$N_{bolts} := \frac{P_u}{\phi r_{nv} \cdot (2)} = 11.1$	use 14 bolts $N_{bolts} := 14$	recall $P_u = 840\ kip$

Check tensile yielding on the brace gross section (J4.1a)

Strength reduction factor	$\phi := 0.9$	recall
Tensile yielding strength of brace	$\phi R_n := \phi \cdot Fy_{brace} \cdot Ag_{brace} = 848.9\ kip$	$Fy_{brace} = 36\ ksi$
Check capacity	if $(P_u < \phi R_n, \text{"OK"}, \text{"NOT OK"}) = \text{"OK"}$	$Ag_{brace} = 26.2in^2$
		$P_u = 840\ kip$

Check tensile rupture on the brace net section (J4.1b)

Net area of brace	$An_{brace} := Ag_{brace} - 4 \cdot t_{brace} \cdot \left(d_{bolt} + \frac{2}{16}in\right) = 22.2\,in^2$	recall
Length of bolted connection	$l_{brace} := \left(\frac{14}{2} - 1\right) \cdot 3\ in = 18\ in$	$Ag_{brace} = 26.2\,in^2$
Shear lag factor (Table D3.1 case 2)	$U := 1 - \frac{x\_bar_{brace}}{l_{brace}} = 0.908$	$t_{brace} = 1\,in$
Effective area of brace	$Ae_{brace} := U \cdot An_{brace} = 20.2\ in^2$	$x\_bar_{brace} = 1.65\,in$
Strength reduction factor	$\phi := 0.75$	$Fy_{brace} = 36\ ksi$
Tensile rupture strength of brace	$\phi R_n := \phi \cdot Fu_{brace} \cdot Ae_{brace} = 877.2\ kip$	$Fu_{brace} = 58\ ksi$
Check capacity	if $(P_u < \phi R_n, \text{"OK"}, \text{"NOT OK"}) = \text{"OK"}$	$Fy_{PL} = 50\ ksi$
		$Fu_{PL} = 65\ ksi$
		$N_{bolts} = 14$
		$d_{bolt} = 0.875\ in$
		$p = 3\,in$
		$t_{PL} = 1\,in$
		$P_u = 840\ kip$

Check block shear rupture on the brace (J4.3)

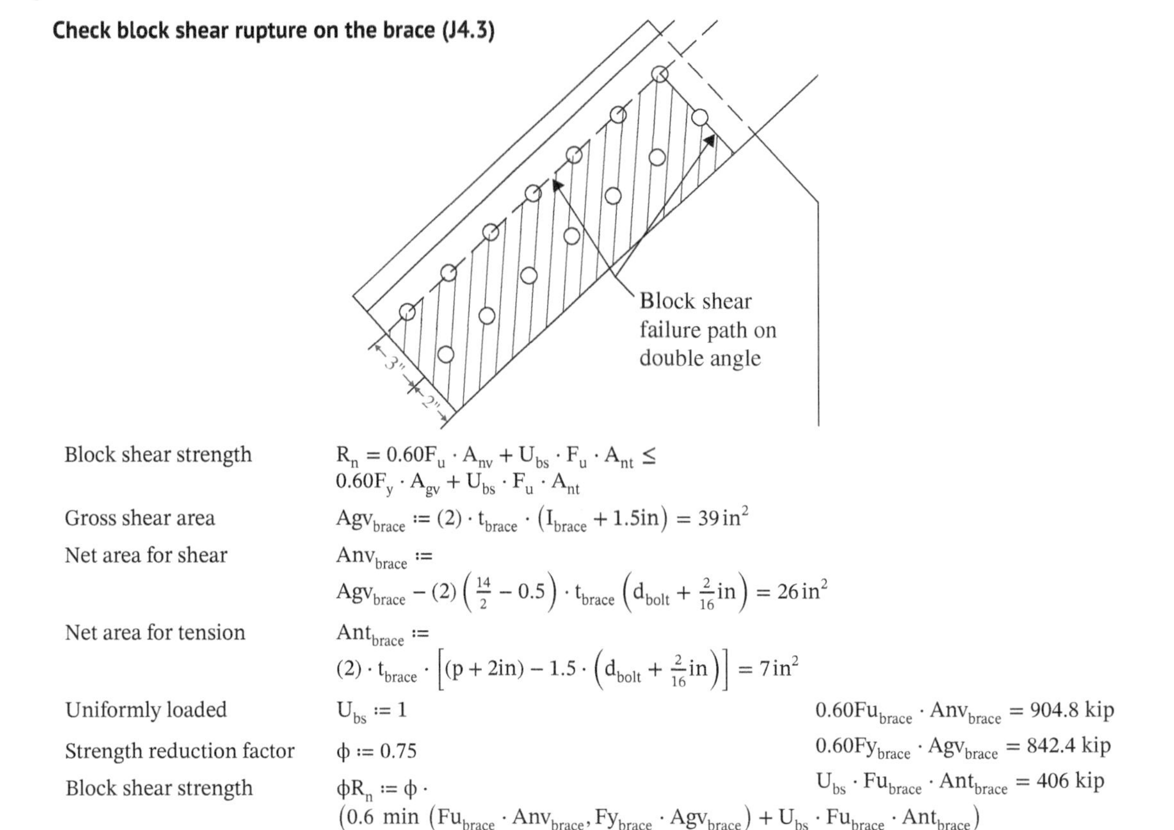

Block shear strength $R_n = 0.60F_u \cdot A_{nv} + U_{bs} \cdot F_u \cdot A_{nt} \leq$
$0.60F_y \cdot A_{gv} + U_{bs} \cdot F_u \cdot A_{nt}$

Gross shear area $Agv_{brace} := (2) \cdot t_{brace} \cdot (l_{brace} + 1.5in) = 39\,in^2$

Net area for shear $Anv_{brace} :=$
$Agv_{brace} - (2)\left(\frac{14}{2} - 0.5\right) \cdot t_{brace}\left(d_{bolt} + \frac{2}{16}in\right) = 26\,in^2$

Net area for tension $Ant_{brace} :=$
$(2) \cdot t_{brace} \cdot \left[(p + 2in) - 1.5 \cdot \left(d_{bolt} + \frac{2}{16}in\right)\right] = 7\,in^2$

Uniformly loaded $U_{bs} := 1$

$0.60Fu_{brace} \cdot Anv_{brace} = 904.8\ kip$

Strength reduction factor $\phi := 0.75$

$0.60Fy_{brace} \cdot Agv_{brace} = 842.4\ kip$

$U_{bs} \cdot Fu_{brace} \cdot Ant_{brace} = 406\ kip$

Block shear strength $\phi R_n := \phi \cdot$
$(0.6\ \min\left(Fu_{brace} \cdot Anv_{brace}, Fy_{brace} \cdot Agv_{brace}\right) + U_{bs} \cdot Fu_{brace} \cdot Ant_{brace})$

$\phi R_n = 936.3\ kip$

Check capacity if $(P_u < \phi R_n$, "OK", "NOT OK" $)$ = "OK"

Check block shear rupture on the gusset plate (J4.3)

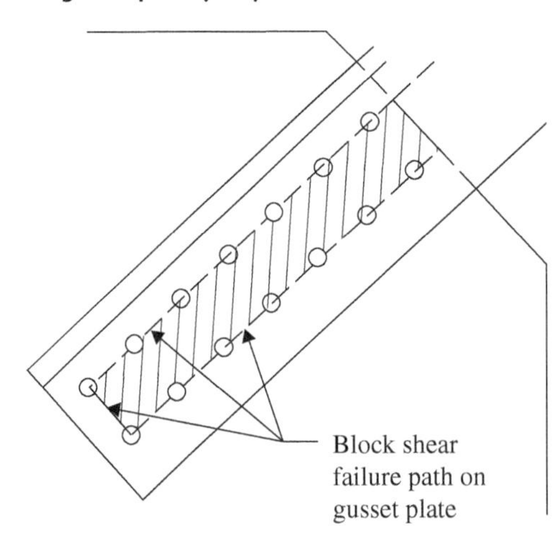

Gross shear area $Agv_{PL} := 2 \cdot t_{PL} \cdot (I_{brace} + 1.5in) = 39\,in^2$

Net area for shear $Anv_{PL} := Agv_{PL} - (2)\left(\frac{14}{2} - 0.5\right) \cdot t_{PL} \cdot \left(d_{bolt} + \frac{2}{16}in\right)$

$Anv_{PL} = 26\,in^2$

Net area for tension $Ant_{PL} := t_{PL} \cdot \left[3\,in - \left(d_{bolt} + \frac{2}{16}in\right)\right] = 2\,in^2$

Uniformly loaded $U_{bs} := 1$

Strength reduction factor $\phi := 0.75$

Block shear strength $\phi R_n := \phi \cdot (0.6\ \min(Fu_{PL} \cdot Anv_{PL}, Fy_{PL} \cdot Agv_{PL}) + U_{bs} \cdot Fu_{PL} \cdot Ant_{PL}) = 858\,kip$

$0.60Fu_{PL} \cdot Anv_{PL} = 1014\,kip$
$0.60Fy_{PL} \cdot Agv_{PL} = 1170\,kip$
$U_{bs} \cdot Fu_{PL} \cdot Ant_{PL} = 130\,kip$

Check capacity if $(P_u < \phi R_n$, "OK", "NOT OK" $)$ = "OK"

Check bolt bearing on the gusset plate (J3.10)

Bolt bearing strength $R_n = 1.2 \cdot I_c \cdot t \cdot F_u < 2.4 \cdot d \cdot t \cdot F_u$

Clear distance (between bolts) $I_{c\_inner} := 3\,in - \left(d_{bolt} + \frac{1}{16}in\right) = 2.06\,in$

Clear distance (between edge) $I_{c\_edge} := 1.5in - 0.5 \cdot \left(d_{bolt} + \frac{1}{16}in\right) = 1.03\,in$

Strength reduction factor $\phi := 0.75$

Edge bolt bearing strength $\phi R_{n\_edge} := \min\left[\phi \cdot 1.2 \cdot I_{c\_edge} \cdot t_{PL} \cdot Fu_{PL}, \phi \cdot 2.4 \cdot d_{bolt} \cdot t_{PL} \cdot Fu_{PL}, (2)\,\phi r_{nv}\right] = 60.3\,kip$

Inner bolt bearing strength $\phi R_{n\_inner} := \min\left[\phi \cdot 1.2 \cdot I_{n\_inner} \cdot t_{PL} \cdot Fu_{PL}, \phi \cdot 2.4 \cdot d_{bolt} \cdot t_{PL} \cdot Fu_{PL}, (2)\,\phi r_{nv}\right] = 75.8\,kip$

Total bolt bearing strength $\phi R_n := (2) \cdot \phi R_{n\_edge} + (N_{bolts} - 2) \cdot \phi R_{n\_inner} = 1029.9\,kip$

Check capacity if $(P_u < \phi R_n$, "OK", "NOT OK" $)$ = "OK"

recall
$\phi r_{nv} = 37.9\,kip$
$d_{bolt} = 0.875\,in$
$Fu_{PL} = 65\,ksi$
$N_{bolts} = 14$
$P_u = 840\,kip$

Check the gusset plate for tensile yielding on the Whitmore section (p. 9-3)

Whitmore section width $I_w := p + \tan(30\,deg) \cdot 2 \cdot I_{brace} = 23.8\,in$

$I_{w\_web} := 4.70\,in$

Effective area of the Whitmore section $A_w := (I_w - I_{w\_web}) \cdot t_{PL} + (I_{w\_web}) \cdot tw_{beam} = 21.5\,in^2$

recall
$I_{brace} = 18\,in$
$tw_{beam} = 0.515\,in$
$t_{PL} = 1\,in$
$Fy_{PL} = 50\,ksi$
$P_u = 840\,kip$

Tensile yielding strength of the gusset plate (J4.1a)

Strength reduction factor $\phi := 0.90$

Tensile yielding strength $\phi R_n := \phi \cdot Fy_{PL} \cdot A_w = 967.7\ kip$

Check capacity $\text{if}\left(P_u < \phi R_n, \text{"OK"}, \text{"NOT OK"}\right) = \text{"OK"}$

Check the gusset plate for compression buckling on the Whitmore section (J4.4)

recall
$t_{PL} = 1in$
$Fy_{PL} = 50\ ksi$
$P_u = 840\ kip$

Effective length factor $K_{PL} := 0.5$

Length of plate to buckle $L_{PL} := 9.76in$

Radius of gyration $r_{PL} := \sqrt{\dfrac{L_{PL} \cdot t_{PL}^{3}}{12 \cdot t_{PL} \cdot L_{PL}}} = 0.289\,in$

$\dfrac{K \cdot L}{r}$ $\dfrac{K_{PL} \cdot L_{PL}}{r_{PL}} = 16.9 \quad < 25$

Gross area $Ag_{PL} := t_{PL} \cdot 20.9in = 20.9\,in^2$

Strength reduction factor $\phi := 0.90$

Compressive strength $\phi P_n = \phi \cdot Fy_{PL} \cdot Ag_{PL} = 940.5\ kip$

Check capacity $\text{if}\left(P_u < \phi P_n, \text{"OK"}, \text{"NOT OK"}\right) = \text{"OK"}$

Connection Interface Forces (Chapter 13)

recall
$d_{beam} = 21.4$
$ind_{column} = 14\,in$
$P_u = 840\ kip$

Beam half depth $e_{beam} := \dfrac{d_{beam}}{2} = 10.7\,in$

Column half depth $e_{column} := \dfrac{d_{column}}{2} = 7\,in$

Slope of brace $\theta_{brace} := \text{atan}\left(\frac{12}{11.125}\right) = 47.2 \cdot deg$

$\alpha - \beta \cdot \tan(\theta) = e_{beam} \cdot \tan(\theta) - e_{column}$ Eq. 13-1

$\beta\_bar := 12in$

$\beta := \beta\_bar$

$\alpha := e_{beam} \cdot \tan(\theta_{brace}) - e_{column} + \beta \cdot \tan(\theta_{brace}) = 17.5\,in$

Plate thickness $t_p := 1\,in$

$\alpha\_bar = \alpha = \dfrac{I_h + t_{clip}}{2} + t_p$

$I_h := 2\left(\alpha - t_p\right) - t_{clip} + 32.2\,in$

$r := \sqrt{\left(\alpha + e_{column}\right)^2 + \left(\beta + e_{beam}\right)^2} = 33.4\,in$ Eq. 13-6

$V_{column} := \dfrac{\beta}{r} \cdot P_u = 301.9\ kip$ Eq. 13-2

$V_{beam} := \dfrac{e_{beam}}{r} \cdot P_u = 269.2\ kip$ Eq. 13-4

$P_u \cdot \cos(\theta_{brace}) - \left(V_{column} + V_{beam}\right) = -0\ kip$

$H_{column} := \dfrac{e_{column}}{r} \cdot P_u = 176.1\ kip$ Eq. 13-3

$H_{beam} := \dfrac{\alpha}{r} \cdot P_u = 439.9\ kip$ Eq. 13-5

$P_u \cdot \sin(\theta_{brace}) - \left(H_{column} + H_{beam}\right) = -0\ kip$

Gusset-to-Beam Connection

Check gusset plate for shear yielding (J4.2a) and tensile yielding (J4.1a) along the beam flange

Weld length	$I_{weld} := I_h - 0.75in = 31.5\,in$	recall
		$I_h = 32.2\,in$
Strength reduction factor	$\phi := 1.00$	$t_{PL} = 1\,in$
Shear yielding strength	$\phi R_n = \phi \cdot 0.60 \cdot Fy_{PL} \cdot t_{PL} \cdot I_{weld} = 944.1\text{ kip}$	$Fy_{PL} = 50\text{ ksi}$
Check capacity	$\text{if}\left(H_{beam} < \phi R_n, \text{"OK"}, \text{"NOT OK"}\right) = \text{"OK"}$	$H_{beam} = 439.9\text{ kip}$
Strength reduction factor	$\phi := 0.90$	$V_{beam} = 269.2\text{ kip}$
Tensile yielding strength	$\phi R_n = \phi \cdot Fy_{PL} \cdot t_{PL} \cdot I_{weld} = 1416.2\text{ kip}$	
Check capacity	$\text{if}\left(V_{beam} < \phi R_n, \text{"OK"}, \text{"NOT OK"}\right) = \text{"OK"}$	

Consider force interaction for gusset plate

interaction Eq plasticity theory (Neal, 1977) and suggested by Astaneh-Asl (1998)

$$\left(\frac{M_{ub}}{\phi M_n}\right) + \left(\frac{V_{ub}}{\phi N_n}\right)^2 + \left(\frac{H_{ub}}{\phi V_n}\right)^4 < 1.$$

recall
$Fy_{PL} = 50\text{ ksi}$
$I_{weld} = 31.5\,in$
$t_{PL} = 1\,in$
$H_{beam} = 439.9\text{ kip}$
$V_{beam} = 269.2\text{ kip}$

Strength reduction factor

$$\phi := 0.90$$

$$\phi Mn_{PL} = \phi \cdot F_y \cdot Z_x$$

Nominal moment strength (Eq. F2-1)

$$\phi Mn_{PL} := \phi \cdot Fy_{PL} \cdot \left(\frac{t_{PL} \cdot I_{weld}^{\,2}}{4}\right) = 11142.1\,in \cdot kip$$

Interaction equation

$$\left(\frac{0}{\phi Mn_{PL}}\right) + \left[\frac{V_{beam}}{0.90 \cdot \left(Fy_{PL} \cdot t_{PL} \cdot I_{weld}\right)}\right]^2 + \left[\frac{H_{beam}}{1.0 \cdot \left(0.60 \cdot Fy_{PL} \cdot t_{PL} \cdot I_{weld}\right)}\right]^4 = 0.08$$

Design weld at gusset-to-beam flange connection

Axial stress in weld $f_a := \dfrac{V_{beam}}{l_{weld}} = 8.6 \cdot \dfrac{kip}{in}$

Shear stress in weld $f_v := \dfrac{H_{beam}}{l_{weld}} = 14 \cdot \dfrac{kip}{in}$

Bending stress in weld $f_b := 0$

Peak stress $f_{peak} := \sqrt{(f_a + f_b)^2 + f_v^{\,2}} = 16.4 \cdot \dfrac{kip}{in}$

Average stress $f_{avg} := \dfrac{1}{2} \cdot \left[\sqrt{(f_a - f_b)^2 + f_v^{\,2}} + \sqrt{(f_a + f_b)^2 + f_v^{\,2}}\right] = 16.4 \cdot \dfrac{kip}{in}$

Load angle $\theta_{load} := \operatorname{atan}\left(\dfrac{f_a}{f_v}\right) = 31.5 \cdot deg$

Additional ductility (p. 13-11) $ductility := 1.25$

Stress on weld $f_{weld} := \max\left(ductility \cdot f_{avg}, f_{peak}\right) = 20.5 \cdot \dfrac{kip}{in}$

Gusset to beam flange weld size $D := \dfrac{f_{weld}}{2 \cdot \left(1.392 \dfrac{kip}{in}\right) \cdot \left(1.0 + 0.50 \cdot \sin\left(\theta_{load}\right)^{1.5}\right)} = 6.2$

$\underline{D} := ceil(D) \cdot in = 7\,in$ 16th of an inch

recall
$l_{weld} = 31.5\,in$
$H_{beam} = 439.9\,kip$
$V_{beam} = 269.2\,kip$

Check beam web local yielding (J10.2)

Strength reduction factor $\phi := 1.00$

$\phi R_n = \phi \cdot F_{yw} \cdot t_w \cdot (2.5k + l_b)$

Web local yielding strength $\phi R_n = \phi \cdot Fy_{beam} \cdot tw_{beam} \cdot (2.5k\_des_{brace} + l_{weld}) = 896.6\,kip$

Check capacity $if\ (V_{beam} < \phi R_n, \text{"OK"}, \text{"NOT OK"}) = \text{"OK"}$

recall
$Fy_{beam} = 50\,ksi$
$tw_{beam} = 0.5\,in$
$k\_des_{brace} = 1.3\,in$
$l_{weld} = 31.5\,in$
$V_{beam} = 269.2\,kip$

Check equivalent normal force, Ne $N_{ue} := V_{beam} + \dfrac{2.0}{(l_{weld} \div 2)} = 269.2\,kip$

recall
$tw_{beam} = 0.5\,in$
$l_{weld} = 31.5\,in$
$d_{beam} = 21.4\,in$

recall
$tf_{beam} = 0.84\,in$
$Fy_{beam} = 50\,ksi$
$E = 29000\,ksi$
$V_{beam} = 269.2\,kip$

Check beam web local crippling (J10.3)

Strength reduction factor $\phi := 0.75$

Web local crippling strength $\phi R_n := \phi \cdot 0.80 \cdot tw_{beam}^{\,2} \cdot \left[1 + 3 \cdot \left(\dfrac{l_{weld}}{d_{beam}}\right) \cdot \left(\dfrac{tw_{beam}}{tf_{beam}}\right)^{1.5}\right] \cdot \sqrt{\dfrac{E \cdot Fy_{beam} \cdot tf_{beam}}{tw_{beam}}} = 765.4\,kip$

Check capacity $if\ (V_{beam} < \phi R_n, \text{"OK"}, \text{"NOT OK"}) = \text{"OK"}$

Gusset-to-Column Connection

Design bolts at gusset-to-column connection (J3.6 and J3.7)

Number of bolts $N_{bolts.end\_PL} := 14$

Ultimate shear force per bolt $r_{uv} := \frac{V_{column}}{N_{bolts.end\_PL}} = 21.6\ kip$

Check capacity $if\ (r_{uv} < \phi r_{nv}, \text{"OK"}, \text{"NOT OK"}) = \text{"OK"}$

Ultimate tensile force per bolt $r_{ut} := \frac{H_{column}}{N_{bolts.end\_PL}} = 12.6\ kip$

Check capacity $if\ (r_{ut} < \phi r_{nt}, \text{"OK"}, \text{"NOT OK"}) = \text{"OK"}$

Strength reduction factor $\phi := 0.75$

Modified nominal tensile Stress factored to include Shear stress effects

$$F'_{nt} = 1.3 \cdot F_{nt} - \frac{F_{nt}}{\phi \cdot F_{nv}} \cdot f_{rv} < F_{nt}$$

$$F'nt_{bolt} := \min\left[1.3 \cdot Fnt_{bolt} - \frac{Fnt_{bolt}}{\phi \cdot Fnv_{bolt}} \cdot \left(\frac{r_{uv}}{A_{bolt}}\right), Fnt_{bolt}\right] = 82.6\ ksi$$

Combined loading strength $\phi R_n := \phi \cdot F'nt_{bolt} \cdot A_{bolt} = 37.2\ kip$

Check capacity $if\ (r_{ut} < \phi R_n, \text{"OK"}, \text{"NOT OK"}) = \text{"OK"}$

recall
$Fnv_{bolt} = 84\ ksi$
$Fnt_{bolt} = 113\ ksi$
$\phi r_{nv} = 37.9\ kip$
$\phi r_{nt} = 51\ kip$
$H_{column} = 176.1\ kip$
$V_{column} = 301.9\ kip$
$A_{bolt} = 0.6\ in^2$

Design gusset-to-end plate weld (Eq. 8-2)

Resultant load on weld $R_u := \sqrt{H_{column}^2 + V_{column}^2} = 349.5\ kip$

Resultant weld load angle $\theta_{resultant} := atan\left(\frac{H_{column}}{V_{column}}\right) = 30.3 \cdot deg$

End plate effective Length of weld $I_{weld.end\_PL} := p \cdot \frac{N_{bolts.end\_PL}}{2} = 21\ in$

End plate weld size

$$D_{end\_PL} := \frac{R_u}{(2) \cdot \left(1.392 \frac{kip}{in}\right) \cdot I_{weld.end\_PL} \cdot \left(1.0 + 0.50 \cdot \sin\left(\theta_{resultant}\right)^{1.5}\right)} = 5.1$$

$$D_{end\_PL} := ceil\left(D_{end\_PL}\right) = 6$$

recall
$H_{column} = 176.1\ kip$
$V_{column} = 301.9\ kip$
$N_{bolts.end\_PL} = 14$
$p = 3\ in$

Check gusset plate tensile and shear yielding at the gusset-to-end-plate interface (J4.1a and J4.2a)

Strength reduction factor	$\phi := 0.90$	recall
Tensile yielding strength	$\phi N_n := \phi \cdot Fy_{PL} \cdot (t_{PL} \cdot I_w) = 1070.3\ \text{kip}$	$I_w = 23.8\ \text{in}$
Check capacity	if $(H_{column} < \phi N_n$, "OK", "NOT OK" $) =$ "OK"	$t_{PL} = 1\ \text{in}$
Strength reduction factor	$\phi := 1.00$	$Fy_{PL} = 50\ \text{ksi}$
Shear yielding strength	$\phi V_n := \phi \cdot 0.60 \cdot Fy_{PL} \cdot (t_{PL} \cdot I_w) = 713.5\ \text{kip}$	$H_{column} = 176.1\ \text{kip}$
Check capacity	if $(V_{column} < \phi N_n$, "OK", "NOT OK" $) =$ "OK"	$V_{column} = 301.9\ \text{kip}$

Check prying action on bolts at the end plate (Part 9 and Figure 5-4a)

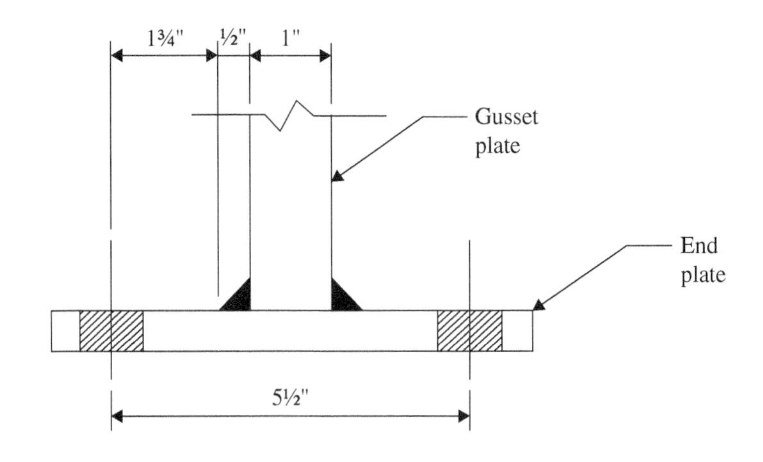

Bolt gage $\quad gage := 5.5\ \text{in}$

$$b_{endPL} := \frac{gage - t_{PL}}{2} = 2.3\ \text{in}$$

Eq. 9-21 $$b'_{endPL} := b_{endPL} - \frac{d_{bolt}}{2} = 1.8\ \text{in}$$

$$a_{endPL} := \frac{h_{endPL} - gage}{2} = 2.25\ \text{in}$$

Eq. 9-27 $$a'_{endPL} := \min\left(a_{endPL} + \frac{d_{bolt}}{2}, 1.25 \cdot b_{endPL} + \frac{d_{bolt}}{2}\right) = 2.69\ \text{in}$$

$$\rho_{endPL} := \frac{b'_{endPL}}{a'_{endPL}} = 0.674$$

$$d' := d_{bolt} + \frac{1}{16}\text{in} = \frac{15}{16}\text{in}$$

Eq. 9-24 $$\delta_{endPL} := 1 - \frac{d'}{p} = 0.69$$

recall
$t_{PL} = 1\ \text{in}$
$d_{bolt} = \frac{7}{8}\ \text{in}$
$h_{endPL} = 10\ \text{in}$
$\phi r_{nt} = 51\ \text{kip}$
$p = 3\ \text{in}$
$\phi r_{nt} = 51\ \text{kip}$
$Fu_{PL} = 65\ \text{ksi}$
$r_{ut} = 12.6\ \text{kip}$

Available tensile Bolt strength — $B := \phi r_{nt} = 51\ kip$

Strength reduction factor — $\phi := 0.90$ — $B := 24.2\ kip$

Thickness to prevent prying action Eq. 9-30 — $t_c := \sqrt{\frac{4 \cdot B \cdot b'_{endPL}}{\phi \cdot p \cdot Fu_{PL}}} = 1\ in$

Eq. 9-35 — $\alpha' := \frac{1}{\delta_{endPL} \cdot (1 + \rho_{endPL})} \cdot \left[\left(\frac{t_c}{t_{PL}}\right)^2 - 1\right] = -0$

Eq. 9-33 — $Q := \left(\frac{t_{PL}}{t_c}\right)^2 \cdot (1 + \delta_{endPL} \cdot \alpha') = 1$

Available strength — $T_{avail} := B \cdot Q = 24.2\ kip$

Check capcity — if $(r_{ut} < T_{avail}$, "OK", "NOT OK") = "OK"

Try t_endPL=5/8″ — $t_{endPL} := \frac{5}{8} in$

$\alpha' := \frac{1}{\delta_{endPL} \cdot (1 + \rho_{endPL})} \cdot \left[\left(\frac{t_c}{t_{endPL}}\right)^2 - 1\right] = 1.35$

$Q := \left(\frac{t_{endPL}}{t_c}\right)^2 \cdot (1 + \delta_{endPL}) = 0.66$

Available strength — $T_{avail} = B \cdot Q = 16\ kip$

Check capcity — if $(r_{ut} < T_{avail}$, "OK", "NOT OK") = "OK"

Check bolt bearing at bolt holes on end plate (J3.10a)

recall

$\phi r_{nv} = 37.9 in$

$d_{bolt} = 0.875 in$

$t_{endPL} = 0.625 in$

$Fu_{PL} = 65\ ksi$

Clear distance — $I_{c\_endPL} := 1.75in - 0.5 \cdot \left(d_{bolt} + \frac{1}{16} in\right) = 1.28\ in$

Strength reduction factor — $\phi := 0.75$

Bolt bearing strength — $\phi R_n := \min\left(\phi \cdot 1.2 \cdot I_{c\_endPL} \cdot t_{endPL} \cdot Fu_{PL}, \phi \cdot 2.4 \cdot d_{bolt} \cdot t_{endPL} \cdot Fu_{PL}, \phi r_{nv}\right) = 37.9\ kip$

Check capcity — if $(\phi R_n \geq \phi r_{nv}$, "bolt shear governs", "check bolt bearing") = "bolt shear governs"

Check block shear rupture of the end plate (J4.3)

recall
$V_{column} = 301.9\ kip$
$N_{bolts.end\_PL} = 14$
$t_{endPL} = 0.625\,in$

Block shear strength $R_n = 0.60F_u \cdot A_{nv} + U_{bs} \cdot F_u \cdot A_{nt} \le 0.60F_y \cdot A_{gv} + U_{bs} \cdot F_u \cdot A_{nt}$

Gross shear area (1 side) $Agv_{endPL} := \left[\left(\frac{N_{bolts.end\_PL}}{2} - 1\right) \cdot p + 1.75\,in\right] t_{endPL} = 12.3\,in^2$

Net shear area (1 side) $Anv_{endPL} := Agv_{endPL} - t_{endPL} \cdot \left(\frac{N_{bolts.end\_PL}}{2} - 0.5\right) \cdot \left(d_{bolt} + \frac{2}{16}in\right) = 8.28\,in^2$

Net tensile area (1 side) $Ant_{endPL} := t_{endPL} \cdot \left[\frac{h_{endPL} - gage}{2} - 0.5 \cdot \left(d_{bolt} + \frac{2}{16}in\right)\right] = 1.09\,in^2$

$U_{bs} := 1$

$0.60Fu_{PL} \cdot Anv_{endPL} = 323\ kip$

$0.60Fy_{PL} \cdot Agv_{endPL} = 370.3\ kip$

$U_{bs} \cdot Fu_{PL} \cdot Ant_{endPL} = 71.1\ kip$

Strength reduction factor $\phi := 0.75$

Block shear strength $\phi R_n := \phi \cdot (2)\left(0.6\ \min\left(Fu_{PL} \cdot Anv_{endPL}, Fy_{PL} \cdot Agv_{endPL}\right) + U_{bs} \cdot Fu_{PL} \cdot Ant_{endPL}\right) = 591.1\ kip$

Check capacity if $\left(V_{column} < \phi R_n, \text{"OK"}, \text{"NOT OK"}\right) = \text{"OK"}$

Check prying action on column flange

recall
$gage = 5.5\,in$
$tw_{column} = 0.44\,in$
$d_{bolt} = 0.875\,in$
$bf_{column} = 14.5\,in$
$tf_{column} = 0.71\,in$
$p = 3\,in$
$Fu_{beam} = 65\ ksi$
$r_{ut} = 12.6\ kip$

$$b_{column} := \frac{gage - tw_{column}}{2} = 2.5\,in$$

$$b'_{column} := b_{column} - \frac{d_{bolt}}{2} = 2.1\,in \quad \text{use}$$

$$a_{column} := \frac{bf_{column} - gage}{2} = 4.5\ in \quad > a_{endPL} = 2.25\ in$$

$$a'_{column} := \min\left(a_{endPL} + \frac{d_{bolt}}{2}, 1.25 \cdot b_{column} + \frac{d_{bolt}}{2}\right) = 2.69\,in$$

$$\rho_{column} := \frac{b'_{column}}{a'_{column}} = 0.78$$

$$\delta_{column} := 1 - \frac{d'}{p} = 0.69$$

$\phi := 0.90$

$t_c := \sqrt{\dfrac{4 \cdot B \cdot b'_{column}}{\phi \cdot p \cdot Fu_{beam}}} = 1.074\,in$ $\quad B = 24.2\,kip$

$\alpha' := \dfrac{1}{\delta_{column} \cdot (1 + \rho_{column})} \cdot \left[\left(\dfrac{t_c}{tf_{column}}\right)^2 - 1\right] = 1.05$

Available strength $\quad Q := \left(\dfrac{tf_{column}}{t_c}\right)^2 \cdot (1 + \delta_{column}) = 0.74$ Available strength

$T_{avail} = B \cdot Q = 17.8\,kip$

Check capacity $\quad$ if$\left(r_{ut} < T_{avail}, \text{“OK”}, \text{“NOT OK”}\right) = \text{“OK”}$

More realistic model

Number of bolts in row $\quad n := N_{bolts.end\_PL} \div 2 = 7$

$a\_bar := \dfrac{bf_{column} - gage}{2}$

$b\_bar := b_{column} = 2.5\,in$

Replace p with p.eff $\quad p_{eff} := \dfrac{(n-1)\,p + \pi \cdot b\_bar + 2 \cdot a\_bar}{n} = 5\,in$

$\delta_{column\_eff} := 1 - \dfrac{d'}{p_{eff}} = 0.81$

recall

$N_{bolts.end\_PL} = 14$

$bf_{column} = 14.5\,in$

$b_{column} = 2.5\,in$

$gage = 5.5\,in$

$Fu_{beam} = 65\,ksi$

$tf_{column} = 0.7\,in$

$T_{avail} = 17.8\,kip$

$t_c := \sqrt{\dfrac{4 \cdot B \cdot b'_{column}}{\phi \cdot p_{eff} \cdot Fu_{beam}}} = 0.833\,in$

$\alpha' := \dfrac{1}{\delta_{column\_eff} \cdot (1 + \rho_{column})} \cdot \left[\left(\dfrac{t_c}{tf_{column}}\right)^2 - 1\right] = 0.26$ $\quad B = 24.2\,kip$

$Q_{eff} := \left(\dfrac{tf_{column}}{t_c}\right)^2 \cdot (1 + \delta_{endPL} \cdot \alpha') = 0.86$ $\quad$ percent change $\dfrac{T_{avail.eff} - T_{avail}}{T_{avail}} = 16.3 \cdot \%$

Available strength $\quad T_{avail.eff} := B \cdot Q_{eff} = 20.7\,kip$

Check capacity $\quad$ if$\left(r_{ut} < T_{avail.eff}, \text{“OK”}, \text{“NOT OK”}\right) = \text{“OK”}$

Check bearing on column flange

$tf_{column} = 0.7\,in > t_{endPL} = \dfrac{5}{8}\,in$

Beam-to-column connection

$$b := \frac{Span}{2} = 150 in$$

$$Height := \frac{Span}{\tan(\theta_{brace})} = 278.1 in$$

$$c := \frac{Height}{2} = 139.1 in$$

DG 29 Eq. 4-12

$$M_{uD} = 6 \cdot \left(\frac{P}{A \cdot b \cdot c}\right) \left(\frac{I_b \cdot I_c}{\frac{I_b}{b} + \frac{2I_c}{c}}\right) \cdot \left(\frac{b^2 + c^2}{bc}\right)$$

$$M_{uD} := 6 \cdot \left(\frac{P_u}{Ag_{brace} \cdot b \cdot c}\right) \left(\frac{Ix_{beam} \cdot Ix_{column}}{\frac{Ix_{beam}}{b} + \frac{2Ix_{column}}{c}}\right) \cdot \left(\frac{b^2 + c^2}{b \cdot c}\right)$$

$$M_{uD} = 1272.8 in \cdot kip$$

$$H_{uD} := \frac{M_{uD}}{\beta\_bar + e_{beam}} = 56.1 kip$$

$$H_u := H_{column} - H_{uD} + Hu_{node} = 220 kip$$

Required shear strength $V_u := V_{beam} + Vu_{beam} = 319.2 kip$

Required axial strength $T_u := H_u = 220 kip$

recall
$Span = 25 \cdot ft$
$\theta_{brace} = 47.2 \cdot deg$
$Ix_{beam} = 1830 in^4$
$Ix_{column} = 999 in^4$
$Ag_{brace} = 26.2 in^2$
$e_{beam} = 10.7 in$
$\beta\_bar = 12 in$
$H_{column} = 176.1 kip$
$P_u = 840 kip$
$V_{beam} = 269.2 kip$

Design bolts at beam-to-column connection (J3.6 and J3.7)

Number of bolts $N_{bolts\_beam} := 12$

Ultimate shear

Force per bolt $r_{uv} := \frac{V_u}{N_{bolts\_beam}} = 26.6 kip$

Check capacity if ($r_{uv} < \phi r_{nv}$, "OK", "NOT OK") = "OK"

Ultimate tensile Force per bolt $r_{ut} := \frac{T_u}{N_{bolts\_beam}} = 18.3 kip$

Check capacity if ($r_{ut} < \phi r_{nt}$, "OK", "NOT OK") = "OK"

Strength reduction factor $\phi := 0.75$

Modified nominal tensile stress factored to

recall
$\phi r_{nv} = 37.9 kip$
$\phi r_{nt} = 51 kip$
$Fnt_{bolt} = 113 ksi$
$Fnv_{bolt} = 84 ksi$
$A_{bolt} = 0.6 in^2$
$V_u = 319.2 kip$
$T_u = 220 kip$

include shear stress effects $F'_{nt} = 1.3 \cdot F_{nt} - \frac{F_{nt}}{\phi \cdot F_{nv}} \cdot f_{rv} < F_{nt}$

$$F'nt_{bolt} := \min\left[1.3 \cdot Fnt_{bolt} - \frac{Fnt_{bolt}}{\phi \cdot Fnv_{bolt}} \cdot \left(\frac{r_{uv}}{A_{bolt}}\right), Fnt_{bolt}\right] = 67.6\ \text{ksi}$$

Combined loading strength $\phi R_n := \phi \cdot F'nt_{bolt} \cdot A_{bolt} = 30.5\ \text{kip}$

Check capacity $\text{if}\left(r_{ut} < \phi R_n, \text{"OK"}, \text{"NOT OK"}\right) = \text{"OK"}$

Try Gr 490 bolts

Nominal tensile stress $Fnt_{bolt\_A490} := 113\ \text{ksi}$ $\phi := 0.75$ factored tensile strength per bolt

Nominal shear stress $Fnv_{bolt\_A490} := 84\ \text{ksi}$ $B := \phi \cdot Fnv_{bolt\_A490} \cdot A_{bolt} = 51\ \text{kip}$

Modified nominal tensile stress factored to include shear stress effects

$$F'nt_{bolt\_A490} := \min\left[1.3 \cdot Fnt_{bolt\_A490} - \frac{Fnt_{bolt\_A490}}{\phi \cdot Fnv_{bolt\_A490}} \cdot \left(\frac{r_{uv}}{A_{bolt}}\right), Fnt_{bolt\_A490}\right] = 67.6\ \text{ksi}$$

Strength reduction factor $\phi := 0.75$

Combined loading strength $\phi R_{n\_A490} := \phi \cdot F'nt_{bolt\_A490} \cdot A_{bolt} = 30.5\ \text{kip}$

Check capacity $\text{if}\left(r_{ut} < \phi R_{n\_A490}, \text{"OK"}, \text{"NOT OK"}\right) = \text{"OK"}$

Design beam web-to-end plate weld (Eq. 8-2)

Resultant force $R_u := \sqrt{{V_u}^2 + {T_u}^2} = 387.7\ \text{kip}$

Angle of load $\theta_{b\_endPL} := \text{atan}\left(\frac{T_u}{V_u}\right) = 34.6 \cdot \text{deg}$

Effective length of connection $I_{b\_endPL} := \frac{N_{bolts\_beam}}{2} \cdot p$

recall

$N_{bolts\_beam} = 12$

$V_u = 319.2\ \text{kip}$

$T_u = 220\ \text{kip}$

End plate weld size

$$D_{b\_endPL} := \frac{R_u}{(2) \cdot \left(1.392 \frac{\text{kip}}{\text{in}}\right) \cdot I_{b\_endPL} \cdot \left(1.0 + 0.50 \cdot \sin\left(\theta_{b\_endPL}\right)^{1.5}\right)} = 6.4$$

$$D_{b\_endPL} := \text{ceil}\left(D_{b\_endPL}\right) \cdot \text{in} = 7\,\text{in}$$

Check the 5.5in. gage with 7/16in. fillet welds (Table 7-15)

Required clearance $C_{3\_req} := \frac{7}{8}\,in$

Clearance $clearance := \frac{gage}{2} - \frac{tw_{beam}}{2} - \frac{D_{b\_endPL}}{16} = 2.05\,in$

Check clearance $if\left(C_{3\_req} < clearance, \text{"OK"}, \text{"NOT OK"}\right) = \text{"OK"}$

recall
$tw_{beam} = 0.515\,in$
$gage = 5.5\,in$
$D_{b\_endPL} = 7\,in$

Check prying action on bolts and end plate (Chapter 9)

Beam end plate thickness $t_{b\_endPL} := \frac{5}{8}\,in$

$$b_{beam} := \frac{gage - tw_{beam}}{2} = 2.5\,in$$

$$b'_{beam} := b_{beam} - \frac{d_{bolt}}{2} = 2.1\,in$$

$$a_{beam} := \frac{h_{endPL} - gage}{2} = 2.3\,in$$

$$a'_{beam} := \min\left(a_{beam} + \frac{d_{bolt}}{2}, 1.25 b_{beam} + \frac{d_{bolt}}{2}\right) = 2.7\,in$$

$$\rho_{b\_endPL} := \frac{b'_{beam}}{a'_{beam}} = 0.76$$

$$\delta_{b\_endPL} := 1 - \frac{d'}{p} = 0.69$$

recall
$tw_{beam} = 0.515\,in$
$gage = 5.5\,in$
$d_{bolt} = 0.875\,in$
$h_{endPL} = 10\,in$
$p = 3\,in$
$Fu_{PL} = 65\,ksi$
$r_{ut} = 18.3\,kip$

$B = 51\,kip$
$B := 30.4\,kip$

$$\phi := 0.90$$

$$t_c := \sqrt{\frac{4 \cdot B \cdot b'_{beam}}{\phi \cdot p \cdot Fu_{PL}}} = 1.19\,in$$

$$\alpha'_{b\_endPL} := \frac{1}{\delta_{b\_endPL} \cdot \left(1 + \rho_{b\_endPL}\right)} \cdot \left[\left(\frac{t_c}{t_{b\_endPL}}\right)^2 - 1\right] = 2.2$$

$$Q_{b\_endPL} := \left(\frac{t_{b\_endPL}}{t_c}\right)^2 \cdot \left(1 + \delta_{b\_endPL}\right) = 0.46$$

$$T_{avail.b\_endPL} := B \cdot Q_{b\_endPL} = 14.1\,kip$$

$$if\left(r_{ut} < T_{avail.b\_endPL}, \text{"OK"}, \text{"NOT OK"}\right) = \text{"NOT OK"}$$

Use thicker plate $t_{b\_endPL} := \frac{3}{4}\,in$

$$\alpha'_{b\_endPL} := \frac{1}{\delta_{b\_endPL} \cdot \left(1 + \rho_{b\_endPL}\right)} \cdot \left[\left(\frac{t_c}{t_{b\_endPL}}\right)^2 - 1\right] = 1.3$$

Available strength $Q_{b\_endPL} := \left(\frac{t_{b\_endPL}}{t_c}\right)^2 \cdot \left(1 + \delta_{b\_endPL}\right) = 0.67$

Check capacity $T_{avail.b\_endPL} := B \cdot Q_{b\_endPL} = 20.3\,kip$

$$if\left(r_{ut} < T_{avail.b\_endPL}, \text{"OK"}, \text{"NOT OK"}\right) = \text{"OK"}$$

Check prying action on column flange

$$t_c := \sqrt{\frac{4 \cdot B \cdot b'_{column}}{\phi \cdot p \cdot Fu_{beam}}} = 1.2\,in$$

$$\alpha'_{column} := \frac{1}{\delta_{column} \cdot (1 + \rho_{column})} \cdot \left[\left(\frac{t_c}{tf_{column}}\right)^2 - 1\right] = 1.5$$

$$Q_{column} := \left(\frac{tf_{column}}{t_c}\right)^2 \cdot (1 + \delta_{column} \cdot 1) = 0.59$$

$$T_{avail.column} := B \cdot Q_{column} = 17.8\,kip$$

$$if\left(r_{ut} < T_{avail.column}, \text{"OK"}, \text{"NOT OK"}\right) = \text{"NOT OK"}$$

recall
$b'_{column} = 2.09\,in$
$a'_{column} = 2.7\,in$
$\rho_{column} = 0.8$
$p = 3\,in$
$\delta_{column} = 0.688$
$Fu_{beam} = 65\,ksi$
$tf_{column} = 0.7\,in$
$r_{ut} = 18.3\,kip$
$B = 30.4\,kip$

Check bolt bearing at end plate

Check block shear rupture on end plate (J4.3)

Block shear strength

$$R_n = 0.60F_u \cdot A_{nv} + U_{bs} \cdot F_u \cdot A_{nt} \leq 0.60F_y \cdot A_{gv} + U_{bs} \cdot F_u \cdot A_{nt}$$

recall
$h_{endPL} = 10\,in$
$t_{b\_endPL} = 0.8\,in$
$N_{bolts\_beam} = 12$
$d_{bolt} = 0.875$
$ingage = 5.5\,in$

Gross area in shear (1 side)

$$Agv_{b\_endPL} := \left[\left(\frac{N_{bolts\_beam}}{2} - 1\right) \cdot p + 4.40\,in\right] \cdot t_{b\_endPL}$$

$$Agv_{b\_endPL} = 14.5\,in^2$$

Net shear area (1 side)

$$Anv_{b\_endPL} := Agv_{b\_endPL} - t_{b\_endPL} \cdot \left[\left(\frac{N_{bolts\_beam}}{2} - 0.5\right) \cdot \left(d_{bolt} + \frac{2}{16}in\right)\right] 10.4\,in^2$$

Net tensile area (1 side)

$$Ant_{b\_endPL} := t_{b\_endPL} \cdot \left[\frac{h_{endPL} - gage}{2} - 0.5 \cdot \left(d_{bolt} + \frac{2}{16}in\right)\right] = 1.31\,in^2$$

$$U_{bs} := 1$$

$$0.60Fu_{PL} \cdot Anv_{b\_endPL} = 406.6\,kip$$
$$0.60Fy_{PL} \cdot Agv_{b\_endPL} = 436.5\,kip$$
$$U_{bs} \cdot Fu_{PL} \cdot Ant_{b\_endPL} = 85.3\,kip$$

Strength reduction factor

$$\phi := 0.75$$

$$\phi R_n := \phi \cdot (2)\left(0.6\ \min\left(Fu_{PL} \cdot Anv_{b\_endPL}, Fy_{PL} \cdot Agv_{b\_endPL}\right) + U_{bs} \cdot Fu_{PL} \cdot Ant_{b\_endPL}\right)$$

Block shear strength

$$\phi R_n = 737.8\,kip$$

recall
$V_u = 319.2\,kip$

Check capacity

$$if\left(V_u < \phi R_n, \text{"OK"}, \text{"NOT OK"}\right) = \text{"OK"}$$

Check beam shear strength (J4.2)

Gross shear area	$Agv_{beam} := d_{beam} \cdot tw_{beam} = 11\,in^2$	recall
Strength reduction factor	$\phi := 1.00$	$d_{beam} = 21.4\,in$
Shear strength	$\phi R_n := \phi \cdot 0.60 \cdot Fy_{beam} \cdot Agv_{beam} = 330.6\,kip$	$tw_{beam} = 0.515\,in$
Check capacity	if $(V_u < \phi R_n$, "OK", "NOT OK" $)$ = "OK"	$Fy_{beam} = 50\,ksi$
		$V_u = 319.2\,kip$

Check column shear strength (J4.2)

Gross shear area	$Agv_{column} := d_{column} \cdot tw_{column} = 6.2\,in^2$	recall
		$d_{column} = 14\,in$
Strength reduction factor	$\phi := 1.00$	$tw_{column} = 0.44\,in$
Shear strength	$\phi R_n := \phi \cdot 0.60 \cdot Fy_{beam} \cdot Agv_{column} = 184.8\,kip$	$Fy_{beam} = 50\,ksi$
Check capacity	if $(H_{column} < \phi R_n$, "OK", "NOT OK" $)$ = "OK"	$H_{column} = 176.1\,kip$

References

AISC (2015). *Vertical Bracing Connections: Analysis and Design*, Design Guide 29. American Institute of Steel Construction, Chicago.

AISC (2016). *Specification for Structural Steel Buildings*. American Institute of Steel Construction, Chicago.

AISC (2017). *Steel Construction Manual*, 15th Edition. American Institute of Steel Construction, Chicago.

Astaneh-Asl, A. (1998), "Seismic Behavior and Design of Gusset Plates," Steel Tips, Structural Steel Educational Council, Moraga, CA.

Neal, B.G. (1977), The Plastic Methods of Structural Analysis, Chapman and Hall, London.

17

Double Web-Angle (DWA) Connections

17.1 Description

In this chapter, the design strength capacities of ten double web-angle (DWA) connection specimens were calculated following the requirements of the AISC *Specification* (AISC, 2016) and the AISC *Manual* (AISC, 2017). Four test specimens were selected from the experimental study performed by McMullin and Astaneh-Asl (1988) in the Department of Civil Engineering at the University of California, Berkeley, and were examined using IDEA StatiCa (V 20.1.3471.1). Six additional models were developed for verification purposes by modifying the parameters based on the available test specimens. Then, the baseline model was analyzed using ABAQUS software (Dassault, 2020) and the results were compared.

17.2 The Experimental Study

Seven full-scale steel beam-column connection specimens were tested, and results were presented in McMullin and Astaneh-Asl (1988). Each connection specimen was bolted to the beam and welded to the column with double angle sections. The schematic of the DWA connection used in these experiments and the test setup are shown in Figure 17.1(a). As shown in Figure 17.1(b), the main goal of these tests is to apply only shear force in the connection with very small bending or moment. To achieve this objective, the actuator "S" near the connection applies the shear force. The actuator "R" near the tip of the cantilever aims to keep the beam horizontal and limit the rotation (bending) of the connection. The properties of the seven DWA connection specimens are provided in Table 17.1.

Beams and columns were made of ASTM A992 steel, and angles were manufactured from ASTM A36 steel. The material properties of the members are presented in Table 17.2. All bolts were A325 bolts with threads excluded from the shear planes. The edge distance of the bolts was 1.25 in. from the top and bottoms of the angle sections, while the bolt spacing was 3.0 in. The weld size of each specimen was 0.25 in. and it was welded to the column using E-70XX electrodes with a nominal strength of 70 ksi.

Four DWA connection specimens, Tests No. 4, 5, 6, and 9, were selected out of the eight specimens tested (Table 17.1). The properties, failure modes, and shear capacities measured during the testing of these four specimens are provided in Table 17.3. Ultimate failure of all four specimens was due to weld failure, including in the heat-affected zone (HAZ).

Steel Connection Design by Inelastic Analysis: Verification Examples per AISC Specification, First Edition.
Mark D. Denavit, Ali Nassiri, Mustafa Mahamid, Martin Vild, Halil Sezen, and František Wald.

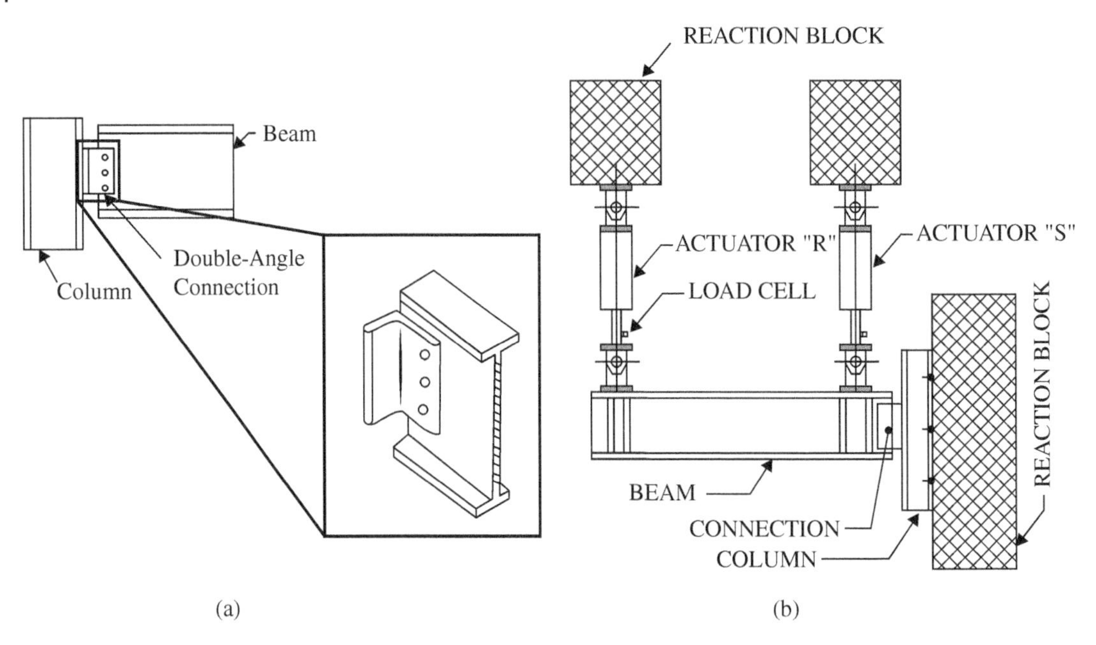

Figure 17.1 (a) DWA connection specimens; and (b) test setup (McMullin and Astaneh-Asl, 1988).

Table 17.1 Properties of DWA connection specimens.

Test no.	Number of bolts	Bolt size	Weld size	Connection length	Weld length	Angle size	Connection detail
4	7	3/4	1/4	20.5	20.5	4 × 3.5 × 3/8	I
5	5	3/4	1/4	14.5	14.5	4 × 3.5 × 3/8	I
6	3	3/4	1/4	8.5	8.5	4 × 4 × 3/8	I
7	7	7/8	5/16	20.5	26.0	4 × 4 × 3/8	III
8	5	7/8	5/16	14.5	20.0	4 × 4 × 3/8	III
9	5	7/8	5/16	14.5	14.5	4 × 4 × 3/8	I
10	5	7/8	5/16	14.5	14.5	4 × 4 × 3/8	II

Shear force-rotation and moment-shear diagrams measured during the experiments (McMullin and Astaneh-Asl, 1988) are shown in Figures 17.2 to 17.5 for each of the four specimens modeled and analyzed in this report.

17.2.1 Instrumentation

The instrumentation used in this experimental study are (see Figure 17.6):

- three linear variable displacement transducers (LVDT)
- three linear potentiometers (LP)
- two load cells

Table 17.2 Material properties.

Members	Strength (ksi)	
Columns and beams (A992)	Yield strength, F_y	50
	Ultimate strength, F_u	65
Angles (A36)	Yield strength, F_y	36
	Ultimate strength, F_u	58
Bolts A325 (threads excluded)	Nominal tensile strength, F_{nt}	90
	Nominal shear strength, F_{nv}	68

Table 17.3 Summary of test results.

	Properties of specimens					Test results		
						At ultimate load		
Test no.	Beam	Column	Double angle (in.)	Number of bolts	Bolt diameter (in.)	Shear (kips)	Rotation (rad)	Failure mode
4	W24 × 68	W10 × 77	4 × 3.5 × 3/8	7	3/4	230	0.0257	Weld sheared along its full length in the HAZ
5	W24 × 68	W10 × 77	4 × 3.5 × 3/8	5	3/4	205	0.0315	Weld cracked in HAZ of angle
6	W24 × 68	W10 × 77	4 × 3.5 × 3/8	3	3/4	117	0.0414	Weld cracked along the top length
9	W24 × 68	W10 × 77	4 × 4 × 3/8	5	7/8	192	0.0332	Weld cracked from top down

LVDT 7 was used to measure the separation of the top of the angle relative to the column flange, while LVDT 5, 6, 8, and 9 measured the relative displacement between the column flange and the beam flange. The small rotations (less than 0.02 rad.) of the beam can be calculated with the following equation:

$$\text{Rotation} = (\text{LVDT } 5 + \text{LVDT } 6 + \text{LVDT } 8 + \text{LVDT } 9) \,/\, (2 \text{ x distance between LVDT centerlines})$$

The deflection at the end of the beam was measured with LP #3. The deflection across from the actuator (actuator "S" in Figure 17.1) was measured using LP #4, while LP #10 was used to measure the displacement at the bolt line in the direction of the applied shear load. The rotation of the beam (larger than 0.01 rad.) can be calculated with the following equation:

$$\text{Rotation} = (\text{LP\#3} + \text{LP\#10}) \,/\text{separation}$$

In this study, it is assumed that the moment at the weld of the connection is the moment transferred into the column. If the location of the inflection point is known, the moment, M, transferred from the

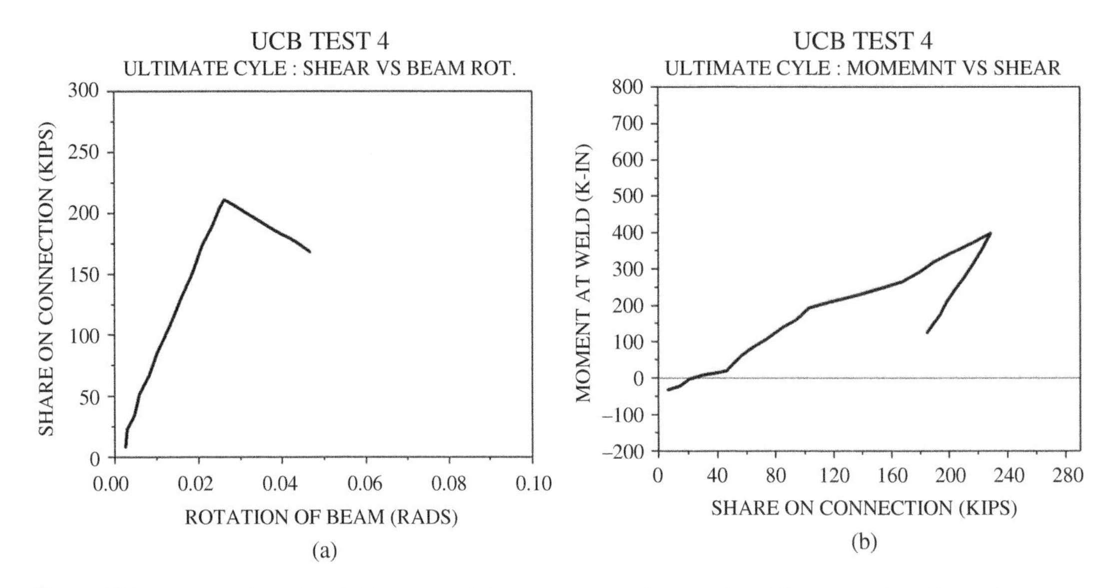

Figure 17.2 Test specimen No. 4: (a) measured shear on connection-rotation of beam relationship; and (b) moment at weld-shear on connection relationship.

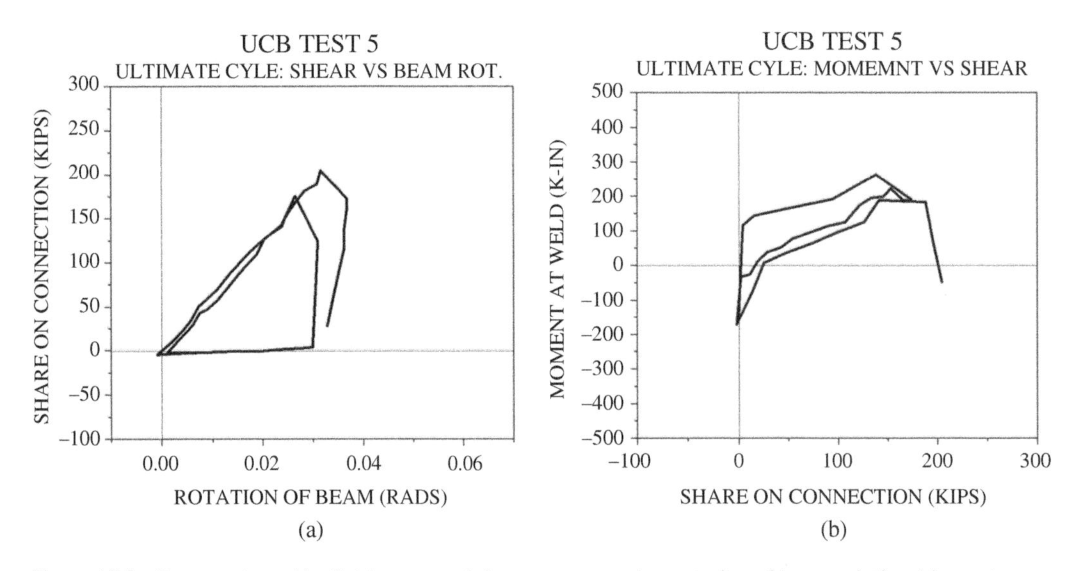

Figure 17.3 Test specimen No. 5: (a) measured shear on connection-rotation of beam relationship; and (b) moment at weld-shear on connection relationship.

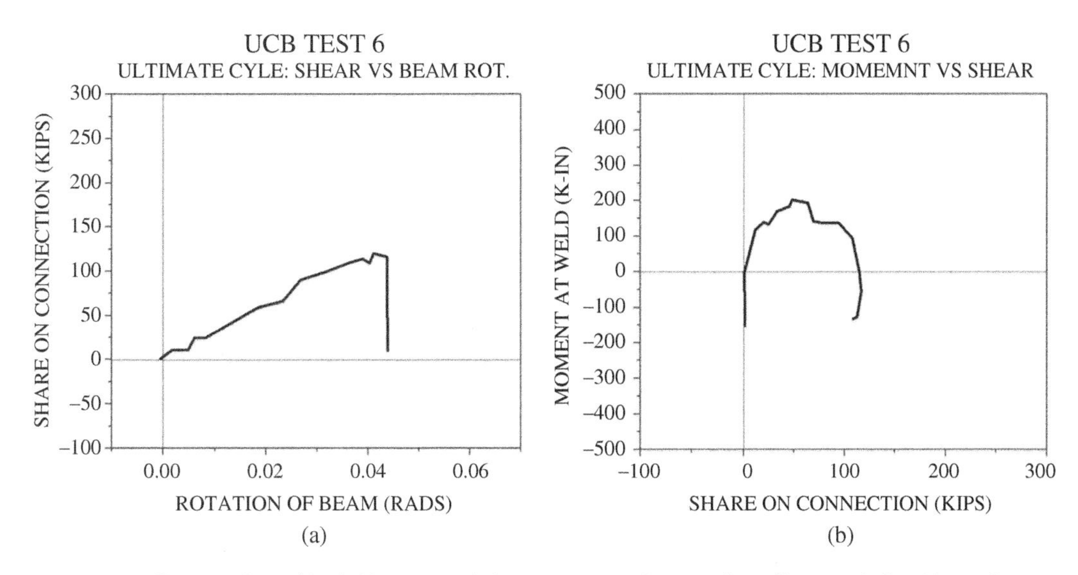

Figure 17.4 Test specimen No. 6: (a) measured shear on connection-rotation of beam relationship; and (b) moment at weld-shear on connection relationship.

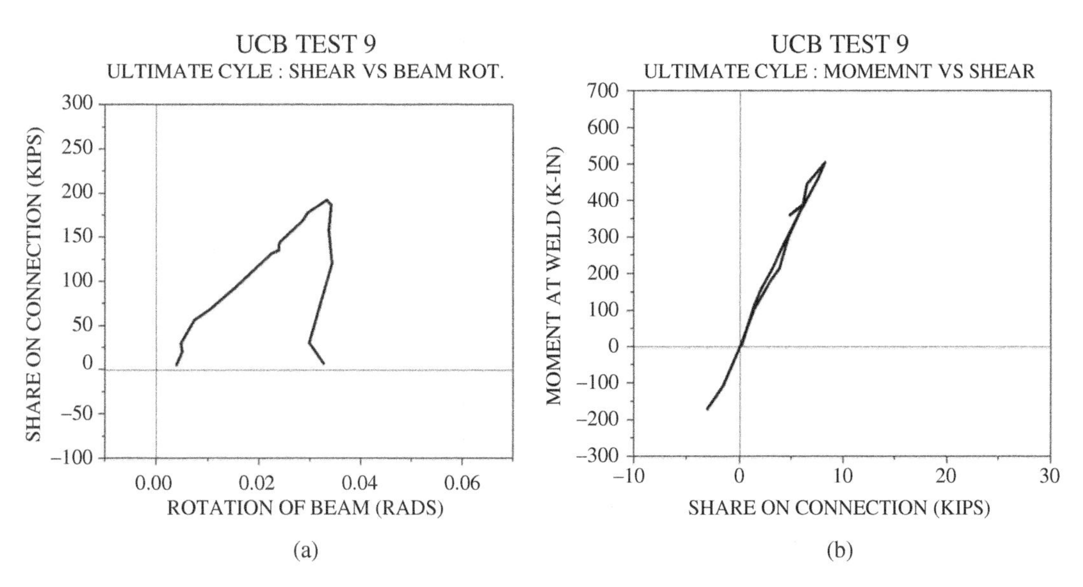

Figure 17.5 Test specimen No. 9: (a) measured shear on connection-rotation of beam relationship; and (b) moment at weld-shear on connection relationship.

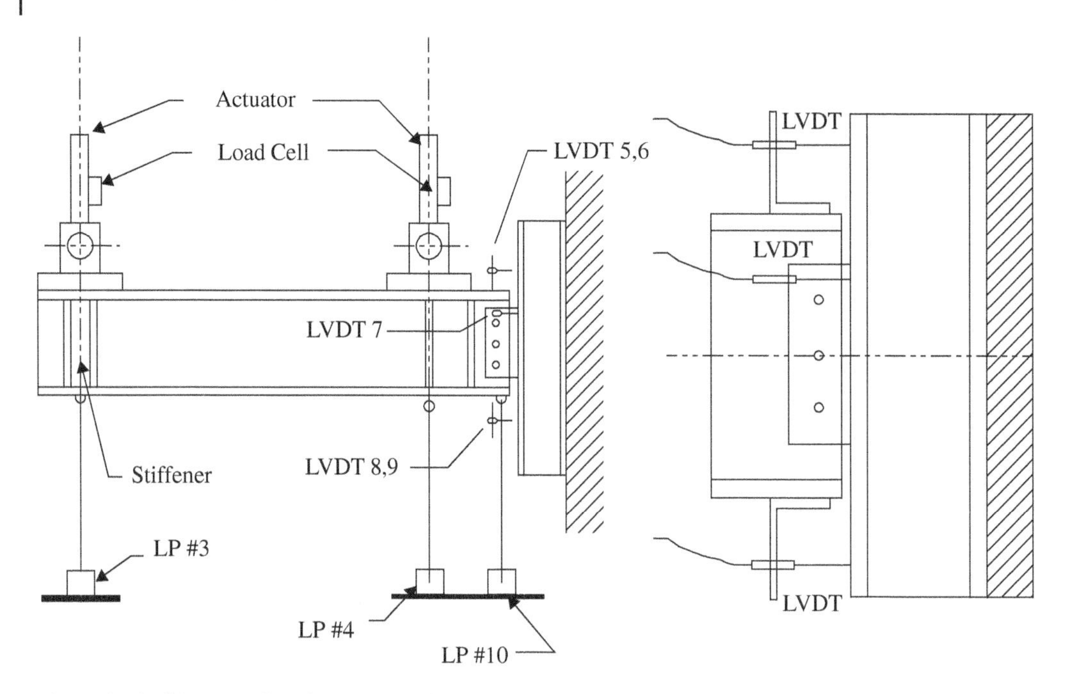

Figure 17.6 Diagram of the instrumentation used during the experiment (McMullin and Astaneh-Asl, 1988).

beam can be calculated by multiplying the applied shear force, V, by the distance between the inflection point and the column, e.

$$M = V \cdot e$$

However, it is not possible to determine the location of the inflection point because of the complexity of the connection and loading. From the results obtained during testing, the actual location of the inflection

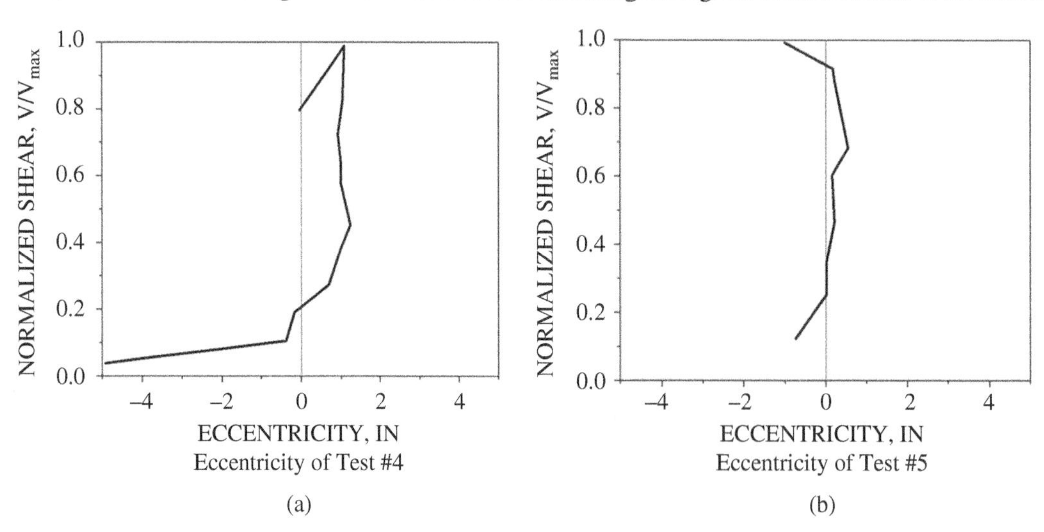

Figure 17.7 Normalized shear-eccentricity: (a) Test No. 4; and (b) Test No. 5 (McMullin and Astaneh-Asl, 1988).

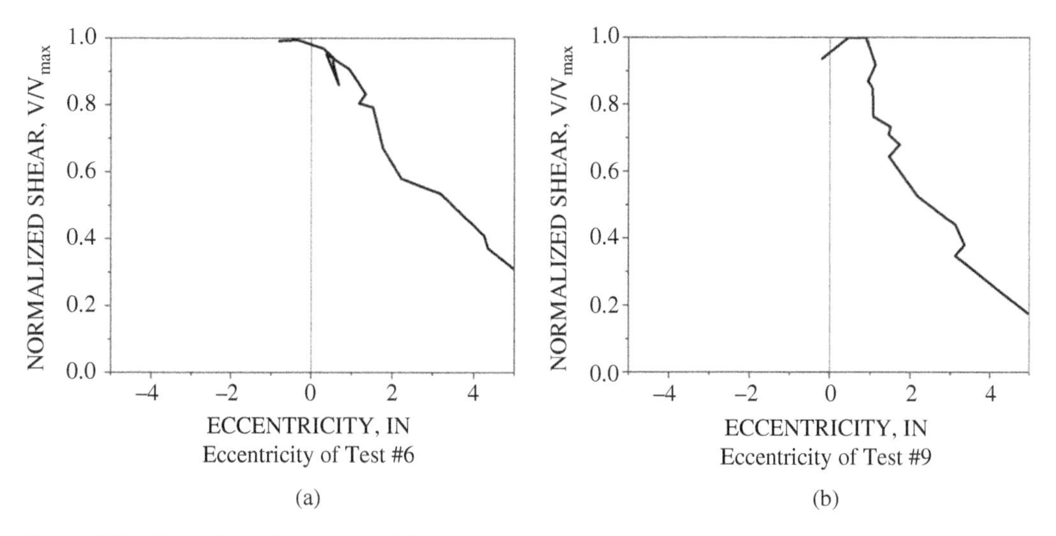

Figure 17.8 Normalized shear-eccentricity: (a) Test No. 6; and (b) Test No. 9 (McMullin and Astaneh-Asl, 1988).

point can be computed by static analysis for any load during the loading history and can be obtained as a function of the normalized shear, V/V_{max} on the connection (see Figures 17.7 and 17.8).

17.3 Code Design Calculations and Comparisons

The design strength capacities (ϕR_n) of the connections were calculated following the requirements of the AISC *Specification* (AISC, 2016) and the AISC *Manual* (AISC, 2017). The nominal strength, R_n, and the corresponding resistance factor, ϕ, for each connection design limit state for load and resistance factored design (LRFD) are provided in Chapter J of the AISC *Specification*. The following 13 design checks were performed according to the LRFD design equations included in the AISC *Specification* or AISC *Manual*.

- Bolt Shear Check (Eq. J3-1, AISC *Specification*)
- Bolt Tensile Check (Eq. J3-1, AISC *Specification*)
- Bolt Bearing on Beam (Eq. J3-6a, AISC *Specification*)
- Bolt Tearout on Beam (Eq. J3-6c, AISC *Specification*)
- Bolt Bearing on Angles (Eq. J3-6a, AISC *Specification*)
- Bolt Tearout on Angles (Eq. J3-6c, AISC *Specification*)
- Shear Rupture on Angles (Beam Side) (Eq. J4-4, AISC *Specification*)
- Block Shear on Angles (Beam Side) (Eq. J4-5, AISC *Specification*)
- Shear Yielding on Angles (Eq. J4-3, AISC *Specification*)
- Shear Yielding on Beam (Eq. J4-3, AISC *Specification*)
- Welds Rupture on Angles (Support Side) (Page 9-5, AISC *Manual*)
- Weld Capacity (Page 10-11, AISC *Manual*)
- Weld Capacity (no eccentricity) (Eq. J4-2, AISC *Specification*)

17.3.1 LRFD Design Strength Capacities of Four Test Specimens

The design strength capacities (ϕR_n) of the four selected test specimens were calculated by following the AISC LRFD code requirements, i.e. the AISC *Specification* (AISC, 2016) and the AISC *Manual* (AISC, 2017). The properties of the four selected test specimens and their design strength capacities are provided in Tables 17.4 and 17.5, respectively. The detailed design strength calculations for Test No. 4 are provided in Appendix A of the full report, published on the IDEA StatiCa website (IDEA StatiCa, 2021).

The minimum yield stress, F_y, and the specified minimum tensile strength, F_u, of the materials of the test specimens are obtained from Table 2-4 in the AISC *Manual* (AISC, 2017). The clear distance between web fillets, T, the thickness of web of the beam, t_w, the depth of the beam, d, the distance from

Table 17.4 Properties of the test specimens.

		Test No. 4	Test No. 5	Test No. 6	Test No. 9
Beam	Type	W24 × 68	W24 × 68	W24 × 68	W24 × 68
	t_w (in.)	0.415	0.415	0.415	0.415
	d (in.)	23.7	23.7	23.7	23.7
	T (in.)	20.75	20.75	20.75	20.75
	k (in.)	17/16	17/16	17/16	17/16
	F_y (ksi)	50	50	50	50
	F_u (ksi)	65	65	65	65
Column	Type	W10 × 77	W10 × 77	W10 × 77	W10 × 77
	t_f (in.)	0.87	0.87	0.87	0.87
	F_y (ksi)	50	50	50	50
	F_u (ksi)	65	65	65	65
Double angle	Dimension (in.)	4 × 3-1/2 × 3/8	4 × 3-1/2 × 3/8	4 × 3-1/2 × 3/8	4 × 4 × 3/8
	leg (in.)	4	4	4	4
	L (in.)	20.5	14.5	8.5	14.5
	F_y (ksi)	36	36	36	36
	F_u (ksi)	58	58	58	58
Bolts	Type	A325	A325	A325	A325
	Diameter	0.75	0.75	0.75	0.875
	Number	7	5	3	5
	Threads	Excluded	Excluded	Excluded	Excluded
	Spacing (in.)	3	3	3	3
	Edge dist. (in.)	1.25	1.25	1.25	1.25
Weld	Size (in.)	0.25	0.25	0.25	0.3125
	Length (in.)	20.5	14.5	8.5	14.5
	Electrode	E70XX	E70XX	E70XX	E70XX

Table 17.5 Calculated LRFD design strength capacities of the test specimens.

Design checks	Test No. 4	Test No. 5	Test No. 6	Test No. 9
Bolt Shear (kips)	315.6	225.4	135.2	306.5
Bolt Tensile (kips)	417.8	298.4	179.0	405.7
Bolt Bearing on Beam (kips)	254.9	182.1	109.3	212.4
Bolt Tearout on Beam (kips)	378.2	344.7	311.2	330.7
Bolt Bearing on Angles (kips)	411.1	293.6	176.2	342.6
Bolt Tearout on Angles (kips)	547.3	375.8	204.4	235.4
Shear Rupture on Angles (kips)	281.4	198.4	114.8	185.8
Block Shear on Angles (kips)	260.5	187.6	114.7	185.4
Shear Yielding on Angles (kips)	332.2	235.0	137.8	235.0
Shear Yielding on Beam (kips)	295.2	295.2	295.2	295.2
Minimum Thickness for Connecting Element Rupture Strength (in.)	0.21	0.21	0.21	0.26
Weld Capacity by AISC *Manual* (2017, pp. 10–11) (kips)	186.8	114.6	48.1	126.6
Weld Capacity (no eccentricity) by Eq. J2.4 (AISC *Specification* (AISC, 2016) (kips)	228.3	161.5	94.7	201.9

the outer face of the flange to the web toe of the fillet, k, and the thickness of the flange of the column, t_f, are obtained from Table 1-1 in the AISC *Manual* (AISC, 2017). The lengths of the double angles, L, are provided in the experiment report (McMullin and Astaneh-Asl, 1988).

From the calculated design capacities for the four test specimens, the lowest shear capacity was selected in Table 17.5. Comparison of the calculated strengths in Table 17.5 shows that the design capacity of each of these four DWA connection specimens was controlled by weld failure according to the AISC *Manual* (AISC, 2017), i.e. the lowest strengths are in the second row from the bottom.

17.3.2 LRFD Design Strength Capacities of Six Additional Connection Models

The design strength capacities (ϕR_n) of six additional models were calculated by following the AISC LRFD code requirements. The properties of these six connections and their calculated design capacities are provided in Tables 17.6 and 17.7, respectively. Test No. 4 was selected as a baseline model. The modified properties are shown in bold italics in Table 17.6.

Comparison of the calculated design capacities of the six additional models in Table 17.7 showed that, in all six additional specimens, the lowest shear capacity corresponds to the weld capacity calculated from the equations in the AISC *Manual* (AISC, 2017).

Table 17.6 Properties of the test specimens.

		Model 1	Model 2	Model 3	Model 4	Model 5	Model 6
Beam	Type	W24×68	W24×68	W24×68	W24×68	W24×68	W24×68
	t_w (in.)	0.415	0.415	0.415	0.415	0.415	0.415
	d (in.)	23.7	23.7	23.7	23.7	23.7	23.7
	T (in.)	20.75	20.75	20.75	20.75	20.75	20.75
	k (in.)	17/16	17/16	17/16	17/16	17/16	17/16
	F_y (ksi)	50	50	50	50	50	50
	F_u (ksi)	65	65	65	65	65	65
Column	Type	W10×77	W10×77	W10×77	W10×77	W10×77	W10×77
	t_f (in.)	0.87	0.87	0.87	0.87	0.87	0.87
	F_y	50	50	50	50	50	50
	F_u	65	65	65	65	65	65
Double Angle	Dimension (in.)	4×3.5×3/8	***4×3.5×1/2***	4×3.5×3/8	4×3.5×3/8	4×3.5×3/8	4×4×3/8
	leg (in)	4	4	4	4	4	4
	L (in.)	20.5	20.5	20.5	20.5	20.5	20.5
	F_y (ksi)	36	36	36	36	36	36
	F_u (ksi)	58	58	58	58	58	58
Bolts	Type	A325-X	A325-X	A325-X	***A325-N***	A325-X	A325-X
	Diameter (in.)	0.75	0.75	***0.5***	0.75	*1*	***0.875***
	Number	7	7	7	7	7	7
	Threads	Excluded	Excluded	Excluded	***Not excluded***	Excluded	Excluded
	Spacing (in.)	3	3	3	3	3	3
	Edge distance (in.)	1.25	1.25	1.25	1.25	1.25	1.25
Weld	Size (in.)	***0.3125***	0.25	0.25	0.25	0.25	***0.3125***
	Length (in.)	20.5	20.5	20.5	20.5	20.5	20.5
	Electrode	E70XX	E70XX	E70XX	E70XX	E70XX	E70XX

17.3.3 Calculated ASD Design Strength Capacities

According to the allowable strength design (ASD), the allowable strength (R_n/Ω) is calculated by dividing the nominal strength, R_n by the safety factor, Ω. The allowable strength capacities of connection specimen Test No. 4 are calculated by following the AISC ASD code requirements. The properties of this test specimen were given in Table 17.4. The calculated ASD design strength capacities (R_n/Ω) of the specimen are

Table 17.7 Calculated LRFD design strength capacities of test specimens.

Design checks	Model 1	Model 2	Model 3	Model 4	Model 5	Model 6
Bolt Shear (kips)	315.6	315.6	139.9	250.6	560.6	429.1
Bolt Tensile (kips)	417.8	417.8	185.2	417.8	741.9	568.0
Bolt Bearing on Beam (kips)	254.9	254.9	169.9	254.9	339.9	297.4
Bolt Tearout on Beam (kips)	378.2	378.2	417.8	378.2	328.7	354.8
Bolt Bearing on Angles (kips)	411.1	548.1	274.1	411.1	548.1	479.6
Bolt Tearout on Angles (kips)	547.3	729.7	407.4	547.3	467.4	514.5
Shear Rupture on Angles (kips)	281.4	375.2	315.7	281.4	238.6	264.3
Block Shear on Angles (kips)	260.5	347.1	264.5	260.5	247.1	258.3
Shear Yielding on Angles (kips)	332.2	442.8	332.2	332.2	332.2	332.2
Shear Yielding on Beam (kips)	295.2	295.2	295.2	295.2	295.2	295.2
Minimum Thickness for Connecting Element Rupture Strength (in.)	0.26	0.21	0.21	0.21	0.21	0.26
Weld Capacity by AISC *Manual* (2017, pp. 10–11) (kips)	233.5	186.8	186.8	186.8	186.8	214.4
Weld Capacity (no eccentricity) by Eq. J2.4 (AISC *Specification* (AISC, 2016) (kips)	285.4	228.1	228.1	228.1	228.1	285.4

provided in Table 17.8. The calculated lowest strength is the weld capacity (124.5 kips) for this specimen. For ASD design purposes, this capacity (124.5 kips) should be compared with the load demand calculated using the ASD design load combinations (AISC *Manual*; AISC, 2017). The detailed design strength calculations for Test No. 4 are provided in Appendix B of the full report accessible on the IDEA StatiCa website (IDEA StatiCa, 2021).

17.4 IDEA StatiCa Analysis

IDEA StatiCa checks four different failure scenarios of this steel connection type: (1) plate failure, (2) bolt failure, (3) weld failure, and (4) buckling. The selected four test specimens (Table 17.4) and the six additional models (Table 17.6) were modeled in IDEA StatiCa and analyzed under a shear force, as shown

Table 17.8 Calculated ASD design strength capacities of Test No. 4.

Design checks	Test No. 4
Bolt Shear (kips)	210.4
Bolt Tensile (kips)	278.5
Bolt Bearing on Beam (kips)	170.0
Bolt Tearout on Beam (kips)	252.2
Bolt Bearing on Angles (kips)	274.1
Bolt Tearout on Angles (kips)	364.9
Shear Rupture on Angles (kips)	187.6
Block Shear on Angles (kips)	173.7
Shear Yielding on Angles (kips)	221.5
Shear Yielding on Beam (kips)	196.8
Minimum Thickness for Connecting Element Rupture Strength (in.)	0.21
Weld Capacity (kips)	124.5
Weld Capacity (no eccentricity) by Eq. J2.4 (AISC, 2016) (kips)	152.0

in Figure 17.9. In the software, the location of the shear force can be arbitrarily selected. Two shear force locations were investigated: (1) in bolts, and (2) at the column face.

The shear force was applied incrementally on the vertical line connecting the bolts and on the vertical welding material in different models until the connections reached their shear capacities in IDEA StatiCa. In this way, the shear capacities of the four tested specimens and the six additional models were computed, as presented in Tables 17.9 and 17.10.

For a new user, modeling the first connection (Test No. 4) takes approximately 8–10 minutes. Since each of the other connections were modeled by modifying the first one, each took 2–3 minutes. The software completed the calculation for each connection in 5–7 seconds on a personal computer. The result screen pointing out the failure mode and the deformed shapes (deformation scale 10) of finite element models from IDEA StatiCa are shown in Appendix C of the full report, published on the IDEA StatiCa website (IDEA StatiCa, 2021).

17.5 ABAQUS Modeling and Analysis

The aim of this section is to compare the results from IDEA StatiCa with those from another commercial finite element code. In this study, ABAQUS software package (Version 2020) (Dassault, 2020) was used. ABAQUS is a robust general-purpose FEA software package suitable for analyzing the whole range of static, dynamic, and nonlinear problems.

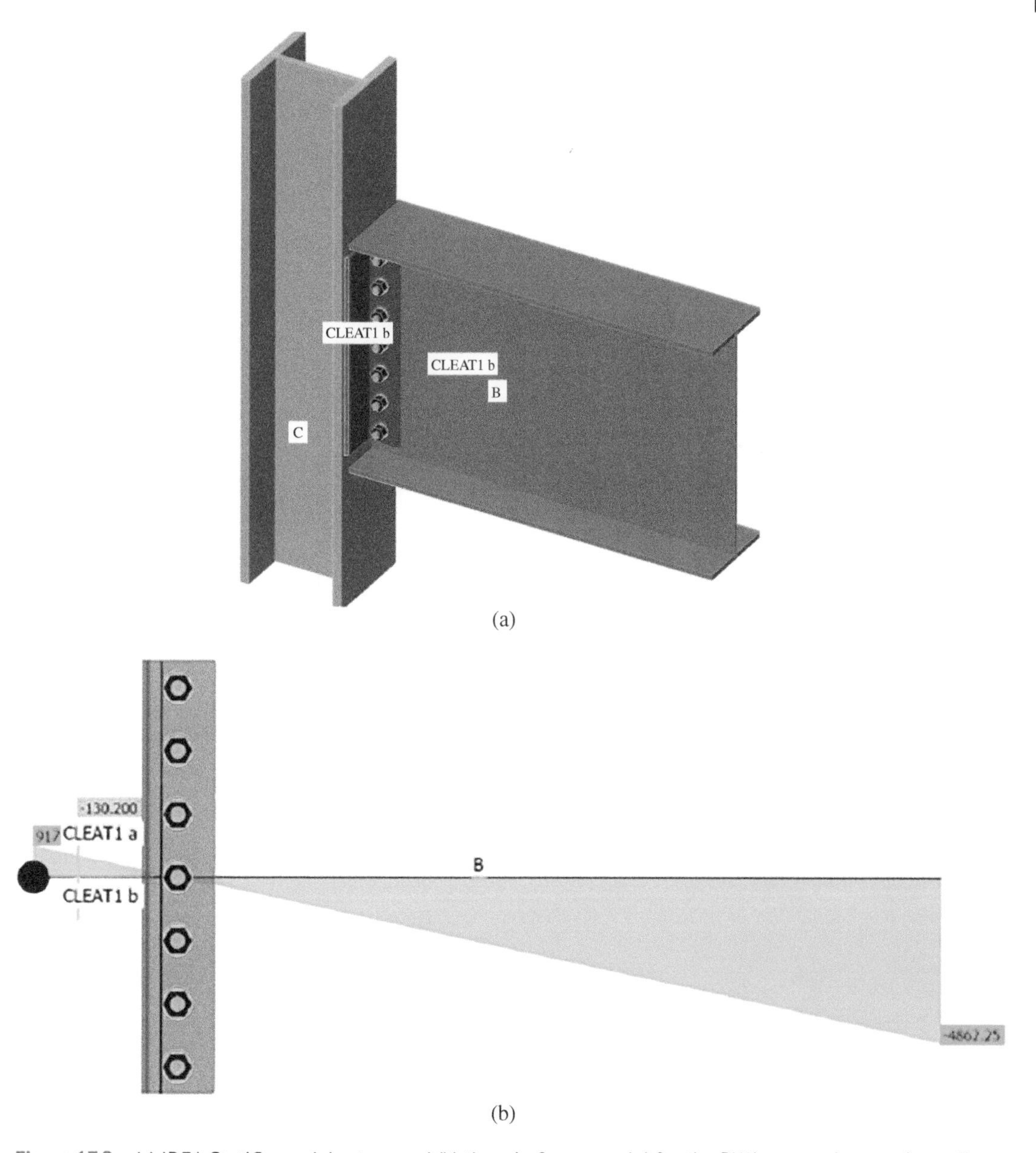

Figure 17.9 (a) IDEA StatiCa model setup; and (b) the wireframe model for the DWA connection specimen, Test No. 4 (force applied on the centroid of the bolts group).

In this study, Test No. 4, as described in Section 17.2, was chosen as a base model. Numerical simulations with almost identical conditions (i.e. in terms of material properties, boundary condition, and loading) were carried out using both IDEA StatiCa and ABAQUS. The model was initially designed in IDEA StatiCa and then the assembly (including beam, column, and double angles) was imported to

Table 17.9 Shear capacities of selected test specimens calculated by IDEA StatiCa.

Strength capacity by IDEA StatiCa	Test No. 4	Test No. 5	Test No. 6	Test No. 9
Strength by IDEA StatiCa – force applied on bolts (kips)	130.2	73.4	31.3	61.3
Failure mode – force applied on bolts	Plate failure (limit plastic strain, 5%)	Plate failure (limit plastic strain, 5%)	Plate failure (limit plastic strain, 5%)	Plate failure (limit plastic strain, 5%)
Strength by IDEA StatiCa when force is applied on welding (kips)	216.6	145.4	74.8	168.0
Failure mode – force applied on welding	Weld failure	Weld failure	Weld failure	Plate failure (limit plastic strain, 5%)

Table 17.10 Shear capacities of the six variation models calculated by IDEA StatiCa.

Strength capacity by IDEA StatiCa	Model 1	Model 2	Model 3	Model 4	Model 5	Model 6
Strength by IDEA StatiCa – force applied on bolts (kips)	127.3	200.1	129.1	130.2	132.3	127.9
Failure mode – force applied on bolts	Plate failure (limit plastic strain, 5%)	Weld failure	Bolt shear failure	Plate failure (limit plastic strain, 5%)	Plate failure (limit plastic strain, 5%)	Plate failure (limit plastic strain, 5%)
Strength by IDEA StatiCa when force is applied on welding (kips)	229.0	226.7	136.0	216.5	213.3	234.1
Failure mode – force applied on welding	Plate failure (limit plastic strain, 5%)	Weld failure	Bolt shear failure	Weld failure	Bolt bearing failure	Bolt bearing failure

ABAQUS using the IDEA StatiCa's viewer platform. Afterward, a simplified model for the bolt and weld were designed and added to the ABAQUS model (see Figure 17.10).

In ABAQUS, the element type of C3D8R (3D stress, 8-node linear brick, reduced integration) was chosen and a total of 29,3294 elements were generated in the model (see Table 17.11 and Figure 17.11 for more details).

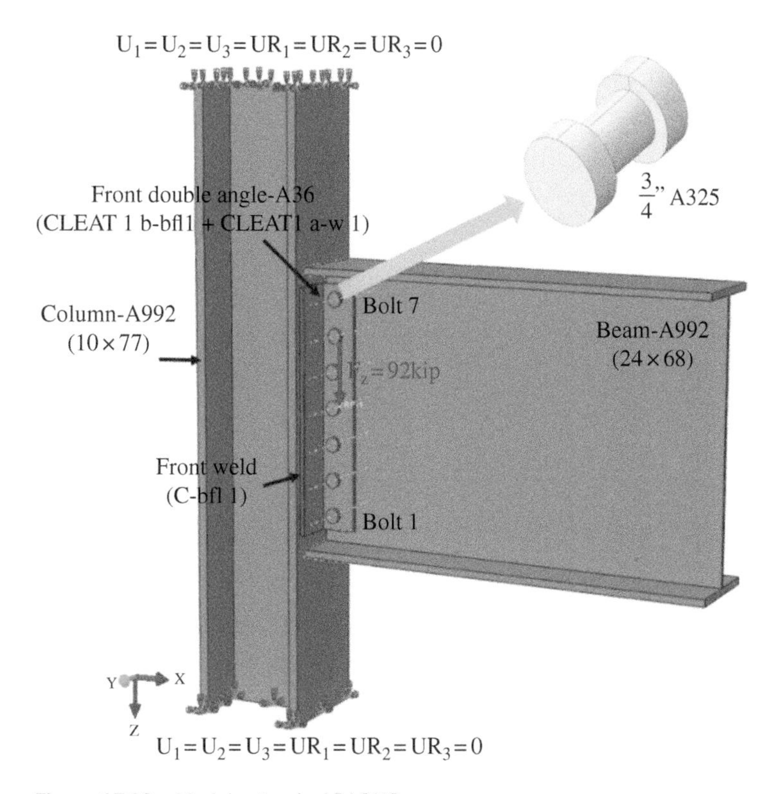

Figure 17.10 Model setup in ABAQUS.

Table 17.11 Number of finite elements in the ABAQUS model.

Item	Number of elements
Column	185,902
Beam	19,430
Bolt	6,304
Weld	1,820
Double angle	40,194

As described in the previous section, different results can be achieved depending on the position of the acting vertical shear force. Therefore, two cases were defined and investigated using the ABAQUS model. In case 1, the vertical shear force of 130.2 kips was applied on the centroid of the bolt group ($x = 7.045$ in., x is the distance from the centerline of the column). In case 2, the vertical shear force of 216 kips was applied on the weld lines ($x = 5.5$ in.). It should be mentioned that in the second case, the beam and column were slightly shorter than the first case to mimic the experimental test. In both cases,

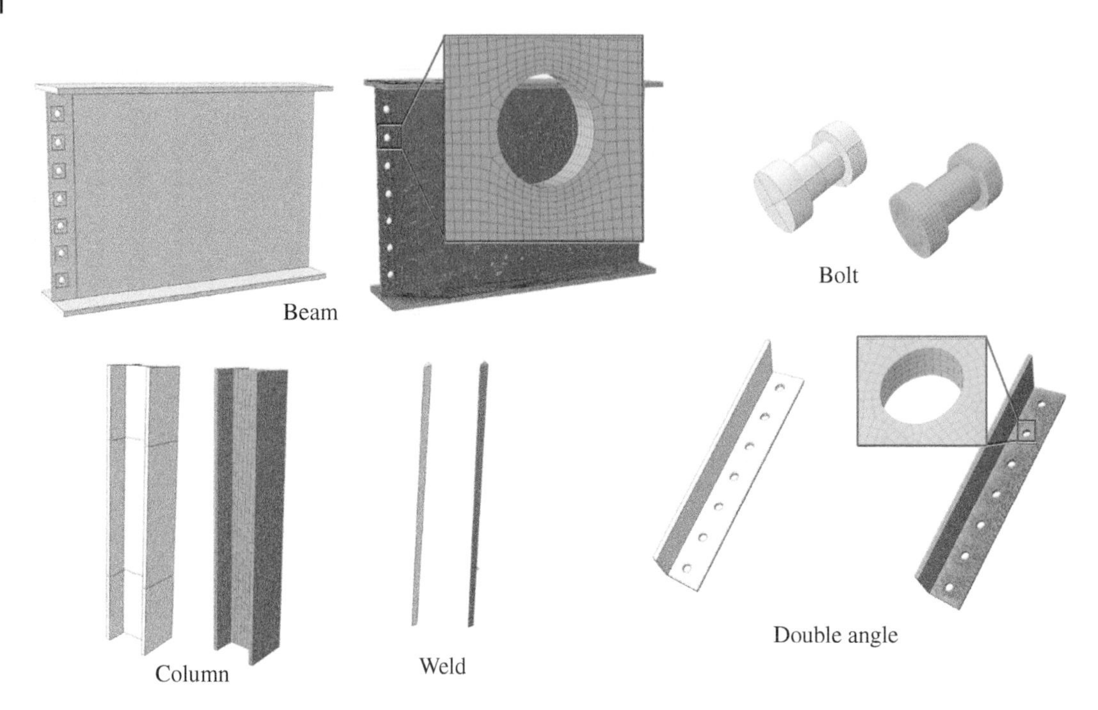

Figure 17.11 ABAQUS model mesh densities.

top and bottom of the column were fixed as a boundary condition (see Figure 17.10). The contact between the parts was defined as surface-to-surface with a finite sliding formulation. Friction was defined with a penalty method, and a Coulomb friction coefficient of $\mu = 0.3$ was used everywhere except between the column face and double angles in which the contact is assumed to be frictionless. Also, a tie constraint was applied between the weld lines and the attaching parts (i.e. column and double angles).

The material behavior was modeled using a bilinear plasticity in ABAQUS. Other parameters, including density, elastic modulus, and Poisson's ratio, were taken from the IDEA StatiCa materials library. The numerical simulations were carried out on two processors (Intel Xenon (R) CPU E5-2698 v4 @ 2.20GHz). Each simulation took approximately 155 minutes. Figure 17.12 depicts the comparison of the predicted von Mises stresses between the IDEA StatiCa and the ABAQUS models for both cases.

17.6 Results Comparison

17.6.1 Comparison of IDEA StatiCa and AISC Design Strength Capacities

Two different weld capacities were calculated for each test specimen following the AISC LRFD design requirements. For the same four test specimens (Table 17.3) and six additional models (Table 17.6), two different weld capacities were calculated from the IDEA StatiCa models by applying the shear force at different locations. In all loading scenarios, it was found that the weakest component of the connections was the welding. The controlling or smallest calculated strengths corresponding to the weld capacities

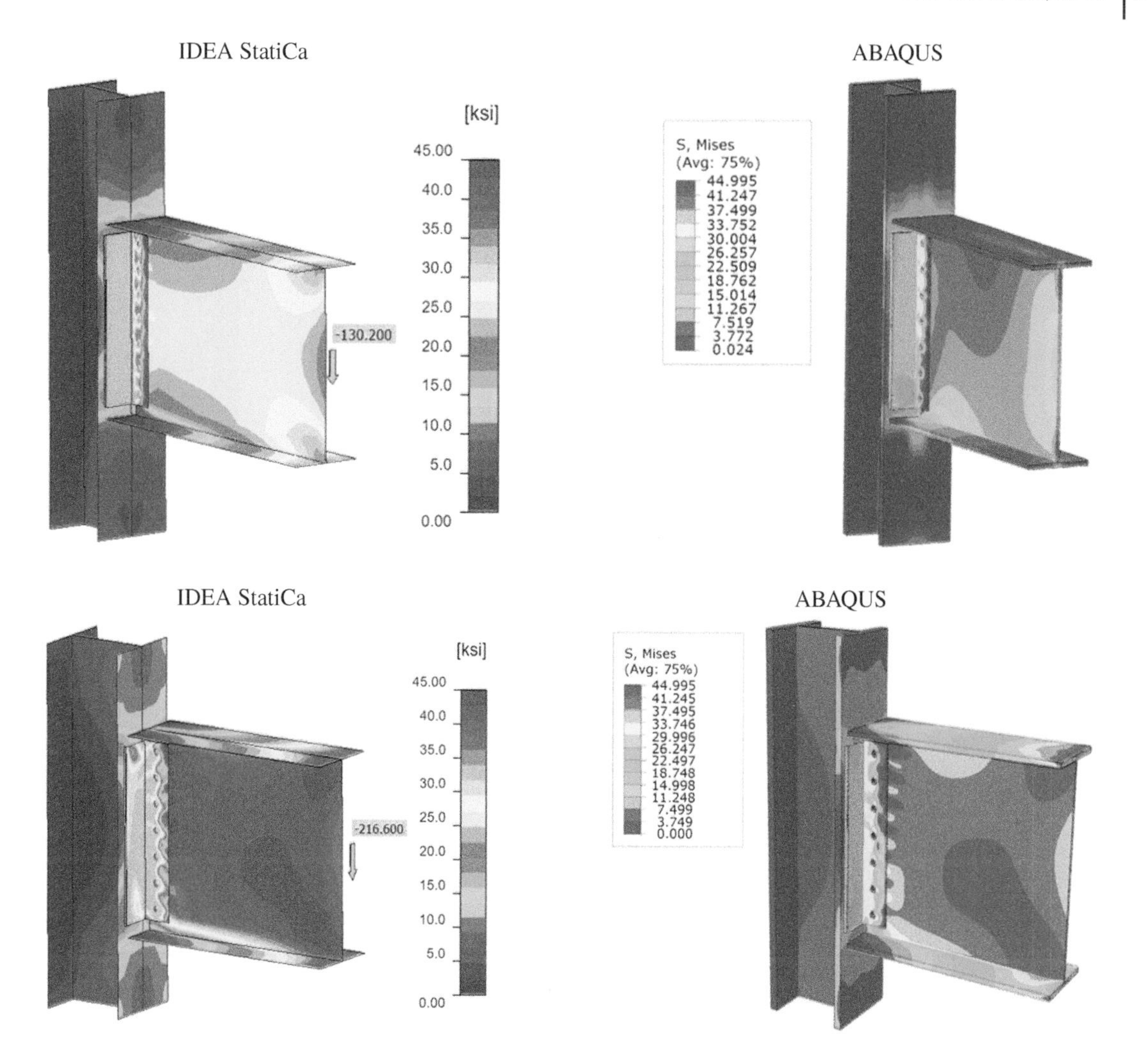

Figure 17.12 Predicted von Mises stress between IDEA StatiCa and ABAQUS models; case 1 (top row): shear load was applied on the centroid of the bolt group; case 2 (bottom row): shear load was applied on the weld lines.

are presented and compared with the ultimate welding shear capacity measured during the experiment in Figure 17.13.

Weld capacities of the test specimens were computed in two different ways by following the AISC LRFD code requirements (AISC *Specification*; AISC, 2016; AISC *Manual*; AISC, 2017). For instance, in Test No. 4, if Equation J2.4 in the AISC *Specification* is followed, the weld design capacity of the specimen is calculated as 228.3 kips. In this solution, no eccentricity is taken into account. To compare this approach with the IDEA StatiCa analysis, the vertical shear force was applied on the welding (parallel to the weld line) and the welding capacity of this specimen was calculated as 216.6 kips, which is very close the one calculated from Equation J2.4 in the AISC *Specification*.

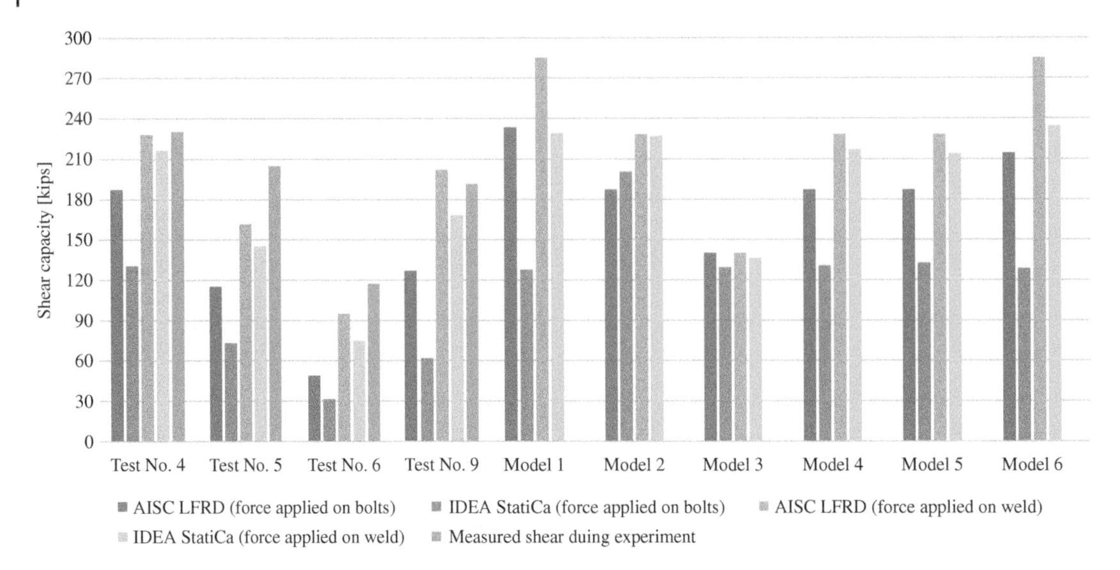

Figure 17.13 Comparison of measured shear capacities with those calculated from AISC LRFD design equations and IDEA StatiCa analysis.

When the shear force is applied on the bolts (external vertical force parallel to the bolt line) in the IDEA StatiCa model, the connection capacity was computed as 130.2 kips. If the welding capacity is calculated by following the LRFD weld strength equation (pages 10–11 of the AISC *Manual* (AISC, 2017)), which considers the eccentricity of the loading on the support side, the welding capacity of the specimen is calculated as 186.8 kips. However, conservatively, this AISC LRFD equation does not account for the eccentricity resulting from the gap between the bolts and the welding. It is believed that this assumption is the reason for the difference between the results calculated from IDEA StatiCa and LRFD strength equation in the AISC *Manual*.

17.6.2 Comparison of IDEA StatiCa and ABAQUS Results

The comparison between the IDEA StatiCa and the ABAQUS results are summarized in Tables 17.12–17.14 for case 1, and Tables 17.15–17.17 for case 2. In these tables:

F_y: Yield strength
σ_{Ed}: Resultant equivalent stress
ε_{pl}: Plastic strain

Check status OK: pass the AISC requirements

F_t: Tension force
V: Result of shear forces in bolt
$\phi R_{n,bearing}$: Bolt bearing resistance
Ut_t: Utilization in tension
Ut_s: Utilization in shear
F_n: Force in weld critical element
ϕR_n: Weld resistance

Table 17.12 Specified yield strengths and calculated stress, strain, and plates check status (case 1).

IDEA StatiCa					ABAQUS				
Item	F_y (*ksi*)	σ_{Ed} (*ksi*)	ε_{pl} (%)	Check status	Item	F_y (*ksi*)	σ_{Ed} (*ksi*)	ε_{pl} (%)	Check status
C-bfl 1	50	25.9	0	OK	Column	50	37.75	0	OK
C-tfl 1	50	4.7	0	OK					
C-w 1	50	9.2	0	OK					
B-bfl 1	50	36.0	0	OK	Beam	50	45.00	0	OK
B-tfl 1	50	36.0	0	OK					
B-w 1	50	45.1	0.2	OK					
CLEAT 1 a-bfl1	36	32.7	1.1	OK	Front double angle	36	32.40	12.7	Not OK!
CLEAT 1 a-w1	36	33.8	4.9	OK					
CLEAT 1 b-bfl1	36	32.7	1.1	OK	Back double angle	36	32.40	12.7	Not OK!
CLEAT 1 b-w1	36	33.8	4.9	OK					

Table 17.13 Calculated tension force, shear force, and bolt bearing resistance when the external load is applied on the bolt line (case 1).

	IDEA StatiCa					ABAQUS				
Item	F_t (*kips*)	V (*kips*)	$\phi R_{n,bearing}$ (*kips*)	Ut_t (%)	Ut_s (%)	F_t (*kips*)	V (*kips*)	$\phi R_{n,bearing}$ (*kips*)	Ut_t (%)	Ut_s (%)
B1	3.10	9.49	16.60	10.4	57.2	3.18	9.64	36.40	9.1	45.9
B2	4.24	9.51	36.28	14.2	53.2	2.47	9.64	36.40	7.1	45.9
B3	4.94	9.46	36.28	16.6	52.9	2.67	9.44	36.40	7.6	45.1
B4	5.63	9.40	36.28	18.9	52.6	5.70	9.53	36.40	16.3	45.4
B5	6.42	9.24	36.28	21.6	51.7	5.01	9.11	36.40	14.3	43.4
B6	7.17	9.10	36.28	24.1	50.9	8.67	8.91	36.40	24.8	42.4
B7	6.88	9.05	36.28	23.1	50.6	10.37	8.92	36.40	29.6	42.5

Table 17.14 Calculated force in weld critical element, weld resistance, and welds check status (case 1).

	IDEA StatiCa				ABAQUS			
Item	F_n (kips)	ϕR_n (kips)	Ut (%)	Check status	F_n (kips)	ϕR_n (kips)	Ut (%)	Check status
C-bfl 1	3.39	4.09	82.9	OK	4.09	4.13	99.1	OK
C-bfl 1	3.39	4.09	82.8	OK	4.09	4.13	99.1	OK

Table 17.15 Specified yield strengths and calculated stress, strain, and plates check status (case 2).

IDEA StatiCa					ABAQUS				
Item	F_y (ksi)	σ_{Ed} (ksi)	ε_{pl} (%)	Check status	Item	F_y (ksi)	σ_{Ed} (ksi)	ε_{pl} (%)	Check status
C-bfl 1	50	23.3	0	OK	Column	50	19.03	0	OK
C-tfl 1	50	4.7	0	OK					
C-w 1	50	9.6	0	OK					
B-bfl 1	50	45.0	0	OK	Beam	50	45.00	3.6	OK
B-tfl 1	50	45.0	0	OK					
B- w 1	50	46.1	3.9	OK					
CLEAT 1 a-bfl1	36	32.8	1.3	OK	Front double angle	36	32.40	1.2	OK
CLEAT 1 a-w1	36	32.4	0.2	OK					
CLEAT 1 b-bfl1	36	32.8	1.3	OK	Back double angle	36	32.40	1.2	OK
CLEAT 1 b-w1	36	32.4	0.2	OK					

Table 17.16 Calculated tension force, shear force, and bolt bearing resistance when the external load is applied on the weld line (case 2).

	IDEA StatiCa					ABAQUS				
Item	F_t (kips)	V (kips)	$\phi R_{n,bearing}$ (kips)	Ut_t (%)	Ut_s (%)	F_t (kips)	V (kips)	$\phi R_{n,bearing}$ (kips)	Ut_t (%)	Ut_s (%)
B7	4.31	16.81	36.28	14.5	94	5.63	16.56	36.40	17.9	92.6
B1	7.31	15.92	36.28	24.5	89.1	7.32	15.76	36.40	24.7	88.1
B2	3.79	16.01	36.28	12.7	89.6	3.51	15.95	36.40	11.9	89.2
B3	2.80	15.67	36.28	9.4	87.6	2.83	15.16	36.40	10.1	84.8
B4	2.60	15.55	36.28	8.7	87	2.75	14.43	36.40	8.8	80.6
B5	2.49	15.64	36.28	8.4	87.5	2.33	14.66	36.40	7.8	82.1
B6	2.64	16.03	36.28	8.9	89.7	2.65	14.97	36.40	8.5	83.7

Table 17.17 Calculated force in weld critical element, weld resistance, and welds check status (case 2).

	IDEA StatiCa				ABAQUS			
Item	F_n (kips)	ϕR_n (kips)	Ut (%)	Check status	F_n (kips)	ϕR_n (kips)	Ut (%)	Check status
C-bfl 1	3.87	3.87	100	Not OK!	4.1	4.12	99.5	OK
C-bfl 1	3.87	3.87	100	Not OK!	4.1	4.12	99.5	OK

17.7 Summary

This chapter aimed to validate the results of Finite Element Analysis (FEA) obtained from the IDEA StatiCa software package for the tested DWA connection. This was achieved by comparing these results with those derived from both US building codes (AISC *Specification*; AISC, 2016; AISC *Manual*; AISC, 2017) and ABAQUS analysis. Overall, IDEA StatiCa provides lower shear strength capacities for DWA connections covered in this study compared to the AISC procedure, meaning that the steel connection design using IDEA StatiCa is more conservative.

Also, there was good agreement between IDEA StatiCa and ABAQUS simulation results. In case 1, in which load was applied on the centroid of the bolt group, more deformation was observed on the double angles in the ABAQUS model. Also, the maximum predicted stress on the beam, column, and weld lines was slightly higher in the ABAQUS model. In addition, a slightly different stress distribution was observed on the beam in the ABAQUS model. While applying the load on the bolt group is not common in traditional finite element software, such a discrepancy could be associated with different contact formulations or element types (i.e. the solid element in ABAQUS versus the shell element in IDEA StatiCa). Also, due to the nature of the tie constraint, larger stresses were obtained on the column in the ABAQUS model. In case 2, in which load was applied on the weld lines, much better agreement was observed between the two models. In both models, it was found that the weakest component of the connections was the weld lines. This is also consistent with the LRFD code design checks (see Section 17.6.1). Stress distributions for each case can be seen in Appendix D of the full report, accessible on the IDEA StatiCa website (IDEA StatiCa, 2021).

References

AISC (2016). *Specification for Structural Steel Buildings*. American Institute of Steel Construction, Chicago.

AISC (2017). *Steel Construction Manual*, 15th Edition. American Institute of Steel Construction, Chicago.

Dassault Systèmes Simulia Corporation (2020). *ABAQUS* (Version 2020) [Computer program]. Dassault Systèmes SE, Providence, RI.

IDEA StatiCa (2021). *Verification of IDEA StatiCa Calculations for Steel Connection Design (AISC)*. https://www.ideastatica.com/support-center/verification-of-idea-statica-calculations-for-steel-connection-design-aisc (accessed October 31, 2023).

McMullin, K.M. and Astaneh-Asl, A. (1988). *Analytical and Experimental Studies of Double-Angle Framing Connections*. Berkeley, CA, University of California.

18

Top- and Seat-Angle with Double Web-Angle (TSADWA) Connections

18.1 Description

In this chapter, the design strength capacities of ten top-and seat-angle with double web-angle (TSADWA) connection specimens were calculated following the requirements of the AISC *Specification* (AISC, 2016) and the AISC *Manual* (AISC, 2017). These specimens were selected from the experimental study performed by Azizinamini et al. (1985) in the Department of Civil Engineering at the University of South Carolina. All specimens were analyzed using IDEA StatiCa, while one of them was analyzed using ABAQUS (Dassault, 2020). Then, the results were compared.

18.2 Experimental Study on TSADWA Connections

Several TSADWA connection specimens comprised of double angles and top- and seat-beam flanges were subjected to static and cyclic loadings to investigate their moment-rotation behavior. A pair of specimens was tested at the same time, as shown in Figure 18.1. One side of the beam sections was bolted to the column and the other side was supported by roller-type seats. The vertical movement of the stub column was allowed by roller guides attached to the top and bottom of the column. The hydraulic actuator was used to apply the load on the column and the connection transferred the load to the beams.

The moment values were calculated from the actuator load cell readings. To obtain the corresponding rotations, the displacements measured by the linear variable differential transducers (LVDT) were converted to relative rotations between the end of the beam and the flange of the column (Figure 18.2).

In this study, ten specimens subjected to static loading were selected to be analyzed. The properties of these ten TSADWA connection specimens are presented in Table 18.1. All connections were bolted to beam and column. The first four beam specimens were framed to a W14 × 38 beam section while W8 × 21 was used for the other six beam sections. The column section of W12 × 96 was used for all ten specimens. The members and connections were made of ASTM A36 steel while the fasteners were ASTM A325 with ¾ in. diameter bolts. The material properties of the members are provided in Table 18.2.

The bolt spacing was 3.0 in. while the edge distance of bolts was 1.25 in. from the top and bottom of the double angle sections. The longitudinal bolt spacing and the edge distance of top and seat angles on beam side were 2.5 in. and 1.25 in., respectively, while those varied on the angles attached to the column flange. Similarly, the transfer bolt spacing, and edge distance of top and seat angles varied, as provided

Steel Connection Design by Inelastic Analysis: Verification Examples per AISC Specification, First Edition.
Mark D. Denavit, Ali Nassiri, Mustafa Mahamid, Martin Vild, Halil Sezen, and František Wald.

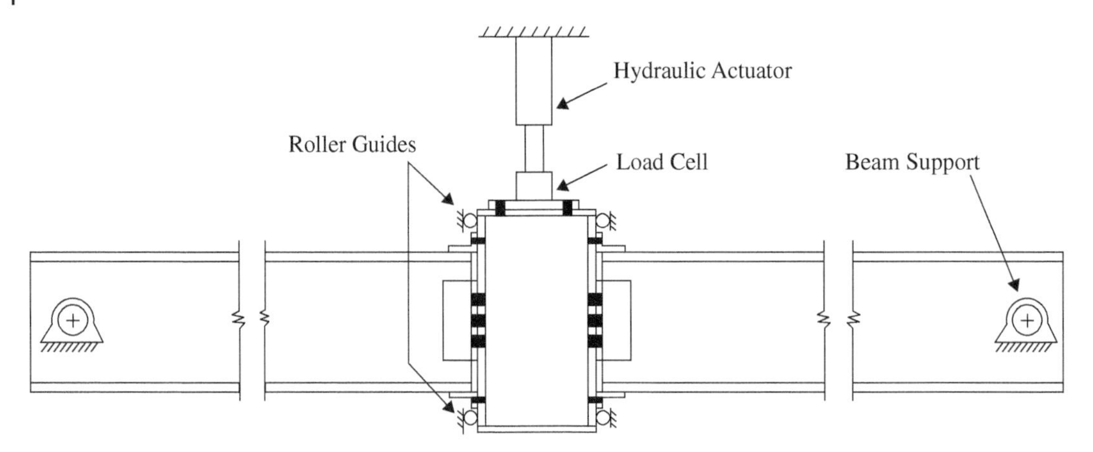

Figure 18.1 Test setup used by Azizinamini et al. (1985).

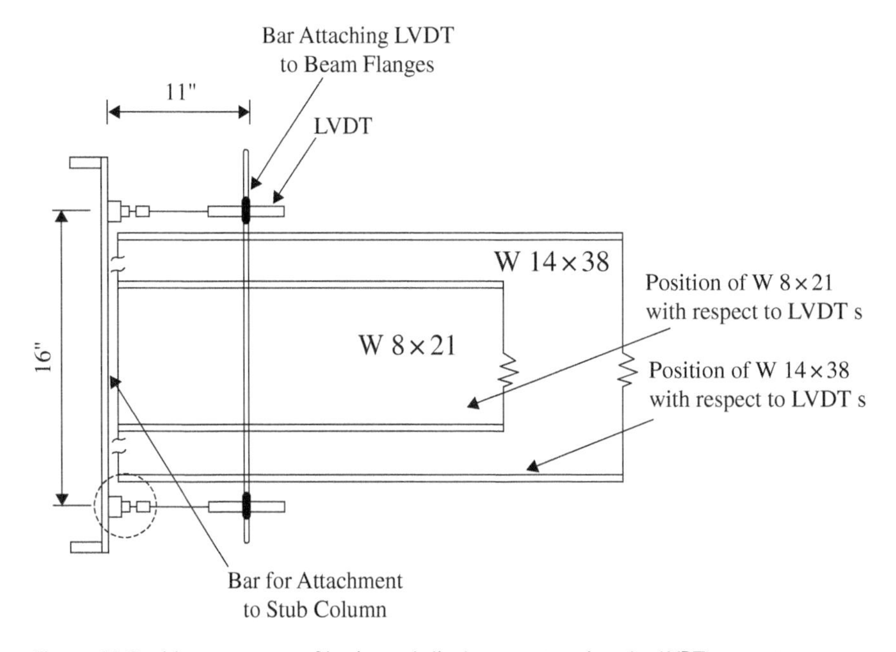

Figure 18.2 Measurements of horizontal displacements using the LVDT apparatus.

in Table 18.1. The geometric details of the connections are shown in Figures 18.3 and 18.4. The summary of the test results measured during static loading are presented in Table 18.3 and Figures 18.5–18.9. Note that in these figures, the blue line shows the resistance determined by the AISC traditional calculation (see Section 18.3) and the orange line depicts the resistance determined by IDEA StatiCa (see Section 18.4).

Table 18.1 Properties of the TSADWA connection specimens.

		Top and bottom flange angles				Web angle		
Specimen number	**Beam section**	**Angle**	**Length (in.)**	**Gage in leg on column flange (in.)**	**Bolt spacing in leg on column flange (in.)**	**Angle**	**Length (in.)**	**Number of the bolt**
14S1	W14×38	L6×4×3/8	8	2.5	5.5	2L4×3.5×1/4	8.5	3
14S2	W14×38	L6×4×1/2	8	2.5	5.5	2L4×3.5×1/4	8.5	3
14S3	W14×38	L6×4×3/8	8	2.5	5.5	2L4×3.5×1/4	5.5	2
14S4	W14×38	L6×4×3/8	8	2.5	5.5	2L4×3.5×3/8	8.5	3
8S1	W8×21	L6×3.5×5/16	6	2	3.5	2L4×3.5×1/4	5.5	2
8S2	W8×21	L6×3.5×3/8	6	2	3.5	2L4×3.5×1/4	5.5	2
8S3	W8×21	L6×3.5×5/16	8	2	3.5	2L4×3.5×1/4	5.5	2
8S4	W8×21	L6×6×3/8	6	4.5	3.5	2L4×3.5×1/4	5.5	2
8S5	W8×21	L6×4×3/8	8	2.5	5.5	2L4×3.5×1/4	5.5	2
8S6	W8×21	L6×4×5/16	6	2.5	3.5	2L4×3.5×1/4	5.5	2

Table 18.2 Material properties of the tested TSADWA connection specimens.

Members	**Strength**	**ksi**
Columns, beams, and angles (A36)	Yield strength, F_y	36
	Ultimate strength, F_u	58
Bolts A325 (threads excluded)	Nominal tensile strength, F_{nt}	90
	Nominal shear strength, F_{nv}	68

18.3 Code Design Calculations and Comparisons

The design strength capacities (ϕR_n) of the connections were calculated following the requirements of the AISC *Specification* (AISC, 2016) and the AISC *Manual* (AISC, 2017). The nominal strength, R_n, and the corresponding resistance factor, ϕ, for each connection design LRFD limit state are provided in Chapter J of the AISC *Specification*. It is assumed that the top- and seat-angles provide moment resistance, and the double web-angle is used for shear resistance for the connection conservatively.

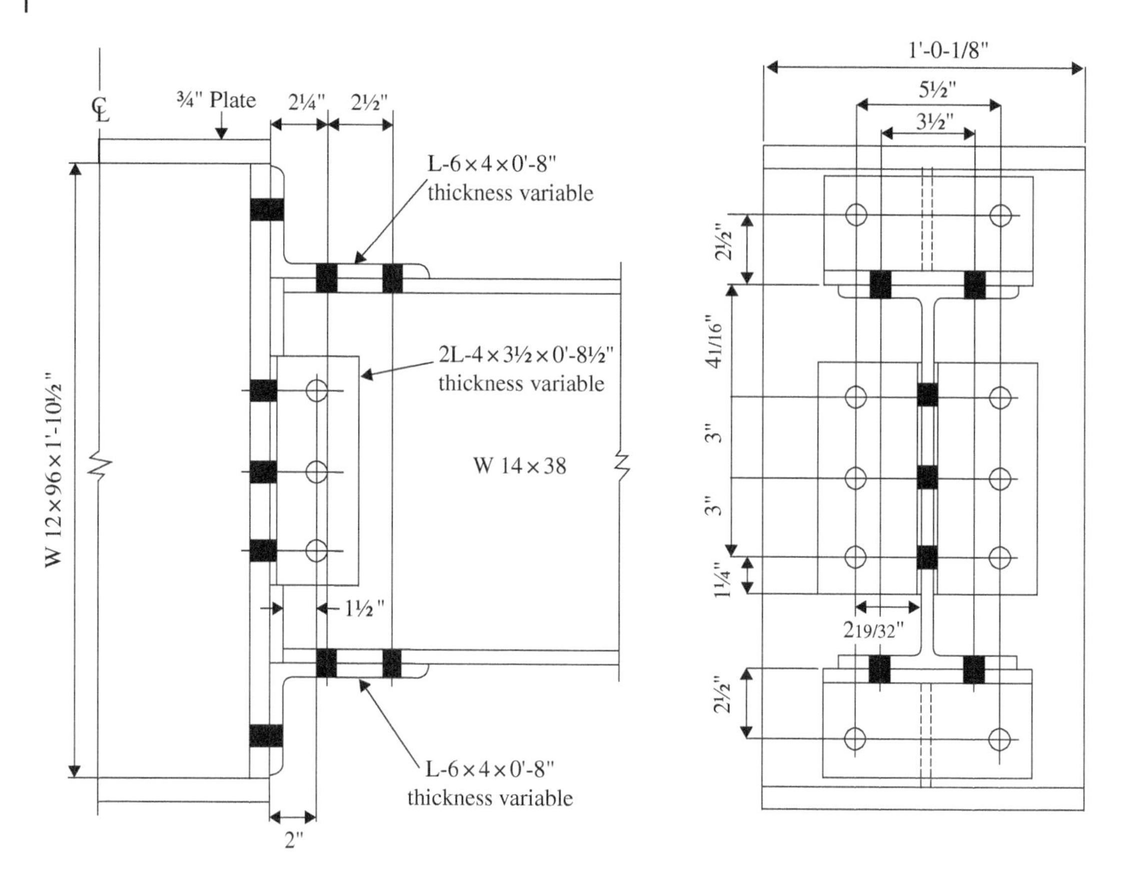

Figure 18.3 Details of connection for W14x38 beam.

18.3.1 Design Strength Capacity of Double Web-Angles

The following 14 design checks were performed according to the LRFD design equations included in the AISC *Specification* (AISC, 2016) or the AISC *Manual* (AISC, 2017) for the design strength capacity of a double web-angle.

1. Angle (beam side)

a. Bolts shear	(Eq. J3-1, AISC *Specification*)
b. Bolt bearing and tearout	(Eq. J3-6, AISC *Specification*)
c. Shear yielding	(Eq. J4-3, AISC *Specification*)
d. Shear rupture	(Eq. J4-4, AISC *Specification*)
e. Block shear	(Eq. J4-5, AISC *Specification*)

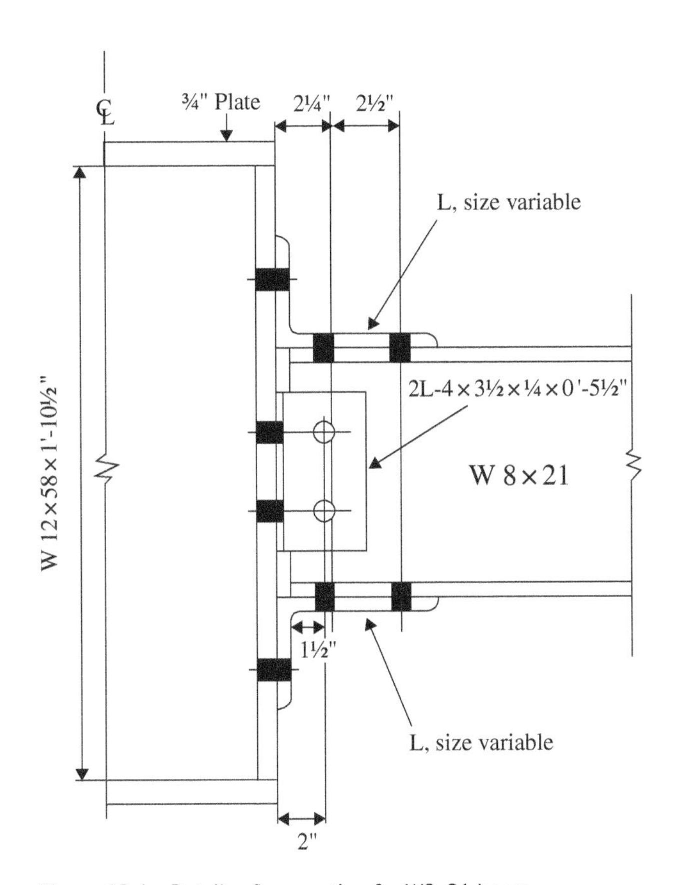

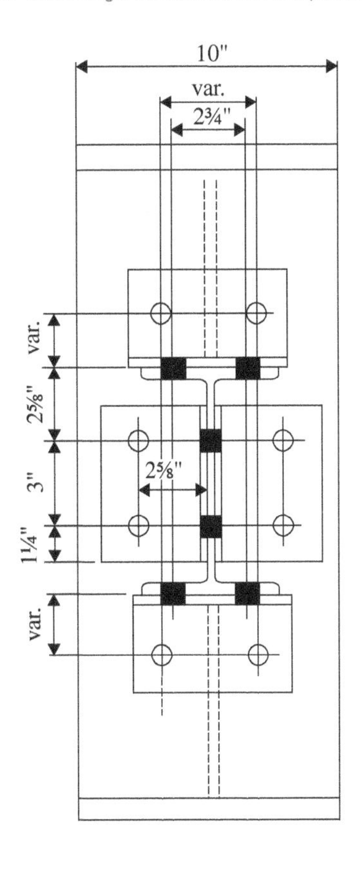

Figure 18.4 Details of connection for W8x21 beam.

2. Angle (column side)

a. Bolts shear	(Eq. J3-1, AISC *Specification*)
b. Bolt bearing and tearout	(Eq. J3-6, AISC *Specification*)
c. Shear yielding	(Eq. J4-3, AISC *Specification*)
d. Shear rupture	(Eq. J4-4, AISC *Specification*)
e. Block shear	(Eq. J4-5, AISC *Specification*)
f. Resulting tension capacity due to prying action	(Part 9, AISC *Manual*)

3. Beam

a. Bolt bearing and tearout	(Eq. J3-6, AISC *Specification*)
b. Shear yielding	(Eq. J4-3, AISC *Specification*)

Table 18.3 Summary of test results.

Specimen number	Initial slope of moment-rotation curve (k-in./radian)	Slope of secant line to moment-rotation curve at 4.0×10^{-3} radians (k-in./radian)	Moment at 4.0×10^{-3} radians (k-in.)	Slope of moment-rotation curve at 24×10^{-3} radians (k-in./radian)	Moment at 24×10^{-3} radians (k-in.)	Remarks
14S1	195.0×10^3	108.7×20^3	435	5.8×10^3	688	
14S2	295.0	151.8	607	12.6	(947)	Major slip at 12×10^{-3} and 20×10^{-3} radians
14S3	115.9	88.8	355	7.2	652	
14S4	221.9	124.0	496	8.3	822	
8S1	66.7	44.3	177	4.1	329	
8S2	123.4	69.0	276	1.5	(384)	Major slip at 16×10^{-3} radians
8S3	104.7	64.3	257	4.0	422	
8S4	15.3	14.4	57.5	2.2	165	
8S5	76.7	47.9	191.5	2.7	337	
8S6	39.5	30.0	120	3.2	244	

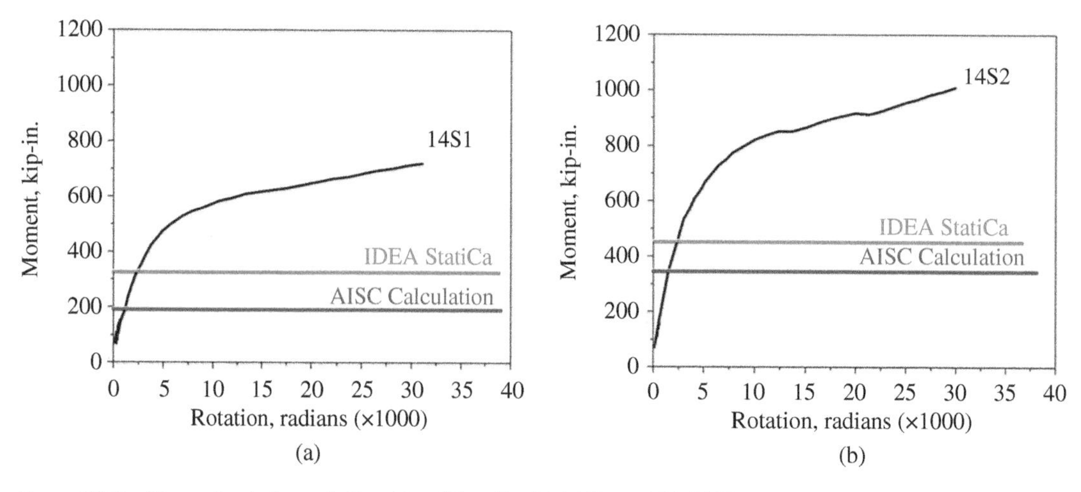

Figure 18.5 Moment-rotation relationship of Test No. (a) 14S1; and (b) 14S2.

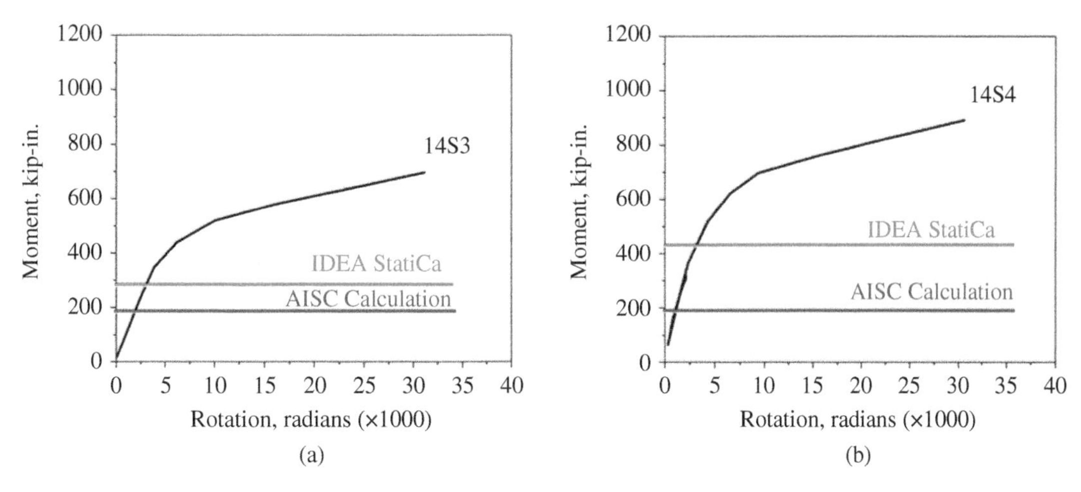

Figure 18.6 Moment-rotation relationship of Test No. (a) 14S3; and (b) 14S4.

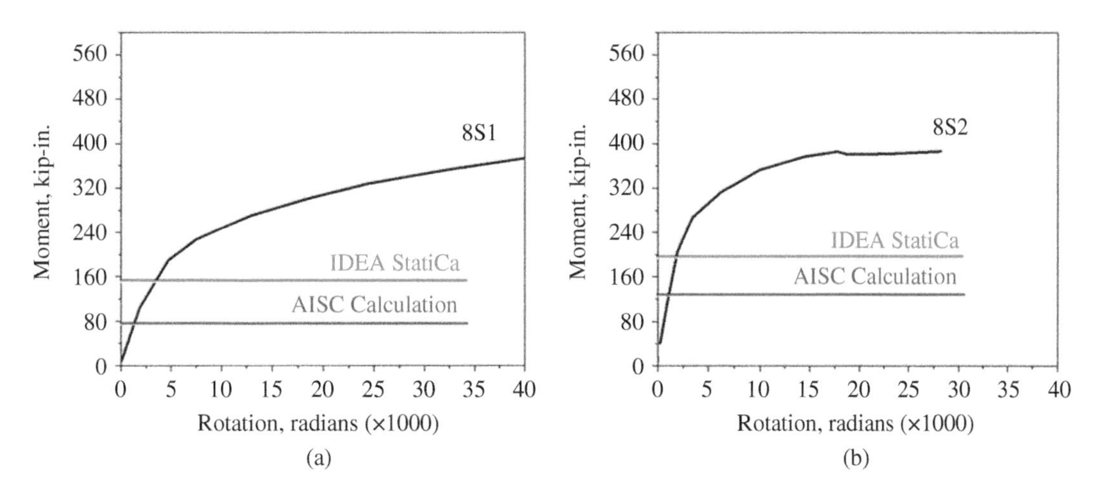

Figure 18.7 Moment-rotation relationship of Test No. (a) 8S1; and (b) 8S2.

4. Column

a. Bolt bearing and tearout	(Eq. J3-6, AISC *Specification*)

The design strength capacities (ϕR_n) of double web-angles for the ten specimens calculated by following the AISC LRFD code requirements are provided in Table 18.4.

From the calculated design capacities for the ten test specimens, the lowest shear capacities are shown in bold and italic in Table 18.4. According to the results, the design capacity of two double web-angles (in specimens 14S1 and 14S2) were controlled by the block shear of the bolts on the angle attached to the beam while the bearing and tearout of the bolts on the beam controlled the shear design capacities of the

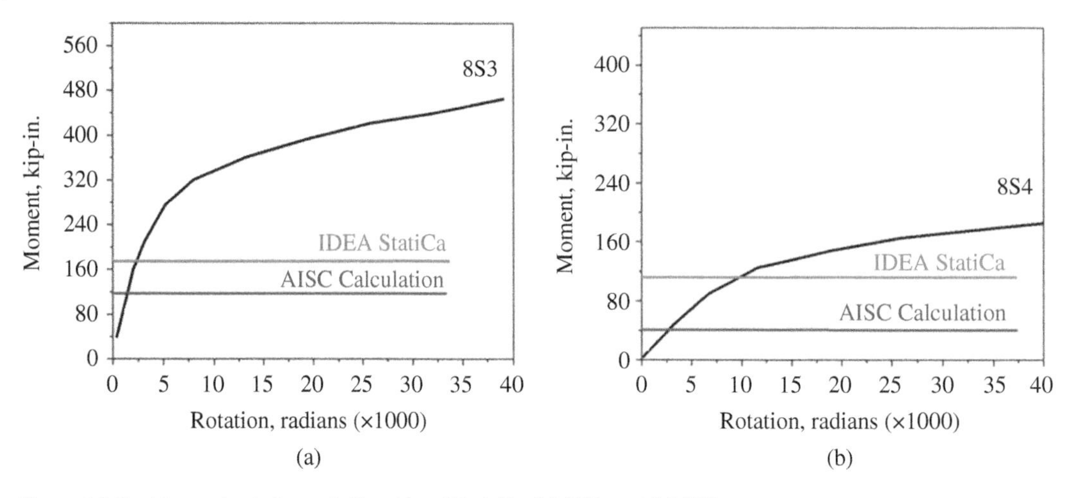

Figure 18.8 Moment-rotation relationship of Test No. (a) 8S3; and (b) 8S4.

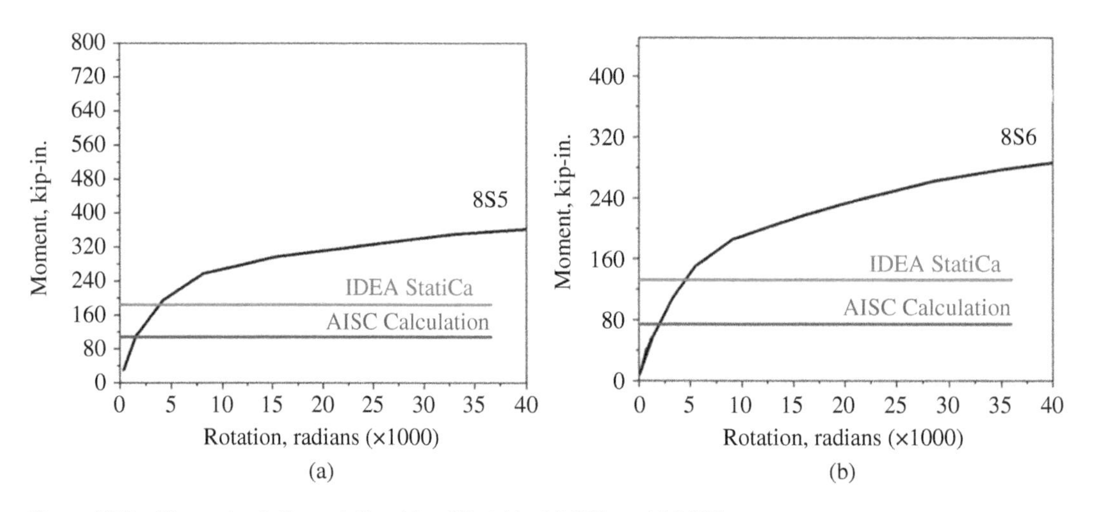

Figure 18.9 Moment-rotation relationship of Test No. (a) 8S5; and (b) 8S6.

other eight specimens. The detailed design strength calculations for Test 14S1 are provided in Appendix E of the full report, published on the IDEA StatiCa website (IDEA StatiCa, 2021).

18.3.2 Design Strength Capacity of the Top- and Bottom Seat-Angles

The following 16 design checks were performed according to the LRFD equations included in the AISC *Specification* (AISC, 2016) or the AISC *Manual* (AISC, 2017) for the design strength capacity of the top- and seat-angle.

Table 18.4 LRFD design strength capacities (ϕR_n) of double web-angles in ten specimens.

Design checks (LRFD)	Design strength capacity (kips)									
				Specimen number						
	14S1	14S2	14S3	14S4	8S1	8S2	8S3	8S4	8S5	8S6
Angles (beam side)										
Bolt shear	135.24	135.24	89.58	134.4	89.58	89.58	89.58	89.58	89.58	89.58
Bolt bearing and tearout	117.45	117.45	67.70	160.30	67.70	67.70	67.70	67.70	67.70	67.70
Shear yielding	91.80	91.80	64.80	145.80	64.80	64.80	64.80	64.80	64.80	64.80
Shear rupture	76.73	76.73	55.46	124.8	55.46	55.46	55.46	55.46	55.46	55.46
Block shear	71.53	71.53	59.56	125.8	59.56	59.56	59.56	59.56	59.56	59.56
Angles (column side)										
Bolt shear	135.24	135.24	89.58	134.4	89.58	89.58	89.58	89.58	89.58	89.58
Bolt bearing and tearout	117.45	117.45	61.17	150.5	61.17	61.17	61.17	61.17	61.17	61.17
Shear yielding	91.80	91.80	64.80	145.80	64.80	64.80	64.80	64.80	64.80	64.80
Shear rupture	76.73	76.73	55.46	124.8	55.46	55.46	55.46	55.46	55.46	55.46
Block shear	71.53	71.53	52.10	114.6	52.10	52.10	52.10	52.10	52.10	52.10
Beam										
Bolt bearing and tearout	72.81	72.81	48.55	72.82	39.15	39.15	39.15	39.15	39.15	39.15
Shear yielding	94.41	94.41	94.41	94.41	44.71	44.71	44.71	44.71	44.71	44.71
Column										
Bolt bearing	211.41	211.41	281.90	422.80	281.90	281.90	281.90	281.90	281.9	281.9

1. Top- and seat-angle (beam side)

a. Tension yielding	(Eq. J4-1, AISC *Specification*)
b. Tension rupture	(Eq. J4-2, AISC *Specification*)
c. Compression	(Sec. J4.4, AISC *Specification*)
d. Bolts shear	(Eq. J3-1, AISC *Specification*)
e. Bolt bearing and tearout	(Eq. J3-6, AISC *Specification*)
f. Block shear	(Eq. J4-5, AISC *Specification*)

2. Top- and seat-angle (column side)

a. Shear yielding	(Eq. J4-3, AISC *Specification*)
b. Shear rupture	(Eq. J4-4, AISC *Specification*)
c. Tension capacity due to prying action	(Pages 9–10, AISC *Manual*)

3. Beam

a. Bolt bearing and tearout	(Eq. J3-6, AISC *Specification*)
b. Flexural strength	(Sec. F13.1, AISC *Specification*)
c. Block shear	(Eq. J4-5, AISC *Specification*)

4. Column

a. Panel web shear	(Eq. J10-9, AISC *Specification*)
b. Flange local bending	(Eq. J10-1, AISC *Specification*)
c. Web local yielding	(Eq. J10-2, AISC *Specification*)
d. Web local crippling	(Eq. J10-4, AISC *Specification*)

The design strength capacities (ϕR_n) of the top- and bottom seat-angles in ten specimens are calculated and provided in Table 18.5. The values highlighted in bold italic in Table 18.5 show the lowest shear capacities in the ten test specimens considered. According to the results, the design capacities of all top- and seat-angles were controlled by tension capacity due to prying action on the angle side bolted to the column. The design capacity of all top- and seat-angles was controlled by tension capacity due to prying action.

The moment capacities of the specimens were calculated by multiplying the tension capacity (which is identical at the top and bottom angles) by the moment arm, which is set equal to the distance from the center of compression to the bolt-row in tension (gage in leg on column flange + beam depth + a half of the thickness of seated-angle), as provided in Table 18.6. This definition may provide slightly larger moment arm because the compressive force is likely to be applied above the bolt-row.

18.3.3 ASD Design Strength Capacities of Test No. 14S1

According to the allowable strength design (ASD), the allowable strength (R_n/Ω) is calculated by dividing the nominal strength, R_n by the safety factor, Ω. The strength of the connection specimen Test No. 14S1 is calculated by following the AISC ASD code requirements. The properties of this test specimen were given in Table 18.1. The calculated ASD strength capacities (R_n/Ω) of the specimen, including double web-angle, and top- and seat-angles, are provided in Tables 18.7 and 18.8, respectively. The calculated lowest strength of the double web-angles in Test No. 14S1 is 48.54 kips due to the bolt bearing and tearout on the beam. The detailed design strength calculations for Test No. 14S1 are provided in Appendix F of the full report, published on the IDEA StatiCa website (IDEA StatiCa, 2021).

The calculated lowest strength of the top- and seat-angle for Test No. 14S1 is 9.15 kips because of the tension capacity due to prying action on the angle bolted to column flange. The moment capacity of the connection (153.63 kips-in.) can be calculated by multiplying by the tension capacity of the angle (9.15 kips) by the moment arm (16.79 in.).

18.4 IDEA StatiCa Analysis

18.4.1 Moment Capacity Analysis Using IDEA StatiCa

The ten test specimens were modeled in IDEA StatiCa with and without web-angles, and analyzed under a shear force applied a certain distance away from the column. The distance was selected to be equal to

Table 18.5 The design strength capacities (ϕR_n) of top- and seat-angle for ten specimens.

Design checks (LRFD)	Design strength capacity (kips)									
	Specimen number									
	14S1	14S2	14S3	14S4	8S1	8S2	8S3	8S4	8S5	8S6
Top- and seat-angles (beam side)										
Tension yielding	97.20	129.60	97.20	97.20	60.75	72.90	81.00	72.90	97.20	60.75
Tension rupture	101.78	135.94	101.78	101.78	57.77	69.33	84.96	69.33	101.78	57.77
Compression	93.13	129.60	93.13	93.13	57.12	69.84	76.16	69.84	93.13	57.12
Bolts shear	90.16	89.58	90.16	90.16	90.16	90.16	90.16	90.16	90.16	90.16
Bolt bearing and tearout	99.03	122.34	99.03	99.03	76.46	79.52	76.46	99.03	99.03	99.03
Block shear	67.25	117.84	67.25	67.25	63.46	64.43	63.46	76.15	76.15	63.46
Top- and seat-angles (column side)										
Shear yielding	64.80	86.40	64.80	64.80	40.50	48.60	54.00	48.60	64.80	40.50
Shear rupture	61.07	81.56	61.07	61.07	34.66	41.60	50.98	41.60	61.07	34.66
Tension capacity due to prying action	***13.73***	***25.01***	***13.73***	***13.73***	***9.00***	***13.25***	***12.47***	***4.84***	***13.62***	***6.72***
Beam										
Bolt bearing and tearout	120.97	120.97	120.97	120.97	118.76	112.23	118.76	118.76	118.76	118.76
Flexural strength	166.05	166.05	166.05	166.05	55.08	55.08	55.08	55.08	55.08	55.08
Block shear	124.46	124.57	124.46	124.46	83.70	83.70	83.70	83.70	83.70	83.70
Column										
Panel web shear	135.79	135.79	135.79	135.79	85.38	85.38	85.38	85.38	85.38	85.38
Flange local bending	164.03	164.03	164.03	164.03	82.94	82.94	82.94	82.94	82.94	82.94
Web local yielding	163.35	168.30	163.35	163.35	88.45	88.45	88.45	88.45	88.45	88.45
Web local crippling	257.31	264.00	257.31	257.31	112.81	112.81	112.81	112.81	112.81	112.81

the one between the column centerline and the beam support. The beam support is assumed to be at 120 in. away from the column centerline for the first four specimens while it was 72 in. for the other six specimens (this beam support is on the right side of the beam while the left side of the beam is bolted to the column as shown in Figure 18.1). The shear force was applied incrementally until the connection models reached their capacities in IDEA StatiCa. All specimens fail because the top-angles attached to the column exceed the plastic strain limit which is defined as 5% by the software. The calculated moment capacities of the connection specimens with and without web angles are shown in Tables 18.9 and 18.10, respectively.

The screenshots from IDEA StatiCa showing the failure modes, and deformed shapes (deformation scale 10) of finite element models are shown in Appendix G of the full report, published on the IDEA StatiCa website (IDEA StatiCa, 2021).

Table 18.6 Design moment calculations for the ten TSADWA connection specimens.

	Specimen number									
	14S1	14S2	14S3	14S4	8S1	8S2	8S3	8S4	8S5	8S6
Tension capacity due to prying action (kips)	13.73	25.01	13.73	13.73	9.00	13.25	12.47	4.84	13.62	6.72
Beam depth, d (in.)	14.10	14.10	14.10	14.10	8.28	8.28	8.28	8.28	8.28	8.28
Gage in leg on column flange, g (in.)	2.5	2.5	2.5	2.5	2.0	2.0	2.0	4.5	2.5	2.5
Flange angle thickness, t (in.)	3/8	1/2	3/8	3/8	5/16	3/8	5/16	3/8	3/8	5/16
Moment arm, z $(d + g + t/2)$ (in.)	16.79	16.85	16.79	16.79	10.44	10.47	10.44	12.97	10.97	10.94
Moment capacity, $M = T{\cdot}d$ (kips-in.)	230.49	421.42	230.49	230.49	93.93	138.69	130.14	62.76	149.38	73.49

Table 18.7 ASD strength capacity (ϕR_n) of double web-angle for Test No. 14S1.

Design checks (ASD)	Design strength capacity (kips)
Angle (beam side)	
Bolt shear	90.18
Bolt bearing and tearout	78.30
Shear yielding	61.20
Shear rupture	51.16
Block shear	55.87
Angle (column side)	
Bolt shear	90.18
Bolt bearing and tearout	78.30
Shear yielding	61.20
Shear rupture	51.16
Block shear	55.87
Beam	
Bolt bearing and tearout	48.54
Shear yielding	62.93
Column	
Bolt bearing	140.94

Table 18.8 Design strength capacity (ϕR_n) of top- and seat-angles in specimen Test No. 14S1.

Design checks (ASD)	Design strength capacity (kips)
Top and seat angle (beam side)	
Tension yielding	64.67
Tension rupture	67.85
Compression	61.96
Bolts shear	60.12
Bolt bearing and tearout	78.30
Block shear	48.04
Top- and seat-angle (column side)	
Shear yielding	43.20
Shear rupture	40.72
Tension capacity due to prying action	*9.15*
Beam	
Bolt bearing and tearout	80.66
Flexural strength	110.48
Block shear	82.98
Column	
Web panel zone shear	90.35
Flange local bending	109.13
Web local yielding	108.90
Web local crippling	181.54

18.4.2 Moment-Rotation Analysis

Moment-rotation analysis for Test No. 14S1 was performed using IDEA StatiCa. To generate the test condition, the mechanical properties of A36 steel provided in the test report were used (see Azizinamini et al., 1985). The mean values of the yielding and ultimate strength of the material were reported in Azizinamini et al. (1985) as 40.65 ksi and 68.43 ksi, respectively. These are the properties used for the materials used in the IDEA StatiCa models. The resistance factors were set to be equal 1.0 and moment-rotation analysis was performed by selecting stiffness analysis option (Figure 18.10).

Table 18.9 Moment capacities of the specimens with web angles calculated using IDEA StatiCa.

Specimen number	Shear force (kips)	Distance (in.)	Moment (kips-in.)
14S1	2.66	120	319.20
14S2	3.75	120	450.00
14S3	2.33	120	279.60
14S4	3.52	120	422.40
8S1	2.13	72	153.36
8S2	2.65	72	190.80
8S3	2.42	72	174.24
8S4	1.54	72	110.88
8S5	2.55	72	183.60
8S6	1.79	72	128.88

Table 18.10 Moment capacities of the specimens without web angles calculated using IDEA StatiCa.

Specimen number	Shear force (kips)	Distance (in.)	Moment (kips-in.)
14S1	1.77	120	212.40
14S2	2.88	120	345.60
14S3	1.76	120	211.20
14S4	1.76	120	211.20
8S1	1.5	72	108.00
8S2	2.03	72	146.16
8S3	1.79	72	128.88
8S4	0.91	72	65.52
8S5	1.92	72	138.24
8S6	1.16	72	83.52

18.5 ABAQUS Analysis

In this section, the output results from IDEA StatiCa were compared to those from ABAQUS software package (Dassault, 2020). In this study, Test No. 14S1 specimen, as described in Table 18.1, was chosen as a base model. Numerical simulations with almost identical conditions (i.e. in terms of material properties, boundary conditions, and loading) were carried out using both IDEA StatiCa and ABAQUS. The model

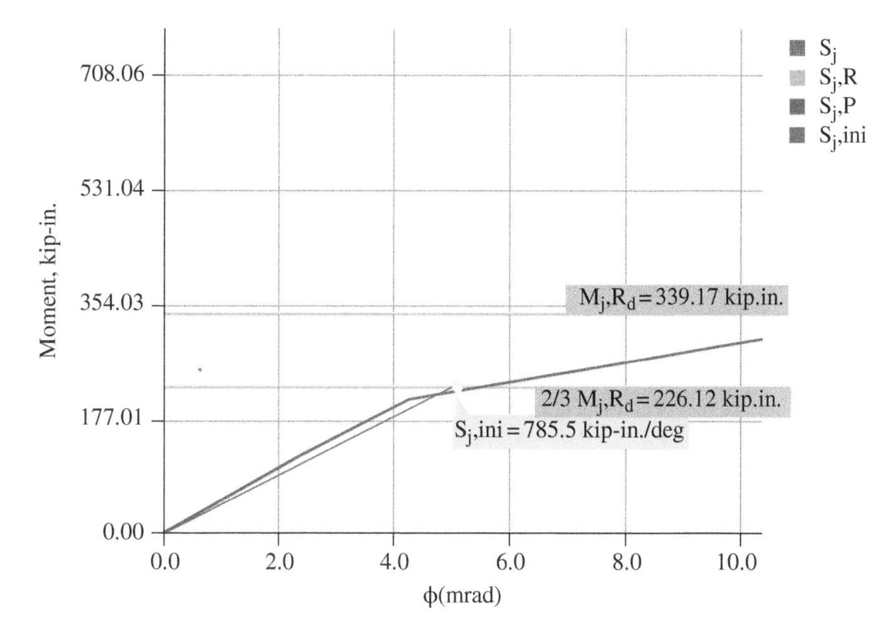

Figure 18.10 Moment-rotation relationship for Test No. 14S1 computed by IDEA StatiCa.

was initially designed in IDEA StatiCa and then the assembly (including beam, column, web-angles, and top- and seat-angles) was imported to ABAQUS using the IDEA StatiCa's viewer platform. Afterward, a simplified model for the bolt was designed and added to the ABAQUS model (see Figure 18.11).

In ABAQUS, the element type of C3D8R (3D stress, 8-node linear brick, reduced integration) was chosen, and a total of 56,2377 elements were generated in the model. More details are provided in Table 18.11 and Figure 18.12.

In the ABAQUS model, the vertical force of 2.66 kips was applied on a reference point (or node) that was defined 10 ft away from the centerline of the column (i.e. x = 10 ft). Then, the coupling constraint (i.e. structural distributing) was defined to connect this reference point to the end section of the beam. The top and bottom of the column were fixed as a boundary condition (see Figure 18.11). The contact between all parts, including column to all angles, was defined as surface-to-surface with finite sliding

Table 18.11 Number of elements in the ABAQUS model.

Item	Number of elements
Column	69,167
Beam	167,574
L-6 × 4 × 3/8	60,550
L-4 × 31/2 × 1/4	33,619
Bolt	6,538

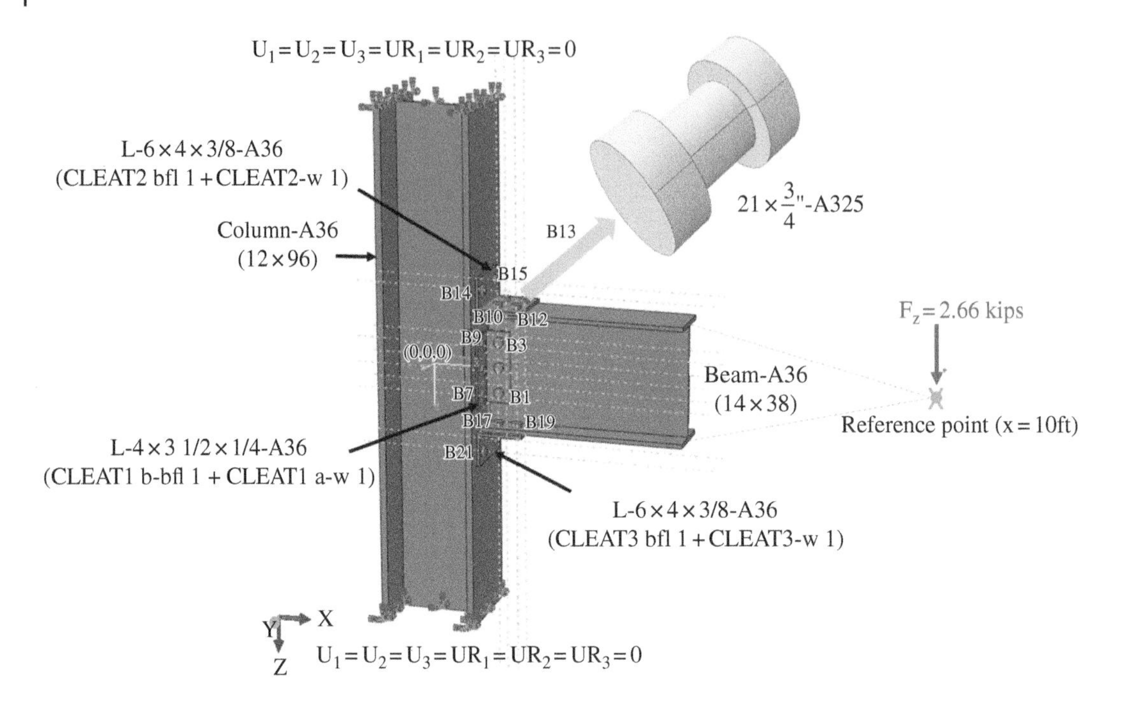

Figure 18.11 TSADWA connection model setup in ABAQUS.

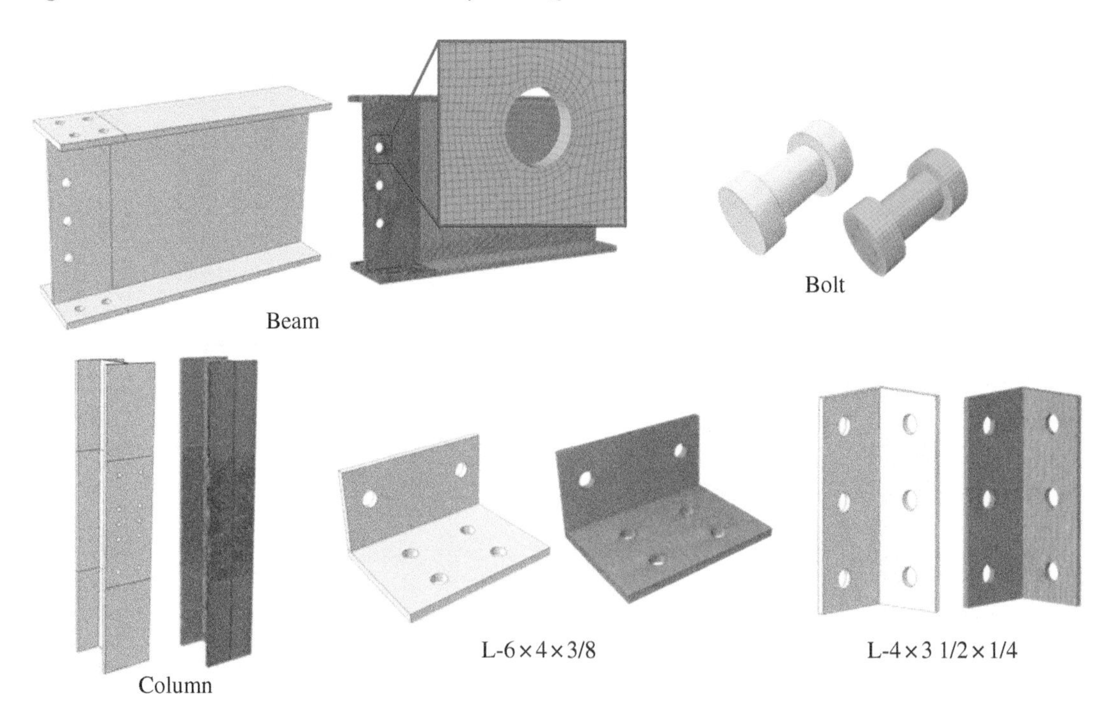

Figure 18.12 ABAQUS model mesh densities.

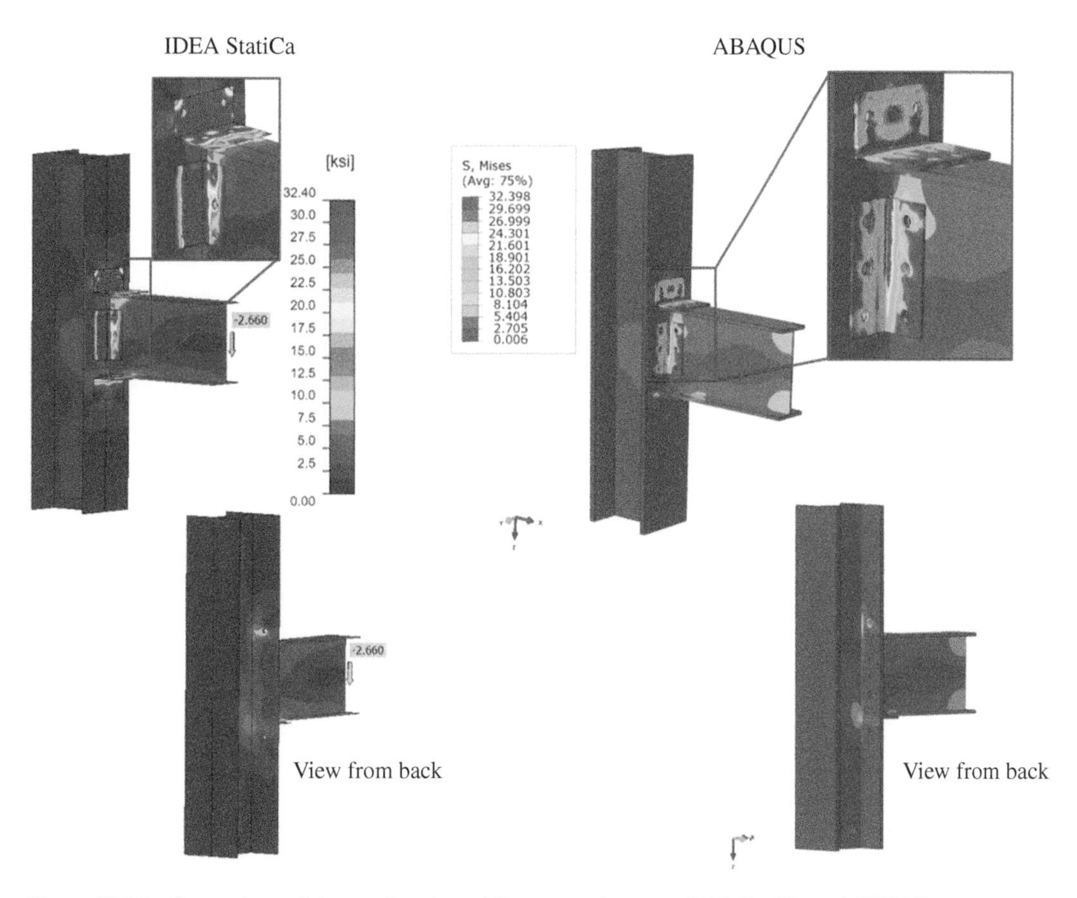

Figure 18.13 Comparison of the predicted von Mises stress between IDEA StatiCa and ABAQUS.

formulation. Friction was defined with a penalty method, and a Coulomb friction coefficient of $\mu = 0.3$ was used everywhere except between the column face and each angle, in which the contact was assumed to be frictionless.

The material behavior was modeled using a bilinear plasticity in ABAQUS. Other parameters, including density, elastic modulus, and Poisson's ratio, were exactly taken from the IDEA StatiCa materials library. The numerical simulations were carried out on four processors (Intel Xenon$^{(R)}$ CPU E5-2698 v4 @ 2.20GHz) and each simulation took approximately 535 minutes to finish. Figure 18.13 compares the calculated von Mises stresses in IDEA StatiCa and ABAQU. Figure 18.14 also shows the side view in which the deformation scale factor of ten was applied to models in both software.

18.6 Results Comparison

18.6.1 Comparison of Connection Capacities from IDEA StatiCa Analysis, AISC Design Codes, and Experiments

The design strength capacities (ϕR_n) of the ten TSADWA connections were calculated using the AISC *Specification* (AISC, 2016) and the AISC *Manual* (AISC, 2017). The moment capacities of the specimens

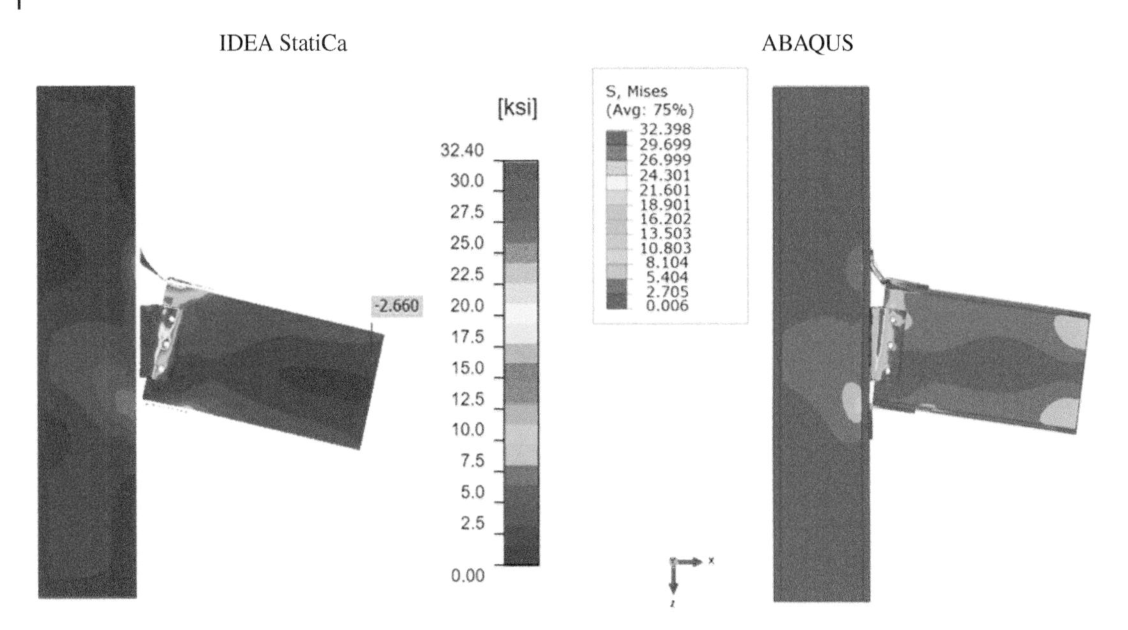

Figure 18.14 Side view comparison between IDEA StatiCa and ABAQUS with deformation scale factor of ten.

were calculated using a conservative approach assuming moment is carried by top- and seat-angles while web-angles resist shear force only. The smallest calculated strengths were determined, and the moment capacities of the connections were obtained corresponding to these controlling strengths.

First, the same specimens were modeled in IDEA StatiCa and analyzed under a shear force applied 120 in. away from the column centerline for the first four specimens (14S1, 14S2, 14S3, 14S4); and 72 in. away for the other six specimens (8S1, 8S2, 8S3, 8S4, 8S5, 8S6). The shear force was increased incrementally until the connections reached their capacities. The moment capacities of the connections were obtained by multiplying by the distance between the shear force application point and column centerline, and the ultimate shear force reached in the incremental loading. In the second part, the web-angles were removed from the specimens, and the moment capacities of the top and seated connections were obtained by following the same procedure to eliminate the resistance of web-angles on the moment capacities of the specimens in IDEA analysis. The results are depicted in Figure 18.15.

The moment-rotation relationship of the Test No. 14S1 was calculated from the IDEA StatiCa analysis using the mean measured material properties (the mean values of the material strengths (F_u, F_y) tested in the lab and provided in the test report) of the tested specimens measured in the experimental study. The calculated response is compared with the moment-rotation relationship provided in the test report (Figure 18.16).

18.6.2 Comparison of IDEA StatiCa and ABAQUS Results

The comparison between the IDEA StatiCa and the ABAQUS results are summarized in Tables 18.12 and 18.13. In general, there was good agreement between the results of the two software packages.

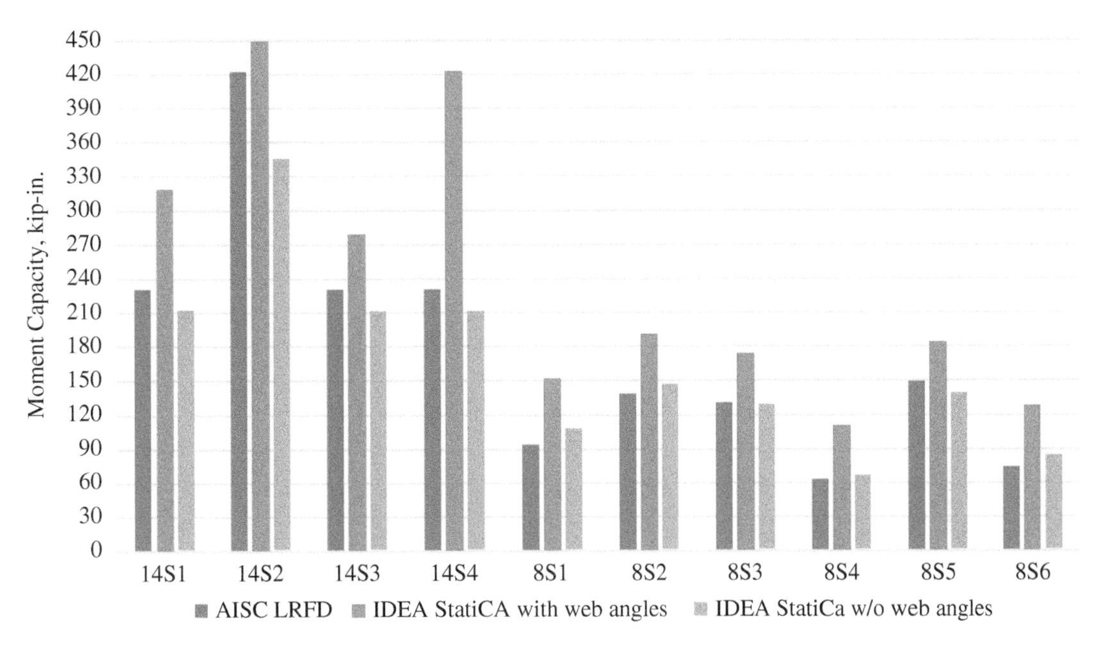

Figure 18.15 Comparison of the moment capacities of the specimens calculated from the AISC design equations and the IDEA StatiCa analysis.

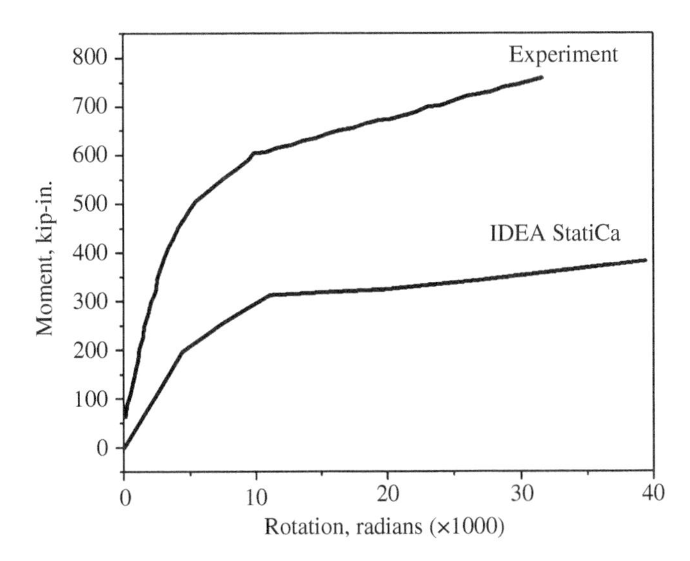

Figure 18.16 Comparison of the moment-rotation relationships of Test No. 14S1 measured during the experiment and calculated from IDEA StatiCa.

Table 18.12 Specified yield strengths and calculated stress, strain, and plates check status ϵ.

	IDEA StatiCa				ABAQUS				
Item	F_y (ksi)	σ_{Ed} (ksi)	ε_{pl} (%)	Check status	Item	F_y (ksi)	σ_{Ed} (ksi)	ε_{pl} (%)	Check status
C-bfl 1	36	13.7	0	Ok	Column	36	15.624	0	Ok
C-tfl 1	36	1.2	0	Ok					
C-w 1	36	8.6	0	Ok					
B-bfl 1	36	17.2	0	Ok	Beam	36	32.398	0	OK
B-tfl 1	36	32.4	0	Ok					
B- w 1	36	29.3	0	Ok					
CLEAT 1 a-bfl1	36	33.3	3.1	Ok	L-4×3 1/2×1/4 (Front)	36	32.398	0.8	OK
CLEAT 1 a-w1	36	33	2.1	Ok					
CLEAT 1 b-bfl1	36	33.2	2.6	Ok	L-4×3 1/2×1/4 (Back)	36	32.398	0.8	OK
CLEAT 1 b-w1	36	33	2.1	Ok					
CLEAT 2 -bfl1	36	33.9	5	Ok	L-6×4×3/8 (Top)	36	32.398	1.8	OK
CLEAT 2 -w1	36	33.8	4.7	Ok					
CLEAT 3 -bfl1	36	26	0	Ok	L-6×4×3/8 (Bottom)	36	32.398	0.1	OK
CLEAT 3 -w1	36	32.5	0.3	OK					

Table 18.13 Calculated tension force, shear force, and bolt bearing resistance.

	IDEA StatiCa					ABAQUS				
Item	F_t (kips)	V (kips)	$\phi R_{n,bearing}$ (kips)	Ut_t (%)	Ut_s (%)	F_t (kips)	V (kips)	$\phi R_{n,bearing}$ (kips)	Ut_t (%)	Ut_s (%)
B1	2.28	1.083	24.359	7.7	8.9	2.169	1.025	26.364	7.2	7.8
B2	3.777	2.250	21.318	12.7	21.1	3.401	2.035	23.146	11.4	17.6
B3	5.633	3.905	18.959	18.9	41.2	5.736	3.621	20.486	19.2	35.4
B4	8.397	2.179	16.596	28.2	13.1	7.536	2.023	17.895	25.2	11.3
B5	14.791	1.841	17.469	49.6	10.5	14.483	1.705	18.869	48.6	9.0
B6	19.713	1.486	19.580	66.2	7.6	18.751	1.526	21.132	62.9	7.2
B7	8.411	2.179	16.585	28.2	13.1	7.555	2.022	17.879	25.3	13.3
B8	14.821	1.842	17.456	49.7	10.5	14.491	1.705	18.859	48.6	9.0
B9	19.731	1.490	19.580	66.2	7.6	18.751	1.526	21.132	62.9	7.2
B10	12.306	3.391	29.293	41.3	15.0	11.89	3.556	31.615	39.9	15.8

Table 18.13 (Continued)

	IDEA StatiCa					ABAQUS				
Item	F_t (kips)	V (kips)	$\phi R_{n,bearing}$ (kips)	Ut_t (%)	Ut_s (%)	F_t (kips)	V (kips)	$\phi R_{n,bearing}$ (kips)	Ut_t (%)	Ut_s (%)
B11	12.276	3.390	29.293	41.2	15.0	11.89	3.556	31.615	39.9	15.8
B12	0.233	3.069	16.285	0.8	18.8	0.450	3.456	17.575	1.5	19.6
B13	0.233	3.069	16.285	0.8	18.8	0.450	3.456	17.575	1.5	19.6
B14	23.861	3.134	22.136	80.1	14.2	23.259	3.222	23.752	78.1	13.5
B15	23.868	3.137	22.138	80.1	14.2	23.259	3.222	23.752	78.1	13.5
B16	2.476	6.092	29.370	8.3	27	2.569	5.957	31.559	8.6	26.4
B17	2.475	6.092	29.370	8.3	27	2.569	5.957	31.559	8.6	26.4
B18	0.423	6.318	29.370	1.4	28	0.445	6.152	31.559	1.5	27.2
B19	0.424	6.318	29.370	1.4	28	0.445	6.152	31.559	1.5	27.2
B20	0.358	1.575	29.370	1.2	7	0.258	1.456	31.559	0.9	6.4
B21	0.364	1.579	29.370	1.2	7	0.258	1.456	31.559	0.9	6.4

18.7 Summary

A comparison between the results from IDEA StatiCa component-based finite element method (CBFEM), the traditional calculation methods used in US practice (AISC *Specification*; AISC, 2016; AISC *Manual*; AISC, 2017), and the traditional finite element analysis using ABAQUS for tested TSADWA connections is presented in this chapter. Overall, a good agreement between the results was observed. The calculated moment capacities following AISC procedure are slightly different than those obtained with IDEA StatiCa. This can be due to the conservative assumption made for TSADWA connections in the AISC *Manual* that ignores the contribution of double web-angles to the overall moment strength of the connections because of its complexity to evaluate them through hand calculations. On the other hand, the IDEA StatiCa intends to capture the real behavior which is expected to be more accurate. In addition, more deformation was captured on the web-angles, top, and bottom flanges in the IDEA StatiCa model compared to the ABAQUS analysis. Also, the stress distributions on the web-angles were slightly different between the two models. This is most likely due to the fact that in the ABAQUS model, solid elements with reduced integration were utilized. In both models, it was found that the weakest component of the assembly was the top flange in tension under the applied shear force pointing downward, which introduces tension in the top flange. Stress distribution for each component can be seen in Appendix H of the full report, published on the IDEA StatiCa website (IDEA StatiCa, 2021).

References

AISC (2016). *Specification for Structural Steel Buildings*. American Institute of Steel Construction, Chicago.

AISC (2017). *Steel Construction Manual*, 15th Edition. American Institute of Steel Construction, Chicago.

Azizinamini, A., Bradburn, J.H., and Radziminski, J.B. (1985). *Static and Cyclic Behavior of Semi-Rigid Steel Beam-Column Connections*. Columbia, SC, University of South Carolina.

Dassault Systèmes Simulia Corporation (2020). *ABAQUS* (Version 2020) [Computer program]. Dassault Systèmes SE, Providence, RI.

IDEA StatiCa (2021). *Verification of IDEA StatiCa Calculations for Steel Connection Design (AISC)*. https://www.ideastatica.com/support-center/verification-of-idea-statica-calculations-for-steel-connection-design-aisc (accessed October 31, 2023).

19

Bolted Flange Plate (BFP) Moment Connections

19.1 Description

In this chapter, the design strength capacities of ten bolted flange plate (BFP) moment connection specimens were calculated following the requirements of the AISC *Specification* (AISC, 2016a) and the AISC *Manual* (AISC, 2017). The base-line specimen was selected from the experimental study performed by Sato et al. (2007) in the Department of Structural Engineering at the University of California, San Diego. The baseline specimen and nine additional variation models were analyzed using IDEA StatiCa while the baseline specimen was also analyzed using ABAQUS (Dassault, 2020). The results were then compared at the end of the chapter.

19.2 Experimental Study on BFP Moment Connections

Three full-scale bolted flange plate (BFP) moment connections were subjected to cycling testing at the University of California, San Diego. All specimens met the requirement of the AISC Prequalified Connections for Special and Intermediate Steel Moment Frames for Seismic Applications (AISC, 2016b) for the beam column connections of special moment frames. The lateral bracing distance for the specimens was determined in accordance with this provision. The vertical displacements were applied by a hydraulic actuator at the tip of the beam, as shown in Figure 19.1.

The loading began at 0.375% drift and the displacement magnitude was increased until the specimen failed. The applied load was measured by the load cell mounted on the actuator. The transducer L1 in Figure 19.2 measured the total displacement of the beam tip while the column horizontal movement was recorded by L5 and L6. The average shear deformation of the column panel zone was measured by L9 and L10 (see Figure 19.2). The moment-rotation relationships at the column face were obtained for all specimens using the data measured by these instruments.

In this study, the Specimen No. BFP was selected as a baseline model. For this specimen, the loading was applied at approximately 177.5 in. away from the column face. The details of this connection are shown in Figures 19.3 and 19.4.

All bolts were A325 bolts with threads excluded from shear planes. The beam and column sections were made of A992 steel, while all the plates were made of A572 Gr. 50 steel. The material properties of

Steel Connection Design by Inelastic Analysis: Verification Examples per AISC Specification, First Edition.
Mark D. Denavit, Ali Nassiri, Mustafa Mahamid, Martin Vild, Halil Sezen, and František Wald.

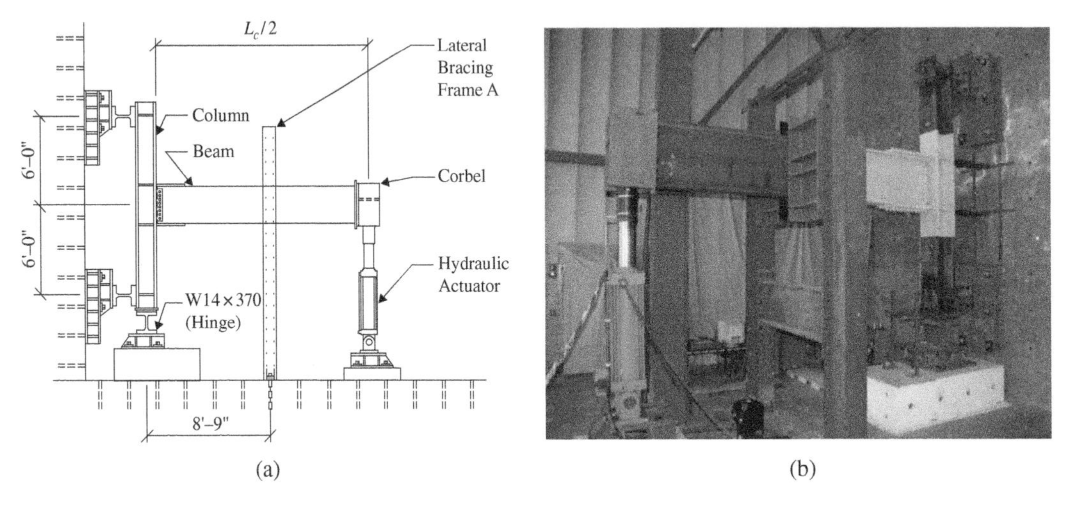

Figure 19.1 Test setup: (a) schematic; (b) photo from the lab (Sato et al., 2007).

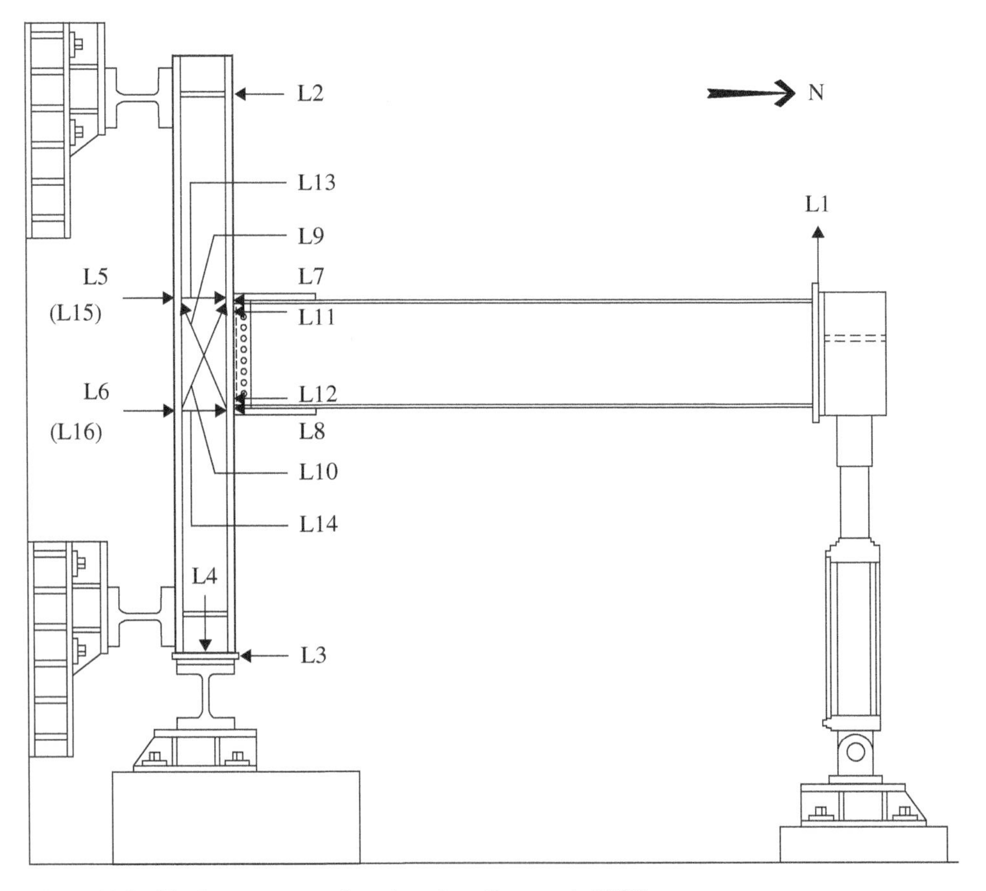

Figure 19.2 Displacement transducer locations (Sato et al., 2007).

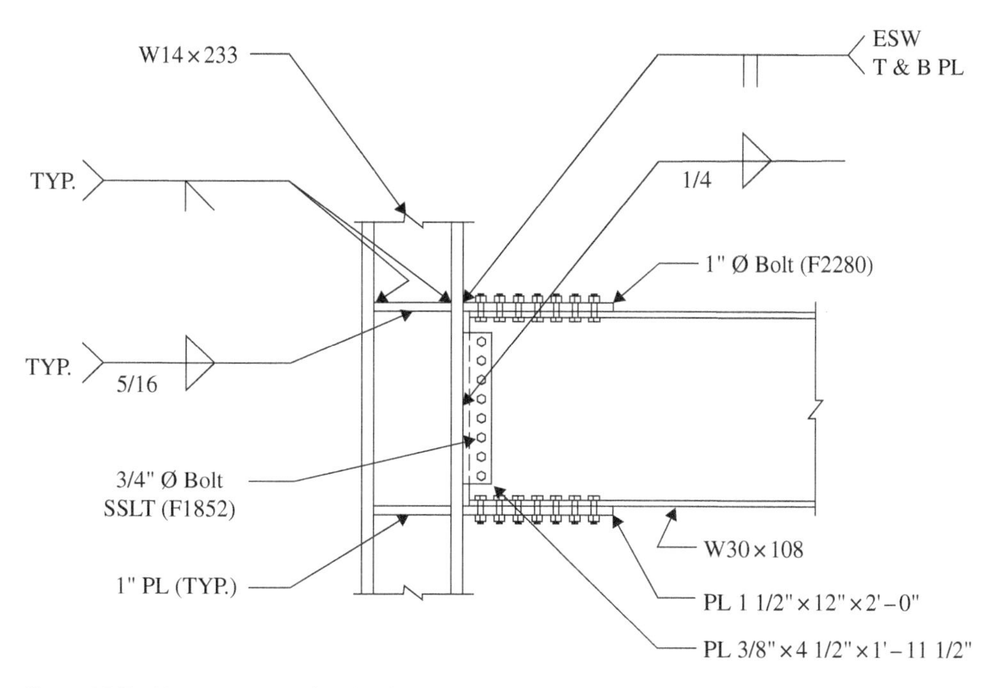

Figure 19.3 Moment connection details for the specimen No. BFP (Sato et al., 2007).

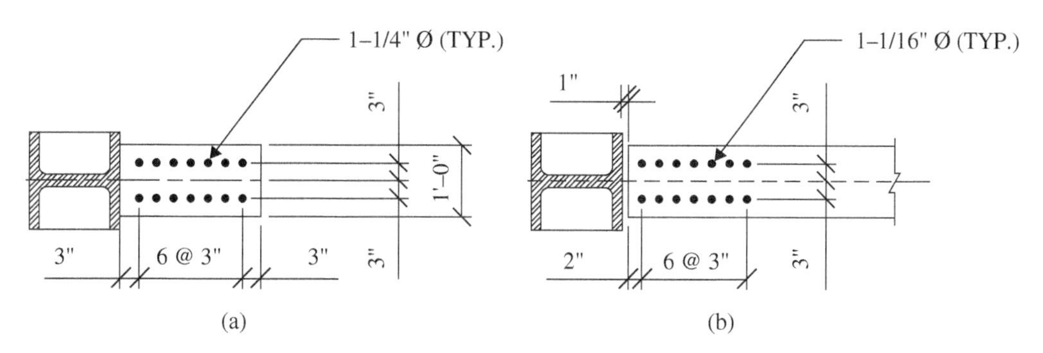

Figure 19.4 Bolt schedule details for the specimen No. BFP: (a) flange plate; and (b) beam flange (Sato et al., 2007).

the members obtained from Colorado Metallurgical Services (CMS) and Certified Mill Test Reports are presented in Table 19.1.

The flange plates were welded to the flange of the column using electroslag welding (ESW) process. Two Arcmatic 105-VMC 3/32 in. diameter electrodes were used. This electrode has a specified minimum Charpy-V Notch Toughness of 15 ft-lbs at -20°F. Flux (FES72) was added by hand per the fabricator's standard procedure.

The Specimen No. BFP failed due to beam flange net section fracture when the interstory drift angle of 0.06 radians was achieved during testing. The applied load-beam tip displacement and the moment at

Table 19.1 Steel mechanical properties.

Member	Steel grade	Yield strength (ksi)	Tensile strength (ksi)
Column	A992	51.5 (57.0)	76.5 (75.5)
Beam	A992	52.0 (57.0)	77.5 (75.0)
Plate	A572 Gr. 50	60.5 (63.0)	87.5 (85.3)

Note: Values in parentheses are based on Certified Mill Test Reports, others from testing by CMS.

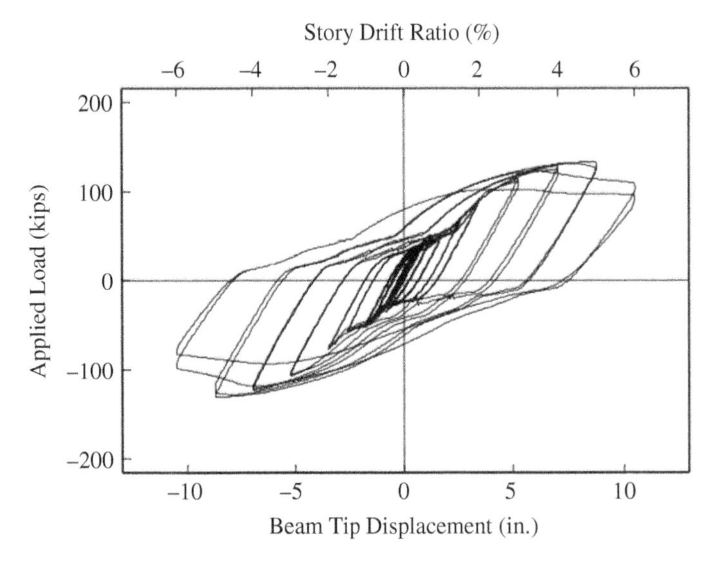

Figure 19.5 Applied load-beam tip displacement (Sato et al., 2007).

column face-beam rotation relationships are provided in Figures 19.5 and 19.6. The fracture location and beam bottom flange net section fracture are shown in Figures 19.7 and 19.8.

19.3 Code Design Calculations and Comparisons

The design strength capacities (ϕR_n) of ten BFP moment connections were calculated following the requirements of the AISC *Specification* (AISC, 2016a) and the AISC *Manual* (AISC, 2017). The nominal strength, R_n, and the corresponding resistance factor, ϕ, for each connection design limit state for load and resistance factored design (LRFD) are provided in Chapter J of the AISC *Specification*.

The Specimen No. BFP was selected as a baseline model from the experimental study and nine additional variation models were generated by changing only one parameter at a time from the baseline

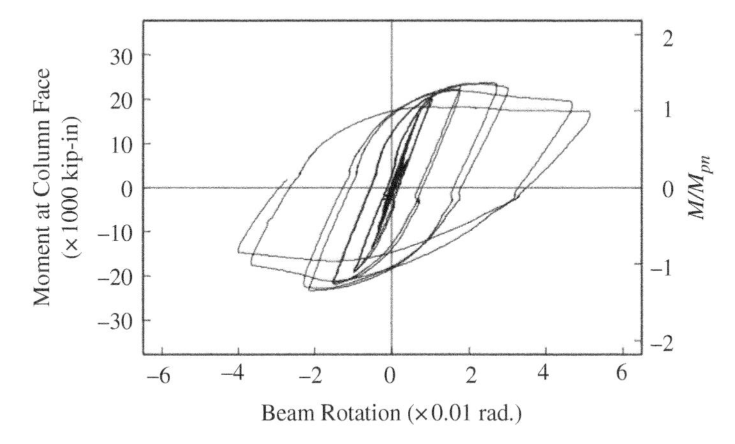

Figure 19.6 Moment at column face-beam rotation (Sato et al., 2007).

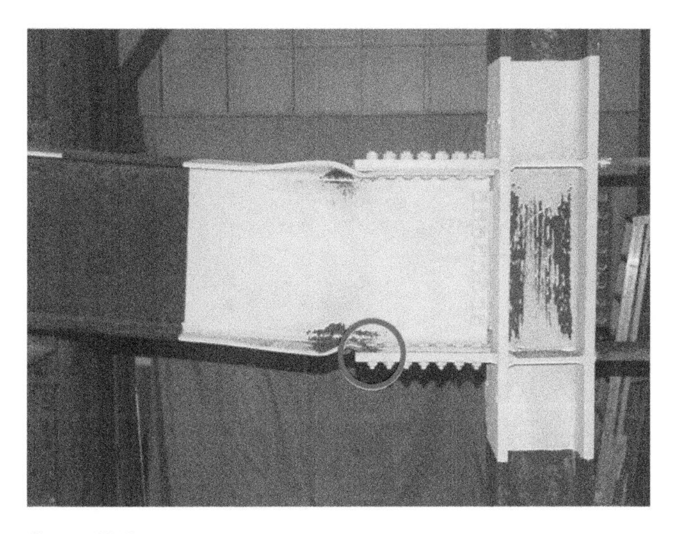

Figure 19.7 Fracture location circled.

model. The properties of the baseline and nine additional variation models are shown in Table 19.2. The changing parameters are in bold italics.

19.3.1 Design Strength Capacity of Single Web Plates

The following eight design checks were performed according to the LRFD design equations included in the AISC *Specification* (AISC, 2016a) and the AISC *Manual* (AISC, 2017) for the design strength capacities of single web plate.

Figure 19.8 Beam bottom flange net section fracture on 2nd cycle at +6% drift (Sato et al., 2007).

Table 19.2 Properties of the ten specimens.

			Single web plate				Flange plates		
Speci-men no.	Beam	Column	Web plate geometry (in.)	Weld size (in.)	Bolt schedule	Bolt dia. (in.)	Flange plates (in.)	Bolt dia. (in.)	Bolt schedule
BFP	W30x108	W14x233	3/8x4.5x23.5	1/4	8x1	3/4	1.5x12x24	1	7x2
Model 1	W30x108	W14x233	***1/4x4.5x23.5***	1/4	8x1	3/4	1.5x12x24	1	7x2
Model 2	W30x108	W14x233	3/8x4.5x23.5	1/4	8x1	***1/2***	1.5x12x24	1	7x2
Model 3	W30x108	W14x233	3/8x4.5x23.5	1/2	***6x1***	3/4	1.5x12x24	1	7x2
Model 4	W30x108	***W14x311***	3/8x4.5x23.5	1/4	8x1	3/4	1.5x12x24	1	7x2
Model 5	W30x108	***W14x370***	3/8x4.5x23.5	1/4	8x1	3/4	1.5x12x24	1	7x2
Model 6	W30x108	W14x233	3/8x4.5x23.5	1/4	8x1	3/4	*1x12x24*	1	7x2
Model 7	W30x108	W14x233	3/8x4.5x23.5	1/4	8x1	3/4	1.5x12x24	***3/4***	7x2
Model 8	W30x108	W14x233	3/8x4.5x23.5	1/4	8x1	3/4	1.5x12x24	1	***5x2***
Model 9	W30x108	W14x233	3/8x4.5x23.5	1/4	8x1	3/4	1.5x12x24	1	***3x2***

1. Web plate
   a. Bolts shear (Eq. J3-1, AISC *Specification*)
   b. Bolt bearing and tearout (Eq. J3-6, AISC *Specification*)
   c. Shear yielding (Eq. J4-3, AISC *Specification*)
   d. Shear rupture (Eq. J4-4, AISC *Specification*)
   e. Block shear (Eq. J4-5, AISC *Specification*)
   f. Weld shear (Eq. 8-2, AISC *Manual*)

Table 19.3 The design strength capacities (ϕR_n) of sing web-plates of the ten specimens.

Design checks (LRFD)	Design strength capacity (kips)									
Specimen number	BFP	Model 1	Model 2	Model 3	Model 4	Model 5	Model 6	Model 7	Model 8	Model 9
Web plate										
Bolt shear	***180.32***	180.32	***80.14***	135.24	***180.32***	***180.32***	***180.32***	***180.32***	***180.32***	***180.32***
Bolt bearing and tearout	263.25	175.50	175.50	197.44	263.25	263.25	263.25	263.25	263.25	263.25
Shear yielding	264.30	176.20	264.30	196.82	264.30	264.30	264.30	264.30	264.30	264.30
Shear rupture	181.06	***120.71***	202.07	***134.41***	181.06	181.06	181.06	181.06	181.06	181.06
Block shear	191.00	127.66	209.44	144.88	191.00	191.00	191.00	191.00	191.00	191.00
Weld shear	261.70	261.70	261.70	194.90	261.70	261.70	261.70	261.70	261.70	261.70
Beam										
Bolt bearing and tearout	382.59	382.59	255.06	286.94	382.59	382.59	382.59	382.59	382.59	382.59
Shear yielding	487.72	487.72	487.23	804.96	487.72	487.72	487.72	487.72	487.72	487.72

2. Beam
   a. Bolts shear (Eq. J3-1, AISC *Specification*)
   b. Bolt bearing and tearout (Eq. J3-6, AISC *Specification*)

The design strength capacities (ϕR_n) of single web plates of the ten specimens calculated by following AISC LRFD code requirements are provided in Table 19.3. The lowest shear capacities are shown in bold italics. Of the calculated design capacities for the ten test specimens. the design capacity of model 2 was controlled by shear rupture while bolt shear led to failure for the other eight specimens.

19.3.2 Design Strength Capacity of Flange Plates

The following 13 design checks were performed according to the LRFD design equations included in the AISC *Specification* (AISC, 2016a) or the AISC *Manual* (AISC, 2017) for the design strength capacities of flange plates.

1. Flange plate
   a. Bolts shear (Eq. J3-1, AISC *Specification*)
   b. Bolt bearing and tearout (Eq. J3-6, AISC *Specification*)
   c. Tensile yielding (Eq. J4-3, AISC *Specification*)
   d. Tensile rupture (Eq. J4-4, AISC *Specification*)
   e. Block shear (Eq. J4-5, AISC *Specification*)
   f. Compression (Sec. J4-4, AISC *Specification*)

e. Beam
   a. Bolt bearing and tearout (Eq. J3-6, AISC *Specification*)
   b. Flexural (Sec. F13.1, AISC *Specification*)
   c. Block shear (Eq. J4-5, AISC *Specification*)
3. Column
   a. Panel web shear (Eq. J10-9, AISC *Specification*)
   b. Flange local bending (Eq. J10-1, AISC *Specification*)
   c. Web local yielding (Eq. J10-2, AISC *Specification*)
   d. Web local crippling (Eq. J10-4, AISC *Specification*)

The design strength capacities (ϕR_n) of flange plates of ten specimens calculated by following AISC LRFD code requirements are provided in Table 19.4. The lowest shear capacities are shown in bold italics.

Of the calculated design capacities for the ten test specimens, the design capacity of seven specimens was controlled by web panel zone shear, two specimens were controlled by bolt shear and one specimen was controlled by block shear. The moment capacities of the specimens were calculated by multiplying by the controlling design capacity by the moment arm as provided in Table 19.5. The moment arm is equal to the depth of the beam for bolt shear while it is equal to the summation of the depth of the beam and the thickness of the plate for the web panel zone shear and block shear strengths (BFP, models 1, 2, 3, 4, 5, 6 and 8). The detailed design strength calculations for Test No. BFP are provided in Appendix I of the full report, published on the IDEA StatiCA website (IDEA StatiCa, 2021).

19.3.3 Calculated ASD Design Strength Capacities of Test No. BFP

According to allowable strength design (ASD), the allowable strength (R_n/Ω) is calculated by dividing the nominal strength, R_n by the safety factor, Ω. The allowable strength capacities of connection specimen Test No. BFP are calculated by following the AISC ASD code requirements. The properties of this test specimen were given in Table 19.2. The calculated ASD design strength capacities (R_n/Ω) of the specimen for the web plate and flange plates are provided in Tables 19.6 and 19.7, respectively.

The calculated lowest strength of the web plate for Test No. BFP is 101.01 kips due to the bolt shear failure while the controlling strength of the flange plate is 307.54 kips because of web panel zone shear failure. The moment capacities of the specimen can be calculated by multiplying by the governing strength of the flange plate (307.54 kips) by the distance of the moment arm (31.33 in.) which is equal to the depth of the beam (29.83 in.) plus one plate thickness (1.5 in.). The detailed design strength calculations for Test No. BFP are provided in Appendix J of the full report, published on the IDEA StatiCA website (IDEA StatiCa, 2021).

19.4 IDEA StatiCa Analysis

19.4.1 Moment Capacity Analysis Using IDEA StatiCa

The ten BFP moment connection specimens were modeled in IDEA StatiCa and analyzed under a shear force applied at 177.5 in. away from the column centerline as in the test report. The shear force was

Table 19.4 The design strength capacities (ϕR_n) of flange plates of ten specimens.

Design checks (LRFD)	Design strength capacity (kips)									
	Specimen number									
	BFP	Model 1	Model 2	Model 3	Model 4	Model 5	Model 6	Model 7	Model 8	Model 9
Flange plate										
Bolt shear	692.44	692.44	692.44	692.44	692.44	692.44	692.44	**389.84**	496.6	**296.76**
Bolt bearing and tearout	2402.60	2402.60	2402.60	2402.60	2402.60	2402.60	1605.20	2760.60	1748.00	1088.1
Tensile yielding	810.00	810.00	810.00	810.00	810.00	810.00	540.00	810.00	810.00	810.00
Tensile rupture	704.00	704.00	704.00	704.00	704.00	704.00	469.30	665.60	704.00	704.00
Block shear	1517.30	1517.30	1517.30	1517.30	1517.30	1517.30	1011.60	1718.40	1199.30	881.20
Compression	810.00	810.00	810.00	810.00	810.00	810.00	540.00	810.00	810.00	810.00
Beam										
Bolt bearing and tearout	1128.18	1128.18	1128.18	1128.18	1128.18	1128.18	1128.18	933.66	1297.50	461.27
Block shear	668.95	668.95	668.95	668.95	668.95	***668.95***	668.95	770.64	507.59	346.42
Flexural	1297.50	1297.50	1297.50	1297.50	1297.50	1297.50	1297.50	1297.50	794.72	1297.50
Column										
Web panel zone shear	***462.24***	***462.24***	***462.24***	***462.24***	***651.00***	802.28	***462.24***	462.24	***462.24***	462.24
Flange local bending	832.05	832.05	832.05	832.05	1436.51	1990.01	832.05	832.05	832.05	832.05
Web local yielding	700.85	700.85	700.85	700.85	1113.90	1477.40	700.85	700.85	700.85	700.85
Web local crippling	1, 93.12	1193.12	1193.12	1193.12	2053.80	2832.00	1193.12	1193.12	1193.12	1193.12

increased incrementally until the connections reached their capacities in IDEA StatiCa. The calculated maximum moment capacities are shown in Table 19.8.

The IDEA StatiCa screenshots showing the failure modes as well as the deformed shapes of finite element models are shown in Appendix K of the full report, published on the IDEA StatiCA website (IDEA StatiCa, 2021).

19.4.2 Moment-Rotation Analysis

Moment-rotation analysis for Test No. BFP was performed using IDEA StatiCa. To be able to generate the test condition, the mean values of yielding and tensile strength of the materials measured by CMS

Table 19.5 The moment capacities of the ten BFP moment connection specimens.

Specimen number	Governing strength (kips)	Length of the moment arm (in.)	Moment capacity (kips-in.)
BFP	462.24	31.33	14481.98
Model 1	462.24	31.33	14481.98
Model 2	462.24	31.33	14481.98
Model 3	462.24	31.33	14481.98
Model 4	651.00	31.33	20395.83
Model 5	668.95	29.83	19954.78
Model 6	462.24	30.83	14250.86
Model 7	389.84	29.83	11628.93
Model 8	462.24	31.33	14481.98
Model 9	296.76	29.83	8852.351

Table 19.6 The design strength capacity (ϕR_n) of the web-plate for Test No. BFP.

Design checks (ASD)	Design strength capacity (kips)
Web plate	
Bolt shear	120.22
Bolt bearing and tearout	175.52
Shear yielding	176.20
Shear rupture	120.71
Block shear	127.34
Weld capacity	174.46
Beam	
Bolt bearing and tearout	227.60
Shear yielding	325.20

and Certified Mill Test Reports were used (Table 19.9). The mean values defined in IDEA StatiCa and the resistance factors were adjusted to 1.0. Then, the moment-rotation analysis was performed by choosing stiffness analysis option (Figure 19.9).

Table 19.7 The design strength capacity (ϕR_n) of the flange plates for Test No. BFP.

Design checks (ASD)	Design strength capacity (kips)
Flange plate	
Bolt shear	461.58
Bolt bearing and tearout	1601.76
Tensile yielding	538.92
Tensile rupture	469.30
Block shear	1011.53
Compression	538.92
Beam	
Bolt bearing and tearout	752.12
Block shear	445.97
Flexural strength	863.27
Column	
Web panel zone shear	307.54
Flange local bending	553.59
Web local yielding	467.23
Web local crippling	795.42

Table 19.8 LRFD moment capacities of the specimens calculated by IDEA StatiCa.

	IDEA StatiCa			
Specimen	Shear force (kips)	Moment arm (in.)	Moment (kips-in.)	Failure mode
BFP	96.70	177.50	17164.25	Beam flange failure (limit plastic strain, 5%)
Model 1	96.05	177.50	17048.88	Bolt shear failure on web plate
Model 2	96.00	177.50	17040.00	Bolt shear failure on web plate
Model 3	96.10	177.50	17057.75	Bolt shear failure on web plate
Model 4	100.20	177.50	17785.50	Beam flange failure (limit plastic strain, 5%)
Model 5	100.40	177.50	17821.00	Beam flange failure (limit plastic strain, 5%)
Model 6	89.70	177.50	15921.75	Flange plate failure (limit plastic strain, 5%)
Model 7	64.00	177.50	11360.00	Bolt shear failure on flange plate
Model 8	92.00	177.50	16330.00	Beam flange failure (limit plastic strain, 5%)
Model 9	61.00	177.50	10827.50	Bolt shear failure on flange plate

Table 19.9 Mean values of measured material properties.

Member	Steel grade	Yield strength (ksi)	Tensile strength (ksi)
Column	A992	54.25	76
Beam	A992	54.5	76.25
Plate	A572 Gr. 50	61.75	86.4

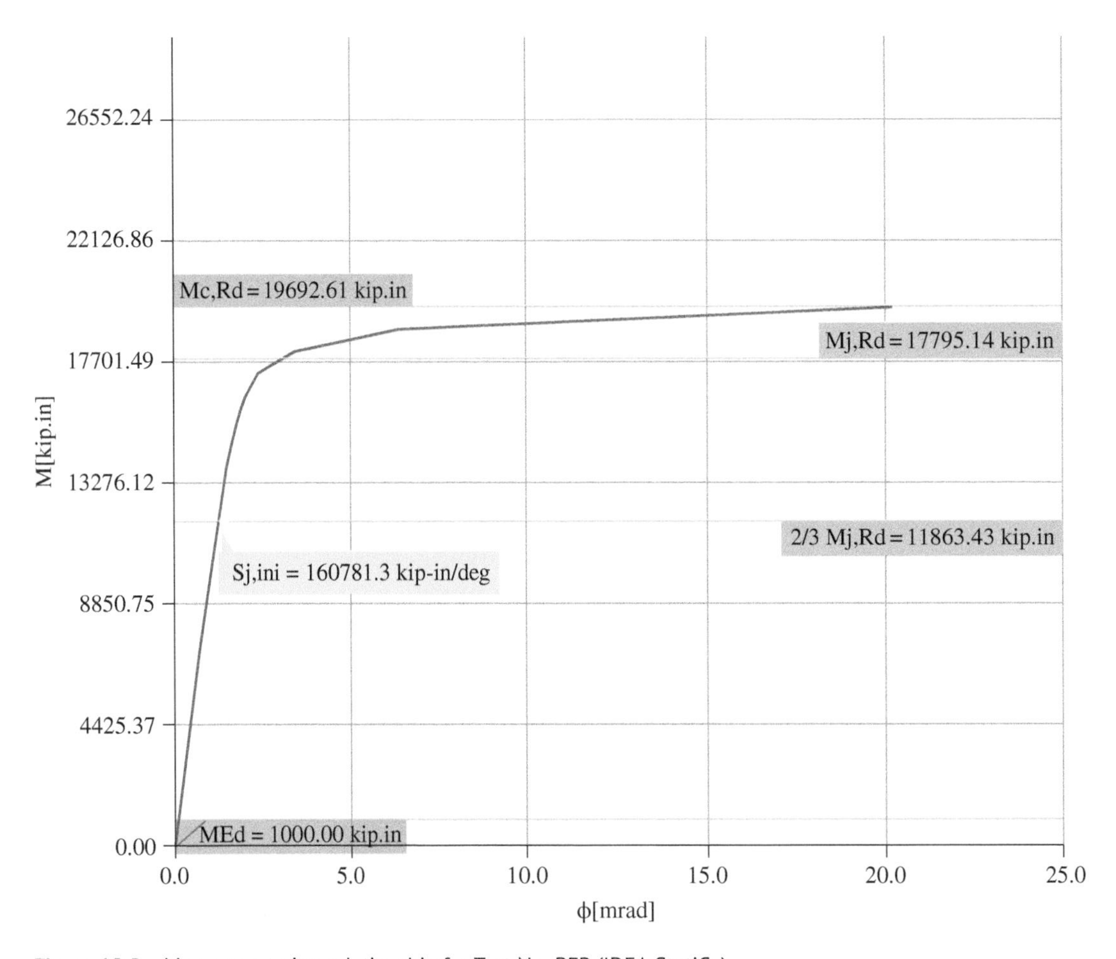

Figure 19.9 Moment-rotation relationship for Test No. BFP (IDEA StatiCa).

19.5 ABAQUS Analysis

In this section, the output results from IDEA StatiCa were compared to ABAQUS software package. In this study, Test BFP, as described in Table 19.2, was chosen as a base model. Numerical simulations with almost identical conditions (i.e. in terms of material properties, boundary condition, and loading) were carried out using both IDEA StatiCa and ABAQUS. The model was initially designed in IDEA StatiCa and then the assembly (including beam, column, and plates) was imported to ABAQUS using the IDEA StatiCa's viewer platform. Afterward, a simplified model for the bolt was designed and added to the ABAQUS model (see Figure 19.10).

In ABAQUS, the element type of C3D8R (3D stress, 8-node linear brick, reduced integration) was utilized, and a total of 681016 elements were generated in the model (see Table 19.10 and Figure 19.11 for more details).

In the ABAQUS model, the vertical shear force of 5.8 kips was applied on a reference point (or node) that was defined 177.5 in. away from the center of the column (i.e. x = 177.5 in.). Then, the coupling constraint (i.e. structural distributing) was defined to connect this reference point to the end section of the beam. The top and bottom of the column were fixed as a boundary condition (see Figure 19.10). The contact between all parts was defined as surface-to-surface with finite sliding formulation. Friction was defined with a penalty method, and a Coulomb friction coefficient of $\mu = 0.3$ was used between all the parts in contact. The top and bottom plates (i.e. Plate 2 and Plate 3) were also welded to the column.

The material behavior was modeled using a bilinear plasticity in ABAQUS. Other parameters including density, elastic modulus, and Poisson's ratio were copied from the IDEA StatiCa materials library. The

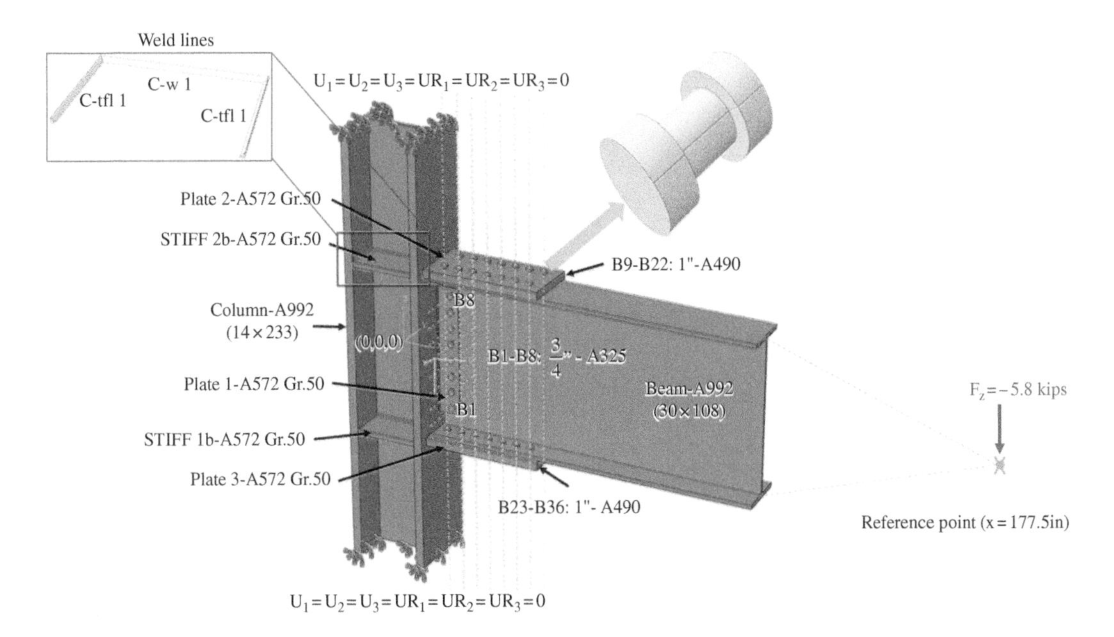

Figure 19.10 Model setup in ABAQUS (Dassault, 2020).

Table 19.10 Number of elements in the ABAQUS (Dassault, 2020) model.

Item	Number of elements
Column	100,569
Beam	38,091
Plate 1	5164
Plates 2, 3	7892
Stiffener	1824
Bolt ¾ in.	7032
Bolt 1.0 in.	16,152
Weld group	700

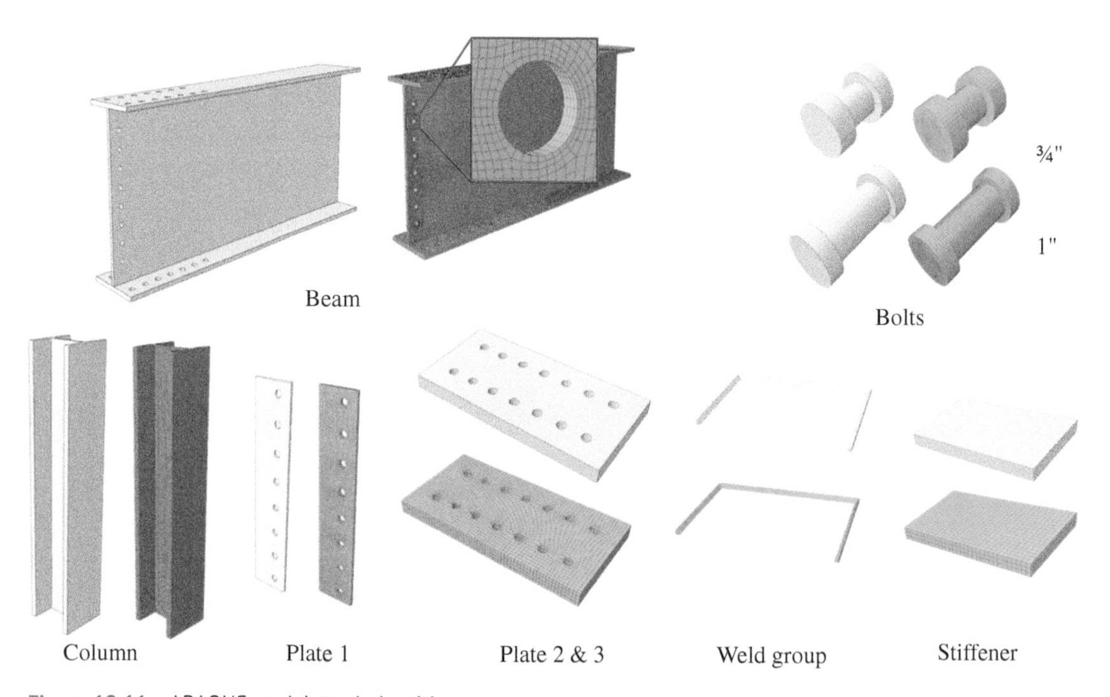

Figure 19.11 ABAQUS model mesh densities.

numerical simulations were carried out on eight processors (Intel Xenon (R) CPU E5-2698 v4 @ 2.20GHz) and the simulation took approximately 685 minutes. Figure 19.12 depicts the comparison between the predicted von Mises stress in IDEA StatiCa and ABAQUS. Figure 19.13 also shows the side view in which the deformation scale factor of 20 was applied to models in both software.

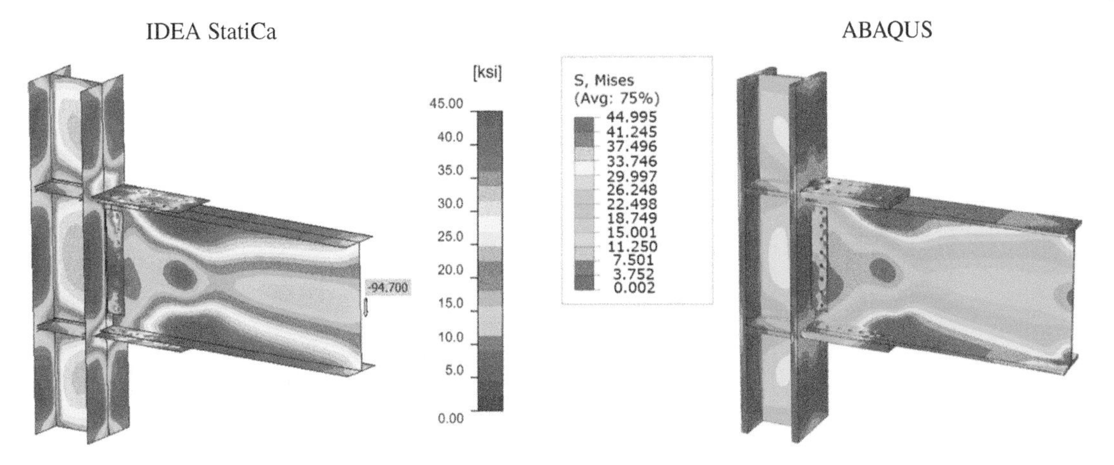

Figure 19.12 Comparison of the predicted von-Mises stress between IDEA StatiCa and ABAQUS.

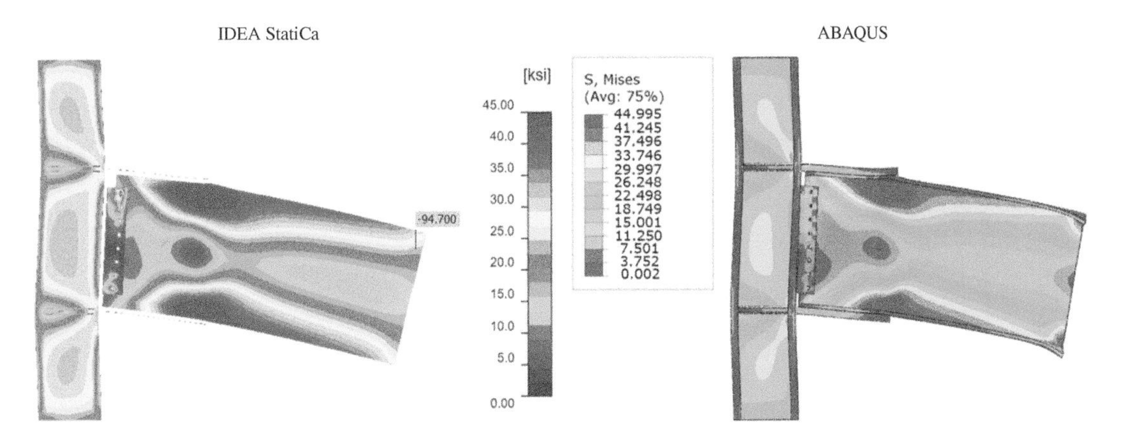

Figure 19.13 Side view comparison between IDEA StatiCa and ABAQUS with deformation scale factor of 20.

19.6 Results Comparison

19.6.1 Comparison of IDEA StatiCa Analysis Data, AISC Design Strengths, and Test Data

The design strength capacities (ϕR_n) of the ten BFP moment connections were calculated following the requirements of the AISC *Specification* (AISC, 2016a) and the AISC *Manual* (AISC, 2017). The smallest calculated strengths were determined, and the moment capacities of the connections were obtained corresponding to these controlling strengths.

The same specimens were modeled in IDEA StatiCa and analyzed under a shear force applied 117.5 in. away from the column centerline. The shear force was increased incrementally until the connections reached their capacities. The moment capacities of the connections were obtained by multiplying by the distance between where the shear force was applied, and the ultimate shear force reached in the incremental loading. The results are compared and shown in Figure 19.14.

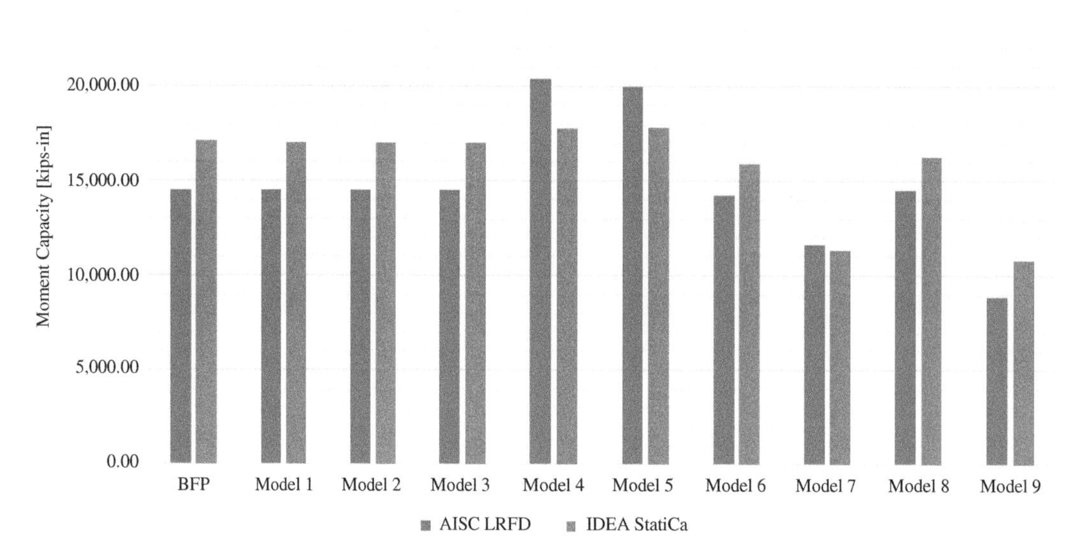

Figure 19.14 Comparison of the moment capacities of the specimens calculated from the AISC design equations and IDEA StatiCa.

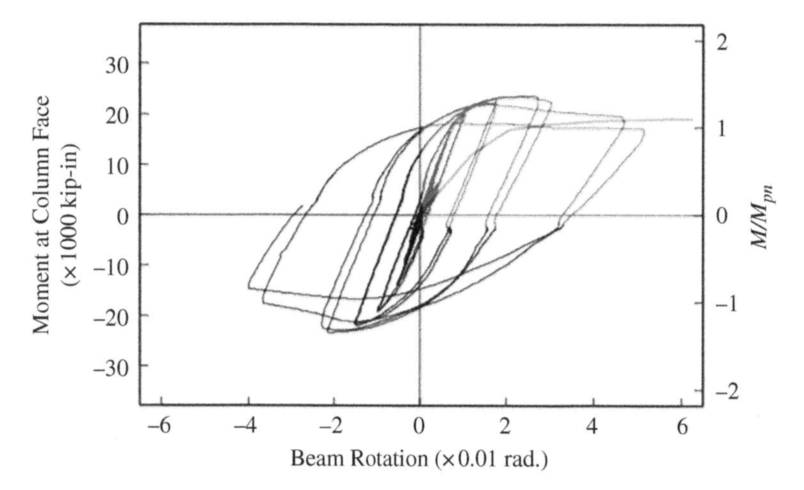

Figure 19.15 Comparison of the moment-rotation relationships of Test No. BFP measured during the experiment and calculated from IDEA StatiCa (blue line).

The moment-rotation relationship of the Test No. BFP was calculated from IDEA StatiCa analysis using the mean values of the material properties of the tested specimens measured in the experimental study and compared with the moment-rotation relationship obtained during static loading provided in the test report (Figure 19.15).

19.6.2 Comparison of IDEA StatiCa and ABAQUS Results

The comparison between the IDEA StatiCa and ABAQUS (Dassault, 2020) results are summarized in Tables 19.11–19.13. As shown in Figure 19.12 and Tables 19.11–19.13, there was a good agreement between the results of the two software packages.

Table 19.11 Specified yield strengths and calculated stress, strain, and plates check status ϵ.

IDEA StatiCa					ABAQUS				
Item	F_y(ksi)	σ_{Ed}(ksi)	ϵ_{pl}(%)	**Check status**	**Item**	F_y(ksi)	σ_{Ed}(ksi)	ϵ_{pl}(%)	**Check status**
C-bfl 1	50.0	37.4	0	OK	Column	50	44.995	0	OK
C-tfl 1	50.0	24.5	0	OK					
C-w 1	50.0	32.9	0	OK					
B-bfl 1	50.0	45.3	1.0	OK	Beam	50	44.995	3.9	OK
B-tfl 1	50.0	46.4	5.0	OK					
B- w 1	50.0	45.1	0.5	OK					
STIFF 1a	50.0	19.5	0.0	OK	Back "bottom stiffener"	50	26.618	0	OK
STIFF 1b	50.0	19.5	0.0	OK	Front "bottom stiffener"	50	26.531	0	OK
FP 1	50.0	39.6	0.0	OK	Plate 1	50	44.995	0.2	OK
FP 2	50.0	45.1	0.2	OK	Plate 2	50	44.995	0.4	OK
FP 3	50.0	45.1	0.2	OK	Plate 3	50	44.995	0.5	OK
STIFF 2a	50.0	19.2	0.0	OK	Back "top stiffener"	50	31.836	0	OK
STIFF 2b	50.0	19.1	0.0	OK	Front "top stiffener"	50	31.736	0	OK

Table 19.12 Calculated tension force, shear force, and bolt bearing resistance.

	IDEA StatiCa					ABAQUS				
Item	F_t (kips)	V (kips)	$\phi R_{n,bearing}$ (kips)	Ut_t (%)	Ut_s (%)	F_t (kips)	V (kips)	$\phi R_{n,bearing}$ (kips)	Ut_t (%)	Ut_s (%)
B1	1.234	10.277	32.906	4.1	45.6	1.156	9.956	34.551	3.6	40.2
B2	0.839	7.145	32.906	2.8	31.7	0.742	6.821	34.551	2.3	27.6
B3	0.412	3.732	32.906	1.4	16.6	0.445	3.863	34.551	1.4	15.6
B4	0.088	0.230	32.906	0.3	1.0	0.132	0.228	34.551	0.4	1.0
B5	0.657	3.156	24.018	2.2	14.0	0.698	3.191	34.551	2.2	12.9
B6	1.292	6.400	24.046	4.3	28.4	1.305	6.556	34.551	3.9	26.5
B7	1.874	6.400	24.072	4.3	28.4	1.945	6.674	34.551	6.0	27.0
B8	2.773	11.864	24.087	9.3	52.6	2.669	11.772	34.551	8.2	47.5

(Continued)

Table 19.12 (Continued)

Item	IDEA StatiCa					ABAQUS				
	F_t (kips)	V (kips)	$\phi R_{n,bearing}$ (kips)	Ut_t (%)	Ut_s (%)	F_t (kips)	V (kips)	$\phi R_{n,bearing}$ (kips)	Ut_t (%)	Ut_s (%)
B9 (IDEA B29)	11.318	35.808	63.899	17.0	72.4	11.058	35.995	93.3471	16.7	72.9
B10 (IDEA B36)	11.140	35.625	63.899	16.7	72.0	11.015	35.995	93.3471	16.6	72.9
B11 (IDEA B28)	3.800	36.199	83.345	5.7	73.2	3.745	36.124	93.3471	5.7	43.2
B12 (IDEA B35)	3.469	36.057	83.345	5.2	72.9	3.562	36.325	93.3471	5.4	73.6
B13 (IDEA B27)	3.071	35.761	83.345	4.6	72.3	3.234	36.034	93.3471	4.9	73.0
B14 (IDEA B34)	2.757	35.688	83.345	4.1	72.1	2.945	35.956	93.3471	4.5	72.8
B15 (IDEA B23)	2.345	35.605	83.345	3.5	72.0	2.567	35.945	93.3471	3.9	72.8
B16 (IDEA B30)	2.203	35.572	83.345	3.3	71.9	2.456	35.857	93.3471	3.7	72.6
B17 (IDEA B24)	2.802	35.943	83.345	4.2	72.7	2.935	36.234	93.3471	4.5	73.4
B18 (IDEA B31)	2.722	35.934	83.345	4.1	72.6	2.845	36.145	93.3471	4.3	73.2
B19 (IDEA B25)	5.883	36.820	83.345	8.8	74.4	5.994	36.945	93.3471	9.1	74.8
B20 (IDEA B32)	5.862	36.820	83.345	8.8	74.4	5.905	36.94	93.3471	8.9	74.8
B21 (IDEA B26)	14.770	38.300	83.345	22.2	77.4	14.95	38.54	93.3471	22.5	78.1
B22 (IDEA B33)	14.770	38.310	83.345	22.2	77.4	14.95	38.54	93.3471	22.5	78.1
B23 (IDEA B57)	0.645	35.969	83.345	1.0	72.7	0.539	36.836	93.3471	0.9	74.6
B24 (IDEA B64)	0.660	35.892	83.345	1.0	72.6	0.634	36.12	93.3471	1	73.2
B25 (IDEA B56)	1.419	35.547	83.345	2.1	71.9	1.525	35.445	93.3471	2.3	71.8
B26 (IDEA B63)	1.472	35.501	83.345	2.2	71.8	1.59	35.966	93.3471	2.4	72.9
B27 (IDEA B55)	1.493	35.320	83.345	2.2	71.4	1.609	35.82	93.3471	2.5	72.6
B28 (IDEA B62)	1.536	35.312	83.345	2.3	71.4	1.635	35.745	93.3471	2.5	72.4
B29 (IDEA B51)	1.390	35.729	83.345	2.1	72.2	1.456	36.09	93.3471	2.2	73.1
B30 (IDEA B58)	1.418	35.743	83.345	2.1	72.3	1.467	36.09	93.3471	2.3	73.1
B31 (IDEA B52)	1.730	36.850	83.345	2.6	74.5	1.837	36.935	93.3471	2.8	74.8
B32 (IDEA B59)	1.732	36.877	83.345	2.6	74.5	1.845	36.894	93.3471	2.8	74.7
B33 (IDEA B53)	1.320	38.689	83.345	2.0	78.2	1.421	38.562	93.3471	2.2	78.1
B34 (IDEA B60)	1.340	38.742	83.345	2.0	78.3	1.456	38.63	93.3471	2.2	78.2
B35 (IDEA B54)	10.395	40.663	83.345	15.6	82.2	10.123	41.345	93.3471	15.3	83.7
B36 (IDEA B61)	10.398	40.682	83.345	15.6	82.2	10.141	41.366	93.347	15.3	83.8

Table 19.13 Calculated force in weld critical element, weld resistance, and welds check status.

		IDEA StatiCa				ABAQUS			
Item	Edge	F_n(kips)	ϕR_n(kips)	Ut(%)	Check status	F_n(kips)	ϕR_n(kips)	Ut(%)	Check status
C-bfl 1	STIFF1a	11.900	15.370	77.4	OK	12.856	16.238	79.2	OK
		12.123	15.414	78.6	OK	12.959	16.238	79.8	OK
C-w 1	STIFF1a	6.939	12.317	56.3	OK	7.234	14.235	50.8	OK
		6.613	12.423	53.2	OK	7.129	14.235	50.1	OK
C-tfl 1	STIFF1a	6.206	13.375	46.4	OK	6.945	15.238	45.6	OK
		5.790	13.171	44.0	OK	6.536	15.238	42.9	OK
C-bfl 1	STIFF1b	12.113	15.415	78.6	OK	13.532	16.235	83.4	OK
		11.907	15.371	77.5	OK	13.134	16.235	80.9	OK
C-w 1	STIFF1b	6.598	12.427	53.1	OK	7.452	13.532	55.1	OK
		6.937	12.314	56.3	OK	7.848	13.532	58.0	OK
C-tfl 1	STIFF1b	5.787	13.172	43.9	OK	5.994	14.235	42.1	OK
		6.207	13.381	46.4	OK	6.345	14.235	44.6	OK
C-bfl 1	FP1	8.010	11.942	67.1	OK	8.556	12.265	69.8	OK
		6.915	11.855	58.3	OK	7.456	12.265	60.8	OK
C-bfl 1	STIFF12a	12.073	15.409	78.4	OK	12.768	17.238	74.1	OK
		11.859	15.359	77.2	OK	12.568	17.238	73.0	OK
C-w 1	STIFF2a	6.534	12.406	52.7	OK	6.876	14.235	48.3	OK
		6.717	12.349	54.4	OK	6.978	14.235	49.0	OK
C-tfl 1	STIFF2a	5.761	13.260	43.4	OK	6.125	16.238	37.7	OK
		6.053	13.346	45.4	OK	6.532	16.238	40.2	OK
C-bfl 1	STIFF12b	11.867	15.360	77.3	OK	12.452	17.265	72.1	OK
		12.054	15.410	78.2	OK	12.645	17.265	73.2	OK
C-w 1	STIFF2b	6.717	12.343	54.4	OK	6.883	14.235	48.4	OK
		6.508	12.416	52.4	OK	6.475	14.235	45.5	OK
C-tfl 1	STIFF2b	6.051	13.364	45.3	OK	6.73	16.238	41.5	OK
		5.760	13.255	43.5	OK	6.134	16.238	37.8	OK

19.7 Summary

In this chapter, a comparison between the results from IDEA StatiCa component-based finite element method (CBFEM), the traditional calculation methods used in US practice (AISC *Specification*; AISC, 2016a; AISC *Manual*; AISC, 2017), and the traditional finite element analysis using ABAQUS for tested BFA connections is presented. Overall, a good agreement between all methods was observed. For the baseline specimen (Specimen BFP), beam flange failure was identified as the failure mode by both the experiment and IDEA StatiCa. In contrast, the AISC procedure identified the web panel zone shear as the governing failure mode. Although the moment strengths computed from IDEA StatiCa for all specimens are higher than those calculated following the AISC's approach, it is believed that this is due to the conservatism in the approach used for the web panel zone shear strength. This can be justified from the measured moment-rotation curve where the inelastic behavior approximately starts around 19000 kip-in (see Figure 19.15). In addition, approximately 17000 kip-in of moment capacity was computed using IDEA StatiCa while 14500 kip-in was calculated from the AISC procedure.

Regarding the comparison between the IDEA StatiCa and ABAQUS results, a minor discrepancy was found where the contacts were defined between the plates and column/beam faces, although the same type of analysis was performed, i.e. small deformation. This could be due to the differences between solid elements and shell elements or contact algorithm(s) that are used in IDEA StatiCa and ABAQUS. Also, the way that IDEA StatiCa code calculates and utilizes the optimum element size was not clear. Additionally, due to the recommended plastic strain limit of 5% by Eurocode (EN1993-1-5 app. C par. C8 note 1) which is defined as a default value in the IDEA StaiCa software, different failure modes were observed. The stress distributions on the beam and column were very close. However, slightly higher stresses were predicted on the column, plate 1, and stiffeners in the ABAQUS model which is most likely due to the nature of the tie constraint. The predicted load on the bolts and weld groups were also very close between the two softwares. Stress distribution for each part can be accessed in Appendix L of the full report, published on the IDEA StatiCA website (IDEA StatiCa, 2021).

References

AISC (2016a). *Specification for Structural Steel Buildings*. American Institute of Steel Construction, Chicago.

AISC (2016b). *Prequalified Connections for Special and Intermediate Steel Moment Frames for Seismic Applications*. American Institute of Steel Construction, Chicago.

AISC (2017). *Steel Construction Manual*, 15th Edition. American Institute of Steel Construction, Chicago.

Dassault Systèmes Simulia Corporation (2020). *ABAQUS* (Version 2020) [Computer program]. Dassault Systèmes SE, Providence, RI.

IDEA StatiCa (2021). *Verification of IDEA StatiCa Calculations for Steel Connection Design (AISC)*. https://www.ideastatica.com/support-center/verification-of-idea-statica-calculations-for-steel-connection-design-aisc (accessed October 31, 2023).

Sato, A., Newell, J., and Uang, C.M. (2007). *Cyclic Testing of Bolted Flange Plate Steel Moment Connections for Special Moment Frames*. La Jolla, CA, University of California, San Diego.

20

Conclusion

Connections are critical to the performance of steel structures, and connection design is critical to the success of structural steel design. The profession has developed an array of design approaches for simple and common connections; however, engineers encounter complex and uncommon connections on a day-to-day basis. The component-based finite element method (CBFEM) is an inelastic finite element analysis where shell elements are used for members and plates and nonlinear springs for bolts, anchors, and welds. It offers an efficient approach to design and evaluation of both typical and atypical connections. Explicitly modeling the flexibility and material response of the connection means that fewer potentially erroneous assumptions need to be made in design. However, as with every method of design, verification and validation are necessary to ensure reliable results.

The CBFEM, as implemented in IDEA StatiCa, has been the subject of rigorous verification and validation to ensure the analysis engine provides accurate results. Comparisons to other finite element software packages performed in this work have further verified the accuracy of the IDEA StatiCa analyses. Similarly, the results of IDEA StatiCa have been compared with European standards by Wald et al. (2020). This work expands on those prior verification studies through comparisons to the provisions of the AISC *Specification for Structural Steel Buildings* (AISC, 2016) and calculations and connections common in the United States.

A variety of connection types were examined, including single plate shear, bolted flange plate moment, extended end-plate moment, top-and-seat angle, corner brace, chevron brace, base plate, bolted wide flange splice, beam-over-column, bracket plate, and T-stub connections. For these common structural steel connection types with established design procedures, the available strength obtained from IDEA StatiCa generally agrees well with traditional calculations per the AISC *Specification* with differences primarily on the conservative side. The conservatism in IDEA StatiCa arises from several sources:

- The maximum stress in connected elements is based on the yield strength and not the tensile strength as is assumed for several calculations per the AISC *Specification*.
- The buckling factor is limited to 3.0 as a simplified means of limiting the effects of geometric nonlinearity and considering elastic stability limit states.
- Minimal inelasticity is permitted in bolts and welds, meaning that the strength of eccentrically loaded bolt and weld groups is governed by the first bolt or segment of weld to achieve 100% utilization.
- IDEA StatiCa does not employ ad hoc assumptions permitted in traditional calculations, such as the use of reduced eccentricity in the design of conventional single plate shear connections.

Steel Connection Design by Inelastic Analysis: Verification Examples per AISC Specification, First Edition.
Mark D. Denavit, Ali Nassiri, Mustafa Mahamid, Martin Vild, Halil Sezen, and František Wald.

- Limitations in the software require some conservative simplifications. For example, when evaluating slip-critical connections subjected to combined tension and shear, IDEA StatiCa conservatively considers only the tension in the bolts and not the contact pressure on the faying surfaces (i.e. the prying force) when determining the available strength.

While the results from IDEA StatiCa were generally observed to be conservative with respect to traditional calculations, there were some cases where the available strength obtained from IDEA StatiCa was greater. These differences were primarily due to conservative assumptions made in the traditional calculations. Examples include:

- It is common in traditional calculations to assume that base plates are rigid. This can overestimate the demands on the base plate. IDEA StatiCa explicitly models the strength and stiffness of the base plates, resulting in lower demands.
- When evaluating web local limit states, the effect of nearby stiffeners is ignored in traditional calculations but can be captured by IDEA StatiCa.
- When evaluating extended end-plate moment connections, the position of the resultant of the contact force with the column is assumed in traditional calculations. Explicit modeling of the connection can result in different positions and increased strength.
- When evaluating top- and seat-angle connections, it is common to ignore the contribution of strength and stiffness provided by web angles. Explicit modeling of the connection naturally includes this contribution.
- Using default parameters, the web panel-zone strength from IDEA StatiCa is similar to the strength from the AISC *Specification* when the effect of inelastic panel-zone deformations on frame stability is accounted for in the analysis to determine required strengths. The lower strength given in the AISC *Specification*, for when the effect of inelastic panel-zone deformations on frame stability is not accounted for in the analysis, to determine required strengths, can be achieved by adjusting the plastic strain limit in IDEA StatiCa.

Additionally, some mesh dependance was observed. While minor in most cases, IDEA StatiCa exhibited reduced strengths when the mesh size was set smaller than the default for some connections and loading conditions.

Overall, IDEA StatiCa and the underlying CBFEM form a rational method of connection design. While not a replacement for engineering judgment, safe designs should be expected for properly defined models using the default analysis and design parameters. Care is required when defining the model to ensure the behavior of the connection is as intended. The authors encourage engineers employing IDEA StatiCa to perform their own studies such as those presented in this work to help understand the software better and gain further confidence in the results.

References

AISC (2016). *Specification for Structural Steel Buildings*. American Institute of Steel Construction, Chicago.

Wald, F., Šabatka, L., Bajer, M., Jehlička, P., Kabeláč, J., Kožich, M., Kuříková, M., and Vild, M. (2020). *Component-Based Finite Element Design of Steel Connections*. Czech Technical University, Prague.

Disclaimer

The information presented in this book has been prepared in accordance with recognized engineering principles and is for general information only. The authors have sought to present accurate, reliable, and useful information on the design of structural steel connections, however, the information contained in this book should not be used or relied upon for any specific project without competent professional assessment of its accuracy, suitability, and applicability. Any person using this information does so at their own risk and assumes all liability arising from such use.

Steel Connection Design by Inelastic Analysis: Verification Examples per AISC Specification, First Edition.
Mark D. Denavit, Ali Nassiri, Mustafa Mahamid, Martin Vild, Halil Sezen, and František Wald.

Terms and symbols

The terms and symbols used in this text are consistent with the 2016 edition of the American Institute of Steel Construction (AISC) *Specification for Structural Steel Buildings*. Where additional terms and symbols were needed, they are defined in the text.

Steel Connection Design by Inelastic Analysis: Verification Examples per AISC Specification, First Edition.
Mark D. Denavit, Ali Nassiri, Mustafa Mahamid, Martin Vild, Halil Sezen, and František Wald.

Index

Steel Connection Design by Inelastic Analysis: Verification Examples per AISC Specification, First Edition.
Mark D. Denavit, Ali Nassiri, Mustafa Mahamid, Martin Vild, Halil Sezen, and František Wald.